T0133674

Computer-Aided Graphing and Simulation Tools for AutoCAD Users

CHAPMAN & HALL/CRC
COMPUTER and INFORMATION SCIENCE SERIES

Series Editor: Sartaj Sahni

PUBLISHED TITLES

ADVERSARIAL REASONING: COMPUTATIONAL APPROACHES TO READING THE OPPONENT'S MIND
Alexander Kott and William M. McEneaney

COMPUTER-AIDED GRAPHING AND SIMULATION TOOLS FOR AUTOCAD USERS
P. A. Simionescu

DELAUNAY MESH GENERATION
Siu-Wing Cheng, Tamal Krishna Dey, and Jonathan Richard Shewchuk

DISTRIBUTED SENSOR NETWORKS, SECOND EDITION
S. Sitharama Iyengar and Richard R. Brooks

DISTRIBUTED SYSTEMS: AN ALGORITHMIC APPROACH, SECOND EDITION
Sukumar Ghosh

ENERGY-AWARE MEMORY MANAGEMENT FOR EMBEDDED MULTIMEDIA SYSTEMS: A COMPUTER-AIDED DESIGN APPROACH
Florin Balasa and Dhiraj K. Pradhan

ENERGY EFFICIENT HARDWARE-SOFTWARE CO-SYNTHESIS USING RECONFIGURABLE HARDWARE
Jingzhao Ou and Viktor K. Prasanna

FUNDAMENTALS OF NATURAL COMPUTING: BASIC CONCEPTS, ALGORITHMS, AND APPLICATIONS
Leandro Nunes de Castro

HANDBOOK OF ALGORITHMS FOR WIRELESS NETWORKING AND MOBILE COMPUTING
Azzedine Boukerche

HANDBOOK OF APPROXIMATION ALGORITHMS AND METAHEURISTICS
Teofilo F. Gonzalez

HANDBOOK OF BIOINSPIRED ALGORITHMS AND APPLICATIONS
Stephan Olariu and Albert Y. Zomaya

HANDBOOK OF COMPUTATIONAL MOLECULAR BIOLOGY
Srinivas Aluru

HANDBOOK OF DATA STRUCTURES AND APPLICATIONS
Dinesh P. Mehta and Sartaj Sahni

HANDBOOK OF DYNAMIC SYSTEM MODELING
Paul A. Fishwick

HANDBOOK OF ENERGY-AWARE AND GREEN COMPUTING
Ishfaq Ahmad and Sanjay Ranka

HANDBOOK OF PARALLEL COMPUTING: MODELS, ALGORITHMS AND APPLICATIONS
Sanguthevar Rajasekaran and John Reif

HANDBOOK OF REAL-TIME AND EMBEDDED SYSTEMS
Insup Lee, Joseph Y-T. Leung, and Sang H. Son

PUBLISHED TITLES CONTINUED

HANDBOOK OF SCHEDULING: ALGORITHMS, MODELS, AND PERFORMANCE ANALYSIS
Joseph Y.-T. Leung

HIGH PERFORMANCE COMPUTING IN REMOTE SENSING
Antonio J. Plaza and Chein-I Chang

HUMAN ACTIVITY RECOGNITION: USING WEARABLE SENSORS AND SMARTPHONES
Miguel A. Labrador and Oscar D. Lara Yejas

IMPROVING THE PERFORMANCE OF WIRELESS LANs: A PRACTICAL GUIDE
Nurul Sarkar

INTRODUCTION TO NETWORK SECURITY
Douglas Jacobson

LOCATION-BASED INFORMATION SYSTEMS: DEVELOPING REAL-TIME TRACKING APPLICATIONS
Miguel A. Labrador, Alfredo J. Pérez, and Pedro M. Wightman

METHODS IN ALGORITHMIC ANALYSIS
Vladimir A. Dobrushkin

MULTICORE COMPUTING: ALGORITHMS, ARCHITECTURES, AND APPLICATIONS
Sanguthevar Rajasekaran, Lance Fiondella, Mohamed Ahmed, and Reda A. Ammar

PERFORMANCE ANALYSIS OF QUEUING AND COMPUTER NETWORKS
G. R. Dattatreya

THE PRACTICAL HANDBOOK OF INTERNET COMPUTING
Munindar P. Singh

SCALABLE AND SECURE INTERNET SERVICES AND ARCHITECTURE
Cheng-Zhong Xu

SOFTWARE APPLICATION DEVELOPMENT: A VISUAL C++®, MFC, AND STL TUTORIAL
Bud Fox, Zhang Wenzu, and Tan May Ling

SPECULATIVE EXECUTION IN HIGH PERFORMANCE COMPUTER ARCHITECTURES
David Kaeli and Pen-Chung Yew

VEHICULAR NETWORKS: FROM THEORY TO PRACTICE
Stephan Olariu and Michele C. Weigle

Computer-Aided Graphing and Simulation Tools for AutoCAD Users

P. A. Simionescu

Texas A&M University
Corpus Christi, USA

CRC Press
Taylor & Francis Group
Boca Raton London New York

CRC Press is an imprint of the
Taylor & Francis Group, an **informa** business

A CHAPMAN & HALL BOOK

CRC Press
Taylor & Francis Group
6000 Broken Sound Parkway NW, Suite 300
Boca Raton, FL 33487-2742

© 2015 by Taylor & Francis Group, LLC
CRC Press is an imprint of Taylor & Francis Group, an Informa business

Printed on acid-free paper
Version Date: 20141020

International Standard Book Number-13: 978-1-4822-5290-3 (Hardback)

Visit the Taylor & Francis Web site at
http://www.taylorandfrancis.com

and the CRC Press Web site at
http://www.crcpress.com

After some thinking, I decided to dedicate this book to the people of Google: Your work makes my work more productive every day and will probably help to sell a few extra copies of this book as well.

Contents

Preface, xv

Legal Notices, xix

Acknowledgments, xxi

Author, xxiii

CHAPTER 1 ▪ Graphical Representation of Univariate Functions
and of (x, y) Data Sets: The **D_2D** Program   1

1.1	ANALYTICAL FUNCTION PLOTS	6
1.2	SHOWING EXTREMA AND ZEROS OF GRAPHS	8
1.3	STEM AND AREA PLOTS: LENGTH OF A CURVE AND AREA UNDER A CURVE	11
1.4	WINDOWING AND PANNING	13
1.5	NUMBERING DATA POINTS	14
1.6	PLOTTING FUNCTIONS WITH SINGULARITIES	15
1.7	CONTROLLING PLOT FEATURES FROM WITHIN THE INPUT DATA FILE	18
1.8	PLOTTING SCATTERED DATA	21
1.9	PLOTTING ORDERED DATA AND HISTOGRAMS	24
1.10	PLOTTING INEQUALITIES	27
1.11	PARAMETRIC PLOTS	28
1.12	ANIMATIONS	33
	REFERENCES AND FURTHER READINGS	40

CHAPTER 2 ▪ Graphical Representation of Functions of Two Variables:
The **D_3D** Program   43

2.1	HOW **D_3D** WORKS?	47
2.2	**D_3D** INPUT DATA STRUCTURE	50
2.3	MESH PLOTS AND THE VISIBILITY PROBLEM	52
2.4	NODE AND STEM PLOTS	56

2.5 EQUALLY SPACED LEVEL-CURVE PLOTS 58

2.6 DEFECT-FREE LEVEL-CURVE PLOTS 60

2.7 LOGARITHMICALLY-SPACED LEVEL CURVES 62

2.8 FILE EXPORT AND **DXF** LAYER ORGANIZATION 65

2.9 AXES REVERSAL AND PLOT ROTATION 67

2.10 GRADIENT PLOTS 68

2.11 TRUNCATED 3D SURFACE REPRESENTATIONS 70

2.12 CONSTRAINED FUNCTION AND INEQUALITY PLOTS 76

2.13 COLOR-RENDERED PLOTS 81

2.14 PLOTTING MULTIPLE SURFACES ON THE SAME GRAPH 85

2.15 IMPLEMENTATION DETAILS OF THE **D_3D** PROGRAM 87

REFERENCES AND FURTHER READINGS 89

CHAPTER 3 ▪ Programs and Procedures for Data Visualization and Data Format Conversion 91

3.1 **LibPlots** SUBROUTINES FOR GENERATING 2D PLOTS 92

    3.1.1 Basic 2D Plotting Using **LibPlots** 92

    3.1.2 Multiple Plots with Markers 93

    3.1.3 Plotting Large Data Sets and Data Read from File 95

    3.1.4 Dynamic Plots with Scan Lines and Scan Points 96

3.2 **Util~TXT** PROGRAM FOR MANIPULATION OF ASCII FILES 98

    3.2.1 Linear Interpolation 99

    3.2.2 Cubic-Spline Interpolation 100

    3.2.3 B-Spline Interpolation 101

    3.2.4 Numerical Differentiation 101

    3.2.5 Angle-Value Rectification 103

    3.2.6 Data Decimation 104

    3.2.7 **DXF** Output of 2D and 3D Polylines 104

3.3 **Util~DXF** PROGRAM FOR VISUALIZATION OF **R12 DXF** FILES 104

    3.3.1 Extracting Polyline Vertex Coordinates 106

    3.3.2 Raster Curve Digitization Using **Util~DXF** and **Util~TXT** 108

    3.3.3 Transferring Level Curves from **D_3D** to **D_2D** 110

3.4 **Util~PLT** PROGRAM FOR MANIPULATING **PLT** FILES 111

    3.4.1 Flattening and Retouching Plots Created with **D_2D** 113

    3.4.2 Alphanumerical Character Discretization 114

3.5   **G_3D.LSP** PROGRAM FOR GENERATING 3D CURVES
AND SURFACES INSIDE **AutoCAD**   116

   3.5.1   3D Polyline Plotting Using **G_3D.LSP**   116

   3.5.2   3D Surface Plotting Using **G_3D.LSP**   119

3.6   **M_3D.LSP** PROGRAM FOR AUTOMATIC 3D MODEL GENERATION
AND ANIMATION INSIDE **AutoCAD**   121

   3.6.1   Animation of **DXF** Files with Multiple Layers Using **M_3D.LSP**   121

   3.6.2   3D Model Generation with Data Read from File   122

   3.6.3   Automatic Insertion of **AutoCAD** Blocks   125

REFERENCES AND FURTHER READINGS   128

CHAPTER 4 ■ Root Finding and Minimization or Maximization
               of Functions   129

4.1   BRENT'S *ZERO* ALGORITHM FOR ROOT FINDING OF NONLINEAR
EQUATIONS   129

4.2   BRENT'S METHOD FOR MINIMIZING FUNCTIONS OF ONE VARIABLE   132

4.3   NELDER–MEAD ALGORITHM FOR MULTIVARIATE FUNCTION
MINIMIZATION   133

4.4   HANDLING CONSTRAINTS IN OPTIMIZATION PROBLEMS   137

4.5   EVOLUTIONARY ALGORITHM FOR BOUNDED-OPTIMUM SEARCH   141

4.6   MULTICRITERIA OPTIMIZATION PROBLEMS   144

   4.6.1   Cantilever Beam Design Example   144

   4.6.2   Design Space and Performance Space Plots   145

   4.6.3   Pareto Front Search   146

REFERENCES AND FURTHER READINGS   149

CHAPTER 5 ■ Procedures for Motion Simulation of Planar Mechanical
               Systems   151

5.1   SAMPLE PROGRAM USING THE **LibMec2D** UNIT
AND SUBROUTINES **Locus** AND **CometLocus**   151

5.2   JOINTS AND ACTUATORS AVAILABLE FOR MECHANICAL SYSTEM
SIMULATION   153

   5.2.1   Kinematic Analysis of Input Rotational Members   154

   5.2.2   Procedures **Crank** and **gCrank**   156

   5.2.3   Kinematic Analysis of Input Translational Members   158

   5.2.4   Procedures **Slider** and **gSlider**   161

5.3 POSITION, VELOCITY, AND ACCELERATION OF POINTS
AND MOVING LINKS — 163

   5.3.1 Procedures `Offset` and `OffsetV` — 166

   5.3.2 Procedures `AngPVA`, `Ang3PVA`, and `Ang4PVA` — 168

5.4 POSITION, VELOCITY, AND ACCELERATION IN RELATIVE MOTION:
SUBROUTINE `VarDist` — 169

5.5 CORIOLIS ACCELERATION EXAMPLE: SUBROUTINE `PutVector` — 171

5.6 MODEL VALIDATION: SUBROUTINE `ntAccel` — 172

5.7 WORKSPACE LIMITS AND INQUIRY SUBROUTINES `PutDist`
AND `PutAng` — 175

5.8 ADDING COMPLEX SHAPES TO SIMULATIONS: SUBROUTINES
`Base`, `Link`, `gShape`, AND `Shape` — 177

5.9 SIMULATIONS ACCOMPANIED BY PLOTS WITH SCAN LINES
AND SCAN POINTS — 179

REFERENCES AND FURTHER READINGS — 181

CHAPTER 6 ■ Kinematic Analysis of Planar Linkage Mechanisms
Using Assur Groups — 183

6.1 ASSUR GROUP–BASED KINEMATIC ANALYSIS OF LINKAGE
MECHANISMS — 183

6.2 INTERSECTION BETWEEN TWO CIRCLES: PROCEDURE `INT2CIR` — 186

6.3 VELOCITY AND ACCELERATION OF THE INTERSECTION POINTS
BETWEEN TWO CIRCLES: PROCEDURE `Int2CirPVA` — 189

6.4 KINEMATICS OF THE RTRTR DOUBLE LINEAR INPUT ACTUATOR:
PROCEDURE `RTRTRc` — 190

6.5 KINEMATICS OF THE RTRTR DOUBLE LINEAR INPUT ACTUATOR
USING A VECTOR EQUATION APPROACH: PROCEDURE `RTRTR` — 196

6.6 MOTION TRANSMISSION CHARACTERISTICS OF RTRTR-BASED
MECHANISMS — 202

6.7 KINEMATIC ANALYSIS OF THE RTRR OSCILLATING-SLIDE ACTUATOR
USING EQUATIONS OF CONSTRAINT: PROCEDURE `RTRRc` — 204

6.8 KINEMATIC ANALYSIS OF THE RTRR OSCILLATING-SLIDE ACTUATOR
USING A VECTOR-LOOP APPROACH: PROCEDURE `RTRR` — 207

6.9 KINEMATIC ANALYSIS OF THE RRR DYAD: PROCEDURES `RRRc`
AND `RRR` — 210

6.10 KINEMATIC ANALYSIS OF THE RRT DYAD USING A VECTOR-LOOP
APPROACH — 214

   6.10.1 RR_T Dyadic Isomer: Procedure `RR_T` — 214

   6.10.2 RRT_ Dyadic Isomer: Procedure `RRT_` — 219

6.11 KINEMATIC ANALYSIS OF THE RTR DYAD USING A VECTOR-LOOP APPROACH: PROCEDURE **RT_R**   222

6.12 KINEMATIC ANALYSIS OF THE TRT DYAD USING A VECTOR-LOOP APPROACH   228

    6.12.1 T_R_T Dyadic Isomer: Procedure **T_R_T**   229

    6.12.2 _TRT_ Dyadic Isomer: Procedure **_TRT_**   235

    6.12.3 T_RT_ Dyadic Isomer: Procedure **T_RT_**   240

6.13 KINEMATIC ANALYSIS OF THE RTT DYAD USING A VECTOR-LOOP APPROACH   245

    6.13.1 R_T_T Dyadic Isomer: Procedure **R_T_T**   245

    6.13.2 RT_T_ Dyadic Isomer: Procedure **RT_T_**   251

    6.13.3 R_TT_ Dyadic Isomer: Procedure **R_TT_**   256

    6.13.4 RT__ Dyadic Isomer: Procedure **RT__T**   261

REFERENCES AND FURTHER READINGS   267

CHAPTER 7 ▪ Design and Analysis of Disk Cam Mechanisms   269

7.1 SYNTHESIS OF FOLLOWER MOTION   269

7.2 SYNTHESIS AND ANALYSIS OF DISC CAMS WITH TRANSLATING FOLLOWER, POINTED OR WITH ROLLER   272

7.3 SYNTHESIS AND ANALYSIS OF DISC CAMS WITH OSCILLATING FOLLOWER, POINTED OR WITH ROLLER   277

7.4 SYNTHESIS AND ANALYSIS OF DISC CAMS WITH TRANSLATING FLAT-FACED FOLLOWER   281

7.5 SYNTHESIS AND ANALYSIS OF DISC CAMS WITH OSCILLATING FLAT-FACED FOLLOWER   286

7.6 SYNTHESIS OF DISC CAMS WITH CURVILINEAR-FACED FOLLOWER   289

    7.6.1 Synthesis of Disk Cams with Arc-Shaped Follower   289

    7.6.2 Synthesis of Disk Cams with Polygonal-Faced Follower   292

REFERENCES AND FURTHER READINGS   295

CHAPTER 8 ▪ Spur Gear Simulation Using **Working Model 2D** and **AutoLISP**   297

8.1 INVOLUTE-GEAR THEORY   297

8.2 INVOLUTE PROFILE MESH   301

8.3 INVOLUTE-GEAR MESH   306

8.4 **WORKING MODEL 2D** SIMULATIONS OF INVOLUTE PROFILE
GENERATION 310

8.5 INVOLUTE PROFILE GENERATION USING `Gears.LSP` 313

REFERENCES AND FURTHER READINGS 318

CHAPTER 9 ■ More Practical Problems and Applications 319

9.1 DUFFING OSCILLATOR 319

9.2 FREE OSCILLATION OF A SPRING–MASS–DASHPOT SYSTEM 321

9.3 FREQUENCY AND DAMPING RATIO ESTIMATION
OF OSCILLATORY SYSTEMS 324

9.4 NONLINEAR CURVE FIT TO DATA 327

9.5 PLOTTING FUNCTIONS OF MORE THAN TWO VARIABLES 329

9.6 RANDOM NUMBER GENERATION AND HISTOGRAM PLOTS 336

9.7 DWELL MECHANISM ANALYSIS 338

9.8 TIME RATIO EVALUATION OF A QUICK-RETURN MECHANISM 341

9.9 EXAMPLES OF ITERATIVE USE OF THE PROCEDURES IN UNIT
`LibAssur` 342

9.10 SIMULATION OF A FOUR-BAR LINKAGE AND OF ITS FIXED
AND MOVING CENTRODES 345

9.11 PLANETARY GEAR KINEMATIC SIMULATION USING **WORKING
MODEL 2D** 346

9.12 IMPLICIT FUNCTION PLOT 349

9.13 INVERSE AND DIRECT KINEMATICS OF 5R AND 2R ASSEMBLY
ROBOTS 351

9.14 INVERSE AND DIRECT KINEMATICS OF THE RTRTR GEARED
PARALLEL MANIPULATOR 355

9.15 KINEMATIC ANALYSIS OF A HYDRAULIC EXCAVATOR
AND OF A ROPE SHOVEL 358

9.16 KINEMATIC ANALYSIS OF INDEPENDENT WHEEL
SUSPENSION MECHANISMS OF THE MULTILINK
AND DOUBLE-WISHBONE TYPE 361

9.17 FLYWHEEL SIZING OF A PUNCH PRESS 367

9.18 A PROGRAM FOR PURGING FILES FROM THE CURRENT DIRECTORY 373

REFERENCES AND FURTHER READINGS 373

APPENDIX A: USEFUL FORMULAE, 377

APPENDIX B: SELECTED SOURCE CODE, 409

INDEX, 599

Preface

O VER THE COURSE OF ALMOST TWO DECADES, I developed a number of Pascal and AutoLISP for AutoCAD programs and Working Model 2D simulations that I used in my publications and presentations. Occasionally, people aware of these computer applications asked for evaluation copies, which I gladly provided them. Such requests encouraged me to spend more time improving and documenting these applications, and ultimately determined me to make these applications and the algorithms behind them available to a wider audience. This is how the idea of writing this book was born.

The intended readership for this book are students, scholars, scientists, and engineers who have access to **AutoCAD** and **Working Model 2D** software and are interested in information visualization, motion simulation of mechanical systems, numerical analysis, optimization, and evolutionary computation. Those who use **AutoCAD LT**, or have access to only a **DXF** viewer, can still make substantial use of this book and of the accompanying programs and simulations.

The first two chapters describe plotting programs **D_2D** and **D_3D**, which have features not yet available in popular software like **MATLAB®**, **Excel**, or **MathCAD**. Some of these features are: showing extrema and zeros of 2D graphs, automatic numbering of data points, controlling the plot appearance from within input data file, plotting inequalities of two variables, trimming the portions of function surface that exceed the plot box, projecting the gradient on the bottom plane in 3D plots, logarithmically spacing level curves, and **DXF** export.

Chapter 3 introduces a collection of Pascal programs and procedures for generating dynamic 2D graphs with scan lines and scan points, for manipulating ASCII files and for viewing **R12 DXF** and **PLT AutoCAD** export files. It also describes two **AutoLISP** applications for plotting curves and surfaces and for generating 3D models consisting of various geometric primitives and predefined blocks using vertex coordinates and model description read from file.

Chapter 4 discusses several algorithms for finding the zeros and minima of functions of one or more variables and for multicriteria optimization. Also presented is a new evolutionary algorithm that explores the boundary between feasible and unfeasible spaces in optimization problems—it is known that in many practical problems the minimum is bounded. Numerical applications of each of these algorithms are accompanied by plots and animations generated using the **D_2D** and **D_3D** programs.

Chapters 5 and 6 introduce a series of procedures, accompanied by examples and the underlying theory, for the kinematic simulation of a wide variety of planar linkage mechanisms.

Chapter 7 deals with the synthesis of the profile of rotating disc cams operating in conjunction with various type followers (pointed, with roller, flat, translating or oscillating). Iterative methods for analyzing the respective cam-follower mechanisms are also presented. In addition, a procedure for synthesizing the follower motion using **AutoCAD** splines is described.

Chapter 8 reviews the theory of planar involute gears and presents a number of **Working Model 2D** simulations and an **AutoLISP** application to illustrate this theory. The **AutoLISP** program is particularly useful because it allows the generation, directly inside **AutoCAD**, of involute gear profiles, internal or external, with any number of teeth.

Chapter 9 is a collection of problems and applications from areas like dynamical systems, vibrations, kinematics, robotics, multidimensional visualization, etc., solved using the software tools presented in the earlier chapters, or using **Working Model 2D**.

Source codes and executables of the programs and simulations discussed in the book are available upon request from the author. The referred animation files can be downloaded from the publisher's website at www.crcpress.com/product/isbn/9781482252903/ or from http://faculty.tamucc.edu/psimionescu/cagstau.html.

While every effort has been made to provide error-free analytical derivations and software implementations of these derivations, in no event shall the author or publisher be liable for any claim, damages, or other liability in connection with the use of the material in this book and of the accompanying computer programs and simulations.

As with any text, the clarity of the writing can be improved and the collection of examples expanded. The **AutoLISP** and Pascal programs provided with this book can also sustain improvements or can be translated into other programming languages. I would therefore appreciate any comments, suggestions, or reports of errors. In particular, I would welcome any serious offer for collaboration on future editions. So my respected reader, before posting critical reviews about this book, please read once again this last paragraph.

Thank you,

Petru A. Simionescu
pa.simionescu@gmail.com

MATLAB® is a registered trademark of The MathWorks, Inc. For product information, please contact:

The MathWorks, Inc.
3 Apple Hill Drive
Natick, MA 01760-2098 USA
Tel: 508-647-7000
Fax: 508-647-7001
E-mail: info@mathworks.com
Web: www.mathworks.com

AutoCAD® is a registered trademark of Autodesk, Inc. For product information, please contact:

Autodesk, Inc.
111 McInnis Parkway
San Rafael, CA 94903 USA
Tel: 415-507-5000
Fax: 415-507-5100
Web: www.autodesk.com

Legal Notices

AUTOCAD, **A**UTOCAD **LT**, AND **AutoLISP** are registered trademarks of Autodesk, Inc. **Working Model 2D**, **WM 2D**, and **WM Basic** are registered trademarks of Design Simulation Technologies, Inc. **Turbo Pascal** and **BGI** are registered trademarks of Borland Software Corporation. All other brand names, product names, or trademarks belong to their respective holders, acknowledged in this book by writing them in boldfaced letters.

Acknowledgments

THE **PCX** RASTER OUTPUT METHOD implemented in some of the programs provided with this book follows an online posting of Bren Sessions of Corvallis, Oregon. Horia Brădău of Vaughan, Ontario, contributed an earlier **AutoLISP** program for the generation of involute gears. Thanks are extended to Dr. Constantin Stăncescu of University Polytechnica of Bucharest, who adapted the **AutoLISP** programs provided with this book to run in any version of **AutoCAD** up to its 2014 release. My appreciation goes to all those that encouraged me to complete this project and to CRC Press for their careful involvement in the preparation of this book for publication.

Author

PETRU AURELIAN SIMIONESCU is on the engineering faculty at Texas A&M University in Corpus Christi. He earned a BSc from the Polytechnic University of Bucharest, a doctorate in technical sciences from the same university, and a PhD in mechanical engineering from Auburn University. Simionescu taught and conducted research at seven Romanian, British, and American universities, and worked for four years in industry as an automotive engineer. His research interests include kinematics, dynamics and design of multibody systems, evolutionary computation, CAD, computer graphics, and information visualization. So far, he has authored over 40 technical papers and has been granted seven patents.

CHAPTER 1

Graphical Representation of Univariate Functions and of (x, y) Data Sets

The D_2D Program

P LOTTING ANALYTICAL FUNCTIONS $y = F(x)$ or simply of (x, y) sets is something that everybody interested in computer graphics most likely has programmed, or at least attempted to do. This is commonly required part of many applications, from mathematics to experimental data analysis. In this chapter, the **D_2D** program available with the book as Pascal source code (**D_2D.PAS** and **UNIT_D2D.PAS**) and as executable file (**D_2D. EXE**) will be introduced. **D_2D** has several features not yet available in popular software like **Excel**, **MathCAD**, **Mathematica**, or **MATLAB** as follows:

1. The size of the plot box can be precisely controlled by the user, advantageous when creating stacked graphs of the same height and/or width.

2. The x-axis can be placed either on the bottom or on the top of the plot box.

3. The divisions on the x- and y-axes can be labeled in fractions or multiples of π, a feature useful when plotting trigonometric functions.

4. When plotting a single curve, **D_2D** adds the lengths of the individual segments that form the graph and displays this number as the length of the curve.

5. Multiple graphs can be plotted simultaneously over four separate y categories named by default **F(x)**, **F2(x)**, **F3(x)**, and **F4(x)**.

6. Markers or glyphs can be inserted one at each data point (the ✱✱✱ marker-placing option) or can be spaced at constant distance along the plot curve (the –✱– marker-placing option). In the latter case, the distance between two successive markers can be specified in marker radii or as an integer number of data points between two glyphs. This integer is read by **D_2D** from a configuration file with the extension **CF2**.

7. Marker types available to distinguish between multiple plot curves are Ø, ○, ●, □, ◇, ▽, △, ✱, ×, +, ¤, ♀, ♂, ¦, and >. The arrow marker > can be used to indicate the order in which the data have been generated (e.g., in time-varying processes), an intuitive still-image substitute to animated comet plots. In nonaccumulated comet plots, the broken-bar ¦ marker will be automatically converted into a vertical scan line. The marker size can be specified either in screen units, or, if the plot is *isotropic*, in the same units as the graph. It will be called *isotropic*, a graph that has the width/height ratio of the plot box equal to the ratio of the x- and y-axis ranges, namely, $(x_{max} - x_{min})/(y_{max} - y_{min})$. Therefore, the isotropic plot of a circle will not be distorted to look like an ellipse. You can make a plot isotropic by manually adjusting its box height and width or the limits over its x- and y-axes. There is also an option where **D_2D** will automatically adjusts the limits over the x- and y-axes so that the graph remains isotropic as the plot-box size is interactively adjusted.

8. The **/#** marker-placing option allows the *minimum* and *maximum* points of the graph to be automatically identified and their coordinates included with the plot, together with the coordinates of the intersection(s) between the graph and the x-axis (the *zeros* of the graph). Both can be exported to ASCII files with extensions **MIN**, **MAX**, and **ZER** and can be used in further calculations and analyses.

9. An alternative to using arrow markers to show the sequence in which data have been generated is to number the points of the graph using the **✱#✱** marker-placing option. The numbering is done automatically by **D_2D** following a pattern specified by the user—by default, every other point will be labeled beginning with 1 that is assigned to the first data point.

10. The **!!!!!!** marker-placing option will generate *stem plots*. A stem plot with the data points connected with a continuous line will be called *area plot*. In case of the latter, when the graph consists of a single curve, the area bounded by the curve and the x-axis will be evaluated by **D_2D** using the *trapezoidal rule of integration* (see Appendix A) and will be automatically displayed on top of the graph together with the length of the curve.

11. Plot-line thickness can be 1 or 3 units (pixels) and either *solid* (options –– or ====), *dashed* (options ––– or ===), or *dotted* (options ··· or :::). Their color can be *blue, green, cyan, red, magenta, brown, gray, light blue, light green, light cyan, light red, light magenta, light blue,* or *yellow.*

Important: In the current implementation of **D_2D**, when large numbers of data points are plotted as a single dotted or dashed line, defects may occur in the form of portions of the graph (or even the whole curve) not being displayed. If exported to **DXF**, however, dotted and dashed lines will be represented correctly. Since the color, thickness, and line type of the graphs can be easily edited from within **AutoCAD**, the **DXF** format is a more advantageous graphic output format of **D_2D**, with the exception of scatter-point plots that may take less disk space if output as raster-images files.

12. For convenient data file management, the (x, y) points belonging to two or more curves of the same y-category can be read from the same file using separators. Color and marker type can be also set or changed from within the data file using separators, as described in the **About** screen of the program (see the following insert) that you can bring up by pressing the <F10> key immediately after launching **D_2D**. Later in this chapter, it will be shown how separators can be used to create animated plots of more than 16 frames or accumulated graphs of more than 16 curves. Note that 16 is the maximum number of data files that can be opened simultaneously by **D_2D**.

13. The graphic screen with the plot can be copied to a **PCX** or to a **DXF** file version **AutoCAD** 12, that is, **R12 DXF**. **PCX** is a common raster graphics format, while **DXF** is a vector format native to **AutoCAD** that can be read by many other graphing and CAD packages. There are also several **DXF** view programs available for download from the Internet as listed in the reference at the end of the chapter. If the **DXF 1:1** export option is selected, **D_2D** will write to **DXF** the plot curves only, and the scale factor will be unity along both axes. If the graph is isotropic, however, then the entire graph (curves and plot box with divisions, values, and labels) will be exported to **DXF** as a one-to-one image.

Important: If you export as **DXF 1:1** a graph trimmed by the plot box, unless it was set previously to be *isotropic*, the limits over either the x- or the y-axis will be displaced outwards from their current positions, and the **DXF** copy of your graph will appear truncated less or not truncated at all.

14. **D_2D** can generate animated graphs and comet plots and the frames in these animations can be exported to **PCX** or **DXF**. You can then assemble these **PCX** files as animated **GIFs** and post them on the Internet or insert them in **Power Point** presentations. I personally prefer the **GIF Animator** program available from www.gif-animator.com because it is affordable, easy to use, and accepts **PCX** files as input. In case you use a different **GIF**-animation software, you might have to convert the **PCX** frames generated by **D_2D** to other raster formats—see the list of graphics format converters at the end of the chapter.

```
┌─────────────────────────────────────────────────────────────────────────┐
│                      About   D_2D ver. 2014                               │
└─────────────────────────────────────────────────────────────────────────┘

┌───────────────────────────────────────────────────────────────────────────┐
│  *  Input binary files D2D (doubles) or R2D (reals) should be structured as:│
│       x1 F(x1) x2 F(x2) x3 F(x3) etc.                                       │
│                                                                             │
│  *  Input ASCII files should be structured as:                              │
│       x1  │ F1(x1) │ F2(x1) │ F3(x1) │ ..                                   │
│       x2  │ F1(x2) │ F2(x2) │ F3(x2) │ ..                                   │
│       x3  │ F1(x3) │ F2(x3) │ F3(x3) │ ..                                   │
│       x4  │ F1(x4) │ F2(x4) │ F3(x4) │ ..                                   │
│       etc.                                                                  │
│  with the column separator '|' any nonnumeric character less the followings:│
│   . - ! ! ~ o x + * > ■ [ ] ∅ % ◆ < > ▲ ^ ▼ υ U δ & ♀ q Q ✳ @            │
│                                                                             │
│  *  D2D files and ASCII files can include control lines as follows:         │
│   1.00E100 1.00E100  '====' or 'empty line' (new line or new animation frame)│
│                                                                             │
│   1.01E100 1.01E100  '%%%%' (change marker to transparent round          )  │
│   1.02E100 1.02E100  'oooo' (change marker to background-color filled round )│
│   1.03E100 1.03E100  '....' (change marker to solid round                )  │
│   1.04E100 1.04E100  '[][]' (change marker to square  -■-                )  │
│   1.05E100 1.05E100  '<><>' (change marker to diamond  -◆-               )  │
│   1.06E100 1.06E100  'υυυυ' (change marker to reversed triangle  -▼-     )  │
│   1.07E100 1.07E100  '^^^^' (change marker to triangle  -▲-              )  │
│   1.08E100 1.08E100  '****' (change marker to  -*-                       )  │
│   1.09E100 1.09E100  'xxxx' (change marker to  -x-                       )  │
│   1.10E100 1.10E100  '++++' (change marker to  -+-                       )  │
│   1.11E100 1.11E100  '@@@@' (change marker to  -✳-                       )  │
│   1.12E100 1.12E100  '&&&&' (change marker to  -δ-                       )  │
│   1.13E100 1.13E100  'qqqq' (change marker to  -♀-                       )  │
│   1.14E100 1.14E100  '!!!!' (change marker to vertical line  -!-         )  │
│   1.15E100 1.15E100  '>>>>' (change marker to arrow  ->-                 )  │
│                                                                             │
│   1.99E100 1.99E100  '~~~~' (suspend marker display                      )  │
│                                                                             │
│   2.01E100 2.01E100  'BLUE'          (change line color to blue          )  │
│   2.02E100 2.02E100  'GREEN'         (change line color to green         )  │
│   2.03E100 2.03E100  'CYAN'          (change line color to cyan          )  │
│   2.04E100 2.04E100  'RED'           (change line color to red           )  │
│   2.05E100 2.05E100  'MAGENTA'       (change line color to magenta       )  │
│   2.06E100 2.06E100  'BROWN'         (change line color to brown         )  │
│   2.07E100 2.07E100  'GRAY'          (change line color to gray          )  │
│   2.08E100 2.08E100  'BLACK'         (change line color to black         )  │
│   2.09E100 2.09E100  'LIGHTBLUE'     (change line color to light blue     )  │
│   2.10E100 2.10E100  'LIGHTGREEN'    (change line color to light green    )  │
│   2.11E100 2.11E100  'LIGHTCYAN'     (change line color to light cyan     )  │
│   2.12E100 2.12E100  'LIGHTRED'      (change line color to light red      )  │
│   2.13E100 2.13E100  'LIGHTMAGENTA'  (change line color to light magenta  )  │
│   2.14E100 2.14E100  'YELLOW'        (change line color to yellow        )  │
│                                                                             │
│   9.99E100 9.99E100  '!!!!'  (return to original color and marker setting )  │
│                                                                             │
│  *  In WINDOWS XP you can automatically open CF2, D2D or R2D files with      │
│  D_2D.EXE by selecting 'Open-With' from the File function of Windows Explorer.│
│                                                                             │
│  *  For the D3D, R3D and G3D formats see the D_3D program.                   │
└───────────────────────────────────────────────────────────────────────────┘
```

15. When you exit **D_2D**, the plot-box dimensions, number of divisions, and limits over the *x*- and *y*-axes, input data file name(s), and, in case of input ASCII file, the column numbers from where the *x* and *y* values were read, will all be saved to configuration file **!.CF2**. If the current plot has been created from scratch, these settings will be written to a new configuration file named generically **!0000001.CF2**, **!0000002. CF2**, and so on. The same will happen if you exit **D_2D** from the **<F1...4>** *screen*, irrespective if the current plot is the result of reading the settings from an existing

CF2 file (other than **!.CF2**) or has been generated new. If you exit **D_2D** at the end of the graphic session and you confirm overwriting the **CF2** file selected as input, the older version of the **CF2** file will be saved with the extension **OLD**, so that it can be restored manually if needed.

16. The **<F1>redo** and **<F2>redoo** options from the **D_2D** *start-up menu* allow to automatically redo the plot associated to the last **CF2** file found by alphabetically searching the current directory. These options are useful when the input data file(s) have been modified and the changes need to be assessed. The difference between them is that **<F2>redoo** will fit the graph to the plot box, while **<F1>redo** will apply the limits over the *x* and *y* axes as they were recorded to the last **CF2** file. The **<F3>CF2** option from the same *start-up menu* allows you to preview and run any **CF2** file on your hard drive. If you want to run a **CF2** file without preview, press the **<F4>** or **<CR>** keys to open it as if it were a regular input data file.

In order for you to make the most out of the aforementioned listed features of the **D_2D** program, a number of examples will be presented next that you can study before solving your own similar problems. You may also want to experiment directly with **D_2D** as the majority of its interactive menus are documented.

Important: There is a small number of plot settings that can be changed only by manually editing the **CF2** configuration file. Note that in a **CF2** file, everything that occurs between curl brackets will be considered comment. These settings are as follows:

(i) Plot-box width and height: Can be changed interactively from within **D_2D**. However, if you do not want these numbers rounded to multiples of 5, you will have to edit line number **4** of the **CF2** file. Note that the plot-box dimensions should be at most 625 × 405 pixels.

(ii) **DXF** polyline coincidence and collinearity parameters: When exporting a plot to **R12 DXF**, the **D_2D** program will eliminate any unnecessary collinear points, by concatenating into polylines as many line segments as possible. If three consecutive data points are found to be *almost* collinear, then the middle point will be eliminated. Similarly, if two separate points of a graph are *almost* coincident, then the second one will be eliminated. These *almost's* are controlled by two parameters read from line **28** of the **CF2** file.

(iii) Marker spacing: This parameter (line number **31** in the **CF2** file) is used with the −*− *marker-placing option*. If positive, then the curvilinear distance between two successive vertices will be measured in multiples of marker radii. If this marker spacing parameter is negative, then the distance between two successive markers will be measured as the number of data points without a glyph.

Important: There is no need to set the *marker spacing parameter* on line **31** of the **CF2** file to zero in order to display a marker at every data point, since this is equivalent to using the ✳✳✳ *marker-placing option*.

(iv) Default file extension: Line number **3** of the **!.CF2** file holds the extension of the files that will be listed for input when you press <F4> from the *start-up menu*. In the copies provided with this book, this line of **!.CF2** reads ***.D2D**. If you want text files to be listed for input instead, then change this line so ***.TXT** (any two or three character file extension is acceptable, including **CF2**). In any other **CF2** file on this third line, it is recorded the title of the plot.

(v) Number of bins in a histogram plot (between 3 and 500): If you read the *x* values from an ASCII file and select column 0 (which is inexistent) from where to read the *y* values, then by default, **D_2D** will plot the data as a histogram. You can set the number of bins in a histogram plot interactively right after you selected the data file. If you want to modify the number of bins and keep all the other settings unchanged, you must edit line number **32** of the configuration file and redo the plot.

Important: When you run **D_2D** with settings from a given **CF2** configuration file, the referred input data files should be either in the folder where they were located when the plot was originally made (i.e., specified by the path that precedes their name in the **CF2** file) or in the same folder where the calling **CF2** file is located. If no configuration file can be found in the same directory with **D_2D.EXE**, the program will not run. Likewise, if a copy of the **DXF.HED** file is not available in the same directory with **D_2D.EXE**, no **DXF** export will be possible.

Important: In **Windows XP**, you can link **CF2** configuration files and data files (extensions **D2D**, **R2D**, etc.) to the **D_2D.EXE** program on your hard drive by editing the *Open With* properties of these files: From **Windows Explorer** select the file you want to link, then from File → Open With → Chose Program menu, select **D_2D.EXE**. Before you click OK, check the option "Always use the selected program to open this kind of file."

1.1 ANALYTICAL FUNCTION PLOTS

Let us begin by graphing the function

$$F(x) = \frac{1}{(x-1)^2 + 0.1} + \frac{1}{(x-3)^2 + 0.2} - 3 \qquad (1.1)$$

over the interval $-1 < x < 5$ like in Figure 1.1. The data file readable by **D_2D** used to do this plot has been generated with the **P1_01.PAS** program listed in Appendix B. **P1_01.PAS** outputs the same data in three different formats, all named **F1_01**, as follows: an ASCII file with the extension **DTA**, a file of doubles with the extension **D2D**, and a file of reals

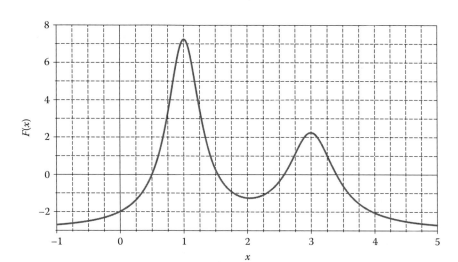

FIGURE 1.1   Plot of the function in Equation 1.1. Configuration file to redo this plot **F1_01.CF2**.

with the extension **R2D**. Since *plot-curve segmentation, color,* and *marker type* cannot be controlled from within a **R2D** file, only the first two data file formats will be emphasized in the remainder of this book.

To recreate the plot in Figure 1.1, launch the **D_2D** program, then press <F3> and open **F1_01.CF2** for preview. Use the <↑>, <↓>, <Page Up>, and <Pg Dn> keys to scroll up and down and inspect this **CF2** file. Press <CR> to confirm your selection or <Esc> to open for preview a different **CF2** file. <CR>, shortcut for Car Return, is the <Enter> key.

Important: To upload a file from the *file-open menu*, you can type its name in the *address line* directly or select it by pressing the <Tab> key first and then navigate the list using the arrow keys. Once highlighted, press <CR> to bring it into the address line, then press <CR> again to confirm your selection and to open it.

D_2D allows you to align the divisions over the *x*- and *y*-axes either with the *origin* or with the *corners* of the plot box: From the *final-graphic screen*, press the <Backspace> key (<Back> in short) to go to the *plot-box edit screen*, then press <Ctrl> and <F1> simultaneously to toggle between the two modes of labeling the *y*-axis. Since the divisions over the *x*-axis are aligned with both the origin and the ends of the axis; the <Ctrl> + <F5> key combination will have no visible effect upon this particular graph.

To change the limits over either *x*- or *y*-axis, press the <Back> key until the program switches to *text mode*. This page will be further referred to as the **<F1...4>** *screen.* Then press <F1> to change limits over *x* and *y*, then type '1' for category *F(x)* and press <CR>. A series of text boxes will let you modify (i) the upper and lower limits over the *y*-axis, (ii) the total number of values that will be written along the *y*-axis (this is equal to the number of major division lines), (iii) the Δy range between two major division lines, and (iv) the number of intervals delimited by inserting minor division lines between two values (i.e., typing '1' will introduce no minor division line). Notice that settings (ii) and (iii) cannot be changed independently. One other option that can be set here is to force the plot to remain isotropic, by automatically adjusting the limits over x and y, as the size of the plot box is being interactively modified.

As you move to the *final-graphic screen* to see the effect of these changes, stop while in the *plot-box edit screen* and press the <G> key several times to toggle between showing and hiding the gridlines. If you press <Ctrl> + <G>, you will be allowed to edit the inner and outer lengths of the *major division lines*. The appearance of the *minor division lines* will also change because they are set by default to 60% of the *major lines* outside length. Towards the inside of the plot box, all division lines will have the same lengths.

Press the <Esc> key from the *final-graphic screen* to exit **D_2D**. This will update the current **CF2** file. You can also exit from the **<F1...4>** *screen*, case in which **D_2D** will generate a new configuration file named **!0000001.CF2** and a temporary file named **D2D0001.$2D**. The latter is a work copy of the plotted data in **D_2D** format. When reading data from an ASCII file, **$2D** will hold copies of the x and y columns that were plotted on the graph. When you read data from multiple files or if you extract more than one (x, y) pairs from the same ASCII files, there will be more than one **$2D** file created.

Important: If you think you could use any of the temporary **$2D** files, change their extension to **D2D** before launching **D_2D** again, or otherwise, they will be erased.

1.2 SHOWING EXTREMA AND ZEROS OF GRAPHS

The function in Equation 1.1 exhibits one *minimum point*, two *maximum points*, and *four zeros*. Finding the zeros of this function requires solving the equation $F(x) = 0$, while finding its minimum and maximum points requires solving the equation $dF(x)/dx = 0$, very unappealing tasks if you are doing them manually. One approach is to approximate the coordinates of these points with the help of the divisions or gridlines of the plot and then use these approximations as initial guesses in some minimization or zero finding iterative schemes.

Instead of solving the equation $F(x) = 0$ of finding the minima and maxima of $F(x)$, you can have **D_2D** inspect the input data and identify any zero or extrema that will be encountered. These can be displayed on the graph as shown in Figure 1.2 and, if desired, can be exported with added decimals to three ASCII files of extensions **ZER**, **MIN**, and **MAX**. Evidently, the precision with which these zeros and extrema are approximated depends on how fine your function has been sampled in the first place.

To redo Figure 1.2 other than by running **D_2D** with setting from the **F1_02.CF2** configuration file, launch **D_2D** and upload the same **F1_01.CF2** file as before (press <F3> and type or select from the list **F1_01.CF2**). After you plot the graph, go back to the **<F1...4>** *screen* and press <F3>. Type 'E' and '1', and then press <CR> twice (you should be under the word 'Line' on the top of the screen). Use the <↑>, and <↓> keys to change the line type from ==== to ——— and press <CR>. Next, select the **/#** marker-placing option using the same arrow keys and press <CR> several times until you get to the *final-graphic screen*. To suppress the gridlines, go back to the *plot-box edit screen* and press the <G> key. Your graph should now look like the one in Figure 1.2.

To write to ASCII the coordinates of these minima, maxima, and zeros, go to the **<F1...4>** *screen* and select option **<F4>**. Scroll up and down through these export options using the

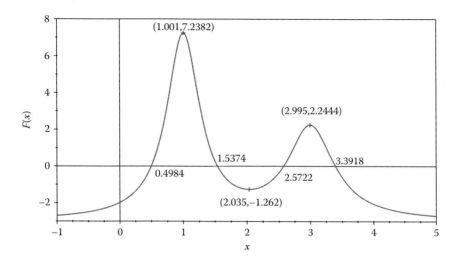

FIGURE 1.2 Plot of the function of Equation 1.1 with 501 data points showing the zeros, minimum, and maximum points automatically introduced by **D_2D**. Configuration file **F1_02.CF2**.

<↑>, and <↓> keys, select Envelopes and then press <CR>. You can accept the default names of the three export files where data will be written or specify your own.

Important: Zeros and extremum points coordinates cannot be exported to ASCII file when the plot originates from multiple data files.

A short note on how the coordinates of these minimum, maximum, and zero points were evaluated: As data are read from file, **D_2D** looks for groups of three successive points with the middle one located above or below the other two (this is called *three-point bracketing*). If it occurs, then between the first and the third point, a local minimum or maximum exists. If you press <F2> from the **<F1...4>** *screen*, you can then choose to either display on your graph the middle point, or the singular point of the parabola that interpolates the three bracketing points. In a similar manner, **D_2D** brackets the zeros of the graph by looking for two successive *y* values of different signs. If this happens, then the intersection of the line that connects these two points with the *x*-axis is an approximation of the respective zero, like in the *secant method* of zero finding. There may be more than one zero between two points that change sign (similar argument can be made about bracketing an extremum). To avoid any ambiguity, the function must be sampled at a rate small enough to capture its 'convolutedness' (see the *Nyquist–Shannon sampling theorem*).

Let us now plot the graph of the first derivative of the function in Equation 1.1, that is,

$$F'(x) = \frac{2x-2}{(x^2-2x+1.1)^2} + \frac{2x-6}{(x^2-6x+9.2)^2} \tag{1.2}$$

The data files used to plot this second function were generated with the **P1_02.PAS** program listed in Appendix B (see lines #2 and #3 in this program). To illustrate additional features of **D_2D**, both function $F(x)$ and its derivative $F'(x)$ were plotted on the same graph, with $F'(x)$ on a secondary *y*-axis (Figure 1.3).

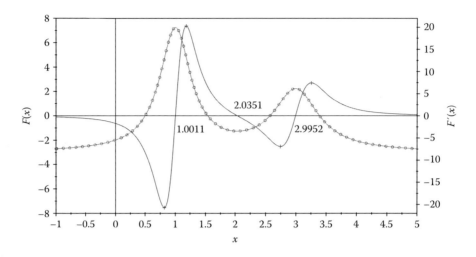

FIGURE 1.3   Plot of the function of Equation 1.1 as a line with markers, and of its derivative $F'(x)$, showing the zeros of the derivative. Configuration file **F1_03.CF2**.

To duplicate this figure, run **D_2D** with settings from **F1_03.CF2**. To make the plot look like Figure 1.4, and then export it to **R12 DXF**, perform the following steps: From the **<F1...4>** *screen*, press <F2>. Change the title to '$F(x)$ & $F'(x)$', then increase the marker size to 4.5 pixels. To make the text colors of the two *y-category names* identical to the color of the respective plot lines or to modify the marker type along the $F(x)$ curve, go to the **<F1...4>** *screen* and press <F3>. Type 'E', then '1', and press <CR> to select the first curve; insert a space in front of the first *y*-category name. Press <CR> three more times, then scroll up to change marker type from transparent round ∅ to diamond ◇, then press <CR> twice. Repeat the procedure and add a space in front of the second *y-category* name to change its color from the default *dark gray* to *blue*, that is, the color of the second plot line.

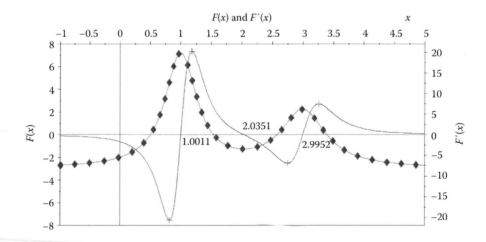

FIGURE 1.4   **R12 DXF** copy of the plot in Figure 1.3, exported after some formatting as explained in the text. This is an **AutoCAD** screenshot after issuing the command *shade*. Configuration file **F1_04.CF2**.

Important: If there are multiple curves per category, the *y*-axis label will take the color of the first curve assigned to that category.

Go to the *plot-box edit screen* and use <Pa Up>, <Pg Dn>, <↑>, and <↓> keys to resize the plot box, then press <Ctrl> + <Pa Up> to move the *x*-axis divisions and values to the top of the graph. When you are satisfied with the appearance of the graph, go back to the **<F1...4>** *screen*, press <F4> to do a **R12 DXF** export, enter the file name, and press the <CR> key. After the **DXF** export is completed, press <Esc> to exit **D_2D**.

To view the **DXF** file that you have just created, open a new drawing in **AutoCAD**, type 'dxfin' at the command line, then type 'hide'. Because the ◊, □, ▽, and △ markers are **AutoCAD** *regions* placed slightly elevated by **D_2D**, they will obscure the plot lines. **AutoCAD** circles behave the same, so opaque round markers will also obscure the plot lines behind then following the *hide* or *shade* commands.

In this previous example, input data were read from two separate files, that is, **F1_01.D2D** and **F1_02.D2D**. Note that it is possible to store the values of both the function and of its derivative in the same ASCII file and simplify data management.

1.3 STEM AND AREA PLOTS: LENGTH OF A CURVE AND AREA UNDER A CURVE

It was mentioned earlier that **D_2D** can generate *stem* and *area plots*. Both are produced by choosing **!!!!!!** from the **Line and markers** *option* of the **<F3> edit add remove lines** *menu*.

You can quickly redo the plot in Figure 1.5 by running **D_2D** with **F1_05.CF2** as input. The data file required is the same **F1_01.D2D** from before.

If you want to create the plot from scratch, launch **D_2D** and press <F4>, then select **F1_01.D2D** (press <Esc> to abort the uploading of additional data files). From the

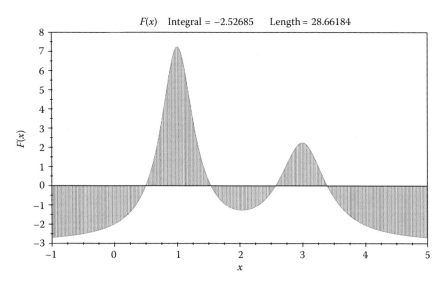

FIGURE 1.5   Area plot of the function in Equation 1.1 with 501 data points. Configuration file **F1_05.CF2**.

<F1..4> *screen*, go to **<F3> edit add remove lines** and using the arrow keys, change the marker-placing option from **-*-** to **!!!!!!**. Also, change the marker size to a smaller value, that is, 1 or even 0.

Finally, for the graph to look exactly like the one in Figure 1.5, you must suppress the *x* = 0 line (the vertical crosshair) by pressing <F5> while in the *plot-box edit screen*.

To create an actual stem plot, input data points should be in smaller number, that is, tens rather than hundreds of points. To generate the plot in Figure 1.6, the number of plot points **nX** inside **P1_01.PAS** was changed from 501 to 61. With this modification and with only a **D2D** file as output (i.e., **F1_03.D2D**), the program was renamed **P1_03.PAS**—see source code in Appendix B. The procedure to generate stem plots is the same as for area plots, with the difference that in the **Line and markers** *section* of the **<F3> edit add remove lines** *menu*, you must choose **no line** instead of the default —— line.

Using this new data file, the area plot in Figure 1.7 has been generated. Notice the differences between the numerically calculated length of the curve and the value of the integral in Figures 1.5 and 1.7. The integral (the area between the curve and the *x*-axis) has been evaluated using the *trapezoid rule of integration* (see Appendix A), which has the benefit that it can handle easily data sampled both at constant and at variable *x* step and the cases where the curve is trimmed by the plot box like in Figure 1.8.

If you want to create your own trimmed-area plot, after generating Figure 1.7, go to the **<F1..4>** screen. Press **<F1>** and modify the lower limit over the *F*(*x*) axis from −3 to −2 and the upper limit from 8 to 3.

You may also want to redo Figure 1.2 using the **F1_03.D2D** file with only 61 data points and observe the effect of *interpolating for extrema*. One very quick way of doing this is to open the **F1_02.CF2** file and on line #**38**, change the input file name from **F1_01.D2D** to **F1_03.D2D**, then run **D_2D** with settings from this modified **CF2** file.

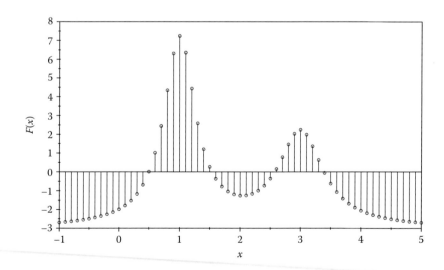

FIGURE 1.6 Stem plot of the function in Equation 1.1 with only 61 data points. Configuration file **F1_06.CF2**.

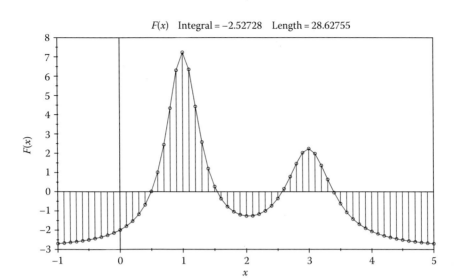

FIGURE 1.7 Area plot of the function in Equation 1.1 with only 61 data points. Configuration file **F1_07.CF2**.

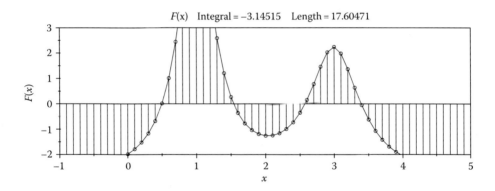

FIGURE 1.8 Same plot in Figure 1.7 restricted to $-2 \leq F(x) \leq 3$. Configuration file: **F1_08.CF2**.

1.4 WINDOWING AND PANNING

Oftentimes, you want to view just a portion of a graph or scroll left and right through your plot. If you export your graph to **R12 DXF** and then open it in **AutoCAD**, you can zoom in or crop the graph any way you like. You can also turn it into a block and then reinsert it scaled at differ-ent rates over *x* and *y*. Similar maneuvers can be done directly from within **D_2D** by modify-ing the limits over the *x*- and *y*-axes from **<F1> change limits over X and Y** *option* as described earlier. To translate your graph to the left or to the right, there is the **<P>an** com-mand available from the *final-graphic screen*. When pressing <P>, you are prompted to type in the amount you want the graph to be displaced horizontally. If you type a positive number, the graph will be translated to the left, while a negative number will translate the graph to the right. A zero input will bring the graph back to its initial location.

Important: If your plot is in a displaced position fallowing a `<P>an` command, then the current *x*-axis limits will be saved to the `CF2` configuration file.

1.5 NUMBERING DATA POINTS

With `F1_03.D2D` file, 61 data points used to plot Figures 1.6 through 1.8 will be used next to demonstrate the capability of `D_2D` to automatically number the points on a graph. Start by replotting Figure 1.7 using the `F1_07.CF2` configuration file. When finished, go to the `<F1..4>` *screen* by pressing <Back> twice and then press <F3>. Type 'E' and then '1' and <CR> to edit the appearance of the graph. Advance horizontally by pressing <CR> and use the arrow keys to change the line type from —— *thin solid* to ≡≡≡ *thick dashed*. Then change the marker-placing option from !!!!!! to *#* and the marker type from *transparent round ∅* to *solid round ○*.

When you get under the `Label pattern`, a box with the text `1:4:61;` will open up. This is the default marker labeling option, interpreted by `D_2D` as "number the first data point as 1, then label every 4th data point up to the last point of the graph, that is, point 61" (point number 61 will be labeled whether it is multiple of 4 or not).

Notice how labels are always placed on the outside of the curve. This requires `D_2D` to estimate the center of curvature around the point to be numbered (i.e., to calculate the center of the circle through this current data point and its two neighbors), and use it as a reference for placing the label.

Here are a few more numbering patters that you may want to experiment with:

`P0:5:60;` this will number every 5th point starting from 0 and will add a *P* in front of every label. You can replace *P* with any other character. Note however that some characters from the extended ASCII set do not have an equivalent in **AutoCAD**.

`1:1:10;50:1:61;` this will number the first 10 and the last 10 data points only.

`1:1:10;2:27;5:40;57;` this will begin with 1, label every point until the 10th, then continue labeling every other point until the 27th point (including), then will label every 5th point until the 40th, and will finally label point number 57.

The plot with this last numbering pattern is available in Figure 1.9. After you run the configuration file `F1_09.CF2`, in order to obtain the exact same appearance as in Figure 1.9, you must remove both the *x* = 0 and *y* = 0 lines by pressing <F1> and <F5> while in the *plot-box edit screen*.

Important: When written to `CF2`, axis division and value placement can be altered due to round offs and may not be recreated exactly. To obtain the exact appearance of the divisions and values along the *x*-axis as in Figure 1.9, after you launch `D_2D` with settings its `CF2` file, go to the `<F1..4>` *screen*, follow option `<F1>`, and modify the interval between two major division lines over the *x*-axis from 0.9 to 1.

After you export the graph to `R12 DXF`, and open it with **AutoCAD**, type 'hide' at the command line so that the glyphs will obscure the line. Also type 'ltscale' and change its value to 10, so that the dashed line will look *tighter*.

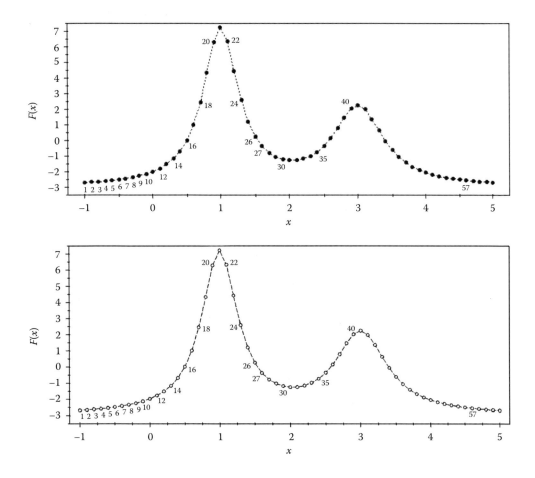

FIGURE 1.9   Plot with automatically labeled data points. Above is a **PCX** copy, below a **DXF** copy showing the effect of the **AutoCAD** *hide* command. Configuration file **F1_09.CF2**.

1.6 PLOTTING FUNCTIONS WITH SINGULARITIES

Functions exhibiting singular points pose additional challenges when it comes to graphical representation. One such example is the function in the following equation:

$$F(x) = x(x^2 - 3)/(x^2 - 4) \tag{1.3}$$

If we are to generate the data to plot this function for $-8 < x < 8$, we must avoid evaluating it for $x = \pm 2$, or otherwise the computer will report *division by zero*. Even when the division by zero is bypassed by checking the value of the denominator, the following two situations are likely to occur: (i) if the function is sampled at a very fine rate, large spikes will occur at singularities (see Figure 1.10a); (ii) irrespective of the sampling rate, the left and right limits at a singular point should not be connected, since the function is not defined here (Figure 1.10b).

In this section, additional features of **D_2D** that were implemented to address such issues will be presented. One is the possibility to suppress the lines that connect two data points located outside the plot box. The other is the possibility of controlling the plot-line interruption directly from within the data file.

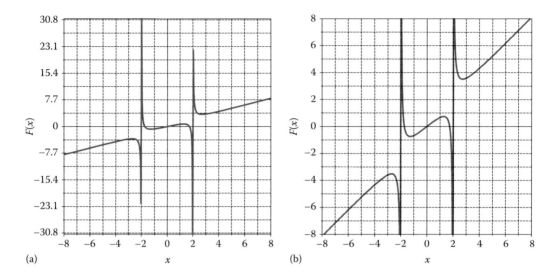

FIGURE 1.10   Plots of Equation 1.3 with 400 data points, done using configuration files **F1_10A.CF2** (a) and **F1_10B.CF2** (b). At $x = -2$ and $x = 2$, the graph should be discontinuous.

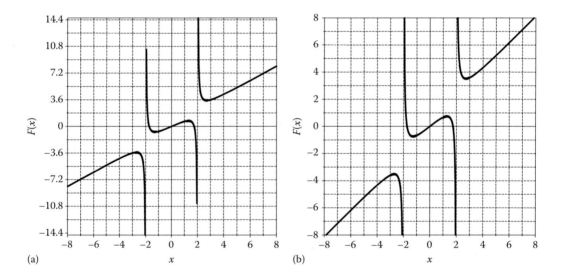

FIGURE 1.11   Graphs of the function in Equation 1.3 with 401 data points and plot-line breakers done using configuration files **F1_11A.CF2** (a) and **F1_11B.CF2** (b). Figure (b) can also be produced using one of the 400 data file, but after editing the limits over the y-axis, you must press the <C> key when in the *final-graphic screen* to disconnect the graph at $x = -2$ and $x = +2$. (i.e., remove curtains).

The **D2D** and **DAT** files used to plot the graph in Figures 1.10 and 1.11 were generated by program **P1_10.PAS** listed in Appendix B. Before the actual function is evaluated, the program checks whether x is *almost* equal to -2 or $+2$ (see line #**14** of the source code), and if found true, then the function is assigned the constant **InfD** defined in unit **LibMath**, which is equal to 10^{100}. Constant **EpsD**, defined in the same unit **LibMath**, is a very small positive number set equal to 10^{-100}.

Important: In all programs provided in this book, 1.0E100 is considered equal to *infinity*, and, together with several of its multiples (1.01E100, 1.02E100, etc.), is used to code additional information about the input data or of the state of the procedure from where the value originates. The **About** screen of **D_2D** (you can bring it up by pressing <F10> right after launching **D_2D.EXE**) explains how these multiples of 1.0E100 can be used to control the color, marker, and interruption of a plot line. According to this protocol implemented in the **P1_10.PAS** program, as the function is sampled at constant step, any time a singular point occurs, a *plot-line breaker* is written to the data file (i.e., the value 1.0E100), which will instruct the **D_2D** program not to connect the two points that the respective *breaker* separates.

P1_10.PAS was run twice: once for **nx** equal to 400 plot points when the files generated were named **F1_10.D2D** and **F1_10.DAT**, and a second time for **nx** equal to 401 plot points, when the same files were named **F1_11.D2D** and **F1_11.DAT**. For **nx=400** points, no division by zero actually occurred, and the corresponding plot looks as shown in Figure 1.10a. After editing *y*-min and *y*-max (see Figure 1.10b), you can eliminate the two extraneous vertical lines that connect the left and right limits at the singular points by pressing the <C> key (C stands for *curtain*) when in the *final-graphic screen*. The lines connecting two points lying outside the plot box, one above and one below, are called *curtains*. With **nx=401** data points, however, division by zero do occur at $x = -2$ and $y = 2$. In this case, the **P1_10.PAS** program wrote to the **D2D** file two 1.0E100 values, while to the **DAT** file, it wrote a control line, that is, ----------. Both the 1.0E100 value pair and the ---------- line are interpreted by **D_2D** as *line breakers*, and the resulting graphs will appear like in Figure 1.11.

Now, it is a good opportunity for you to experiment with the **R12 DXF** copies of Figure 1.10 or 1.11 (notice that Figure 1.10b and b are of *isotropic* type). The **DXF 1:1** exports of Figures 1.10a and 1.11a will include only the curve, while the same of Figures 1.10a and 1.11a (which are isotropic) will include the graph together with the plot box scaled 1 to 1. In both cases, the origin of the drawing will coincide with the origin of the graph. If you perform a regular **DXF** output of these figures and open them in **AutoCAD**, you will notice that the origin of the graph will be located somewhere outside the plot box and that the dimensions of the plot box will be equal to those from **D_2D**.

One unnatural thing about Figure 1.11a is that at the singular points the plot line does not extend all the way to the plot-box edge, that is, the graph should look like Figure 1.11b irrespective of the minimum and maximum limits set over the *y*-axis. To remedy this, we must differentiate between the $-\infty$ and $+\infty$ limits at a point. The solution implemented in the **P1_12A.PAS** program (see Appendix B) was to evaluate the sign of the function *slightly left* and *slightly right* of the *x* point at which 1.0E100 is returned (see the **DX** variable calculated on line #**23** of program **P1_12A.PAS**) and write to file either −1.0E100 or +1.0E100 as limits of the function to the left and to the right of the singular point. Files **F1_12A.D2D** and **F1_12A.DAT** were generated this way and have been used to produce the graphs in Figure 1.12a. Note that the *plot-line breakers* are not essential since the *curtains* can be removed from within the **D_2D** program.

In the aforementioned examples, the singular points were assumed known. A fully capable function-plotting program should be able to identify these automatically and

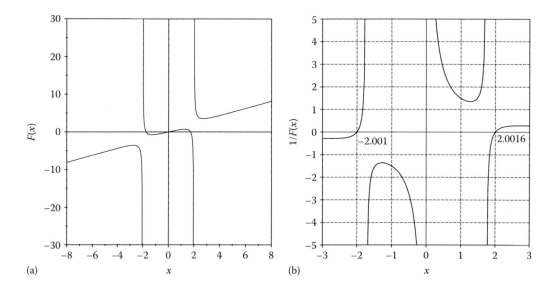

FIGURE 1.12 (a) Same graph as in Figure 1.11a, with ±1.0E100 assigned to the function value at the singular points. Note that irrespective of the limits over the *y*-axis, the plot lines will extend up to the plot-box edge. (b) Graph of 1/*F*(*x*) in Equation 1.3 with the *curtains* removed. Configuration files **F1_12A.CF2** and **F1_12B.CF2**.

increase the sampling rate around them. Remember that for 400 data points, the singular points were not detected.

As Figure 1.12b indicates, the singular points of *F*(*x*) coincide with the roots of equation 1/*F*(*x*) = 0. The data files needed to generate this last figure were produced with **P1_12B. PAS** available with the book, which was straightforwardly obtained by modifying earlier program **P1_10.PAS**, where function *F* was replaced with 1/*F*.

1.7 CONTROLLING PLOT FEATURES FROM WITHIN THE INPUT DATA FILE

In addition to type and color, **D_2D** allows marker occurrence to be controlled from within the data file, that is, they can be turned *off* and back *on*. However, their style cannot be set or changed, say from '*******' to '**-*-**' or '***#***', and a line without markers cannot be turned into a line with markers from within the data file.

Important: You will have to assign some type of markers to your graph from within **D_2D** in order for the input data file control lines to have an effect.

To exemplify, open the ASCII data file **F1_11.DAT** using **Notepad** or other ASCII editor and insert two empty lines right before the first plot-line breaker '**====**' (this should be on line **153** from the top). Then type '**<><><><>**' on one of these lines, and on the other one, type the word 'Red'. These will change, from that point over, the marker type to diamond and the color to *red*. To limit these changes only to the middle portion of the plot, scroll down to line **253** and insert above the second plot-line breaker an empty line on which type '**!!!!!!!!!!**'—this will restore the original marker type and line color (see Figure 1.13a). Save the **F1_11.DAT** file twice: once

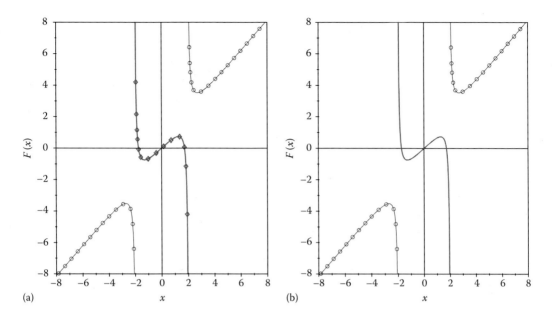

(a)

(b)

FIGURE 1.13 Plots of data files (a) **F1_13A.DAT** and (b) **F1_13B.DAT** having additional control lines inserted as described in the text. Configuration files: **F1_13A.CF2** and **F1_13B.CF2**.

under the name **F1_13A.DTA** and a second time under the name **F1_13B.DTA**. Open this second ASCII file and change the line of '<><><><>' you inserted earlier, into '~~~~~~'. This will suspend the marker display until reaching the *reset* line '!!!!!!!!!!' (see Figure 1.13b).

If you want to redo the plots in Figure 1.13, launch **D_2D.EXE** and press the <F3> key to load one of the configuration files **F1_13A.CF2** and **F1_13B.CF2**.

There are several other instances where controlling graph-line interruptions from within data file can become useful. For example, Figure 1.14 consists of over 130 distinct polylines, the vertices of which are read from a single data file, that is, **F1_14.XY**.

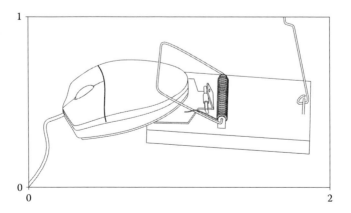

FIGURE 1.14 Example of a plot created with **D_2D** where line-break controls were used multiple times. Configuration files: **F1_14.CF2**.

The *plot-line breakers* in this file play similar to the 'pen up' and 'pen down' commands a plotter receives when in operation. This figure was generated staring from a photograph that was opened inside **AutoCAD**, and its contours traced with polylines. The drawing was then exported to **R12 DXF**, and using the **UTIL~DXF** program described in Chapter 3, and the vertices of these polylines were then written to ASCII file **F1_14.XY**.

Another application of the line segmenting capability of **D_2D** is on plotting *families of curves*, with data read from a single file. Let us consider the *amplitude ratio* of a damped forced linear oscillator function:

$$H(\zeta\Omega/\Omega_n) = \left[\left(1-\left(\Omega/\Omega_n\right)^2\right)^2 + \left(2\zeta\cdot\Omega/\Omega_n\right)^2\right]^{-0.5} \tag{1.4}$$

and plot 10 separate curves corresponding to the *damping ratio* ζ, between 0.1 and 1, and for the *frequency ratio* Ω/Ω_n between 0 and 2.5 like in Figure 1.15.

One possibility is to write the data to 10 separate files (one file per each ζ value) and plot them on the same graph (**D_2D** can read data from up to 16 different files). Alternatively, an ASCII file with 11 columns can be generated: one column for the independent variable, that is, Ω/Ω_n, and 10 columns for each damping ratio value.

The third possibility is to write a program with two nested *for loops* that will output data to the same file. The points belonging to one curve must be separated from those of other curves using line breakers. The **P1_15.PAS** program (see Appendix B) implements such an approach and serves to create data files **F1_15.D2D** and **F1_15.DAT** available with the book. The plot in Figure 1.15 has been generated using the first of these files, and then it was exported to **R12 DXF**.

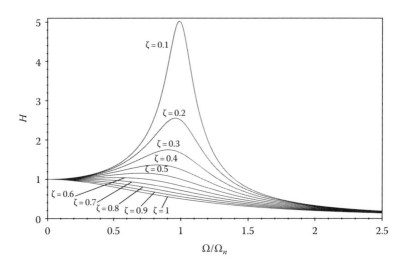

FIGURE 1.15 Family of curves plotted using either **F1_15.D2D** or **F1_15.D3D**. Configuration files **F1_15D2D.CF2** or **F1_15D3D.CF2**. **D3D** are specific to program **D_3D** discussed in Chapter 2. Curve and axis labeling has been done inside **AutoCAD**.

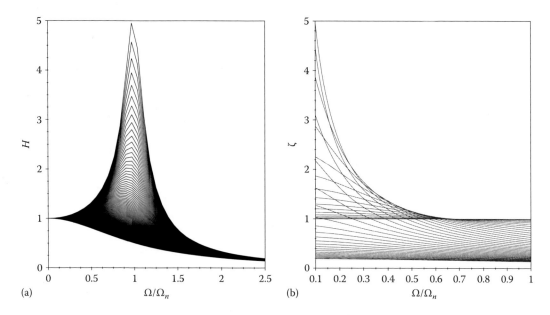

(a) Ω/Ω_n (b) Ω/Ω_n

FIGURE 1.16   2D projections of function $H(\zeta,\Omega/\Omega_n)$ in Equation 1.4, done by plotting the **F1_16.D3D** data file using **D_2D** with input options **f(x,y) vs x** (a) and **f(x,y) vs y**, (b).

The fourth solution to the same problem is to generate a **D3D** data file, that is, **F1_15. D3D** using program **P1_156.PAS** (see Appendix B), and then plot this file using **D_2D**. The **D3D** files are specifically formatted to be read using the **D_3D** program described in Chapter 2—see Figures 2.1, 2.6, and 2.7, which are plots of the function $H(\zeta,\Omega/\Omega_n)$ in Equation 1.4. When a **D3D** data file is opened using the **D_2D** program, a temporary file of double (extension **$2D**) is first created, which employs *plot-line breakers* to separate the individual x = constant or y = constant lines. Note that **D_2D** allows two separate plotting options for **D3D** files, that is, **f(x,y) vs x** and **f(x,y) vs y**, which correspond to the *side view* and *front view* of the $f(x,y)$ function surface, respectively (see Figure 1.16).

1.8 PLOTTING SCATTERED DATA

Plotting scattered data, illustrated by an example in this section, is common to experimental data analysis and statistics. Inequalities of two variables can also be represented graphically as large collections of scattered points as will be shown later.

The plot in Figure 1.17 has been generated using the **F1_17.DTA** file and represents the life of six groups of bearings subjected to various operating conditions. Notice that group sizes are not identical, so in order to keep the file structured orderly, dots were used as place holders (see the **F01_17.DTA** insert) although any nonnumerical character can be employed. Alternatively, you can rearrange the columns from longest to shortest as visible in the **F01_18.TXT** insert.

F01_17.DTA

G1	L1	G2	L2	G3	L3	G4	L4	G5	L5	G6	L6
1	35.4	2	40.1	3	49.3	4	51.7	5	50.6	6	54.0
1	34.8	2	38.6	3	48.3	4	51.3	5	50.3	6	53.3
1	33.5	2	38.3	3	47.7	4	50.5	5	50.1	6	52.7
1	32.4	2	37.6	3	47.2	4	50.2	5	50.0	6	52.4
1	31.6	2	37.3	3	47.0	4	49.4	5	49.7	6	51.8
1	31.4	2	36.5	3	47.0	4	49.2	5	49.0	6	51.3
1	31.3	2	36.5	3	46.9	4	49.1	5	48.5	6	50.8
1	30.7	2	35.9	3	45.9	4	49.1	5	48.3	6	50.1
1	30.4	2	35.5	3	44.5	4	48.3	5	47.7	6	48.7
1	30.2	2	35.3	3	44.0	4	48.0	5	46.8	6	48.1
1	30.0	2	34.6	3	43.3	4	47.7	5	46.8	6	47.6
1	29.3	2	34.4	3	42.0	4	47.5	5	46.2	6	47.5
1	28.3	2	33.5	3	41.9	4	47.2	5	45.0	6	47.5
1	27.5	2	33.3	3	41.6	4	46.7	5	44.3	6	47.5
1	27.1	2	33.0	3	40.1	4	46.3	5	43.1	6	45.3
1	26.6	2	32.5	3	38.9	4	45.2	5	42.9	6	44.2
1	26.3	2	32.3	3	37.3	4	44.4	5	42.4	6	43.8
1	25.1	2	31.7	3	36.4	4	43.6	5	41.8	6	43.6
1	24.7	2	31.6	3	35.8	4	42.7	5	40.6	6	43.3
1	24.1	2	33.1	3	35.4	4	42.5	5	39.9	6	43.1
1	22.1	2	31.4	3	35.0	4	42.0	5	39.2	6	42.7
1	21.8	2	31.0	3	34.8	4	41.3	5	38.5	6	42.3
1	18.0	2	30.1	3	33.5	4	40.7	5	38.0	6	42.2
.	.	2	29.3	3	31.5	4	39.9	5	36.3	6	40.8
.	.	2	28.8	3	30.6	4	38.9	5	34.0	6	40.0
.	.	2	24.5	3	29.2	4	37.7	5	32.2	6	37.2
.	.	2	21.9	3	26.3	4	37.2	5	27.5	6	36.6
.	.	2	19.3	3	23.4	4	35.4	5	24.2	6	33.9
.	.	.	.	.	.	4	34.3	.	.	6	29.4
.	.	.	.	.	.	4	32.5	.	.	6	47.6
.	.	.	.	.	.	4	31.1	.	.	.	.
.	.	.	.	.	.	4	29.8	.	.	.	.
.	.	.	.	.	.	4	25.7	.	.	.	.
.	.	.	.	.	.	4	23.5	.	.	.	.

F01_18.TXT

G4	L4	G6	L6	G2	L2	G3	L3	G5	L5	G1	L1
4	51.7	6	54.0	2	40.1	3	49.3	5	50.6	1	35.4
4	51.3	6	53.3	2	38.6	3	48.3	5	50.3	1	34.8
4	50.5	6	52.7	2	38.3	3	47.7	5	50.1	1	33.5
4	50.2	6	52.4	2	37.6	3	47.2	5	50.0	1	32.4
4	49.4	6	51.8	2	37.3	3	47.0	5	49.7	1	31.6
4	49.2	6	51.3	2	36.5	3	47.0	5	49.0	1	31.4
4	49.1	6	50.8	2	36.5	3	46.9	5	48.5	1	31.3
4	49.1	6	50.1	2	35.9	3	45.9	5	48.3	1	30.7
4	48.3	6	48.7	2	35.5	3	44.5	5	47.7	1	30.4
4	48.0	6	48.1	2	35.3	3	44.0	5	46.8	1	30.2
4	47.7	6	47.6	2	34.6	3	43.3	5	46.8	1	30.0
4	47.5	6	47.5	2	34.4	3	42.0	5	46.2	1	29.3
4	47.2	6	47.5	2	33.5	3	41.9	5	45.0	1	28.3
4	46.7	6	47.5	2	33.3	3	41.6	5	44.3	1	27.5
4	46.3	6	45.3	2	33.0	3	40.1	5	43.1	1	27.1
4	45.2	6	44.2	2	32.5	3	38.9	5	42.9	1	26.6
4	44.4	6	43.8	2	32.3	3	37.3	5	42.4	1	26.3
4	43.6	6	43.6	2	31.7	3	36.4	5	41.8	1	25.1
4	42.7	6	43.3	2	31.6	3	35.8	5	40.6	1	24.7
4	42.5	6	43.1	2	33.1	3	35.4	5	39.9	1	24.1
4	42.0	6	42.7	2	31.4	3	35.0	5	39.2	1	22.1
4	41.3	6	42.3	2	31.0	3	34.8	5	38.5	1	21.8
4	40.7	6	42.2	2	30.1	3	33.5	5	38.0	1	18.0
4	39.9	6	40.8	2	29.3	3	31.5	5	36.3	1	
4	38.9	6	40.0	2	28.8	3	30.6	5	34.0	.	.
4	37.7	6	37.2	2	24.5	3	29.2	5	32.2	.	.
4	37.2	6	36.6	2	21.9	3	26.3	5	27.5	.	.
4	35.4	6	33.9	2	19.3	3	23.4	5	24.2	.	.
4	34.3	6	29.4	.	.	.	.	.	.	.	.
4	32.5	6	47.6	.	.	.	.	.	.	.	.
4	31.1	.	.	.	.	.	.	.	.	.	.
4	29.8	.	.	.	.	.	.	.	.	.	.
4	25.7	.	.	.	.	.	.	.	.	.	.
4	23.5	.	.	.	.	.	.	.	.	.	.

Rather than using **D_2D** to preview the input ASCII data file and select its column interactively, you can create a configuration file that will allow you to plot the same data (i.e., **F1_18.TXT**) as scattered points, but with flipped axes as shown in Figure 1.18.

Begin by opening the master configuration file **!.CF2** using **Notepad** and save it under a different name (this is file **F1_18.CF2** available with the book). Change line **36** of this file to '6' (i.e., the number of groups or pairs of data), then edit the remaining 13 lines according to Table 1.1.

Copy and paste these 13 lines five times at the end of the file and then change the column numbers from where the *x* and *y* values are read (i.e., those commented with **{x column}** and **{y column}**, respectively) to **4 3, 6 5, 8 7, 10 9,** and **12 11,** respectively.

Modify the **{marker radius in screen units}** from 2.5 to 10. Leave line **27** unchanged since it will be ignored, and change line number **33**, which currently reads

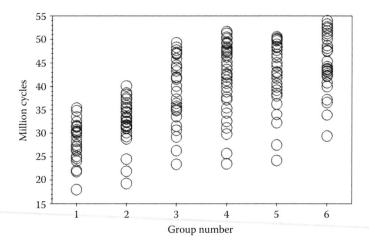

FIGURE 1.17 Plot of the **F1_17.DTA** experimental data file. Configuration file: **F1_17.CF2**.

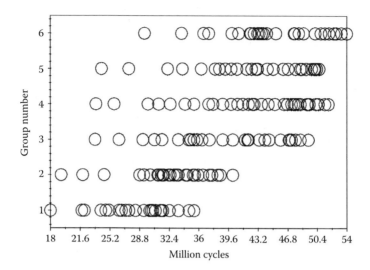

FIGURE 1.18   Plot of the **F1_18.TXT** files with settings from configuration file **F1_18.CF2** created by hand according to the text.

TABLE 1.1   Modifications to Obtain Configuration File **F1_18.CF2**

	Line to Append	Comments
1	F1_17.TXT	Name of the file from where data are read.
2	N	N (No) because the curve is not a background curve of an animation (see next paragraph on producing animations).
3	0	In case of **D3D** data files, this should be 1 or 2, otherwise it must remain zero.
4	2 {x column}	Column number for the x value of the (x, y) pair.
5	1 {y column}	Column number for the y value of the (x, y) pair.
6	2	Read data from file beginning with the 2nd row. *Important*: Inserting 1 here can result in erroneous plots because portions of the header may be interpreted as data.
7	1000	You can insert any number greater/equal than the total number of rows in the file, i.e., 35. To ensure that <F2>redo option will capture all data from future versions of the file, it is advisable to insert a safely large value here.
8	1	y-axis category, i.e., 1 for F1(x), 2 for F2(x), 3 for F3(x), and 4 for F4(x).
9	7	Line type 1 through 7 for ——, ---, ···, ====, ====, :::, and no line.
10	1	Color (1 through 8): blue, green, cyan, red, magenta, brown, gray, black.
11	2	Marker pattern 1 through 6 for -*-, ***, *#*, ! ! ! ! ! !, /#\, and no marker.
12	2	Marker type from 1 to 14 for Ø, ○, ●, □, ◇, ▽, △, ∗, ×, +, ¤, ♀, ♂, ¦, >.
13	1:1:1000;	Since this information is not used, any numbering pattern is acceptable.

1E100 -1E100 F1(x) with **0.5 6.5 group number**. These will be the minimum and maximum limits over the vertical axis and its new name 'group number'. The name of the x-category on line **32** currently reads **x**. Before saving and closing your configuration file, change this to **million cycles** and include six spaces at the end, while leaving the limits as they are, that is, **1E100 -1E100**. The **D_2D** will recalculate them such that the box will tightly fit the plot.

Important: You can selectively reset the limits over *x*- or any of the *y*-axes by editing the corresponding lines **32** through **36** and making the lower limit bigger than the upper limit, or by inserting a **1E100 -1.0E100** pair. If you want to reset the limits on all axes, it might be more convenient to just use the **<F2>redo⁑** option from the *start-up menu*.

Launch **D_2D** and select the configuration file that you have just created. You should obtain a plot similar to that in Figure 1.18. To make it look even nicer, consider editing the number of divisions and values over the horizontal axis (either from inside **D_2D** or by further editing its configuration file).

Important: In case of an incorrect **CF2** file input, an error message will be issued by **D_2D**. The debugging information provided is not complete however. It is therefore advisable to always save under a different name the last functional copy of the **CF2** file that you are editing on.

1.9 PLOTTING ORDERED DATA AND HISTOGRAMS

One capability of **D_2D** is to autogenerate the *x*-coordinate values as 1, 2, 3, 4, etc., useful when plotting one column only from an ASCII file. You can instruct **D_2D** to do so by setting to zero the column number from where the *x* values are read.

To exemplify, let us edit the **CF2** file of the plot in Figure 1.17 and set to zero the **column # for x** six times. In addition, assign different marker types to each data set (i.e., **10, 9, 8, 7, 6, 5**) and change their width from 10 to 4 screen units. Next, on line **33**, enter the text: **0 35 specimen #** (the *x* label) followed by several spaces to offset it to the left. Edit line number **6** so that it reads **8 5 {no. of values & divisions over x axis}**—this will make the horizontal axis of the graph look nicer. Save the file under a new name (**F1_19.CF2** is the name of the one available with the book) and open it with **D_2D** to produce the graph in Figure 1.19. This is actually a **DXF** copy of the plot, where markers ◇, □, △, ▽ are transparent rather than opaque. If you issue the *hide*

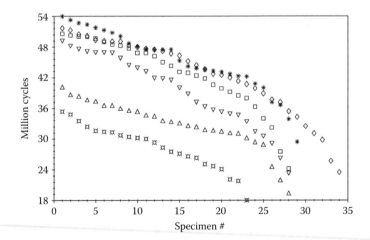

FIGURE 1.19 **DR12 DXF** copy of the plot of the **F1_17.TXT** file, with the *x* values generated automatically. Configuration file **F1_19.CF2**.

or *shade* commands inside **AutoCAD**, the markers will obscure each other, similarly to the **D_2D** screen.

If you read the x values from ASCII and set to zero the column number for the y values, **D_2D** will generate a histogram and not a scatter plot with y the independent variable (if you need one of those instead, your only option is to rotate the graph 90° inside **AutoCAD**). A histogram is a graph of adjacent vertical bars showing what proportions of data fall into each of the given intervals or bins. In case of **D_2D**, these bins are equal width, and their number must be specified immediately after setting the column for the y-axis to zero. According to Bendat and Piersol (2010), for N data points the number of bins n_b should be

$$n_b = 1.87(N-1)^{0.4} + 1 \tag{1.5}$$

For n_b equal size bins, the left and right limits of a current bin I will be

$$f_{min} + (i-1)(f_{max} - f_{min})/n_b \quad \text{and} \quad f_{min} + i(f_{max} - f_{min})/n_b \tag{1.6}$$

where f_{min} and f_{max} are the lower and upper range of the data series read from file. Figure 1.20 shows a seven-bin histogram of the most numerous bearing test group in **F1_17.TXT**.

Important: The limits f_{min} and f_{max} introduced earlier appear on the histogram centered with the leftmost and rightmost bins. Currently, **D_2D** does not allow the user to directly modify them and is independent of the horizontal-axis minimum and maximum limits.

The appearance of a histogram will depend on the number of bins n_b and on the f_{min} and f_{max} values in Equation 1.5. The former can be set at the beginning when data are read from file or by editing line number **32** of the **CF2** and running **D_2D** with these new settings. However, the latter can be modified only indirectly, for example, by adding two properly selected values to the input data file and trimming the graph to the left and to the right as it will be exemplified next.

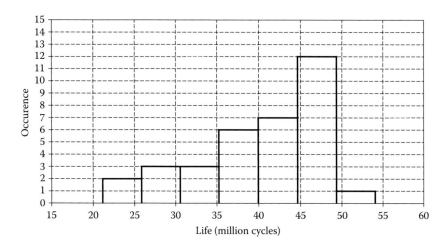

FIGURE 1.20   Seven-bin histograms of the most numerous (34 samples) bearing group in **F1_17. TXT** file. Configuration file: **F1_20.CF2**.

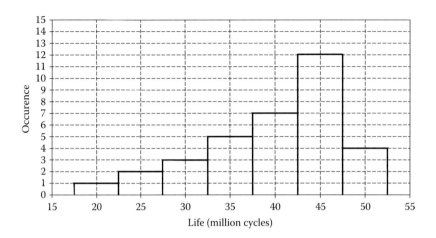

FIGURE 1.21 Histogram of the data used for Figure 1.21 modified such that the bins are centered at rounded values. Configuration file: **F1_21.CF2**.

In order to center the bins of the histogram with 20×10^6 to 50×10^6 cycles as shown in Figure 1.21, values 10.0 and 60.0 have been added to the original data file, and the file was then saved under the new name **F1_20.TXT**. The total number of bins of the histogram was then changed to **11** = 7 (original number of bins) + **2** (empty bins, one to the left and one to the right) + **2** (bins for two new entries). With these modifications, four new bins have been added to the graph (see Figure 1.22) as follows: two bins each with only one data point (one centered at 10×10^6 cycles and the other centered at 60×10^6 cycles), separated by the rest of the histogram by two empty bins (one centered at 10 million and the other centered at 55×10^6). To obtain the correct appearance corresponding to the original data, the extraneous bins must be eliminated by setting the x-axis range from 15 to 55 million cycles.

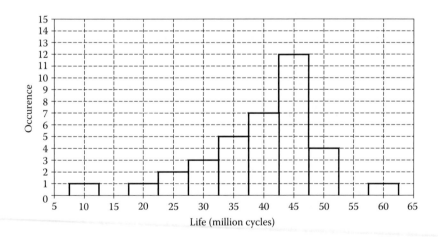

FIGURE 1.22 Histogram in Figure 1.21 before trimming the extraneous bins to the left and to the right. Configuration file: **F1_22.CF2**.

Note that both the **Integral** and **Length** of the curve displayed at the top of the graph are both related to the number of samples in the input data file according to the following equations:

$$n_{\text{samples}} = \text{Integral} \cdot \frac{n_b - 1}{f_{\max} - f_{\min}}$$

$$2n_{\text{samples}} = \text{Length} - \frac{n_L \left(f_{\max} - f_{\min} \right)}{n_b - 1}$$

(1.7)

where n_L is the number of horizontal lines of length equal to the width of one bin, which are visible on the graph. Also note that for a histogram with empty bins, if you make the lower limit of the y-axis less than zero, the number of horizontal segments increases (and so does the total length of the graph).

1.10 PLOTTING INEQUALITIES

Plotting scattered data and inequalities are actually related issues. To exemplify, let us look at the problem of graphing the inequality:

$$\left(\sin(x) + \sin(y) \right)^2 - (x \cdot y + 0.5) \geq 0$$

(1.8)

with $-\pi \leq x \leq \pi$ and $-\pi \leq y \leq \pi$. This is the top view of the intersection between the surface $z = (\sin x + \sin y)^2$ and the hyperboloid of equation $z = x \cdot y + 0.5$.

The **P1_23.PAS** program (see Appendix A) that generates the **F1_23.D2D** file used to plot Figure 1.23 has a very simple structure: It essentially evaluates inequality (1.8) over a 406×406 grid, and if it is not satisfied, then the corresponding (x, y) pair is written to file.

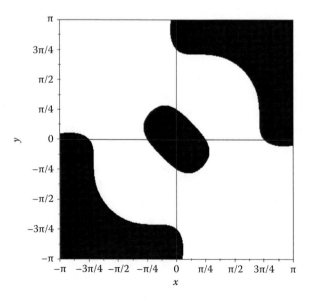

FIGURE 1.23   Plot of the inequality 1.8. Configuration file: **F1_23.CF2**.

If you save the plot in Figure 1.20 as **R12 DXF** and open it with **AutoCAD**, you will be able to see its doted structure. If you increase the point size, then the **AutoCAD** drawing will begin resembling the raster figure. To do this, type '_ddptype' at the command line, change the point type from dot (which is not scalable) to circle, and increase its size. Because of the large number of graphic entities (i.e., dots), inequality plots are recommended to be saved and manipulated as raster graphics.

Important: Both the point size and the grid size will influence the appearance of an inequality plot. For raster graphics output, set **D_2D** point size to zero and sampling size over x and y slightly bigger than the dimensions in pixels of the plot box.

If you redo Figure 1.23 with settings from **F1_23.CF2**, you will notice that the values over the x- and y-axes are in decimal numbers, not in fractions of π, like in Figure 1.23. This is caused by the round offs that occur when recalculating the interval between two values with settings from the configuration file. To fix this, go to the *plot-box edit screen* and press <Ctrl> + <F1> then <Ctrl> + <F5>. If it has no effect, you will have to go to the **<F1> change limits...** *menu* and type 'pi/4' where it says **Write a value every** for both the x- and y-axes. To enter the actual character π, hold the <Alt> key and type 227 or type 'pi' without quote marks.

1.11 PARAMETRIC PLOTS

2D parametric curves are defined by separate equations for the x and y coordinates of their points, that is,

$$\begin{cases} x = F_x(t) \\ y = F_y(t) \end{cases} \tag{1.9}$$

where t is an independent variable parameter. In many cases, t is associated to time and can assume only positive values, including zero. For polar curves written in Cartesian form (like the *Archimedean spiral* considered next), the independent parameter is sometimes noted θ and represents an angle measured in radians.

For certain parametric curves, if data are generated at a constant increment of the independent variable, rapid jumps in the F_x and F_y function values can occur, and the graph will look nonsmooth in those areas (see Figure 1.24a). This is less likely to happen in case of single-valued functions $y = F(x)$, because their graph does not turn over itself, and the total length of the curve remains short. A remedy proposed by Reverchon and Duchamp (1993) is to evaluate the distance ΔL between every two consecutive points $x(t)$, $y(t)$ and $x(t + \Delta t)$, $y(t + \Delta t)$ and if this distance is greater than a given *maximum length* $\Delta L_{\min}$, then increment Δt is reduced, and the second point is recalculated. Conversely, if the distance between these two points is smaller than a given *minimum length* $\Delta L_{\max}$, then step Δt will be increased. The procedure is repeated until there is no need to adjust Δt, and only then the new point $x(t + \Delta t)$, $y(t + \Delta t)$ is written to file.

If in an animated comet plot you make $\Delta L_{\min}$ very close to $\Delta L_{\max}$, the graph will appear to *grow* at a constant speed, because all segments of the polygon that approximates the curve will be about equal. In other instances, however, like in projectile or robot

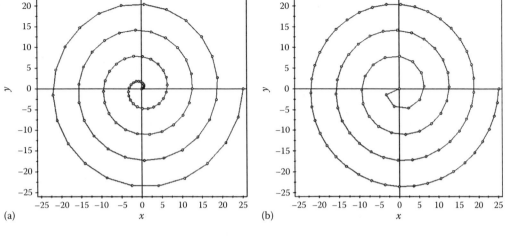

FIGURE 1.24   Plot of the curve in Equation 1.10 with $0 \le \theta \le 8\pi$ generated (a) for constant increment Δt and (b) for an adjustable increment Δt such that $\Delta L = 31 \pm 0.001$. Both plots have 91 data points. Configuration files **F1_24A.CF2** and **F1_24B.CF2**.

end-effector path problems, we would explicitly want the plot points to be displayed at a constant time interval Δt, so that a comet plot animation will appear realistic.

Archimedean spiral of equations

$$\begin{cases} x = \theta \cdot \cos(\theta) \\ y - \theta \cdot \sin(\theta) \end{cases} \tag{1.10}$$

is a classical parametric curve (see Figures 1.24 and 1.26). The data used to produce Figure 1.24a have been generated with the program **P1_24A.PAS** (see Appendix B), where parameter θ increases at a constant step between 0 and 8π.

The companion plot in Figure 1.24b was produced with the program **P1_24B.PAS** (see Appendix B), which implements the variable step-size algorithm discussed earlier. It ensures a 31 pixels long with 0.1% accuracy to each segment of the polygon that approximates the graph (the plot-box size was assumed to be 405 × 405 pixels). For the particular function in Equation 1.7, neither approach appears to be satisfactory: The plot in Figure 1.24a looks properly sampled close to the origin of the spiral, while the plot in Figure 1.24b looks better towards its outer end. In addition, the number of function calls inside **P1_24B.PAS** required to attain the specified polygon segment accuracy was considerable (over 95,000). The same result can be obtained in fewer function evaluations, and with improved accuracy, if a rapidly converging zero finding procedure is employed (see Chapter 4). On the other hand, the overall appearance of the graph will not change, unless the number of plot points is increased.

A different, more efficient strategy of curve polygonalization was implemented in the **P1_25.PAS** program (see Appendix B). Here, the function is evaluated at a constant parameter θ step, but not all points are written to file, that is, the program verifies (i) if the current point coincides within a given tolerance with previous point (see line #**61** of the program) and (ii) if the current point and the previous two points are collinear, the same with a given tolerance (see line #**64**).

The coincidence and collinearity conditions mentioned earlier are verified by the **Coinc2Pts2D** and **Colin3Pts2D** functions of the *Boolean* type that are called from unit **LibGe2D**. For conformity, these two functions are listed next:

```
function  Coinc2Pts2D(xA,yA, xB,yB, Eps2: double): Boolean;
{Check if points A and B coincide with precision Eps2}
BEGIN
   Coinc2Pts:=TRUE;
   if (xA = xB) AND (yA = yB) then Exit;
   if (Sqr(xA-xB)+Sqr(yA-yB) > Eps2) then Coinc2Pts:=FALSE;
END;
function Colin3Pts2D(x1,y1, x2,y2, x3,y3, Eps3: double): Boolean;
{Check if points 1, 2 and 3 are collinear with precision Eps3}
var ReqEps3, D_13, Max_123: double;
BEGIN
   Colin3Pts:=FALSE;
   D_13:=Sqr(x1-x3)+Sqr(y1-y3);   {distance between 1st and 3rd point}
   Max_123:=Max3(Sqr(x1-x2)+Sqr(y1-y2), D_13, Sqr(x2-x3)+Sqr(y2-y3));
   if (D_13 = Max_123) then BEGIN   {triangle 3-2-1 is obtuse}
      ReqEps3:=4.0*Sqr((x2-x1)*(y3-y1)-(y2-y1)*(x3-x1))/(D_13+1e-100);
      if (ReqEps <= Eps3) then Colin3Pts:=TRUE;
   END;
END;
```

where function **Max3** called by **Colin3Pts2D** from unit **LibMath** returns the maximum of three numbers.

Note that in both procedures, in order to eliminate the repeated calling of the mathematical function **Sqrt**, the squared rather than the actual distances between two points were used in calculations. Also, notice that in order to avoid a division by zero when evaluating **ReqEps3**, a very small positive number was added to the denominator.

It is visible that the plot in Figure 1.25, produced with the data from **P1_25.PAS** program in Appendix B, ensures a better distribution of the vertices of the approximating polygon. Parameters **EPS2** in the **Coinc2Pts2D** and **EPS3** in **Colin3Pts2D** functions were set to 8.3e-3 and 4.2e-3, respectively. You may want to experiment with **P1_25.PAS** and see how these two values and the number of initial data points (this was considered equal to 1000 for Figure 1.26—see line #**10** of the program) affect the number and disposition of the vertices of the approximating polygon.

D_2D program employs the same two functions **Coinc2Pts2D** and **Colin3Pts2D** for optimizing polyline vertices before they are written to the **DXF** file. The corresponding **EPS2** and **EPS3** values are the **DXF** polyline coincidence and collinearity parameters that are read from line **25** of the **CF2** configuration file. When you export a graph to **DXF**, the **D_2D** program will report at the end the maximum required coefficients **EPS2** (first

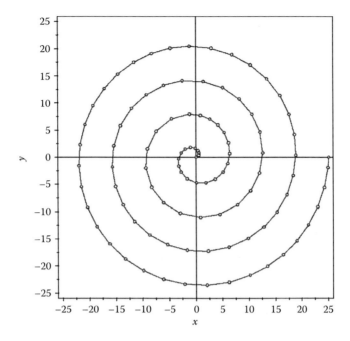

FIGURE 1.25 Plot of the curve in Equation 1.10 with $0 \le \theta \le 8\pi$ generated based on an initial pool of 1000 data points of constant $\Delta\theta$ increment, decimated to 99 points using the **Coinc2Pts** and **Colin3Pts** functions. Configuration file: **F1_25.CF2**.

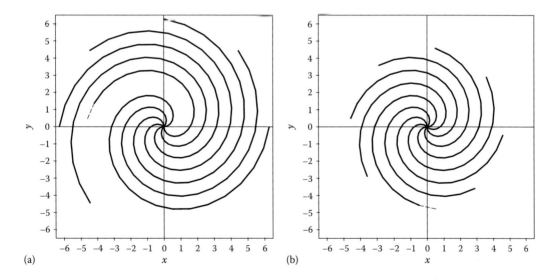

FIGURE 1.26 Polar array of *Archimedean spirals* (Equation 1.11) with $n = 8$ and $0 \le \theta \le 2\pi$ divided between 31 data points of constant θ increment. Both figures are *plots in progress*, showing how data have been generated (i.e., last segment drawn is shown in dashed line) as (a) one spiral at a time and (b) all spirals are grown simultaneously. Configuration files: **F1_26A.CF2** and **F1_26B.CF2**.

number) and **EPS3** (second number), for which at least one plot point will be eliminated because of its coincidence or collinearity with its neighbors. In order to control the amount of vertex removal the polylines written to **DXF** will be subject to, you can modify the default **EPS2** and **EPS3** values on line **25** of the **CF2** file.

Important: When a **DXF 1:1** export is performed, coefficients **EPS2** and **EPS3** are automatically set to zero, so that all plot points are preserved.

Another useful capability of **D_2D**, other than eliminating points that are near coincident and near collinear, is that it concatenates into polylines line segments that are placed head to tail. This occurs even if these segments were generated out of sequence or with their ends flipped. To exemplify, let us look at the problem of plotting a polar array of *n Archimedean spirals* (see Figures 1.26 and 1.27) of equations:

$$\begin{cases} x = \theta \cdot \cos[\theta + 2\pi(i-1)/n] \\ y = \theta \cdot \sin[\theta + 2\pi(i-1)/n] \end{cases} \tag{1.11}$$

with $i = 1$ to n and $0 \le \theta \le 2\pi$. We will generate the data points in two ways: (i) the spirals will be generated one at a time and (ii) all n spirals will grow simultaneously. The data files used to plot Figure 1.26a and b were produced with programs **P1_26A.PAS** and **P1_26B.PAS** (see Appendix B), respectively, which consist of the same two *for loops* but nested in different order. In both programs, line breakers are used to separate the individual spirals. In addition, program **P1_26B.PAS** inserts line breakers to separate the individual segments that approximate the spirals. Evidently, in the latter case, data file organization is less effective because of the increased number of separators and because each plot point (except of end points) is written to file twice.

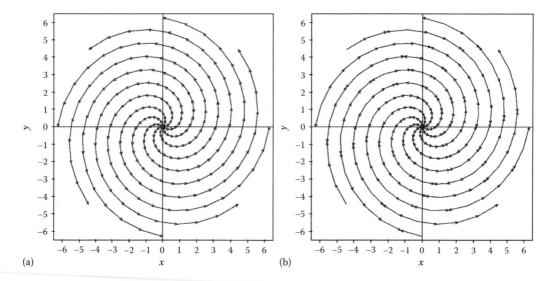

FIGURE 1.27 Polar array of *Archimedean spirals* (Equation 1.11) with $n = 8$ and $0 \le \theta \le 2\pi$ divided between 31 data points of constant θ increment and with the plot points (a) orderly oriented and (b) oriented at random. Configuration files: **F1_27A.CF2** and **F1_27B.CF2**.

In both cases, however, the final graphs will look the same. The differences are visible only when these plots are shown as comet plots (see Section 1.12 on animations). Additional differences between these two graphs will become apparent when you export the respective plots to **DXF**, that is, the line segments the endpoints of which are identical (or coincide within tolerance **EPS2**) will be concatenated into polylines, even if they were plotted out of sequence. To verify, run **D_2D** with settings from **F1_26A.CF2** and **F1_26B.CF2** and export each of the graphs to **DXF** once with the option to **separate graphs into DXF layers** *on*, and the second time with this option turned *off*—you can toggle between these two export variants from option **<F2>** of the **<F1..4>** *screen*.

When opening inside **AutoCAD** the **DXF** file generated with settings from **F1_26A.CF2**, you will notice that each spiral belongs to a separate layer titled 'Line1section1' to 'Line1section8'. In case of the **DXF** file created with settings from **F1_26B.CF2**, each segment of the eight spirals belongs to 240 separate layers titled 'Line1section1' to 'Line1section240'. If you chose not to separate the lines into **DXF** layers, go to the **<F1..4>** *screen*, option **<F2>** in case of either **F1_26A.CF2** or **F1_26B.CF2** configuration files. In this case, a single layer named 'Line1section1' will contain all plot lines. Moreover, some spirals will appear joined into a single polyline because their initial points coincide. Also notice that for a **DXF** file generated with settings from **F1_26B.CF2**, even if the vertices were originally plotted out of sequence, they were concatenated correctly when converted to **DXF** polylines.

A third program available with the book named **P1_27.PAS**, which is a modification of Pascal program **P1_26B.PAS** (source code not included in appendix), generates the same type of data point structure as for Figure 1.26b, with the difference that the line segments are written to file at random, that is, either the outer point first followed by the inner point or vice versa. If you do **DXF** export and you chose to **separate the graph into DXF layers** from option **<F2>** of the **<F1..4>** *screen*, then the individual segments that form the spirals will still be connected into polylines, following a proper reordering of vertices. This helps reducing the size of output **DXF** files and also makes it easier to edit the plots generated by **D_2D** using **AutoCAD**, because the graphs consist of polylines rather than separate line segments.

Important: If a plot has been generated with data from two or more separate files, then a **DXF** layer will be created for each curve and their names will be 'Line1section1', 'Line2section1', etc. (see the **DXF** output of Figure 1.4). If line breakers are used, however, and if you chose to **separate graphs into DXF layers**, then additional layers will be created and their names will be 'Line1section1', 'Line1section2', etc. If the plot was generated with data from multiple input files, then layers 'Line2section1', 'Line2section2', etc., will also be created.

1.12 ANIMATIONS

D_2D allows you to create *comet plots* and *animations*, with or without having some of the curves plotted as *background* images and with or without accumulating frames. A *comet plot* is a regular plot where displaying the line segment that connects the current plot point with the rest of the graph or displaying the next marker of the plot is delayed a certain amount of time. In nonaccumulating-frame comet plots that consist of broken-bar markers only (i.e., option ✶✶✶ with ¦ markers), each of these markers will be drawn as a vertical scan line.

In *multiple-frame animations*, a new frame is created and will be delayed any time a line breaker is encountered in the input data file or when the end of the input file is reached. In plots with data read from more than one input file, a new frame will be generated any time the data from a new **$2D** file is plotted on the screen. After the last **$2D** data file is plotted, everything is repeated.

If the plot consists of multiple **$2D** files, the curves generated using one or more of these files can be defined as *background*, and only the remaining curves will be animated.

You can choose to accumulate the frames in an animation or refresh the screen every time a new frame is displayed. In both cases, if one or more background curves have been specified, these will be displayed in each frame.

If you choose to number the vertices of a curve that originates from a single file using the **✶#✶** marker option, and if line breakers are used to separate the graph into frames, then the numbering will be restarted every new frame, unless you choose to accumulate them by setting to 'Y' the *accumulate graphs* from option **<F2>** of the **<F1..4>** *screen*.

The amount of delay between frames can be changed interactively, including holding the current frame indefinitely, that is, the next frame will be displayed only after pressing a key. When in the *frame-hold mode*, the current screen can be copied to **PCX**. A number of such **PCX** screenshots can be assembled into a stand-alone animated **GIF** or a movie file. Stream **PCX** export and export to multilayer **DXF** is possible from the **<F4>** option of the **<F1..4>** *screen*.

Important: If the animation rate is set to 0 or 1, stream **PCX** export will copy all frames to **PCX**. If the frame rate is 2, 3, or higher, only every 2nd, 3rd frame, and so on will be exported to **PCX**.

Multiple-layer **DXF** files can then be animated inside **AutoCAD** using the **M_3D.LSP** program discussed in Chapter 3. Currently, **D_2D** allows only line/section and nonaccumulated comet plots (i.e., scan lines or scan point plots) to be exported to multiple-layer **DXF** files, provided that the total number of animation frames is less than 1000.

In the remainder of this chapter, these features of **D_2D** will be exemplified. Of the plots discussed earlier, some will be changed into animations, and a couple of new examples will be presented. Since animations cannot be printed on paper, you will have to run **D_2D.EXE** with settings from the respective **CF2** configuration files or play the animated **GIFs** available with the book.

- **Comet plots**: Once you have created a line graph, you can easily animate it as comet plot. To do so with the plot in Figure 1.1, run **D_2D** with settings from **F1_01.CF2**, then choose option **<F2>** from the **<F1..4>** *screen* and select the **Animate graph** and **Comet plot** options. For this particular case, it is irrelevant if **separate graphs into animation frames** is set to 'Y' or 'N'. Alternatively, you can directly edit the **F1_01.CF2** configuration file and set to 'Y' lines **24** and **25**— (see the **F1_30.CF2** file available with the book). While the animation is running, you can adjust the frame rate using the up and down arrow keys (see Figure 1.28).

When the frame rate is zero (see Figure 1.29), you can copy the current screen to **PCX** by pressing the <F10> key. The name assigned by default to the **PCX** file will start with **D2D00000.PCX** (this is a copy of the background plot) and will be incremented by one with every new **PCX** export.

If **Label animation frames** from the **<F2>** option of the **<F1..4>** *screen* is set to 'Y', then the number of the last line segment or plot point (these are called *sections*) that was added to the graph will be printed to the bottom of the screen (see Figure 1.30).

Important: To ensure that **PCX** screenshot file export is consistent, remove from the current directory all preexisting **PCX** files.

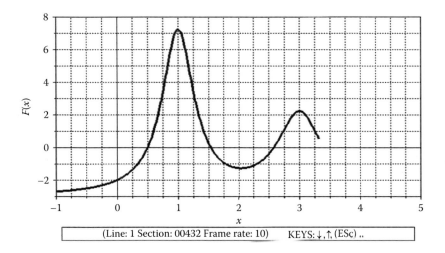

FIGURE 1.28   Screenshot of a comet plot in progress when in the *automatic screen refreshing* mode. Configuration file: **F1_28.CF2**.

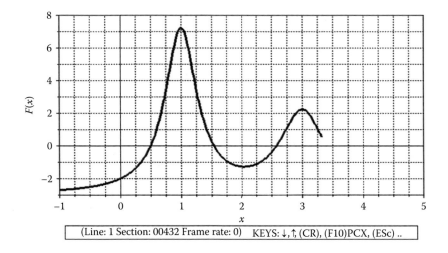

FIGURE 1.29   Screenshot of a comet plot in progress when in the *frame-hold mode*. Configuration file **F1_29.CF2** (remember to repeatedly press <CR> or the space bar for the animation to occur).

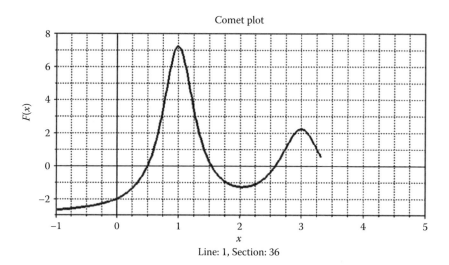

FIGURE 1.30   **PCX** screenshot of one frame of the comet plot in Figure 1.29 showing frame labeling (configuration file **F1_30-00.CF2**). See also animation file **F1_30-00.GIF** generated with every 10th section of the plot.

Examples of scan point and scan line plots with a background the same curve in Figure 1.30 can be produced using configuration files **F1_30-01.CF2** and **F1_30-02. CF2**. See also animated **GIFs F1_30-01.GIF** and **F1_30-02.GIF** available with the book.

- **Line/section animations**: If you animate the previous graph under the option **line/ section animation,** the plot will only flicker because it consists of only one frame. To obtain a meaningful multiple-frame animation, the plot must either (i) originate from several files, each providing data for plotting one frame; (ii) originate from a single file that contains several line breakers; or (iii) originate from multiple files of which some of the files have line breakers inserted within.

The plot in Figure 1.3 originates from two files. In order to animate it, launch **D_2D** with settings from **F1_03.CF2** and change to 'Y' the **animate graph** and set animation option to **line/section animation**. Run **D_2D** twice: once with the **Accumulate graphs in an animation** set to 'N' and a second time set to 'Y'. Because there are only two animation frames, their rate must be reduced to clearly observe the difference. The configuration files that will let you play these two animations are **F1_30-03.CF2** (frames accumulate, i.e., the second frame is added to the first frame) and **F1_30-04.CF2** (frames do not accumulate). See also the corresponding animation files **F1_30-03.GIF** and **F1_30-04.GIF**.

Configuration files **F1_30-05.CF2, F1_30-06.CF2, F1_30-07.CF2, F1_30-08. CF2, F1_30-09.CF2**, and **F1_30-10.CF2** provide several more examples based on Figures 1.15, 1.26a and 1.26b. These are *multiple-frame animations* where data are read from a single

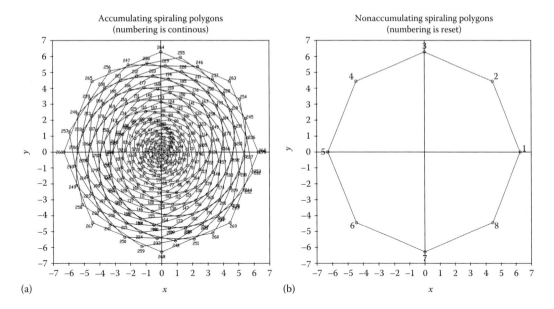

FIGURE 1.31 Last frame of the animations generated with configuration files (a) **F1_31A.CF2** (vertex numbering continues from previous frame) and (b) **F1_31B.CF2** (vertex numbering is restarted with each frame). The same animations are available as **GIF** files **F1_31a.GIF** and **F1_31b.GIF**.

file and line breakers are used as frame separators. The corresponding animated **GIF** files **F1_31-05.GIF** to **F1_30-10.GIF** are also available with the book.

- **Vertex numbering**: This example shows how vertex numbering is affected by the separation of the graphs into animation frames (see Figure 1.31). Data files **F1_31.D2D** and **F1_31.DTA** used in this application were generated with program **P1_31.PAS** (see source code in Appendix B). This new program originates from the one used to produce Figure 1.26, with the difference that the points placed at equal distance from the origin are connected together to form a closed polygon. With proper line breaking inserted into the input data file, when plotted using **D_2D**, it results in an array of spiraling polygons that can be also animated (see configuration files **F1_35A.CF2** and **F1_35B.CF2** and the corresponding animated **GIF** files **F1_35A.GIF** and **F1_35B.GIF**).

Important: In an accumulated-frame graph, data point numbering is continuous. When the frames are plotted separately, point numbering is restarted every frame. Currently, there is no interactive way, nor via **CF2** editing, to continue vertex numbering from the previous frame.

- **Background-curve animations**: When plotting data from multiple files, you can select the curve(s) originating from one or more of these files to be displayed as background curves and display the remaining file(s) animation frames. A couple of comet-plot animations have already been mentioned (see **F1_30-01.GIF** and **F1_30-02.GIF**).

To animate as accumulated graph with a background the plot in Figure 1.3, launch **D_2D** with settings from **F1_03.CF2**, change to 'Y' the **animate graph** and select **comet plot.** At this time, you have two plots that will be animated as comets, and depending whether **accumulate graphs into animation** is set to 'N' or 'Y', the graph of $F'(x)$ will be either displayed on the top of $F(x)$ or it will replace it.

In order to turn the graph of $F'(x)$ into a background curve, go to the **<F1..4>** *screen* and press <F3>, then type 'B' for background and '2', for the 2nd curve, then press <CR> four times (see also **F1_32-1.CF2** and animation file **F1_32-1.GIF**).

To display $F(x)$ as background curve and animate as comet the derivative $F'(x)$, go to the same **<F3> edit add remove lines** *menu*, type 'B' for background curve then type '1'. At this point, both curves are background curves, so the plot will be motionless. To animate as comet the $F'(x)$ graph, type 'B' again, then '2'. Then go to the *final-graphic screen* to watch the result. See also configuration files **F1_32-1.CF2** and **F1_32-2.CF2** and animated **GIF** files **F1_32-1.CF2** and **F1_32-2.GIF**.

Two additional background-curve animations of an increased visual appeal have been generated using the data files generated earlier as follows: file **F1_26A.D2D** provides the background curve, that is, the spirals in Figure 1.26a, while the polygons in Figure 1.31 read from data file **F1_31.D2D** are animated as separate frames (see Figure 1.32a and animation file **F1_32A.GIF**) or as accumulated frames (see Figure 1.32b and animated **GIF** file **F1_32B.GIF**).

Important: If you are in the animation mode and you exit **D_2D** from the **<F1..4>** *screen*, the program will leave behind a file named **D2D00000.PCX**. This is a copy of the plot box and, if it is the case, of the background curve(s). In case you want to utilize it, save this

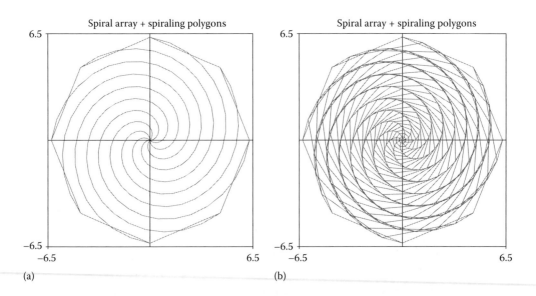

FIGURE 1.32   Last frames of the animations generated using configuration files **F1_32A.CF2** (a) and **F1_32B.CF2** (b). See also animated **GIF**s **F1_32a.GIF** and **F_32b.GIF**.

FIGURE 1.33   Last animation frames of randomly colored spiraling polygons with 2, 3, 4, and 5 sides. Configuration files **F1_33-2.CF2** through **F1_33-5.CF2**. See also **F1_33-2.GIF** through **F1_33-5.GIF**.

PCX file under a different name because the next time you launch **D_2D**, it will be erased together with all **$2D** temporary data files.

Before ending this chapter, one more program will be introduced, that is, **P1_33.PAS** (see Appendix B). This program generates the data files to animate polygons that spiral both forward and backward, which in addition are randomly colored in groups of 10 (see Figure 1.33). Spiraling direction change has been attained by assigning a negative initial value to the parameter, that is, **Tmin=-2*Pi**. You may want to experiment with other initial values and number of vertices and observe the effect. Ideas for more such animations are available from the references listed at the end of the chapter.

***

In this first chapter, the capabilities of **D_2D** plotting program have been explained and illustrated with examples so that you can solve your own similar problems. Further applications of the **D_2D** program are presented in the remainder of the book. The source codes of **D_2D.PAS** and of **Unit _ D2D.PAS** it uses are both available with the book. The comments provided with the code will help you understand how the features discussed throughout this chapter have been implemented.

REFERENCES AND FURTHER READINGS

For raster graphic format convertors, see
www.online-utility.org/image_converter.jsp.
www.nchsoftware.com/imageconverter/index.html.

To download **DXF** viewers, go to
www.edrawingsviewer.com.
www.bravaviewer.com/viewers.

To download **GIF**-animation software, go to
www.gif-animator.com.
www.blumentals.net/egifan.
www.snapfiles.com/get/msgifanimator.html.

For more information on *Nyquist–Shannon sampling theorem* and on histogram plots, see
Alciatore, D. G. and Histand, M. B. (2007). *Introduction to Mechatronics and Measurement Systems.* Boston, MA: McGraw Hill.
Bendat, J. S. and Piersol, A. G. (2010). *Random Data: Analysis and Measurement Procedures.* Hoboken, NJ: John Wiley & Sons.

For trapezoidal rule of integration, see Appendix A and
Kreyszig, E. (2011). *Advanced Engineering Mathematics.* Hoboken, NJ: John Wiley & Sons.

For more details on the center and radius of a circle through three points, see
http://mysite.verizon.net/res148h4j/zenosamples/zs_circle3pts.html.

For polygonal representation of curves and other related topics, see
Douglas, D. H. and Peucker, T. K. (1973). Algorithm for the reduction of the number of points required to represent a digitized line or its caricature. *Journal Cartographica: The International Journal for Geographic Information and Geovisualization, 10*(2), 112–122, University of Toronto Press.
Feschet, F. (2005). Fast guaranteed polygonal approximations of closed digital curves. *Proceedings of the Scandinavian Conference on Image Analysis, in Lecture Notes in Computer Science, 3540,* Berlin, Germany: Springer, pp. 910–919.
Hosur, P. I. and Ma, K.-K. (1999). Optimal algorithm for progressive polygon approximation of discrete planar curves. *Proceedings of the International Conference on Image Processing, ICIP 99, 1,* October 24–28, 1999, Kobe, Japan, pp. 16–20.
Kolesnikov, A. and Fränti, P. (2007). Polygonal approximation of closed discrete curves. *Pattern Recognition, 40*(4), 1282–1293.
Kolesnikov, A. and Fränti, P. (2007). Data reduction of large vector graphics. *Pattern Recognition, 38*(3), 381–394.
Lee, K. H., Woo, H. and Suk, T. (2001). Data reduction methods for reverse engineering. *International Journal of Advanced Manufacturing Technology, 17*(10), 735–743.

Marji, M. and Siy, P. (2003). A new algorithm for dominant points detection and polygonization of digital curves. *Pattern Recognition, 36*(10), 2239–2251.

Masood, A. and Haq, S. A. (2007). A novel approach to polygonal approximation of digital curves. *Journal of Visual Communication and Image Representation, 18*(3), 264–274.

Pikaz, A. and Dinstein, I. (1995). An algorithm for polygonal approximation based on iterative point elimination. *Pattern Recognition Letters, 16*(6), 557–563.

Reverchon, A. and Duchamp, M. (1993). *Mathematical Software Tools in C++.* Hoboken, NJ: John Wiley & Sons.

Sato, Y. (1992). Piecewise linear approximation of plane curves by perimeter optimization. *Pattern Recognition, 25*(12), 1535–1543.

For minimum bracketing and for the secant method of finding zeros of functions, see
Press, W. H., Flannery, B. P., Teukolsky, S. A., and Vetterling, W. T. (1989). *Numerical Recipes in Pascal: The Art of Scientific Computing.* Cambridge, MA: Cambridge University Press.

For more on spiraling polygons and other recursive graphics ideas, see www.physics. emory.edu/~weeks/ideas/.

Graphical Representation of Functions of Two Variables

The D_3D Program

THIS CHAPTER IS ABOUT A PROGRAM named **D_3D** that allows $z = F(x, y)$ single-valued functions of two variables to be represented graphically as *surfaces, level curves, color-coded nodes,* and *stem plots*. **D_3D** also allows the gradient of the function to be graphed, alone or combined with other type of plots. The program was briefly introduced in Chapter 1 (see Figure 1.16) and is available with the book as source code (**D_3D.PAS** and **UNIT_D3D.PAS**) and executable file (**D_3D.EXE**). Its main capabilities are as follows:

1. 3D surface plots can be represented as *lines of constant x, lines of constant y, lines of constant z* (i.e., *raised level curves*), or *node points* mapped on the function surface. These lines or nodes can be plotted alone or combined, while the function surface on which they are mapped can be made *transparent* or *opaque*.

2. When plotted alone, the nodes can be connected with a vertical line to the base of the plot box. These will be called *stem plots* and are useful in representing graphically functions of discrete or integer arguments.

3. Both the nodes and the level curves can be monochrome or can be colored according to their height.

4. When set to opaque, the patches that approximate the function surface can be colored in *light gray* (symbol **G**), background color (*white* - symbol **W**), *yellow* (symbol **Y**) or can be colored according to their elevation (symbol **:**). These symbols must be entered on *chime menu 1* of **D_3D**. The x = constant and y = constant lines of the surface mesh can be set to 1 or 3 pixel thick from the same menu, while their color can be set only by editing the **CF3** configuration file.

5. Any of the previously mentioned plots can be represented in *parallel* or *perspective projections*, as well as in *top view* and *side views*. Side-view plots can also be generated using the **D_2D** program (starting with the same input data file) as explained in Chapter 1 with reference to Figure 1.16.

6. Level curves (i.e., z = constant lines) can be mapped on the function surface or can be projected on the bottom plane of the bounding box. For a given vertical axis range $z_{min}...z_{max}$ and number of level curves, you can choose to distribute the level curves either *evenly spaced* or *logarithmically (log) spaced*. *Log-spaced level curves* can be accumulated towards z_{min} (the **Log spaced down** option), towards z_{max} (the **Log spaced up** option), or towards $z = 0$ (the **Log spaced from zero** option) whether or not $z_{min} < 0 < z_{max}$. Level-curve elevation can also be edited interactively and can be optionally saved to the **CF3** configuration file, from where they will be read next time you run **D_3D**. If saved to **CF3**, level-curve heights can also be modified using a text editor.

7. Similarly to level curves, for any view other than the side views, it is possible to represent the *gradient* of the function as a set of arrows projected on the bottom plane. The gradient is evaluated by **D_3D** through finite differences using the already available plot data. By contrast, the gradient plotting functions in **MATLAB** and **Scilab** (www.scilab.org) require the components of the gradient to be supplied separately.

8. The orientation of the z-axis can be reversed from within **D_3D**, which is more intuitive than viewing the function surface from below.

9. The upper and lower limits of the z-axis can be modified by the user, and if it is the case, the function surface will be truncated where it intersects the top and/or bottom planes of the plot box. These intersections between the function surface and the bounding box can be shown either opaque or transparent or can be plotted alone, without the main body of the function.

10. The patches that approximate the function surface can be selectively displayed based on their location, that is, if they are situated inside the plot box or outside the plot box, or if they intersect the upper and/or lower planes of the plot box. This feature of **D_3D** is useful in representing constrainted functions and inequalities of two variables.

11. The original input data file can be *scattered* (decimated), so that fewer points will be utilized in plotting the function (see *chime menu 3* options).

12. Plots can be exported to file in **PCX** format or **DXF AutoCAD** 12 vector format (i.e., **R12 DXF**). When in top view, the level curves can be exported as **DXF 1:1**, that is, the scale factor will be equal to one on both x- and y-axes.

13. When exiting **D_3D**, the plot-box size, orientation, divisions and values over the three axes, limits over the z-axis, input data file name, and, in case of ASCII

input, the column numbers from where data were read will all be saved to a configuration file with the extension **CF3**. If you exit **D_3D** from the **<F1..4>** *screen*, or if a new plot has been created from scratch, these settings will be written to a new configuration file named automatically **!0000001.CF3, !0000002.CF3,** and so on. If you exit **D_3D** in graphic mode, these settings will be saved to the active **CF3** file, and the original configuration file will have its extension changed to **OLD**.

14. Similarly to **D_2D**, the **<F1>redo** and **<F2>redoo** *options* from the *startup menu* allow the user to recreate the plot with its settings recorded in the **CF3** file found last in the current directory. The **<F1>redo** option will apply the limits $z_{min}...z_{max}$ as they were recorded to the configuration file, while **<F2>redo‡** will reset these limits such that the function surface will exactly fit the bounding box. The **<F3>CF3** option from the *startup menu* allows the user to inspect (but not edit) the **CF3** file before passing it to **D_3D**. To run a **CF3** file without preview, press <F4> or <CR> at *startup* and open it as if it were a data file.

Important: There are a few settings that can be modified *only* by manually editing the **CF3** file. The text between curl brackets serves as comments and should not be deleted because **D_3D** will report an error. These settings are as follows:

(i) Plot window width **w** and height **h**: These refer to the rectangular viewport of the computer screen that fits the projected plot box. Can be changed interactively, but if for any reason you do *not* want these dimensions to be multiples of 5, you must edit the first two numbers on line **6** of the **CF3** file. Remember that they cannot exceed 625 and 430 units, respectively.

(ii) Plot-box orientation and perspective parameters **kH**, **kV**, **tan(Gamma)** and **tan(Delta)**: Are listed on line 6 of **CF3** files, together with parameters **w** and **h** mentioned earlier. Coefficients **kH** and **kV** define the horizontal and vertical locations of the origin of the 3D plot inside the viewport. In turn, **tan(Gamma)** and **tan(Delta)** are the *shear* and *taper* angles that allow *parallel* and *perspective projections* to be emulated. These parameters can be changed in discrete increments from within **D_3D**, but editing the **CF3** file can be assigned any value. Do not exceed $-0.97 \leq$ **kH** ≤ 0.97, $0 \leq$ **kV** ≤ 1, $0 \leq$ **tan(Gamma)** ≤ 0.95, and $0 \leq$ **tan(Delta)** ≤ 0.3, or otherwise the plot will appear distorted.

(iii) Mesh line and node color: Can be changed only by modifying the code on line number **5** of the **CF3** file, that is, 1, blue; 2, green; 3, cyan; 4, red; 5, magenta; 6, *brown*; 7, light gray; 8, *dark gray*; 9, light blue; 10, light green; 11, light cyan; 12, light red; 13, light magenta; 14, yellow; and 0 or 15, *white* (the recommended mesh colors that do not cause confusion with the elevation color scale were italicized). Note that a white color mesh will be visible only on a colored patch, but not as wireframe views or white patches (see *chime menu 1*).

(iv) Node type: Can be changed only by editing line number **7** of the **CF3** file. Acceptable values are 0, 1, and 2 for opaque, solid, and transparent circles, and between 5 and 9 for ∗, ◇, □, ▽, and △, respectively.

(v) Division lines outside and inside lengths: Major division lines outside length on all three axes and inside lengths on the *z*-axis only can be modified by editing line number **8** of the **CF3** file. Along the *x*- and *y*-axes, the inside division line lengths will always be zero. Outside lengths on all three axes can be set to any value, but for esthetical reasons, they should not exceed 10 units. The length of the minor division lines is by default to 60% the length of the major division lines.

(vi) **DXF** polyline coincidence and colinearity parameters: They have the same meaning as in **D_2D** and are read from line number **27** of the **CF3** file. These are required when optimizing, prior to **R12 DXF** export, wireframe plots, level curves projected on the bottom plane, or top-view plots. For polyline optimization to take place, these two parameters must be less than the values reported by **D_3D** after completing a **R12 DXF** export.

Important: When a plot with the hidden lines removed is saved to **DXF** and then opened inside **AutoCAD**, to change the wireframe appearance of the plot, the *hide* or *shade* command must be issued. In order for these commands to have effect, the segments that form the function-surface mesh are drawn along the borders of identically shaped **AutoCAD** *regions*, and as the plotting advances, both the regions and their border segments are progressively elevated a small amount. This is the reason why the line segments of plots created in hide mode inside **D_3D** cannot be concatenated into polylines after they have been exported to **AutoCAD**.

(vii) Level-curve heights: If you choose to manually write them to file, you must add them one per line at the end of the current **CF3** file (after the line that reads ∗∗∗ **Level curve heights** ∗∗∗) and change to 5 the value on line number **15**. This latter change is not essential, however, because choosing to read level-curve heights from file can be set interactively from within **D_3D**.

(viii) Default file extension: When starting a new plot, **D_3D** will extract from line **4** of the master configuration file **!.CF3** the extension of the files that will be listed for input when pressing <F4> at startup. You can write a different extension on line 4 or you can edit it blank, in which case **D_3D** will assume the default extension to be .**CF3**. In a regular **CF3** file, line number **4** holds the title of the plot.

Important: In **Windows XP**, you can link **CF3** files and input data files of extensions D3D, R3D, T3D, and G3D to the **D_3D.EXE** program available on your hard drive by editing the *Open With* properties of these files. From **Windows Explorer**, select the file you want to link. Then from the File → Open With → Chose Program menu, select **D_3D.EXE**. Make sure you check the option "Always use the selected program to open this kind of file" before pressing OK to confirm the setting.

Important: When you redo a plot by running its configuration file, the input data file should be either in the same folder with **D_3D.EXE** or in the directory that precedes its name on line **3** of the **CF3** file. If it is in neither place, an error message will be issued.

Important: A copy of the master configuration file **!.CF3** or of any other valid **CF3** file should be available in the folder from where you launch **D_3D.EXE**. If none is available, **D_3D** will report error. Likewise, in order to be able to export your plot to **R12 DXF**, a copy of the **DXF.HED** file should be available in the current directory.

The remainder of this chapter explains in detail the capabilities of **D_3D**. Its user interface is similar to that of **D_2D** discussed in Chapter 1, so you may find it easy to directly experiment with the program.

2.1 HOW **D_3D** WORKS?

The function surface to be plotted is approximated by an array $Z_{m \times n}$ of height values $z_{ij} = F(x_i, y_j)$ in a regular grid of samples x_i, y_j (where $i = 1 \ldots m$ and $j = 1 \ldots n$). Every row of this array will correspond to a single x coordinate, and every column of the array will correspond to a single y coordinate. Knowing the grid values x_i and y_j used to generate the $Z_{m \times n}$ array, (x_i, y_j, z_{ij}) triplets can be formed, each corresponding to a single point on the function surface.

Instead of using these (x_i, y_j, z_{ij}) triplets to generate the screen coordinates of the projected points of the function surface as other plotting programs do, **D_3D** performs all calculations in the 2D image space. Therefore, the input data can be limited to only a number of height values z_{ij} equally spaced over the $[x_{min}, x_{max}] \times [y_{min}, y_{max}]$ domain. In addition to these z_{ij} values, the grid size $m \times n$ and the limits over x and y must be specified to **D_3D**. These limits are required only to properly place division and values along the horizontal axes.

Important: The maximum $m \times n$ size that a data file can have is 501×501. Evidently, the larger the total number of points z_{ij}, the longer it will take **D_3D** to generate a plot or to save it to **DXF**.

The way **D_3D** generates *oblique projections* is by diagonally offsetting a family of curves (see Figure 2.1). Additional *parallel* and *perspective projections* can be obtained by *shearing* and/or *tapering* an initial oblique projection as illustrated in Figure 2.2. Using *Painter's algorithm* (Foley et al. 2013) and the polygon scan conversion procedure **FillPoly** available in Pascal, the hidden-line removal can be done conveniently in the 2D image space. The components of the gradient, as well as the intersections between the function surface and the lower and/or upper plane of the plot box (occurring in level curve and truncated surface plots), can all be evaluated in the image space, without the need for complicated 3D calculations.

To illustrate how various projections can be generated through separate or combined shear and taper transformations, launch **D_3D**, press <F3> at startup, and open configuration file **F2_01DN.CF3**. Your plot should look similar to the front portion of Figure 2.1. Press the <Backspace> key twice to go to the screen showing the plot box with its axes oriented and labeled **x**, **y**, and **z** as in Figure 2.3. This will be called *deformable-box screen*. Notice the

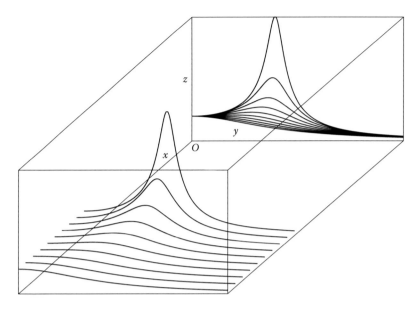

FIGURE 2.1   Oblique projection shown as a diagonal offset of a family of x = constant curves. The figure has been generated inside **AutoCAD** by combining the plots with settings from configuration files **F2_01DN.CF3** and **F2_01UP.CF3**.

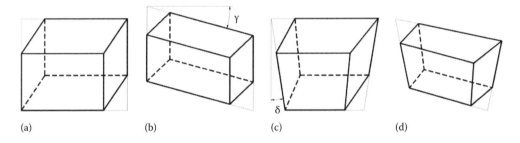

(a)   (b)   (c)   (d)

FIGURE 2.2   (a) An initial oblique projection and various other projections obtained through (b) shearing by angle γ, (c) tapering by angle δ, or (d) combining shearing and tapering.

numbers on the top of the view window and how their values change as you reshape/reorient the plot box. These are the already mentioned parameters **w**, **h**, **kH**, **kV**, **tan(Gamma)**, and **tan(Delta)** read from the **CF3** file.

To resize the viewport that fits the plot, hold the <Ctrl> key and press either <Pg Up>, <Pg Dn>, <←>, or <→>. As you do this, the first two numbers (i.e., **w** and **h**) on the top of the screen will change in increments of 5.

To modify the location of the origin of the projected reference frame inside the viewport, use the four arrow keys. The effect will be equivalent to a left–right, up–down displacement of the viewpoint relative to the plot box. Observe how parameters **kH** and **kV** displayed 3rd and 4th on the top of the screen change their values between −1 and 1 and 0 and 1, respectively.

To obtain more realistic parallel projections (Figures 2.2b and 2.4a), press the <Pg Dn> key several times. This will increase the value of shear angle γ. To undo, press the <Pg Up>

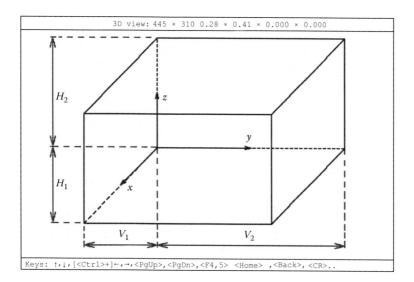

FIGURE 2.3 *Deformable box* that fits a viewport of width $H_1 + H_2 = 445$ and height $V_1 + V_2 = 310$. The origin of the plot is located by the **kH** = H_1/H_2 and **kV** = V_1/V_2 coefficients. For a view point from the 4th quadrant, **kH** < 0, while for a view point from the 1st quadrant, **kH** > 0. As shown **kH** = 0.28 and **kV** = 0.41.

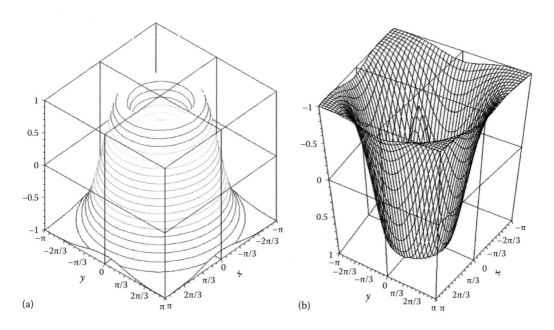

(a)   (b)

FIGURE 2.4 Plot of the *orange-squeezer function* in Equation 2.2 shown as (a) raised level curves and as (b) crosshatched surface with the *z*-axis reversed. Configuration files **F2_04A.CF3** and **F2_04B.CF3**.

key repeatedly. Note that tan(γ) is being displayed as the 5th number on the top of the screen (see Figure 2.3).

To *taper* the plot box as shown in Figure 2.2c (i.e., to increase angle δ) and provide a pseudo-*perspective projection* of your plot, press several times <F5>. To undo, hold the <Ctrl> key and press <F5> again. Notice how the last number on the top of the screen, that is, tan(δ), changes its value.

If a shear and a taper transformation are combined on the same graph, it will result in additional perspective views as illustrated in Figures 2.2d and 2.4a.

Important: You can restore the plot-box orientation to (approximately) an isometric view by pressing the <Home> key. If you press simultaneously the <Ctrl> and <Home> keys instead, the orientation will change to top view, like for a level-curve plot.

2.2 **D_3D** INPUT DATA STRUCTURE

To demonstrate the capabilities of the **D_3D** program, several functions will be considered as follows:

The function in Figure 2.1 known from Chapter 1, with a simplified description, that is,

$$F_1(x,y) = \frac{1}{\sqrt{\left(1-y^2\right)^2 + \left(2x \cdot y\right)^2}} \tag{2.1}$$

and $0.1 \leq x \leq 1$ and $0 \leq y \leq 2.5$

Function

$$F_2(x,y) = 2e^{-\left(\sqrt{x^2+y^2}-1.5\right)^2} - 1 \tag{2.2}$$

plotted in Figure 2.4 for $-\pi \leq x \leq \pi$ and $-\pi \leq y \leq \pi$, which will be further called the *orange squeezer function*

Function with two minima and two maxima of equation

$$F_3(x,y) = 10\left[e^{-(2x+1)^2-(y+1)^2} - e^{-(x-1)^2-(y+1)^2}\right]^2 + 15\left[e^{-(x-1)^2-(y-1)^2} - e^{-(2x+1)^2-(2y-1)^2}\right]^2 \tag{2.3}$$

graphed in Figure 2.5 for $-1.5 \leq x \leq 2.5$ and $-2.5 \leq y \leq 2.5$.

Because of its appearance, this third function will be further referred to as the *four-hump function*. Additional functions will be introduced later.

The data files required to plot Figures 2.1, 2.4, and 2.5 have been generated using program **P2_123.PAS** listed in Appendix B (you may also want to review program **P1_156. PAS** discussed in Chapter 1, which generates a file similar to **F2_1.D3D**). Note how function names **F1**, **F2**, and **F3** can be one by one assigned to variable **F** of the **argF2** type

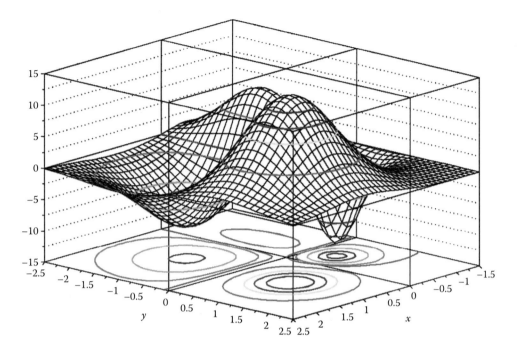

FIGURE 2.5 Mesh plot of the function in Equation 2.2 (the *four-hump function*) with 61 × 61 data points, featuring raised and projected z = constant *level curves*, also known as *contour lines*. Configuration file **F2_05.CF3**.

declared in unit **LibMath**, so that these functions can be activated or inactivated, depending on which file is to be generated (lines #**10**, #**12**, #**34**, #**40**, and #**46**).

The companion program **P2_3.PAS** (see Appendix B) generates four different data file types, all named **F2_3**, that can be used to plot function F_3 in Equation 2.3. The formats of these files are as follows: (i) a file of doubles with the extension **D3D** and identical with the one output by program **P2_123.PAS**; (ii) a file of reals with the extension **R3D**; (iii) one ASCII file with the extension **T3D**; and (iv) one ASCII file with the extension **G3D**. Although of different types, the **D3D**, **R3D**, and **T3D** files have the same structure: the first six entries in these files are the grid size and the limits over the x- and y-axes, that is, m, n, x_{min}, x_{max}, y_{min}, y_{max}, followed by the z_{ij} components with $i = 1...m$ and $j = 1...n$ of the function surface mesh. In turn, the **G3D** file is structured as (x_i, y_j, z_{ij}) rows—including parentheses—which is a format intended primarily for the **G_3D.LSP AutoLISP** program that allows true 3D surfaces and 3D curves to be generated inside **AutoCAD** as explained in Chapter 3.

If you press the <F10> key right after you launch **D_3D**, an **About** screen will come up where these four file formats are explained (see the **About** insert on next page), and the possible input data errors that **D_3D** may report. As mentioned in the **About** screen, if the number of z_{ij} components read by **D_3D** from data file is less than the grid size $m \times n$ recorded at the beginning of the same file (for reasons like accidently or intentionally aborting a lengthy data generating run), you will still be able to graph the available data, but the plot will appear incomplete.

```
┌──────────────────────────────────────────────────────────────────────┐
│                      About   D_3D ver. 2014                            │
├──────────────────────────────────────────────────────────────────────┤
│  < < < < <   D_3D plotting program.    (c) P.A. Simionescu 2014.  > > > > >  │
│                                                                        │
│  Expected input file extensios are D3D and R3D.  Any other extenssion will be │
│  assumed of an ASCII file.                                             │
│                                                                        │
│  D3D (doubles) R3D (reals) or ASCII files on columns should be structured as: │
│  m n Xmin Xmax Ymin Ymax Z11 Z12 Z13..Z1m Z21 Z22 Z23..Z2m...Zn1 Zn2 Zn3..Znm │
│  with m grid size over X axis and n grid size over Y axis.             │
│                                                                        │
│  ASCII files can also be structured as (Xi Yj Zij) ordered rows with   │
│  i=1..n and j=1..m.  This is the format readable by G3D.LSP.           │
│                                                                        │
│  Values Abs(Zij) > 1.0E+30 will be truncated automatically by the plot box. │
│                                                                        │
│  Possible input data error messages:                                   │
│    1: empty file;                                                      │
│    2: malformed file or premature end of data;                         │
│    3: m < 2;                                                            │
│    4: m > 501;                                                          │
│    5: n < 2;                                                            │
│    6: n > 501;                                                          │
│    7: invalid Zij components, or Zij fewer than n·m                     │
│                                                                        │
│  Note that data files with Zij fewer than n·m can still be graphed truncated │
│  to the nearest n.                                                     │
│                                                                        │
│  In WINDOWS XP you can automatically open CF3, D3D, R3D or G3D files with │
│  D_2D.EXE by selecting 'Open-With' from the File function of Windows Explorer │
└──────────────────────────────────────────────────────────────────────┘
```

2.3 MESH PLOTS AND THE VISIBILITY PROBLEM

This section explains how the visibility problem is solved by **D_3D**. You will also learn how to reduce the number of plot points (i.e., how to *scatter* the original data) and how to switch between a *mesh plot* (also known as *crosshatch plot*) and an x = constant only or y = constant only plot. Also explained is how to edit the appearance of the division and values over the x-, y-, and z-axes and the gridlines along the sides of the plot box.

Begin by running the **D_3D** program with settings from configuration file **F2_06A.CF3**, and redo Figure 2.6a. The companion plot in Figure 2.6b has been generated using the same data file as input, but the points along the y-axis were *scattered* from within **D_3D**. Additional modifications over Figure 2.6a are plotting the y = 0 and y = 2.5 boundaries of the function surface and a new layout of the side gridlines.

After launching **D_3D** with settings from **F2_06A.CF3**, to modify the gridlines and z-axis divisions, do the following: From the *final graphic screen*, press the <Backspace> key once to go to what will be further called the *graphic edit screen*. Here press <Insert> then <Z> and respond by typing '0.5' to change the interval between two z values over the vertical axis. Then type '4' to change the number of small intervals between two values; typing '1' instead will display no minor division line. Changing the division and value placements along the x- and y-axes can be done in the same manner. Next, press <Insert> then <CR> to confirm the default option, and then type '6'. This will reduce the number of side gridlines from 11 to 6. You can turn the gridlines completely off by typing '0' instead of '6' or by pressing <G> when in the *graphic edit screen*. Pressing <G> a second time will turn the gridlines back *on*.

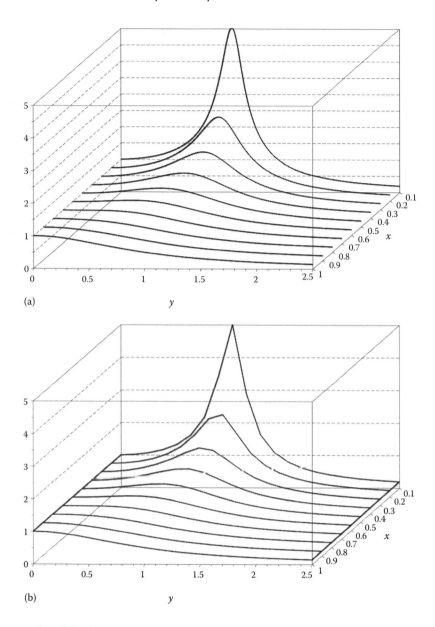

(a)

(b)

FIGURE 2.6   Plot of the function F_1 in Equation 2.1 as lines of constant x. Figure (a) has 10×261 data points, while figure (b) has 10×21 data points. Configuration files **F2_06A.CF3** and **F2_06B.CF3**.

Note that the z-axis division and value editing can be also done following option **<F1>** of the **<F1..4>** *screen*. From here, you will in addition be able to modify the z_{min} and z_{max} limits of the plot, including resetting them to their original values.

Important: If the level curves mapped on the 3D function surface are equally spaced, you can align them with the gridlines so that extracting data manually from the plot becomes more convenient. When doing so, remember that the top and bottom edges of the plot box count as gridlines.

If you press the <Y> key while in the *graphic edit screen*, you will activate the displaying of the *y* = constant lines, in addition to the already existing *x* = constant lines. However, because there are too many points (i.e., 261) along the *y*-axis, it will be hard to distinguish them. One solution is to run **P2_123.PAS** again and generate a new data file having fewer points. Alternatively, you can use the capability of **D_3D** to *scatter* the data and reduce the number of points along the *y*-axis. Note, for example, how the original input files with 481 × 481 points have been scattered to only 41 × 41 points in case of Figure 2.4 and to 61 × 61 points in case of Figure 2.5. This is done inside **D_3D** by eliminating every other point or every two, three, or more data points along the respective axis.

In case of the **F2_1.D3D** data file, the 261 points along the *y*-axis can be reduced to 131, 66, 53, 27, 21, 14, 11, 6, 5, and 3 points. To scatter the number of *y* points of the plot at any of these levels, go to the **<F1..4>** *screen* and press <F3>. A new screen will open with what will be called a *chime menu*—this particular one will be referred to as *chime menu 3*. Change to *Y* the first of the six characters of the menu, and press <CR>. Press <CR> again to leave unchanged the number of plot points over the *x*-axis (i.e., **10 of 10**), and then use the <↑> and <↓> keys to set the new data-point resolution along the *y*-axis to **21 of 261**. Press <CR> several more times until you get to the *final graphic screen*. What you should obtain must be similar to Figure 1.4b, less the *y* = constant lines to the left and to the right, that is, the *surface borders*. To turn these borders *on*, go to the **<F1..4>** *screen* and press <F2>. Then press <CR> twice to leave unchanged the title and the *chime menu 1* settings (the one on top). On *chime menu 2*, change the last **N** into **#** to activate the displaying of the *borders* of the graph. When you are finished, go back to the *final graphic screen* to see the change appearance of your plot.

Important: The editing mode of *chime menus* is *write-over*, so do not use the delete keys. If you want to restore the initial settings of a *chime menu*, just press the <Esc> key.

Important: The settings input on *chime menu 1, 2*, and *3* are saved to lines **30** to **32** of the **CF3** file. Any time you edit *chime menu 3*, the new settings will be added to the current **CF3** file above the lines that reads **∗∗∗ Level curve heights ∗∗∗**. Also recorded in line with *chime menu 3* are two numbers representing the *depth* at which data were scattered over the *x*- and *y*-axes, respectively. For example, in the configuration file **F2_06A.CF3**, line 32 reads: **YNNNNN 1 6**. These are interpreted as "scatter the input data at depth 1 along the *x*-axis (i.e., will remain unchanged) and at depth 6 along the *y*-axis (i.e., will retain only 21 points out of 261 since 21 is the 6th number in the row 261, 131, 66, 53, 27, 21, 14, 11, 6, 5, and 3)." If you replace 6 with 12 or bigger number, nothing will occur because 11 is the deepest *y*-axis *scatter level* possible for the given input data file. Note that *chime menu 3* settings are not recorded to the master configuration file **!CF3**.

Important: Every time *chime menu 3* is activated, a different **$3D** temporary file (in **D3D** format) is generated. The most recent of these files holds the data used to generate your final plot, and depending on the *chime menu 3* history, it could have fewer points, or the data could be recorded in a different order than in the original file. If you exit **D_3D** from the **<F1..4>** *screen* rather than the *final graphic screen*, these **$3D** files will be preserved. If you want to use any of these temporary files, modify their extension to **D3D**, or otherwise they will be deleted next time you launch **D_3D**.

Important: A built-in *scatter calculator* is available inside **D_3D** and can be launched by pressing the <F9> key from the startup menu. This will list the possible scatter options for number up to 501. For example, if the grid size over *x*- or *y*-axis is 481, it can be scattered to 241, 161, 121, 97, 81, 61, 49, 41, 33, 31, 25, 21, 17, 16, 13, 11, 9, 7, 5, 4, 3. In case of an actual plot, the same numbers will be available from *chime menu 1*.

To remove the invisible lines and plot both the *x* = constant and *y* = constant lines as shown in Figure 2.7 starting from the plot in Figure 2.6b, run **D_3D** with settings from configuration file **F2_6B.CF3**. Then go back to the *graphic edit screen* and press <H> to change visibility from *wireframe* to *hide*. Also press <Y> to turn the *y* = constant lines *on*. To modify the mesh color, you must edit line number **5** of the **F2_6B.CF3** file. Use 6 for *brown*, 8 for *dark gray*, 14 for *yellow*, and 15 for *white*.

D_3D employs the scan conversion procedure **FillPoly** in Pascal to fill with color the patches that form the function surface and thus remove the invisible lines. The order in which the polygons are drawn and filled with color is from *back to front*. For first-quadrant views, this succession is from *left to right*, while for fourth-quadrant views, the succession is from *right to left* (see Figure 2.8).

Only after the current patch is generated and filled with color are the *x* = constant, *y* = constant, and/or the elevated *z* = constant lines drawn. This approach, called

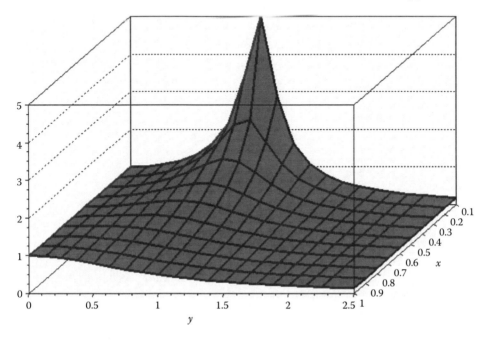

FIGURE 2.7 **PCX** output of the crosshatch (mesh) plot of the function of Equation 1.4. Configuration file **F2_07.CF3**.

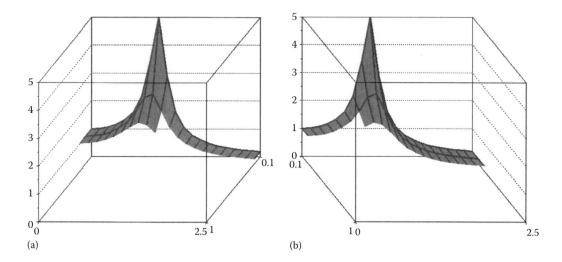

FIGURE 2.8 Plot in progress showing the order in which the hidden lines are removed, (a) for views from the first quadrant and (b) from the fourth quadrant. Configuration files **F2_08A.CF3** and **F2_08B.CF3**.

Painter's algorithm, is also applied when a **DXF** copy of your plot is generated by drawing each patch slightly more elevated than the previous one, so that the **AutoCAD** *hide* command will have the intended effect.

To better understand how *Painter's algorithm* is applied by **D_3D** to solve the visibility problem, a second executable file named **D_3Dslow.EXE** was prepared and is available with the book. This program is identical to **D_3D**, with the difference that you will have to press the <CR> key for every patch or line segment of the function surface to be drawn on the screen. Run **D_3Dslow.EXE** with settings from **F2_08A.CF3** to observe the order in which the patches and lines that form the function surface are plotted on the screen (see Figure 2.8a). Switch to a fourth-quadrant view from within the program, or simply run the **F2_08B.CF3** configuration file, and note the changed order of plotting the gridlines and surface patches (see Figure 2.8a).

2.4 NODE AND STEM PLOTS

D_3D allows you to place a node at every data point, whether or not x = constant, y = constant, or z = constant lines are mapped on the function surface. In turn, the function surface can be set to either *transparent* or *opaque*. Figure 2.9a shows an example of an opaque function surface with nodes filled with background color. The companion Figure 2.9b displays colored *stem plot* for 10 × 11 points of the original data file. In the remainder of this paragraph, it will be explained how these two plots were generated, starting from the configuration file of the plot shown in Figure 2.7.

To recreate the plot in Figure 2.9a, launch **D_3D** with settings from **F2_07.CF3**, then choose <F2> from the <F1..4> *screen*, and on *chime menu 2* change the node size from **0** to **5**. Before you move to the *final graphic screen* to view your plot, press <G> to suppress the gridlines.

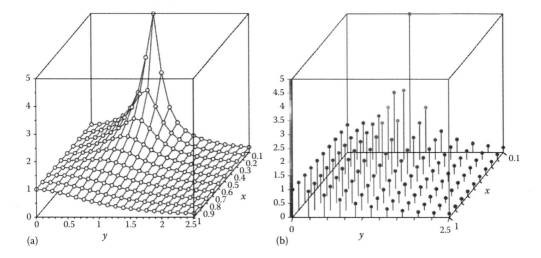

FIGURE 2.9 The function F_1 in Equation 2.1 represented as (a) opaque crosshatched-surface mapped with empty round nodes and (b) colored node stem plot—notice the color scale band integral part with the z-axis labeling. Configuration files **F2_09A.CF3** and **F2_09B.CF3**.

Creating the stem plot in Figure 2.9b requires additional interaction with the program: First go to **<F1..4>** *screen*, press <F2>, and change the last entry of *chime menu 2* from '#' to '!' - —this will activate the *stems*. Also press <F3> from the same **<F1..4>** *screen* and edit *chime menu 3* to reduce to 11 the number of points along the y-axis. Stems will not be displayed unless you press the <W> key when in the *graphic edit screen* to change visibility from *hide* to *wireframe*. With this same occasion, suppress the displaying of the x = constant and y = constant lines by pressing the <X> and the <Y> keys. Next edit the divisions and values over x- and y-axes so that they will look as shown in Figure 2.9b. Finally, set the number of values over the z-axis to 10 and the number of minor intervals between two values to 23 (the maximum allowable in this case). Note that if the total number of minor division lines over the z-axis is greater than 127, **D_3D** will thicken the division lines over the z-axis and color them according to elevation on a 10-color scale. Therefore, even with 23 minor intervals, the z-axis will be color coded since $10 \times 23 > 127$.

Important: *Stem lines* and *level curves* share the same thickness, controlled by the first entry on *chime menu 2*. This menu also controls the color of the *level curves* and of the *nodes*, that is, if you want these colored according to their height, then set the third entry of *chime menu 2* to 'Y'.

To change the node type from empty round to solid round and to widen the z-axis color band, you must exit **D_3D** and apply these changes to the last **CF3** file created as explained next: Open the **CF3** file with **Notepad**; on line number 7, change the node type from 0 to 1; and on line **8**, change the outside length of the division lines to 9 and their inside length to 0 (the latter change will be applied only to the division lines placed along the z-axis). Save your configuration file under a different name (i.e., **F2_08B. CF3**) and open it with **D_3D**. You should obtain a plot similar to the one in Figure 2.9b.

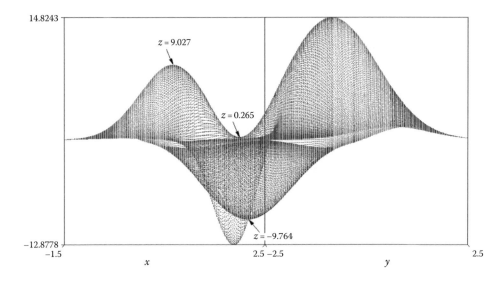

FIGURE 2.10 The *four-hump function* represented as 481 × 481 pixel-size nodes mapped on a transparent surface. The local minimum, local maximum, and saddle-point elevations were extracted through inspection inside **AutoCAD**. Configuration file **F2_09.CF3**.

Another example of a node plot is given in Figure 2.10. This is a zero-elevation diagonal view of the *four-hump function* represented as *point cloud*, that is, nodes are mapped on a transparent function surface. Using **AutoCAD**, the local minimum and maximum points were estimated as −9.746 and 9.027, and the height of the saddle point was estimated as 0.265. Evidently, the accuracy with which these values were determined depends on the grid size at which the function has been sampled. If you are interested only in the global minimum and global maximum values, there is no need to export your graph to **AutoCAD** since these can be displayed on the graph by fitting the function surface to the plot box, that is, go to the **<F1..4>** *screen*, press <F1>, and type '.' (a dot) for the upper and lower limits of the *z*-axis in the respective boxes (this will reset the limits over the *z*-axis in **D_3D**).

Important: The *z*-axis scale can be displayed in two ways: with the values and divisions starting from zero or aligned with the ends of the respective axis. To toggle between the two alignment modes, press <F5> from the *graphic edit screen*. For the plot in Figure 2.10, this will have no effect, however, since along the *z*-axis there are only two values, that is, the end values.

2.5 EQUALLY SPACED LEVEL-CURVE PLOTS

In this paragraph, it will be explained how to match the elevation of the side *gridline* with the height of the *level curves*. You will also learn how you can modify the level-curve heights from within **D_3D** and how to append these values to the current **CF3** file and use them in later plots. Once saved to file, you can further edit these height values, delete some of them, or append new ones. *Bridge-like defects* that may occur when producing level-curve plots, or when the function surface is trimmed by the plot box, are also discussed in this section.

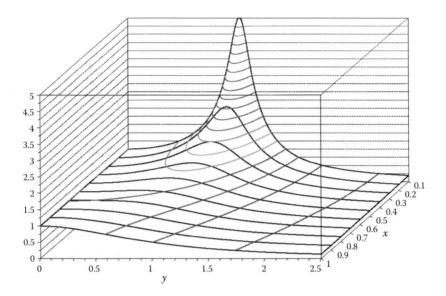

FIGURE 2.11 **R12 DXF** copy of a raised level-curve plot with gridlines and level curves placed at the same heights. This is a wireframe plot, the visibility problem of which has been solved by hand from within **AutoCAD**. Configuration file **F2_11.CF3**.

First, we will add elevated level curves to the plot in Figure 2.6 and then make their heights coincide with the elevations of both the gridlines on the side of the pot box and the division lines along the z-axis—see Figure 2.11. To do so, launch **D_3D** with settings from **F2_06A.CF3**, then go to the *graphic edit screen*, press <Z> to turn the z = constant lines *on*, then press <W> for a wireframe plot. Then press <Insert> followed by <Z> to change the interval between two values along the z-axis to 0.5 units and the number of minor division lines per interval to 2. The total number of divisions, both short and long, that will be placed along the z-axis will now be 21. To make the number of gridlines equal to 21 as well, press <Insert> then <G>. **D_3D** will suggest several numbers to choose from, including 21, which will ensure (although not always) that each gridline is an extension of a z-axis division line.

Go to the **<F1..4>** *screen* and press <F2>, then change the last entry of *chime menu 2* to '**#**' (this will turn the border of the function surface *on*). Finally, go back to the **<F1..4>** *screen* and press <F4>. Leave unchanged the z-axis settings, and type 'Y' when asked to **Update level curves?** Scroll through the available options using <↑> and <↓> and select **Evenly spaced**, then type '21'. Continue pressing the <CR> key until you get to the *final graphic screen*.

If you want the level-curve heights to be appended to the **CF3** file that is created when you exit **D_3D**, go back to the *graphic edit screen* and press <E>. **D_3D** will prompt you to edit one by one the level-curve heights or just press the <CR> key to confirm an existing values. When asked if you want them saved to file, respond by typing 'Y', then go to the *final graphic screen* and exit the program. Use **Notepad** to open the latest **CF3** file in the current directory and see that indeed these level-curve heights were appended after the line that reads ***** Level curve heights *****.

You can edit, delete, or add more values to this list any time you want. These height values do not have to be ordered, nor the corresponding level curves have to actually intersect the function surface.

Launch **D_3D** and open the same configuration file—if you have not changed its default name, just press <F1>; otherwise, press <F3> and select its name. After redoing the plot, go back to the **<F1..4>** *screen*, and press <F1> and <CR> until asked to update level curves. Answer 'Y' then scroll down using the arrow keys. Notice that there is now a fifth option available, that is, to read the level-curve heights from the configuration file. Select this option and see how **D_3D** identifies and uploads all level curves that can potentially intersect the function surface.

Important: If they are automatically generated, the total number of level curves cannot exceed 999. If their height values are read from file, or to edit them interactively from the *graphic edit screen*, they cannot be more then 50.

2.6 DEFECT-FREE LEVEL-CURVE PLOTS

Level curves are the intersections between the function surface and a horizontal cutting plane placed at various elevations z_k. **D_3D** uses the already available *four-sided patches* of the function surface to evaluate the intersection with this horizontal cutting plane, rather than triangulating the surface as other level-curve plotting programs do (Bourke 1987). In addition to the corners of the current patch, **D_3D** also uses the height value of four of its neighboring points to estimate the sign of the curvature of the surface over the area of the respective patch. This way, the level curves will have a correct appearance without exhibiting defects that look like bridges, even if the function surface is undersampled (see Figure 2.12).

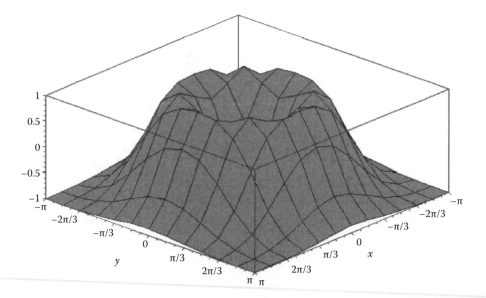

FIGURE 2.12 Plot of the function in equation F_2 with 13 × 13 data points. Configuration file **F2_12.CF3**.

Let us consider the *orange-squeezer function* in Figure 2.4, with only 13 × 13. At such a low resolution, a 3D graph of the function surface will turn very 'choppy' (see Figure 2.12). When the intersection between this function surface and a horizontal cutting plane is evaluated in the process of extracting level curves, connectivity defects may occur around the areas close to the top of the graph.

The level-curve plot in Figure 2.13a produced with the following **MATLAB** commands

```
x=-pi:2*pi/12:pi;
y=x;
[X,Y]=meshgrid(x,y);
Z=2*exp(-(sqrt(X.^2+Y.^2)-1.5).^2)-1;
contour(X,Y,Z, 17);
```

exhibits bridge-like defects when the number of equally spaced level curves is 23 or more. For the same number of level curves and sampling size, a plot produced with **D_3D** is defect-free (Figure 2.13b) and remains that way even for 40 equally spaced curves.

A detailed explanation on how these intersections are evaluated and how they are corrected based on the sign of local curvature of the function surface is available in Simionescu (2003). Some of these aspects will be also discussed in paragraph 2.11.

Important: The bridge-like defects of level curves in top view may occur in different locations or may not occur at all if you press the <F4> key while in the *graphic edit screen* or in the *deformable-box screen*. This will reverse the *z*-axis of the plot, and as a consequence, the points used to evaluate the function curvature will be different, with possible favorable effect.

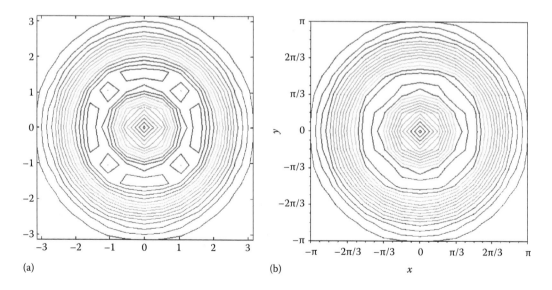

(a)   (b)

FIGURE 2.13   17-level-curve plot (also known as contour plot) of the *orange-squeezer function* with 13 × 13 data points produced (a) with **MATLAB** and (b) with **D_3D**. Configuration file **F2_13B. CF3**.

2.7 LOGARITHMICALLY-SPACED LEVEL CURVES

This paragraph refers to plotting *logarithmically* or *log-spaced level curves*. This is a unique feature of **D_3D**, useful when you want to concentrate the level curves around certain points of interest. The less elegant approach to this problem is to manually alter the heights of a set of equally space level curves, until the details of interest are revealed.

D_3D can automatically create *log-spaced level* curves that are either

1. Accumulated towards z_{min} as shown in Figure 2.14 or
2. Accumulated towards z_{max} as shown in Figure 2.15 or
3. Accumulated towards zero from both above and below the $z = 0$ plane as shown in Figure 2.16

(i) For n_{LC} the total number of level curves and z_k the height of the kth curve with $1 \le k \le n_{LC}$, the following formula ensures accumulating the level curves around z_{min} as *log-spaced down* curves:

$$z_k = z_{min} + \exp\left[(k-1)\cdot ln\left(\frac{z_{max} - z_{min} + 1}{n_{LC} - 1}\right)\right] - 1 \tag{2.4}$$

(ii) To get the same number of level curves converging to z_{max} as *log-spaced up* level curves, Equation 2.5 should be used instead, that is,

$$z_k = z_{max} - \exp\left[(k-1)\cdot ln\left(\frac{z_{max} - z_{min} + 1}{n_{LC} - 1}\right)\right] + 1 \tag{2.5}$$

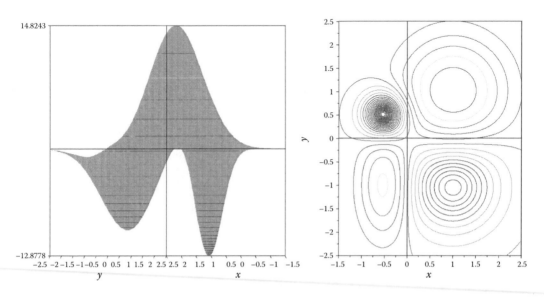

FIGURE 2.14 *Log-spaced down* level-curve plot of the *four-hump function*. Configuration files **F2_14A.CF3** and **F2_14A.CF3**.

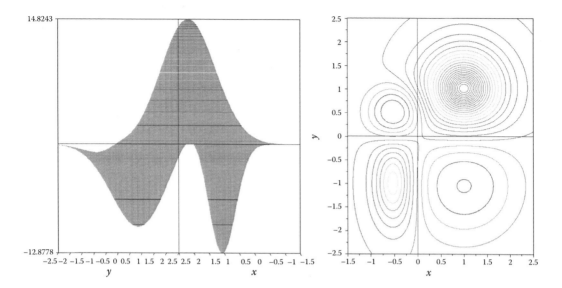

FIGURE 2.15 *Log-spaced up* level-curve plot of the *four-hump function*. Configuration files **F2_15A.CF3** and **F2_15B.CF3**.

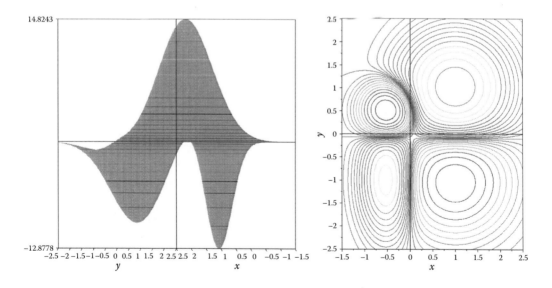

FIGURE 2.16 *Log-spaced from zero* level-curve plot of the *four-hump function*. Configuration files **F2_16A.CF3** and **F2_16B.CF3**.

(iii) For the level curves to be *log-spaced from zero* in both directions, the total number of curves n_{LC} has to be divided into curves located below the $z = 0$ plane:

$$n_{LCdn} = Round \left| \frac{n_{LC} \cdot z_{min}}{(z_{max} - z_{min})} \right| \qquad (2.6)$$

and level curves located above the $z = 0$ plane:

$$n_{LCup} = n_{LC} - n_{LCdn} \qquad (2.7)$$

With these notations, the heights of *log-spaced from zero* level curves can be calculated by setting $z_{max} = 0$ and $n_{LC} = n_{LCdn}$ in Equation 2.4 and setting $z_{min} = 0$ and $n_{LC} = n_{LCup}$ in Equation 2.5. The **D_3D** program actually implements a more general approach, where this type of level curves can be plotted even if zero is not an inner point of the $[z_{min}, z_{max}]$ interval.

When producing Figures 2.14 through 2.16, the lower and upper limits in Equations 2.4 through 2.7 were considered $z_{min} = -12.8778$ and $z_{max} = 14.8243$, as extracted from among the 481×481 data points of file **F2_3.D3D** (see Figure 2.10). If you modify z_{min} and z_{max} from the values extracted by the program, then these new limits will be utilized in applying Equations 2.7 through 2.7. Note that by saving them to file, you can further modify the z-axis limits while maintaining the same level-curve appearance.

An extension of procedure (iii) where that level curves are *log-spaced from any point z_0* within the interval $[z_{min}, z_{max}]$ will be explained next. The number of level curves located below z_0 will be calculated with

$$n_{LCdn} = Round \left| \frac{n_{LC} \cdot (z_0 - z_{min})}{(z_{max} - z_{min})} \right| \qquad (2.8)$$

while the number of level curves located above the z_0 is given by the same Equation 2.7. We then apply Equation 2.4 with $z_{max} = z_0$ and $n_{LC} = n_{LCdn}$ to calculate the height values of the level curves located below z_0 and apply Equation 2.5 with $z_{min} = z_0$ and $n_{LC} = n_{LCup}$ to calculate the height values of the level curves located above z_0.

Program **P2_ZLC.PAS** in Appendix B implements this procedure to generate an ASCII file named **Z.LCS** having **NrLC** = 28 height values that are *log-spaced from* **z0** inside the interval **zmin**= 12.8243 and **zmax**=14.8243 (see line #**6** of this program). The 28 height values generated using program **P2_ZLC.PAS** for **z0**=0.265 (i.e., the saddle point) of the *four-hump function* in Figure 2.10 have been appended manually at the end of the configuration file **F2_17.CF3**. Using this modified configuration file, the plot in Figure 2.17 has been generated.

To generate the same level-curve elevations using **D_3D**, you must follow these steps: First, calculate n_{LCdn} and n_{LCup} using Equations 2.8 and 2.7. Then run **D_3D** with z_{min} set to z_0 and n_{LCdn} level curves *log-spaced up*. Save these height values to the current configuration file by selecting **<E>** from the graphic edit screen (you must confirm each of height value before writing them to file). Then run **D_3D** again with z_{max} set to z_0 and n_{LCup} level curves *log-spaced down*. Again save the resulting height values to file. Combine these two sets of level-curve heights in the same **CF3** file and use this file to generate your plot.

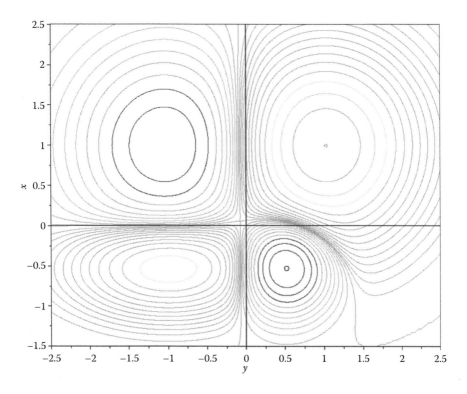

FIGURE 2.17   *Log spaced from* $z_0 = 0.265$ level-curve plot of the *four-hump function* with 481×481 data points. Configuration file **F2_17.CF3**.

2.8 FILE EXPORT AND **DXF** LAYER ORGANIZATION

Any plot can be exported to **PCX** by pressing <F10> from the graphic screen. For other types of exports, you must go to the **<F1..4>** *screen* and select **<F4>**, then scroll down using the arrow keys, and choose either **G3D**, **DXF**, or (if available) **DXF 1:1**. The **DXF 1:1** export mode is available only for level curves in top view, when **D_3D** will export to **R12 DXF** the curves only (without the plot box, divisions, labels etc.), scaled one-to-one. When exporting wireframe plots to **DXF**, **D_3D** will concatenate, inasmuch as possible, their x = constant, y = constant, and z = constant lines into single contiguous polylines.

The **G3D** export option causes the points of the function surface to be written to an ASCII file that can be read by the **G_3D.LSP** application, which allows true 3D surfaces to be generated inside **AutoCAD**. Note that to the **G3D** file the function surface is recorded at the current resolution and without trimming the function surface by the bounding box (assuming that the input data file has been scattered or the limits over the z-axis have been modified from their original values).

If you want to manually generate a top-view level-curve plot scaled one-to-one and have the axes, divisions, and values included, perform the following steps: Adjust the plot-box size until its height over width ratio **h/w** equals $(x_{max} - x_{min})/(y_{max} - y_{min})$, and then export the graph to **DXF**. It is assumed that x_{max}, x_{min}, y_{max}, and y_{min} are the spans over the respective axes of the level-curve plot and that x is the vertical axis and y the horizontal axis of the plot. Open this **DXF** file inside **AutoCAD** and then scale the plot with the factor $(x_{max} - x_{min})/$**h**. Alternatively, you can export to **DXF** the plot as is, then open it inside **AutoCAD**, and copy it to a *block*. Reinsert that block scaled by a factor $(y_{max} - y_{min})/$**w** over horizontal axis of the drawing and by a factor $(x_{max} - x_{min})/$**h** over vertical axis of the drawing. The only problem with this second method is that the text will be distorted, and the division lines will have different lengths over the two axes.

The plot in Figure 2.5 is shown again in Figure 2.18a as an **AutoCAD** drawing obtained through **DXF** import. From within **D_3D**, you can turn *on* and *off* the raised level curves by pressing the<Z> key while in the *graphic edit screen*. Similarly, by pressing the <F1>, <F2>, or <F3> from the same *graphic edit screen*, you can turn *on* and *off* the lines representing the *OXY*, *OXZ*, and *OYZ* planes (these are called *zero lines*). To turn *off* the level curves projected on the bottom plane, or to draw both these and the raised curves in thick line, you must go to the **<F1..4>** *screen*, press <F2>, and edit the first and second entry of *chime menu 2*. The third entry of *chime menu 2* controls the color of the level curves, that is, they can be either monochrome or colored according to their elevation.

In the current implementation of **D_3D**, you cannot edit the thickness and color of the projected and raised level curves separately. However, you can do it easily inside **AutoCAD**, because various entities of the plot are conveniently placed in separate layers as explained next (see also Figure 2.18b):

Layer '0' contains the plot-box lines, less the *zero lines* and *gridlines* that are placed in their own layers. The long and short division lines are placed in layer 'divisions.' The values along the three axes are placed in layer 'text' together with the plot *header*. Layer '$_body' contains the **AutoCAD** regions, one for each patch of the function surface, which serve to obscure the invisible lines when the **AutoCAD** *hide* command is issued. Level curves of the same height elevation (whether projected on the bottom plane or mapped on the function surface) are placed in separate layers, named 'C' followed by their elevation with the decimal point replaced with '_' (the underscore sign). Additional layers are 'const_x' and 'const_y' that host the x = constant and y = constant lines, less the lines that form the edges of the function surface that are assigned to layers 'border_x' and 'border_y'.

Important: The *plot header* visible in Figure 2.18a lists the **D_3D** parameters **w**, **h**, **kH**, **kV**, **tan(Gamma)**, and **tan(Delta)**, followed by the name of the file from where the data originate. This information becomes useful if you want to recreate a plot for which you no longer have a copy of its **CF3** configuration file.

Important: If a wireframe plot is desired, it is best to export it as such to **DXF**. This way, all x = constant, y = constant, and raised z = constant lines will be saved as contiguous polylines. Sometimes, it is possible to manually hide the invisible lines by trimming them inside **AutoCAD**.

Important: The outer ends of the two short oblique lines on the top left and bottom right of every **DXF** copy of a plot (see Figure 2.18a) form the corners of a 640 × 480 rectangle. These two lines are useful in scaling a plot back to its original size following a **PLT** export (see the **Util~PLT** program in Chapter 3 for details).

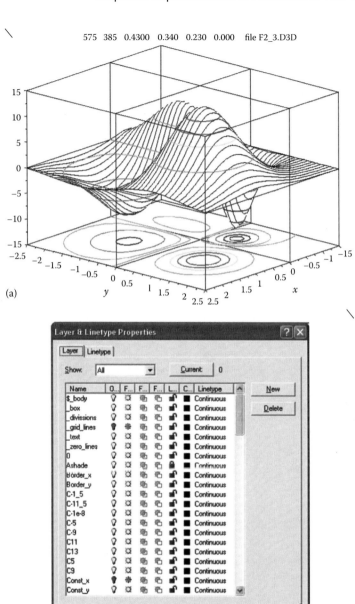

(a)

(b)

FIGURE 2.18   (a) **R12 DXF** copy of the plot in Figure 2.5 and (b) its **AutoCAD** layer settings. Note that the x = constant line and side grid lines have been suppressed by turning the respective layers off.

2.9 AXES REVERSAL AND PLOT ROTATION

Sometimes, it is of interest to reverse the orientation of one or more of the plot axes. Because **D_3D** does not generate true 3D plots, x- and y-axes reversal provides additionally a means to rotate the entire plot about their axis. For top-view plots, x- and y-axes reversal can be done inside **AutoCAD** using the *rotate* and *mirror* commands (remember to first set the **AutoCAD** parameter *mirrtext* to 0, to prevent the text from being mirrored together with the rest of the graph).

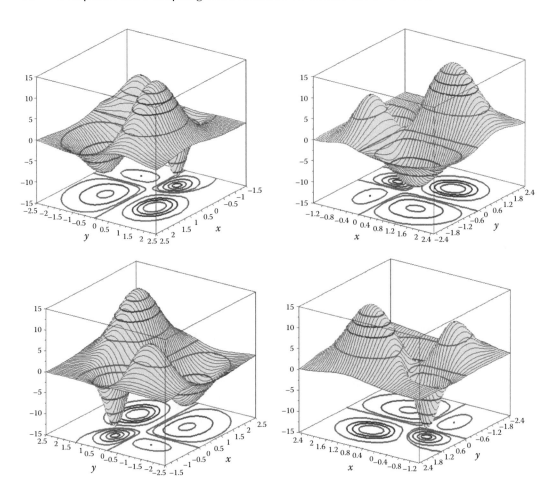

FIGURE 2.19 Rotated views of the *four-hump function*. Configuration files **F2_19A.CF3**, **F2_19B.CF3**, **F2_19C.CF3**, and **F2_19D.CF3**.

From inside **D_3D**, the orientation of the *z*-axis can be reversed by pressing the <F4> key while in the *deformable-box screen* or in the *graphic edit screen*. *z*-axis reversal can sometimes remedy the bridge-like defects of level-curve plots. **D_3D** is also capable of reversing and swapping the *x*- and *y*-axes, without the need for the user to produce a new input data file. These transformations can be induced from the *chime menu 3* accessible through option **<F3>** of the **<F1..4>** *screen*.

Rotating the entire plot in 90° increments (see Figure 2.19) is done by **D_3D** as combinations of *x*- and *y*-axes swap, followed by reversing the orientation of one of them. If after swap the *y*-axis is reversed, the graph will rotate 90° counterclockwise. If the *x*-axis is reversed instead, the graph will rotate 90° clockwise. These two types of rotations are controlled from the same *chime menu 3* of **D_3D**.

2.10 GRADIENT PLOTS

The gradient of a scalar function $f(x, y, z)$ is a vector field that, at a given point (x, y, z), is oriented in the direction of the greatest rate of increase and has its magnitude equal to the slope of the function along this direction. The components of the gradient are the partial derivatives of the function, that is,

$$grad(f) = \nabla f = \left(\frac{\delta f}{\delta x}, \frac{\delta f}{\delta y}, \frac{\delta f}{\delta z} \right) \tag{2.9}$$

Being able to visualize the gradient can reveal some important characteristics of the function at hand, and it is therefore a feature available in a number of function plotting programs.

D_3D is capable to represent the gradient as a set of arrows mapped on the bottom of the plot box, either in top view (Figure 2.20) or in 3D view (Figure 2.21). In both situations, **D_3D** estimates the components of the gradient through finite differences, using the image space coordinates of the corners of the patch, rather than their original 3D coordinates (Simionescu 2011).

As visible in Figure 2.20, the arrows representing the gradient are placed at the projected center of each of the four-sided patches that approximate the function surface. The relative size of these arrows can be controlled by the user from *chime menu 2*, accessible through option **<F2>** of the **<F1..4>** *screen*.

D_3D does not allow you to plot the gradient if the mesh size of the function surface is too dense, specifically if $n + m < 160$. Therefore, in order to generate Figures 2.20 and 2.21, the original file **F2_3.D3D** with 481×481 data points had to be scattered to only 21×25 points.

Note that it is possible to manually combine on the same representation a gradient plot with a higher-resolution surfaces and/or level-curve plot. To do so, you have to generate

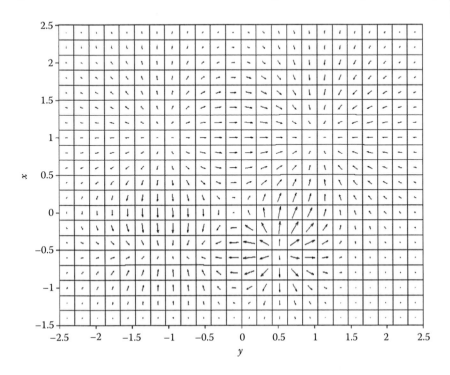

FIGURE 2.20 Gradient plot of the *four-hump function* projected on the bottom plane, overlapped with the mesh grid (resolution 21×25 points). Configuration file **F2_20.CF3**.

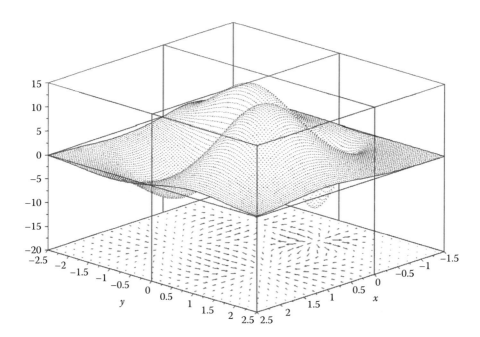

FIGURE 2.21   Projected gradient combined with a 3D plot of the *four-hump function*. Configuration files **F2_21DN.CF3** and **F2_21UP.CF3**.

them separately using different data file resolutions, export them to **PCX** or **DXF**, and then overlap them using **Paint** or **AutoCAD**, respectively. For example, Figure 2.21 shows the gradient mapped on the bottom plane at a 21 × 25 point resolution, while the node-on-opaque surface representation of the function has a 97 × 121 points. As an exercise, you can overlap the level-curve plot in Figure 2.17 generated using 481 × 481 data points, with the gradient plot in Figure 2.20 generated using 21 × 25 data points only.

2.11 TRUNCATED 3D SURFACE REPRESENTATIONS

One of the main reasons I developed **D_3D** was to generate 3D plots where the limits over the z-axis have been reduced from their original values and the function surface is truncated by the upper and/or lower planes of the plot box. Penalized objective functions encountered in optimization problems are the prime example where such a plotting feature is useful (Simionescu 2011), as well as functions with singularities and inequalities in two variables.

In recent years, several commercial software programs were enhanced with such capabilities, for example, **SigmaPlot**, **Mathematica**, and **MATLAB**. The only major software lagging behind is **Excel**. Their truncated 3D plots appear to be rather alterations of the function values (see Figure 2.22), where data points with z greater than the imposed z_{max} are simply forced equal to z_{max} (see also the spreadsheet file **Fig2_22.XLS** available with the book). A 36 × 36 point data file named **F2_2.TXT** that was imported in **Excel** to generate Figure 2.23 has been produced with the **P2_2T.PAS** program listed in Appendix B.

Now contrast Figure 2.22 with Figure 2.23a and b, both generated using **D_3D**. Notice the accurate intersection between the function surface and the upper plane of the plot box in Figure 2.23 and the possibility of representing this intersection either opaque or

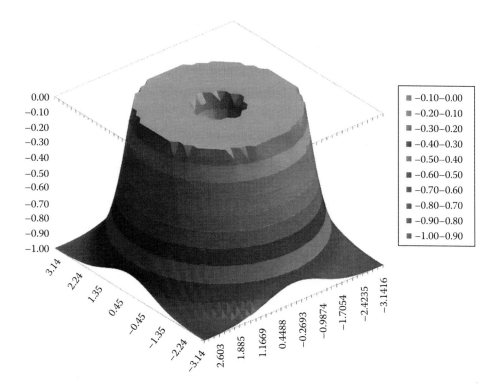

FIGURE 2.22   Truncated plot of the *orange-squeezer function* generated with **Office Excel**, starting from ASCII file **F2_2.TXT** produced with program **P2_2T.PAS**.

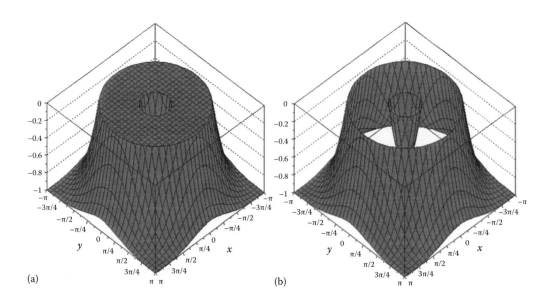

(a)

(b)

FIGURE 2.23   Truncated plots of *orange-squeezer function* with (a) opaque and (b) transparent intersections between the function surface and the bounding box. Configuration files **F2_23A.CF3** and **F2_23B.CF3**.

transparent. In order to generate truncated function plots, you will have to narrow the initial $[z_{min}, z_{max}]$ interval following option **<F1>** of the **<F1..4>** *screen* of **D_3D**. Additional settings refer to displaying or not the top land, controlled from the *chime menu 2* accessible by pressing <F2> from the same **<F1..4>** *screen*.

Figure 2.24 shows another truncated function-surface plot, this time of the *four-hump function*. Notice the hidden-line removal imperfections due to **AutoCAD**. These will not occur if the surface is more coarsely approximated (i.e., the patches are bigger and in fewer number). In Chapter 3, it will be explained how you can manually correct these defects by saving the plot to a **PLT** file and then export it back to **DXF** for editing. The same *four-hump function* is pictured in top view in Figure 2.25, this time with the x = constant and y = constant lines turned *off*.

Since the intersection of the function surface with a horizontal cutting plane is an actual level curve, the same approach is implemented inside **D_3D** to produce truncated function-surface plots.

Important: The borders of the intersections between the function surface and the upper and/or lower horizontal planes of the plot box are treated as level curves mapped on the function surface. If you want to display them on your graph, you must turn the elevated level curves *on* by pressing the <Z> key while in the *graphic edit screen*. You must also update the level-curve heights after modifying the z_{min} and z_{max} limits, possibly setting their number to only two if you do not want additional level curves in between mapped on the function surface.

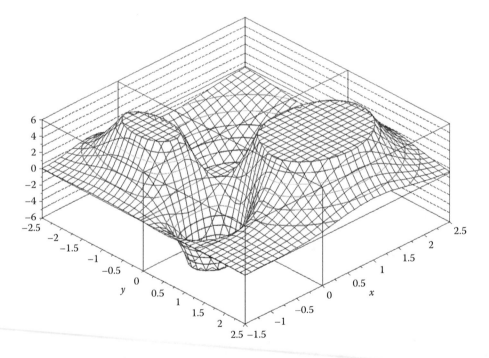

FIGURE 2.24   Truncated plot of the *four-hump function*. Note the hidden-line removal artifacts on the edges of the top and bottom lands due to **AutoCAD**. Configuration file **F2_24.CF3**.

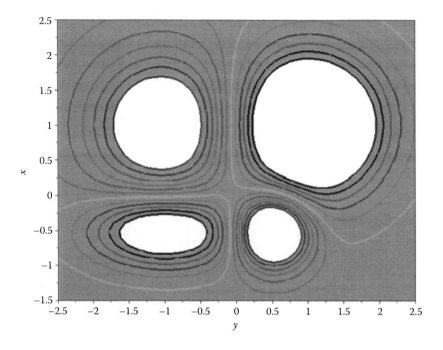

FIGURE 2.25   Top view of the plot in Figure 2.25. Configuration files **F2_25.CF3**.

The intersection between a patch of the function surface and a horizontal plane (also required when generating z = constant level curves) is handled as follows (see Figure 2.26): An initial four-sided patch (0000) of the function surface is modeled as an eight-vertex polygon with two by two of its vertices coincident. If this patch is intersected by one or both horizontal planes of the plot box, its originally coincident vertices can split or merge with other vertices, depending on their position relative to the intersecting plane. All intersection variants between an initial four-sided patch (shown in gray) and a horizontal cutting plane can be handled using a single eight-vertex polygon and four auxiliary five-vertex polygons noted a, b, c, and d shown in white in Figure 2.26. These intersection variants are symbolized (0111), (1011), (1110) etc., where the first binary digit corresponds to node 1–2 of the initial patch, the second digit to node 2–3, and so on. If one of these four nodes is located outside the plot box, then the corresponding digit will be set to 1; otherwise, it will be set to 0. The last two variants symbolized (1010) and (0101) correspond to a saddle point occurring over the current patch of the function surface. These last two variants are treated correctly by **D_3D** only in 50% of the cases, the other 50% causing *bridge-like-defects* as shown in Figure 2.27.

To further explain the accuracy of these ambiguities, let us consider the *hyperbolic paraboloid* of equation:

$$F_4(x, y) = 0.1\, xy \tag{2.10}$$

for $-\pi \le x \le \pi$ and $-\pi \le y \le \pi$, which exhibits a saddle point for $x = 0$ and $y = 0$ (see Figure 2.28). This figure was produced with data file **F2_4.D3D** output by program **P2_4.PAS** in Appendix B.

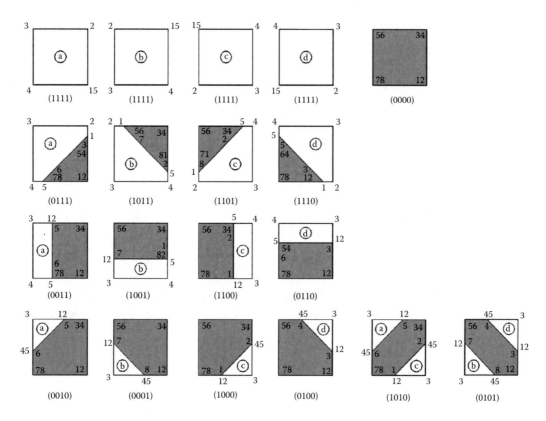

FIGURE 2.26 Intersection variants of an initial four-sided patch (gray) with a horizontal plane. The gray polygons correspond to the portion of the function surface located inside the bounding box, while the white polygons a, b, c, and d correspond to the outside areas.

As visible in Figure 2.29a, the intersection variant (1010) with the saddle point located inside the plot box is analyzed correctly by **D_3D** (see the higher resolution in Figure 2.29b for comparison). The same (1010) variant but with the saddle point located outside the plot box would require the use of two disjointed 'gray' polygons (see Figure 2.29c). These two cases are not differentiated by **D_3D** and are the reason for the 'bridge-like defects' occurring in half the cases as explained earlier.

When generating level-curve plots, the saddle-point variants (1010) and (0101) are dealt with correctly, because they require plotting lines only, and not of function-surface patches. In the process, use is made of the sign of the curvature of the function surface along the two diagonals of the current patch. These signs are estimated using the elevations of the four nodes of the original patch (0000), together with the elevations of the previous patch nodes 5–6 and 7–8 and elevations of the immediately following patch nodes 1–2 and 3–4 (see Figure 2.27 and Simionescu [2003]).

Remember that by increasing the number of sampling data points, these defects can be reduced or eliminated entirely. Alternatively, you can export your plot to **PCX** or **DXF** and retouch it using **Paint** or **AutoCAD**.

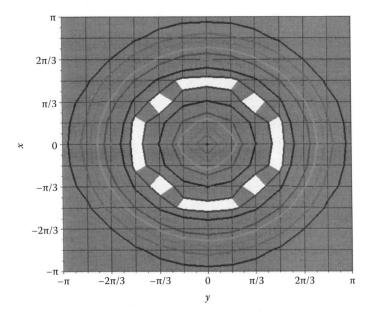

FIGURE 2.27 Defects occurring at the intersection of the *orange-squeezer function* with 13 × 13 data points and the $z = 0.88$ plane. These defects are corrected in the equivalent level-curve plot, based on the local curvature of the function surface. Configuration file **F2_27.CF3**.

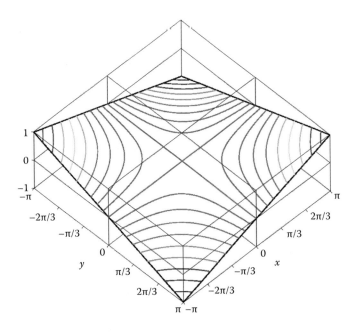

FIGURE 2.28 Elevated *log-spaced from zero* level-curve plot of the function in Equation 2.10, showing the saddle point at (0,0,0). Configuration file **F2_28.CF3**.

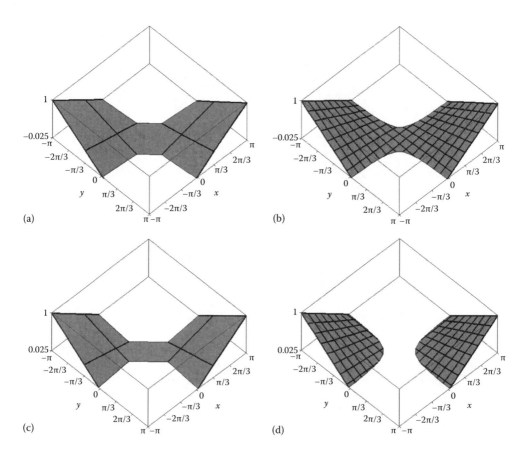

FIGURE 2.29 Truncated plot around the saddle point of the function in Equation 2.10 at (a and c) low and (b and d) high resolutions. Figures (a) and (b) correspond to the saddle point being located inside the plot box, and (c) and (d) correspond to the saddle point being located outside the plot box, of which figure (c) is incorrect. Configuration files **kF2_29A.CF3**, **F2_29B.CF3**, **F2_29C.CF3**, and **F2_29D.CF3**.

Important: When not of zero length, sides 3–4 of the white polygons in Figure 2.27 will always be part of a x = constant line to the left, sides 1–2 will be part of a x = constant line to the right, sides 4–5 will be part of a y = constant line to the rear of the plot, and sides 1–8 will be part of a y = constant line to the front. Similar ordering holds for the five-vertex polygons a, b, c, and d.

2.12 CONSTRAINED FUNCTION AND INEQUALITY PLOTS

This paragraph discusses how you can display or hide the surface patches located completely inside the plot box (referred together as *body*) independent from the patches intersected by the top or bottom planes of the plot box. The portions of the intersected patches located inside the plot box will be called *curtain*, while the outside portions will be called *top* and *bottom land*, depending on whether they are mapped on the top and bottom planes of the plot box (Figure 2.30). Such capabilities of **D_3D** allow you to plot surfaces with discontinuities and of inequalities of two variables. These visibilities are controlled by editing the *chime menu 1* accessible by choosing **<F2>** from the **<F1..4>** *screen*.

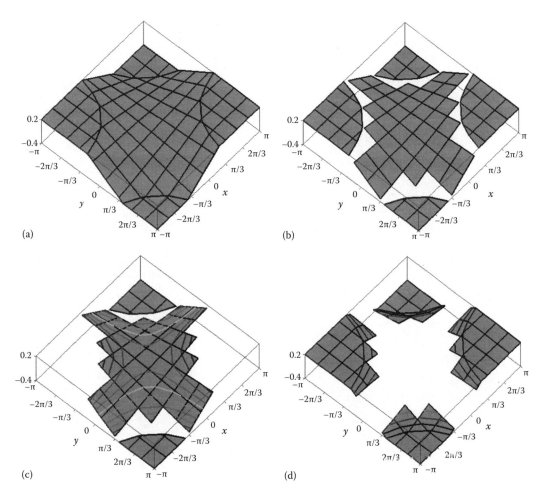

FIGURE 2.30 Various z-axis truncated plots of the function in Equation 2.10: (a) complete plot, (b) curtain missing, (c) top land and curtain missing, (d) body missing. Configuration files **F2_30A.CF3**, **F2_30B.CF3**, **F2_30C.CF3**, and **F2_30D.CF3**.

Let us consider the problem of plotting the surface of the function in Equation 2.10, less a circular hole of radius 1.5 centered at $x = 0$ and $y = 0$. This new function can be described analytically as

$$F_5(x, y) = \begin{cases} 0.1\,xy & \text{for } x^2 + y^2 \geq 1 \\ K & \text{for } x^2 + y^2 < 2.25 \end{cases} \qquad (2.11)$$

where K is a large constant value, assigned to the *infeasible* area, that is, the region where the function is not defined. In the Pascal program used to generate the data for this plot, this constant was set to -10^{30} when output to file **F2_5N.D3D** and to 10^{30} when output to files **F2_5P.D3D** and **F2_5.D3D** (see Figure 2.32a and b and the source code **P2_5.PAS** listed in Appendix B).

Important: When evaluating the limits over the z-axis, **D_3D** will ignore any z_{ij} value read from the input data file that is less than -10^{30} or greater than 10^{30}. Consequently,

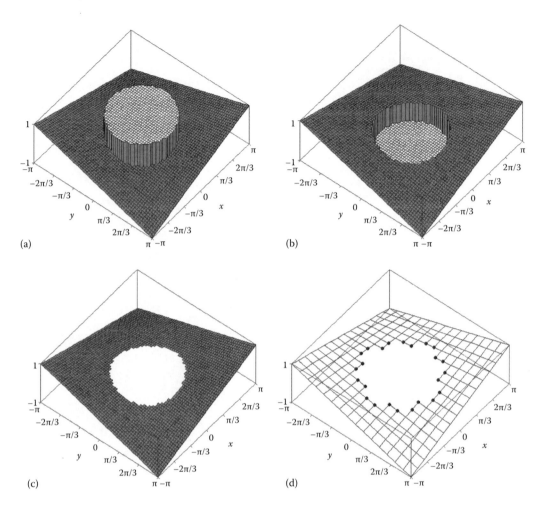

FIGURE 2.31 Plots of the function in Equation 2.11: (a) with $K = -10^{30}$ and 61×61 data points, (b) with $K = 10^{30}$ and 61×61 data points, (c) with 61×61 data points and the curtain and the top/bottom of the plot removed, and (d) with 16×16 data points and nodes placed at the perforation edges. Configuration files **F2_31A.CF3**, **F2_31B.CF3**, **F2_31C.CF3**, and **F2_31D.CF3**.

in Figures 2.32a and b, the points equal to -10^{30} or 10^{30} were automatically trimmed out by the upper and lower planes of the plot box. If in any of these two plots the visibility of the *curtain* and of the *top* and *bottom lands* are turned off, the function surface will look as shown in Figure 2.31c. As the companion Figure 2.12d illustrates, **D_3D** has the ability to place glyphs or markers at the nodes where the function surface is interrupted. This feature is activated by changing the last entry of *chime menu 2* to 'E'. The edge glyphs size and type are controlled by the 5th entry of the same *chime menu 2* and by line **7** of the **CF3** file (same as in a regular node plot like the one in Figure 2.9). Evidently, the accuracy with which these glyphs approximate together the singularity of the function depends on how fine the function has been sampled.

Important: Note that the nodes labeled 1–2, 3–4, 5–6, and 7–8 of the gray polygons (0010), (0001), (1000), and (0100) in Figure 2.26 are not assumed to be *edge nodes* (see also Figure 2.31d).

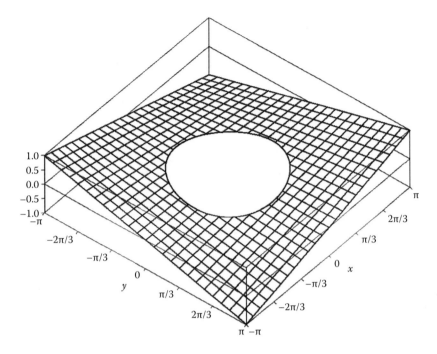

FIGURE 2.32   Mesh plot of the pricewise continuous function in Equation 2.11 obtained by combining and further editing inside **AutoCAD** the main body of the function surface (configuration files **F2_32-1.CF3**), with a polyline that connects the edge nodes (configuration files **F2_32-2.CF3**).

Important: When regular nodes (with or without stems) are plotted on a graph, the nodes below or above the upper and lower planes of the plot box will not be represented nor their stems. If such a plot is exported to **DXF**, the regular nodes and the edge nodes will be placed on layers 'nodes' and 'edge_nodes,' respectively. Because distinguishing between regular nodes and edge nodes was inconvenient to code inside **D_3D**, some edge nodes occur both in the 'edge_nodes' and the 'nodes' layers.

The fact that the edge nodes are placed in separate layers makes it easy to connect them with polyline(s) and manually retouch using **AutoCAD** a constrained function plot (see Figure 2.33).

The same approach described earlier can be applied to representing graphically inequalities of two variables. To exemplify, the same inequality 1.5 in Chapter 1 will be considered equivalent in terms of surface plotting with the following piecewise continuous function:

$$F_6(x, y) = \begin{cases} 0 & \text{for } \left(\sin(x)+\sin(y)\right)^2 - (x \cdot y + 0.5) < 0 \\ 10^{30} & \text{for } \left(\sin(x)+\sin(y)\right)^2 - (x \cdot y + 0.5) \geq 0 \end{cases} \qquad (2.12)$$

For $-\pi \leq x \leq \pi$ and $-\pi \leq y \leq \pi$ and 101×101 data points, a z-axis truncated plot of this function will look as shown in Figure 2.34. The input data file **F2_6.D3D** used to generate this figure has been produced with program **P2_6.PAS** (see Appendix B).

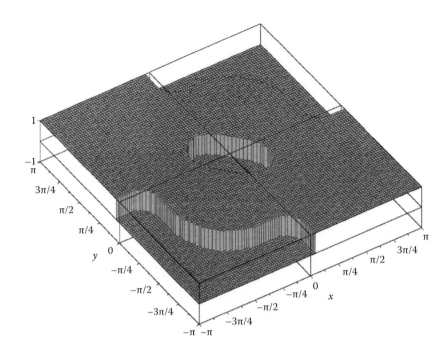

FIGURE 2.33   Plot of inequality 2.12 with 101 × 101 data points. This is a screenshot of a **DXF** copy of the plot taken after issuing the **AutoCAD** *shade* command. Configuration files **F2_33.CF3**.

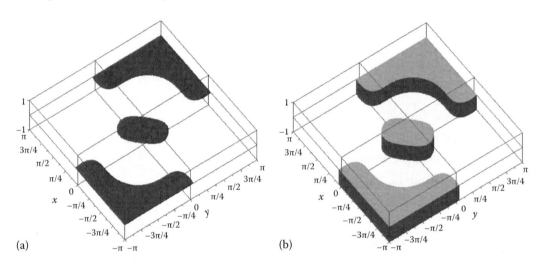

FIGURE 2.34   501 × 501 data point plot of inequality 2.11 as (a) body of the surface only and (b) as stem plot with colored nodes. Configuration files **F2_34A.CF3** and **F2_34B.CF3**.

Note that the large values of 10^{30} were assigned to the cases where the original inequality holds. You can obtain alternative representations by changing the 10^{30} value into -10^{30} and by switching between plotting the top land, the bottom land, or the body of the surface only.

Figure 2.34a is a 3D view of the body of the surface only, shown as x = constant and y = constant lines. For the given resolution (501 × 501 data points), the exact same graph can be obtained as node plot. Another way of representing inequality 2.12 is as stem plot with size-one nodes (Figure 2.34b). The nodes must be colored according to height to distinguish the top land of the graph. Either plot in Figure 2.34, when viewed from the top, will result in a 2D representation similar to Figure 1.20 in Chapter 1.

2.13 COLOR-RENDERED PLOTS

The use of color increases the appeal of a graph and can add more information to a plot. Their only drawback, oftentimes overlooked, is that some information is lost when you print or photocopy them in black-and-white. **D_3D** has the ability of representing color-rendered surface plots in any view, including top view. **D_3D** can produce colored plots by mapping the function surface with nodes and/or level curves or by filling the patches with color according to the elevation of the respective entities.

Figure 2.35 is a top view of a 481 × 481 node plot (the node size was set to one pixel) of **F2_3.D3D** data file. To eliminate the occurrence of voids in the pixel plot, the height and width of the box was made slightly smaller than the grid size, that is, 480 × 385 pixels. Overlapped with this is a gradient plot generated from only 21 × 25 data points of the same file **F2_3.D3D**. The color scale box to the right has been created separately as an *empty plot* viewed from the front, with the number of side gridlines set to 200. When the number of side gridlines exceeds 200 (option **<Insert>** then **<G>** from the *graphic edit screen*), they will be colored according to their elevation, similarly to the *z*-axis division lines. An *empty plot* can be generated by turning off its *top land*, *curtain*, *body*, and *bottom land* from *chime menu 1*.

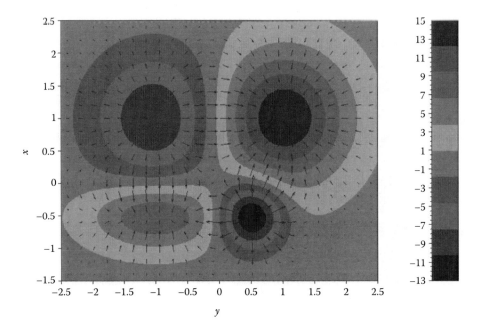

FIGURE 2.35 Gradient plot manually overlapped with a colored node plot in top view. The color scale to the right has been generated separately. Configuration files **F2_35DN.CF3**, **F2_35UP.CF3**, and **F2_35CS.CF3**.

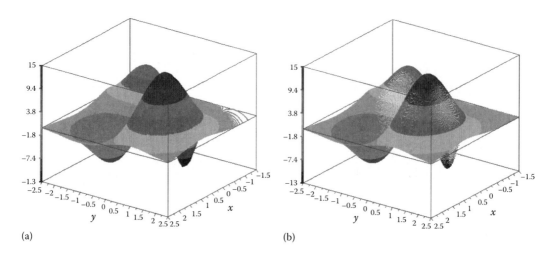

(a)   (b)

FIGURE 2.36 Rendered surface plots of the *four-hump function* (a) with 999 color-coded level curves over a 26 × 26 data-point transparent surface and (b) with 481 × 481 color points mapped on an opaque, white surface. Configuration files **F2_36A.CF3** and **F2_36B.CF3**.

Figure 2.36 shows two types of 3D color plots. When a surface in a general 3D view is rendered with raised level curves (Figure 2.36a) or with colored nodes (Figure 2.36b), voids are more likely to occur than in a top-view plot. Since the number of nodes cannot exceed 501 × 501 and the number of level curves cannot exceed 999, one possible remedy is to reduce the size of the plot before exporting it to **PCX**. Any remaining voids can be then corrected manually using **Paint** or other raster image editing software.

A raised level-curve plot takes longer to generate in **D_3D**, even for a moderately dense grid size. For this reason, in Figure 2.36b where the number of level curves is 999 (the maximum possible), the function surface has only 26 × 26 data points.

Note that the surface in Figure 2.36a was set to *hide* and their color to *white* (i.e., type 'w' to last entry of *chime menu 1*). In Figure 2.36b, the nodes were mapped on a transparent function surface, that is, the plot was in *wireframe* mode set from the *graphic edit screen*. Different appearances can be achieved with thin or thick level curves, with node sizes bigger than 1 (see Figure 2.37) and of other shapes, empty or solid—available node shapes are ○, □, ◇, ▽, △, and ∗.

As visible in Figure 2.36, raised level curves provide better color rendering over the steep regions of the function surface, complementing a node plot that renders better the flat portions of the surface. The appearance of a rendered surface can be improved by overlapping a high-density node plot with a raised level-curve plot using **Paint.**

Remember that in a plot exported to **AutoCAD** via the **R12 DXF** format, round nodes will always be drawn as empty circles and not as solid doughnuts. Also remember that the z-axis color coding in a **PCX** screenshot has 10 colors, versus 20 colors in **DXF**.

The occurrence of voids as described previously is eliminated if the surface patches are filled with color according to their elevation (see Figures 2.38 through 2.40). Because the colors available for scaling are twice as many in a **DXF** copy of the plot, it is better to export your graphs to **AutoCAD**, rather than doing a raster copy straight from **D_3D**. Note that the **AutoCAD** *shade*, *hide*, and *render* commands have effect upon 3D plots

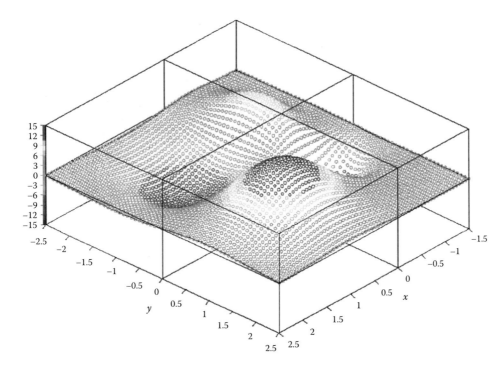

FIGURE 2.37 **DXF** copy of the *four-hump function* plot with 61 × 61 nodes mapped on an opaque surface. Configuration file **F2_37.CF3**.

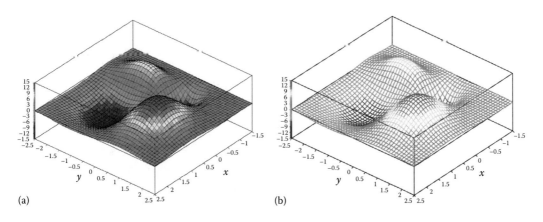

(a)

(b)

FIGURE 2.38 Plot of the *four-hump function* with 25 × 25 colored patches (a) with and (b) without a mesh grid, exported to **AutoCAD**, showing the effect of the (a) *shade* command and (b) *hide* command. Configuration files **F2_38A.CF3** and **F2_38B.CF3**.

generated with **D_3D**, even if they are not truly 3D. Also note that the exact same plot in Figure 2.38b can be obtained if you turn off layers 'border_x', 'border_y', 'color_x', and 'color_y' of the plot in Figure 2.38a and apply the **AutoCAD** *hide* command again.

Important: Plots with color-filled patches will appear different when exported to **PCX** than when opened in **AutoCAD** following a **DXF** export. Differences may also occur on the **D_3D** screen at the end of a **DXF** export.

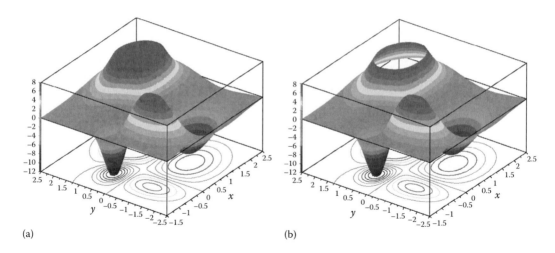

(a)   (b)

FIGURE 2.39 **AutoCAD** rendered plots of the truncated *four-hump function* with 481 × 481 color points, (a) with and (b) without the top land in place. Configuration file **F2_39.CF3**.

Regarding the plots in Figure 2.39, notice that the **AutoCAD** *render* command has the effect of obscuring all line and text entities, with only the actual function surface remaining visible. To compensate, take two screenshots: one on the rendered surface and one on the plot box, then overlap them inside **Paint**. Before taking the second screenshot, turn layers 'colormesh_body' and 'colormesh_topbtm' off, and then issue the *hide* command. This way, the portions of the plot box hidden by the surface will not show on the screen. You can copy to clipboard the active window on your computer screen by pressing simultaneously <Alt> and <Prnt Scrn>.

The plot in Figure 2.39b has the intersections with the plot box removed. You can make these changes inside **D_3D** and export it to **DXF** anew, or you can turn layers 'colormesh_body' and 'phantom_body' off and generate two new screenshots. Considering the

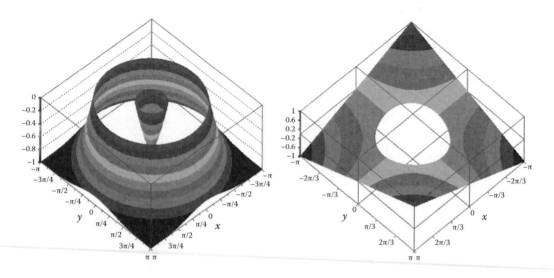

FIGURE 2.40 **PCX** copy of the color-rendered plot version of Figures 2.23b and 2.31c. Configuration file **F2_40A.CF3** and **F2_40B.CF3**.

significant extent of time it takes **D_3D** to output a **DXF** file with this many entities (over 500,000), the latter approach is evidently preferable.

Figure 2.40 provides two more examples of truncated surface plots selected from those already discussed in this chapter (i.e., Figures 2.23b and 2.31c). You may want to compare the appearance of the same plots in Figure 2.40, when exported to **DXF** and then *rendered* or *shaded* inside **AutoCAD**.

2.14 PLOTTING MULTIPLE SURFACES ON THE SAME GRAPH

Since it is not a true 3D graphing program, **D_3D** is not the best tool to represent parametric surfaces or multiple surfaces that intersect each other. In certain cases, however, it is possible to generate plots of surfaces that fold over themselves or combined plots of two or more single-valued functions as discussed next.

The first example is that of plotting a sphere of radius 1.7, centered at (0,0,0). The way this problem was solved was to plot the bottom and top hemispheres separately (see Figure 2.41a), then export them to **PCX**, and then overlap them using **Paint**. The **P2_7.PAS** program (see Appendix B) was used to generate an ASCII file named **F2_7. T3D** having two columns: one for the bottom hemisphere (the lower sign in Equation 2.13) and the other column for the top hemisphere (the upper sign in Equation 2.13):

$$F_7(x,y) = \begin{cases} \pm\sqrt{2.89 - x^2 - y^2} & \text{for } 2.89 \le x^2 - y^2 \\ \pm 10^{30} & \text{for } 2.89 \ge x^2 - y^2 \end{cases} \tag{2.13}$$

When plotting the individual hemispheres, the top and bottom lands must be turned off from *chime menu 1* of **D_3D**. Note that the lower hemisphere in Figure 2.41a can be obtained from the upper hemisphere by flipping the z-axis of the plot, or vice versa. As shown in Figure 2.41b, additional shapes can be obtained starting from the same data file by simply editing the limits over the z-axis.

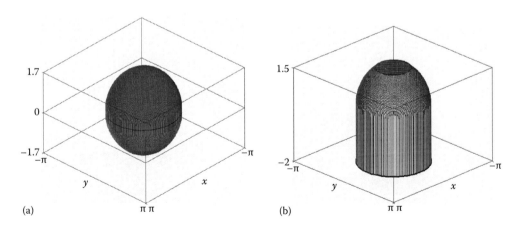

FIGURE 2.41 (a) A sphere produced as the overlap of two hemispheres plotted separately and (b) plot of cylinder a extending with a trimmed hemisphere. Configuration files **F2_41A-1.CF3**, **F2_41A-2.CF3**, and **F2_41B.CF3**.

As an example of graphing two intersecting surfaces, we will consider overlapping the surfaces $Z = F_2(x, y)$ and $Z = F_4(x, y)$ in Equations 2.2 and 2.10. If we are drawing vertical lines at each grid point, these will intersect the combined surfaces in two points, one higher and one lower. If the lower points of these pairs are separated from the upper points, and if you plot them as distinct graphs, you can then combine them using **Paint** or **AutoCAD**, same as we did with the sphere in Figure 2.41. Additional editing, prior or after overlapping these graphs, might be required in areas of incorrect or incomplete visibility.

Figure 2.42 shows two representations of the same combined 3D plot. The difference between them is that Figure 2.42a has been obtained as the overlap of two surfaces (see Figure 2.43), while the one in Figure 2.42b is the overlap of four separate plots (see Figure 2.44). One single file provides the data source of all constituent plots in Figures 2.43 and 2.44. This file named **F2_8.T3D** is organized on six columns and has been generated using program **P2_8.PAS** (see Appendix B).

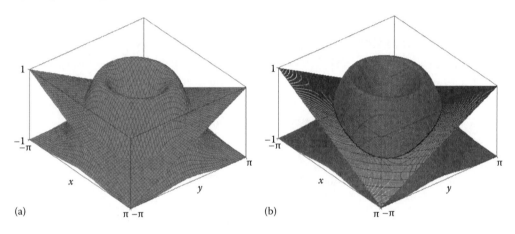

(a)   (b)

FIGURE 2.42 Combined plot of the *orange-squeezer function* and the *paraboloid* in Equation 2.10, obtained as the overlap of separately drawn entities assembled as shown in Figures 2.43a and 2.44b.

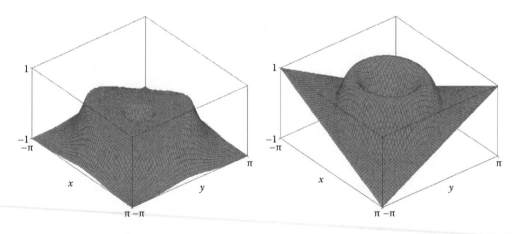

FIGURE 2.43 Constituent plots of Figure 2.42a. Configuration files **F2_43-1.CF3** and **F2_43-2.CF3**.

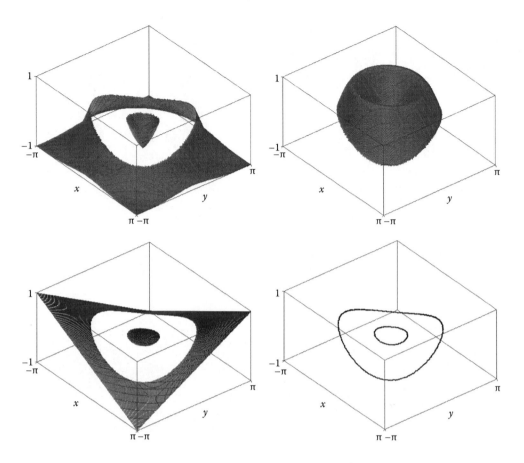

FIGURE 2.44   Constituent plots of Figure 2.42b. Configuration files **F2_44-1.CF3**, **F2_44-2. CF3**, **F2_44-3.CF3**, and **F2_44-4.CF3**.

The two functions F_2 and F_4 are evaluated at a current grid point (x,y), their values are then compared (lines **#37** to **#42** of program **P2_8.PAS**) and then written to file. Depending on their elevation and whether they belong to F_2 or F_4, the two values are of the same program written to columns 1 and 2 or columns 3 to 6 of file **F2_8.T3D** (see lines **#43** and **#44**). Data on columns 1 and 2 served to plot Figures 2.43, while the values on columns 3 to 6 were used to plot Figures 2.44. Note that in case of columns 3 to 6 of file **F2_8.T3D**, use has been made of value 1.0E30 to indicate 'curtains' and 'top lands' that can be selectively plotted or suppressed from within **D_3D**.

2.15 IMPLEMENTATION DETAILS OF THE **D_3D** PROGRAM

This section is provided for those who wants to understand in more detail how the **D_3D.PAS** program and its accompanying unit **UNIT_D3D.PAS** work. Additional useful information (outside of the comments provided with **D_3D.PAS**) is available in Simionescu (2003, 2011).

As explained in paragraph 2.1, all graphic operations are performed by **D_3D** in the 2D image space. The coordinates of the corners of the plot box relative to the computer-screen reference frame OXY are noted as **Xc[..]**, **Xc[..]** for corners C_1 through C_4 and **Xcp[..]**, **Ycp[..]** for corners C_1' through C_4' (Figure 2.45).

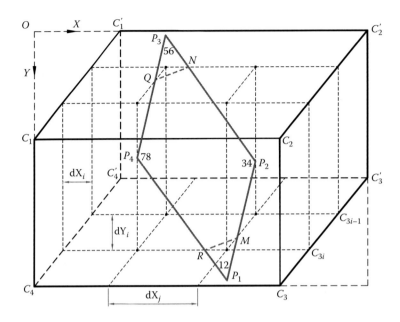

FIGURE 2.45 A surface patch in an (1010) instance (see also Figure 2.26).

The viewport coordinates of a point P_1 of a single patch as shown in Figure 2.46 equivalent to point P_{ij} of the surface is given by equation

$$X_{P_1} = j \cdot dX_j + (n-i) \cdot dX_i$$

$$Y_{P_1} = i \cdot dY_i + \frac{z_{max} - z_{ij}}{z_{max} - z_{min}} \left(Y_{C_4} - Y_{C_1} \right) \tag{2.14}$$

where z_{ij} the function value at P_1, z_{min} and z_{max} are the limits over the z-axis of the graph, and $m \times n$ is the grid size.

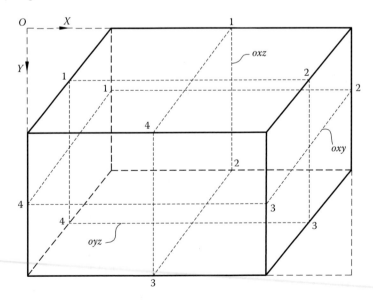

FIGURE 2.46 Outlines of the *oxy, oxz* and *oyz* reference planes.

The dX_i, dX_j, and dY_i increments in Equation 2.14 are as follows:

$$dX_i = \frac{\left(X_{C'_1} - X_{C_1} \right)}{\left(m - 1 \right)} \quad dX_j = \frac{\left(X_{C_3} - X_{C_4} \right)}{\left(n - 1 \right)} \quad dY_i = \frac{\left(Y_{C_4} - Y_{C_1} \right)}{\left(n - 1 \right)} \tag{2.15}$$

The coefficients **kH** and **kV** that position reference corner C'_4 of the plot box (Figure 2.45) are given by equations:

$$\mathbf{kH} = \frac{\left(X_{C'_4} - X_{C_4} \right)}{\left(X_{C'_3} - X_{C_4} \right)} \quad \mathbf{kV} = \frac{\left(Y_{C'_4} - Y_{C_4} \right)}{\left(Y_{C'_1} - Y_{C_4} \right)} \tag{2.16}$$

The outlines of the horizontal reference plane *oxy*, and the two vertical reference planes *oxz* and *oyz*, (Figure 2.46) are represented by **D_3D** using points 1, 2, 3, 4 the coordinates of which are stored in variables **Xoxy[1]**, **Yoxy[1]** through **Xoyz[4]** and **Yoyz[4]** of **D_3D.PAS**.

✳✳✳

The **D_3D** plotting program subject of this chapter combines an offsetting of the lines of constant *x*, with a shear transformation. Solving the visibility problem, the intersection of the function surface with the horizontal planes of the bounding box and level-curve generation is done entirely in the 2D image space. Consequently, the amount of input data and CPU resources per plot is reduced to a minimum. Executable **D_3D.EXE** and source codes **D_3D.PAS** and **Util_D3D.PAS** are available with the book, together with all configuration and data files used in this chapter. Additional examples of **D_3D** use are available in Chapters 3, 4, and 9.

REFERENCES AND FURTHER READINGS

Bourke, P. D. (July, 1987). A contouring subroutine. *Byte Magazine, 12(6)*, 143–150 (also available at http://paulbourke.net/papers/conrec/).

Simionescu, P. A. (2003). Improved display methods of single-valued functions of two variables. *ASME Journal of Computing and Information Science in Engineering, 3(2)*, 136–143.

Simionescu, P. A. (2011). Some advancements to visualizing constrained objective functions and inequalities of two variables. *ASME Journal of Computing and Information Science in Engineering, 11(1)*, 014502 (7p.).

Additional useful titles are:

Fuller, G. and Tarwater, D. (2013). *Analytic Geometry*, Upper Saddle River, NJ: Pearson.

Hughes, J. F., van Dam, A., McGuire, M., Sklar, D. F., Foley, J. D., Feiner, S. K., and Akley, K. (2013). *Computer Graphics: Principles and Practice*, Boston, MA: Addison-Wesley.

Shirley, P., Ashikhmin, M., and Marschner, S. (2009). *Fundamentals of Computer Graphics*, Natick, MA: A K Peters/CRC Press.

Vince, J. (2013). *Mathematics for Computer Graphics*, London, UK: Springer.

CHAPTER 3

Programs and Procedures for Data Visualization and Data Format Conversion

IN THIS CHAPTER, several programs that you may find useful will be presented. These include a collection of Pascal procedures for generating 2D line plots; three programs for manipulating ASCII, **R12 DXF**, and **HP-GL PLT** files; and two **AutoLISP** applications for automatically generating 3D entities from within **AutoCAD** with description read from file.

LibPlots.PAS is a unit with procedures that allows you to write programs that generate graphs very similar to those done with **D_2D**.

The **Util~TXT.PAS** program can be used to add between every two data points read from a file, additional points interpolated linearly, and spline or B-spline. It can also evaluate numerically the first and second derivatives and can transfer to an output file every certain row of the original data. The program can also make continuous a series of angle values restricted, for example, to $[-\pi\ldots\pi]$ and can apply a logarithm transform to the input data safe from crashing when encountering a number that is negative or zero. By directly editing its code, additional transformations are possible, like scaling, offsetting of data, and custom functional transforms.

Util~DXF.PAS is a **DXF** viewer that can display 2D and 3D lines and polylines, circles, and arcs of circle read from an **R12 DXF** file. In addition, the program can be used to extract to ASCII the x, y or x, y, z coordinates of selected polyline(s), a feature useful for transferring level-curves plots from **D_3D** to **D_2D** or for digitizing curves available only as raster images. Figure 1.14 has been produced this way, that is, a picture was imported into **AutoCAD** and polylines were drawn over. When completed, the drawing was exported to **R12 DXF**, and the x, y coordinates of the vertices of these polylines were then extracted to file using **Util~DXF**. This file then served as input to the **D_2D** program when producing Figure 1.14.

Util~PLT.PAS can open **PLT** files exported from **AutoCAD** using the **Hewlett-Packard Graphics Language** (HP-GL) **ADI 4.2** by **Autodesk** #7550 driver. The polylines in these **PLT** files can then be exported to **R12 DXF** while simultaneously the x, y coordinates of their vertices will be saved to an ASCII file. The **Util~PLT** program can be used to "flatten" in the hide mode 3D drawings and surface plots generated using **D_3D** or to digitize alphanumeric characters, arches of circles, and spline curves created inside **AutoCAD**.

G_3D.LSP is an **AutoLISP** application that allows you to generate inside **AutoCAD** true 3D curves and meshed surfaces with vertices read from file.

M_3D.LSP is the second **AutoLISP** program that can automatically generate and animate lines, cylinders, spheres, tori, and cylindrical helixes with dimensions and orientations read from file. It can also insert blocks at locations and with orientations read from the same input file. (These blocks must preexist in the **DWG** file from where **M_3D.LSP** is being run.)

3.1 **LibPlots** PROCEDURES FOR GENERATING 2D PLOTS

Available in this book, there are several Pascal units for user interfacing in text and **BGI** graphical mode, for mathematical calculation, and for 2D plotting. Of these, the **LibPlots** unit will be discussed in more details here. By calling its procedures, you can generate 2D graphs similar to those produced with **D_2D** and export them to **R12 DXF** and **PCX**. Additional features not available in **D_2D** that **LibPlots** allow are (i) assigning different size markers to different curves of the same plot and (ii) having the x-axis intersect the y-axis at $y = 0$ and vice versa. A number of programs that implement these new features will be discussed in the remainder of this section.

3.1.1 Basic 2D Plotting Using **LibPlots**

P3_01A.PAS listed in Appendix B is a simple example of **LibPlots** procedure use, namely, of **PlotCurve**, **PlotXaxis**, and **PlotYaxis**. The program calls several procedures from units **Unit_PCX**, **LibGraph**, and **LibDXF** and uses the **VDp** vector type and the **Pmax** constant, both declared in the **LibMath** unit.

Lines #**19** to #**22** of **P3_01A.PAS** serve to generate the (t, Y) pairs that will be plotted on the graph (Figure 3.1a). The actual plot has been produced by executing lines #**25**, #**26**, and #**27** of the program. The default size and location of the plot on the computer screen can be changed by calling procedure **NewPlot** and assigning different values to the corners of the box as it has been done in the companion program **P3_01B.PAS**. Note that the procedures responsible for drawing the x- and y-axes are called only after all curves have been plotted. This is because the limits stored by variables **xmin**, **xmax**, **ymin**, and **ymax** are assigned meaningful values only after vectors **t** and **Y** are inspected inside the **PlotCurve** procedure. The first parameter (i.e., **1**) in procedures **PlotCurve**, **PlotXaxis**, and **PlotYaxis** specifies the plot number. The second parameter in procedures **PlotXaxis** and **PlotYaxis** controls axis location (possible values are **0**, **1**, or **2**, where **1** will place the axis at $y = 0$ or $x = 0$, respectively), while the third and fourth parameters in these same two procedures represent the number of values and the number of minor intervals that will be placed along the respective axis (see Figure 3.1a).

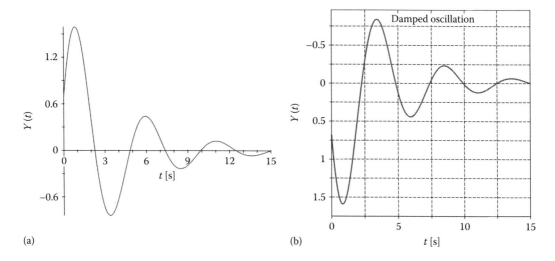

FIGURE 3.1   Plots created with programs (a) **P3_01A.PAS** and with (b) **P3_01B.PAS**.

Important: Currently, **LibPlots.PAS** does not allow more than four separate plots to be opened simultaneously in the same program. These can be drawn on the computer screen in four separate plot boxes, in two boxes having each a primary and a secondary vertical axis, or all four overlapped in the same plot box.

In the companion program **P3_01B.PAS** (see Appendix B), procedure **NewPlot** sets the corners of the **ViewPort** where the plot will be drawn, that is, (150, 50) the top-left corner and (500, 430) the bottom-right corner. If you call **NewPlot** with its second parameter set to the constant **IsoPlot** or TRUE, instead of **FitBox**, the limits of the graph will be adjusted so that it becomes isotropic. The last parameter of **NewPlot** is a character string that will be written at the top of the plot box as title.

The limits over the *x*- and *y*-axes can be extracted from vectors **X** and **Y** by calling procedures **UpdateLimitsX** and **UpdateLimitsY**. These limits can then be accessed by the main program through functions **GetXmin**, **GetXmax**, **GetYmin**, and **GetYmax**. On line #32 of **P3_01B.PAS**, the last two of these *getter* functions are used as parameters in the **NewLimitsY** procedure to extend the range of the *y*-axis and also to reverse it.

The border around the plot box has been produced by calling procedure **DrawBorder** (line #30) from unit **LibGIntf**, while the plot curve was set to **ThickWidth** by calling Pascal's **SetLineStyle** procedure. In order to turn the gridlines *on* (see Figure 3.1b), procedure **SetDivLine** on line #35 has been called with its second parameter set to a value greater than nine. Note that if you call this procedure after **PlotYaxis**, only the vertical gridlines will be plotted.

Note the use of the **WaitToGo** procedure from unit **LibInOut**, which will suspend the program execution until the user presses a key (line #**39**).

3.1.2 Multiple Plots with Markers

When a plot consists of multiple curves, you can assign them different colors and different line types (i.e., normal width or thick, solid, dashed, or dotted) or add markers to

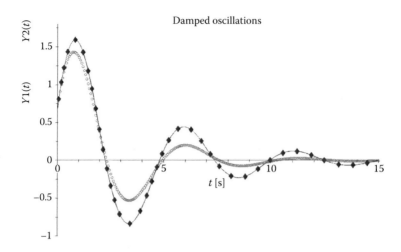

FIGURE 3.2 Plot generated with **P3_02.PAS** showing equally spaced diamond markers and round markers placed one at each data point.

them (see Figure 3.2). Program **P3_02.PAS** (see Appendix B) is an example of marker utilization. If you want to employ different line types, you must insert the adequate **Graph** command(s) prior to calling the **PlotCurve** procedure, as it has been done on line #**33** of the **P3_01B.PAS** program.

Lines #**20** to #**26** in this new program **P3_02.PAS** (see Appendix B) serve to generate data vectors **t**, **Y1**, and **Y2**. In order to encompass both **Y1** and **Y2** components within the *y*-axis limits, procedure **UpdateLimitsY** is called first with **t**, then with **Y1** and with **Y2** as arguments (lines #**30**, #**31**, and #**32**). Procedure **ResizeY** called on line #**33** has the effect of expanding the *y*-axis range by about 0.2 (i.e., reduces **ymin** by 10% and increases **ymax** by 10%); in addition, the new limits will be adjusted so that the associated numbers will be rounded (see also the **P3_03B.PAS** program in Appendix B).

Line #**35** in **P3_02.PAS** sets the marker type placed along the first curve to diamond, and their size to 2. Signaled by the ':' character, these markers will be equally spaced along the plot curve. The distance measured along the curve between every two successive markers will be about six times the marker size (value hardcoded in procedure **PlotCurve**). If on this line #**35** you change ':<>' into '|<>', then the diamond markers will be placed at every data point (see also line #**39**). If the last parameter of the **PlotCurve** is set to a negative value, then markers only will be plotted without the curve. If this parameter is set to a positive value, then the markers will be drawn as the plot curve progresses. Similarly to **D_2D**, if markers are polygonal or round, only the first half of them will obscure the curve (same as in **D_2D**). To plot a curve without markers, insert the command **SetMarker(0, ' ')** right before the **PlotCurve** procedure is called. The allowed second arguments in **SetMarker** are '%' 'o' '.' '[]' '<>' 'v' '^' '*' 'x' '+' '@' '&' 'q', that is, the same as in **D_2D** less the arrow marker that is not available in **LibPlots**.

Note that both *y* category names are written in the same color as the curves for which they stand for (Figure 3.2). This feature is activated by adding a space to the

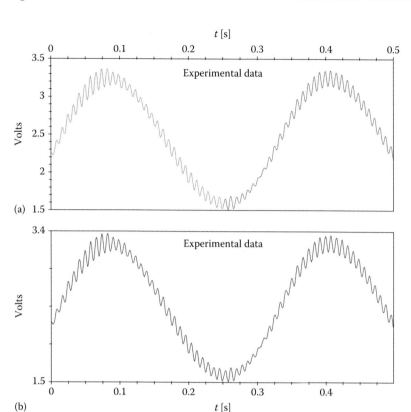

FIGURE 3.3 Plots created with programs (a) **P3_03A.PAS** and (b) **P3_03B.PAS**.

left of the character string parameter (i.e., the category name) of the **PlotYaxis** procedure—see lines #**38** and #**41**.

3.1.3 Plotting Large Data Sets and Data Read from File

Program **P3_03A.PAS** (see Appendix B) shows how to read data from a multiple column ASCII file with more than and 502 rows. 502 is the maximum size of a **VDp** type vector, which means that data has to be plotted as a series of concatenated curves as shown in Figure 3.3.

Procedure **Extract_V** called once on line #**44** and a second time on line #**45** of program **P3_03A.PAS** accepts one row from the input file (or, in general, a character string consisting of groups of numbers separated by one or more nonnumerical characters, including spaces) and returns the value corresponding the specified column. Also, note the random colors assigned to the individual sections of the plot curve (line #**50**) and the *x*-axis placed at the top of the plot box.

The companion program **P3_03B.PAS** (source code not included in Appendix A) is very similar to **P3_03B.PAS**, with the difference that the limits over the *x*- and *y*-axes are established prior to plotting the curves, rather than being provided by the user. This is useful when you want the curves to tightly fit the plot box (see Figure 3.3) or when you want to add flexibility to your program and make the *x* and *y* limits self-adjusting.

3.1.4 Dynamic Plots with Scan Lines and Scan Points

Program **P3_04.PAS** available with the book solves the direct dynamics problem of a two degree-of-freedom *elastic pendulum*. It also provides an example of **PlotScanLine** and **PlotScanPoint** procedure use (see Figure 3.4). In addition, it is also a first introduction to the procedures in the **LibMec2D** and **LibMecGr** units also available with the book.

P3_04.PAS consists of three parts. Firstly, the differential equations of motion of a two degrees of freedom *elastic pendulum* with no damping are solved numerically, and vectors **_t**, **_Theta**, **_Rho**, **_xA**, and **_yA** are generated. These vectors are then used to plot the time response graphs **θ(t)**, **Rho(t)** and the parametric curve **y(t)** versus **x(t)**. In the third part of the program, a *scan line* and a *scan point* are animated synchronously with the motion of the spring. To represent the spring and its fix-end attachment, procedures **Spring** and **PutGPoint** are called form unit **LibMec2D**. The locus of the pendulum bob is then plotted by calling procedure **CometLocus** from the same unit.

The dynamic equilibrium equations about the center of mass of the pendulum bob are (see Figure 3.5a):

$$\begin{cases} \Sigma F_\theta = ma_\theta \\ \Sigma F_\rho = ma_\rho \end{cases} \tag{3.1}$$

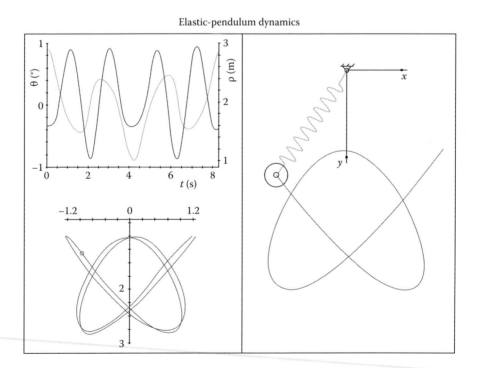

FIGURE 3.4 One of the animation frames generated by program **P3_04.PAS**. See also **F3_04.GIF**.

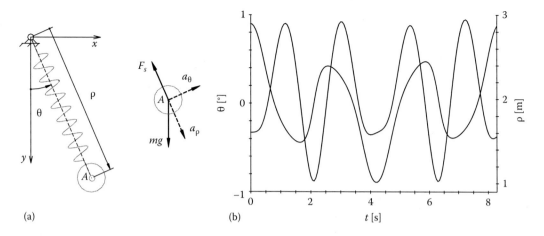

(a)   (b)

FIGURE 3.5   Elastic pendulum geometry and free-body diagram of the bob (a), and plot of the time response curves θ and ρ vs. t written to **DXF** by program **P3_04.PAS** (b).

Applying the tangential and radial accelerations equations of a particle moving in polar coordinates (Meriam and Kraige 2006), we get

$$\begin{cases} -mg\sin(\theta) = m(2\dot{\rho}\dot{\theta} + \rho\ddot{\theta}) \\ mg\cos(\theta) - F_s = m(\ddot{\rho} - \rho\dot{\theta}^2) \end{cases} \tag{3.2}$$

After a few transformations, the equations of motion are derived as

$$\begin{cases} \ddot{\theta} = -\dfrac{g\sin(\theta) + 2\dot{\rho}\dot{\theta}}{\rho} \\ \ddot{\rho} = g\cos(\theta) - \dfrac{F_s}{m} + \rho\dot{\theta}^2 \end{cases} \tag{3.3}$$

where F_s is the force developed by the spring

$$F_s = (\rho - l_0)k \tag{3.4}$$

The free length of the spring was considered $l_0 = 1$ m, its constant $k = 10$ N/m, the mass of the bob $m = 1$ kg, and the acceleration due to gravity $g = 9.81$ m/s^2. With these values, Equations 3.3 are integrated inside **P3_04.PAS** using Euler's method (see Appendix A). For initial conditions $x_A(0) = 1.25$ m, $y_A(0) = 1$ m, and the pendulum at rest, that is, $d\theta(0) = d\rho(0) = 0$, the response of the system for the first 8.3 s of the simulation is obtained (Figure 3.4).

Program **P3_04.PAS** generates, in addition to the **PCX** animation frames, three **R12 DXF** files and a text file named **F3_04.TXT** where the time response data is written. One of these **DXF** files records the polyline representing the spring, the pendulum bob, and the locus of its center (also a polyline), each layer representing a separate animation frame.

The other two **DXF** files record overlapped the graphs $\theta(t)$ and $\rho(t)$ (see Figure 3.5b), and the plot of the locus of the pendulum bob $y_A(x_A)$—see Figure 3.4.

3.2 **Util~TXT** PROGRAM FOR MANIPULATION OF ASCII FILES

Many known graphing programs, like **Excel**, are capable of fitting an interpolated curve to data. Others, like **MATLAB**, have dedicated functions that can add interpolated points to an initial data set, which can then be represented graphically. **D_2D** does not have interpolating capabilities, so the **Util~TXT.PAS** program and the companion **Unit_TxT** are provided in compensation. The program can add up to 100 points between each two original data points read from file that are interpolated either (i) linearly, (ii) cubic spline, (iii) quadratic B-spline, or (iv) cubic B-spline. Other features of **Util~TXT** include (v) making continuous a series of angle values that were forced within $[-\pi...\pi]$, $[-\pi/2...3\pi/2]$, $[0...2\pi]$, or similar interval by some inverse trigonometric function; (vi) decimating a given input file by extracting to the output file every kth row, where k is specified by the user; (vii) scaling and translating the points extracted from a **R12 DXF** file using the **Util~DXF** program; (viii) evaluating the logarithm; and (ix) calculating numerically the first and second derivatives of the input data. Regarding this last transformation, in order to apply the more accurate *centered difference formula* to the end points same as to the interior ones, cubic extrapolated pairs x, y are added, one at the beginning and one at the end of the data series.

Util~TXT can be used in two ways: as executable file with settings read from a configuration file of extension **CON**, or modified and recompiled, case in which additional transformations can be coded into the program. Only the transformations that can be controlled via a configuration file will be discussed here.

Important: When you launch **Util~TXT** or if you press the <F10> key after loading the **CON** file, you can read about the restrictions and limitations that apply to the input data (see the following screenshot). Note that **Util~TXT** is capable of performing linear, spline, and B-spline interpolations to 2D data only.

```
                   About   Util~TXT version 2014

» Input file must be less than 1000 rows. Any additional row will be ignored.
» To apply log10(..) or ln(..) the input data must be positive.
» Deriv1(..) and Deriv2(..) cannot be used together with ContnAng(..).
» Spline and B-spline interpolation cannot be done simultaneously.
» The number of knots to be added between two points by InterpSpline3(..),
   InterpBSpline2(..) and InterpBSpline3(..) cannot exceed 100.
» To calculate the 1st or the 2nd derivative, or to do spline interpolation,
   the X data must be monotonic and the number of input points greater than 3.

» For XY to D_2D format conversion you must add manually on the 3rd line
   Xmin Ymin and on the 4th line Xmax Ymax of your plot in engineering units.
   The next 3 lines must be the x,y coordinates of 3 corners of the plot box,
   (the box should not be tilted), followed by a line breaker i.e. '========'
```

3.2.1 Linear Interpolation

Let us consider a linear interpolation example first, where **Util~TXT** reads several x, y data pairs from an ASCII file named **F3_06.TXT** and then adds six points interpolated linearly between each of these pairs. To do this, edit the master configuration file **!.CON** such that the first two lines read **F3_06.D2D** and **F3_06.TXT** (these are the output and input file names). Set to 'Y' (i.e., Yes) the first character on the **{Interpolate linear}** line and change the number of points to **6**. Save the configuration file under the name **F3_06.CON**, launch **Util~TXT**, and select as input the **CON** file that you have just created. Confirm the remaining default options by pressing the <Enter> key, although you could modify these defaults if you want to. Figure 3.6 is a combined plot of the original data (the transparent rounds Ø) and of the linearly interpolated points read from **F3_06.D2D** (the * markers).

Important: By default, **Util~TXT** assumes that the output is to an ASCII file. Alternative output file formats are **D2D** readable by the **D_2D** program and **DXF**.

A different type of linear interpolation **Util~TXT** can do is to add evenly spaced points along the curve, useful when some of these points are too far apart. To add equally spaced points, modify line number **9** of the previous **CON** file so that instead of '6', it reads '−0.5'. Also change the output file name to **F3_07.D2D** and save the modified configuration file as **F3_07.CON**. The corresponding graph will now look as shown in Figure 3.7.

Adding points interpolated linearly to a graph is useful when plotting the path of a mill cutter using **D_2D**, with round markers representing the actual tool (marker diameter can be accurately set from within **D_2D**). Synthesizing the motion program of the follower of a cam mechanism or the path of the end effector of robot may also require adding interpolated points.

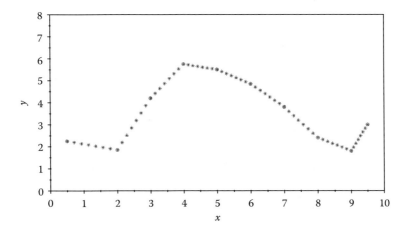

FIGURE 3.6   Plot of initial data (the round markers) and of linearly interpolated points placed in groups of 6 between 2 data points (the asterisk markers). Configuration files **F3_06.CON** and **F3_06.CF2**.

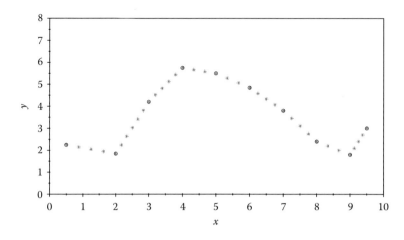

FIGURE 3.7   Same plot as in Figure 3.6 with the interpolated points (asterisk markers) placed at a distance of about 0.5 units along the graph. Configuration files **F3_07.CON** and **F3_07.CF2**.

3.2.2 Cubic-Spline Interpolation

To add cubic-spline interpolated points between the same control points as before, open the last **CON** file and turn the linear interpolation option *off* and the cubic-spline interpolation *on*. Also, change the name of the output file to **F3_08.D2D**. Save your file as **F3_08.CON** and run **Util~TXT** with settings read from it. A plot of the resulting interpolated points read from the new data file **F3_08.D2D**, overlapped with the control points is shown in Figure 3.8.

Remember that for a cubic-spline interpolation to be possible, the x components of the original data must be strictly increasing (Press et al. 1989), that is, $x_j > x_{j+1}$ for any j between 1 and the rank of the second last point. Note that the curve passes smoothly through each of the given point. Contrast this to a B-spline interpolated curve that never passes through the given points as shown in the next section.

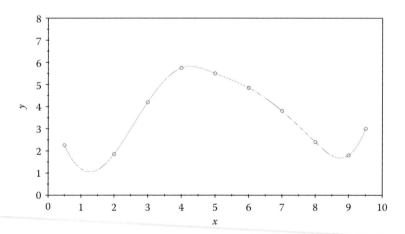

FIGURE 3.8   Plot of the initial data points as round markers and of a cubic-spline interpolated curve through these points. Configuration files **F3_08.CON** and **F3_08.CF2**.

3.2.3 B-Spline Interpolation

Util~TXT can add both quadratic and cubic B-spline interpolation points to a set of control points read from file. The cubic B-spline interpolation is more frequently used in practice, as the degree of smoothness of the resulting curve is higher (Zecher 1993). A quick comparison is available in Figure 3.9, where both a quadratic and a cubic B-spline interpolated curve (the solid line and the dashed lines, respectively) were plotted, together with their control points. The number of points between every two data points in Figure 3.9 has been set to 6 inside configuration files **F3_09-1.CON** and **F3_09-2.CON**.

One advantage of the B-spline curves over splines is that they can be fit through closed *control polygons* or through *control points* that are arranged in neither increasing nor decreasing order. It is called *control polygon*, the polyline that connects the control points of a B-spline curve. To exemplify, two different inputs were assume, both written to file **F3_10.TXT**, that is, an open control polygon (columns 1 and 2) and a close control polygon (columns 3 and 4). The cubic and quadratic B-spline curves through these nodes are shown in Figure 3.5. In order to obtain the intended results, the column numbers from where the *x* and *y* coordinates of the two sets of control points are read must be correctly specified inside configuration files **F3_10-1.CON** through **F3_10-4.CON**. Same about the row numbers where the transformation begins and where it ends (Figure 3.10).

3.2.4 Numerical Differentiation

The possibility of numerically calculating the first and second derivatives of a data set is another capability of the **Util~TXT** program (see Appendix A for the underlying theory). You can use this feature to check if the symbolically calculated derivative of a given function is correct (suspecting hand calculation or computer coding errors) by comparing its graph with the graph of the numerical derivative values generated using **Util~TXT**.

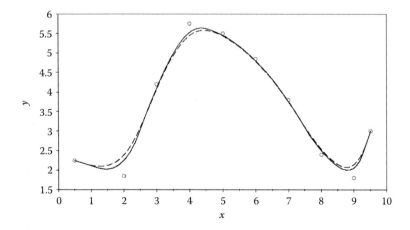

FIGURE 3.9   Plots of quadratic B-spline (solid line) and cubic B-spline (the dashed lines) curves with increasing (monotonic) control points. Configuration files **F3_09A.CF2** and **F3_09B.CF2**.

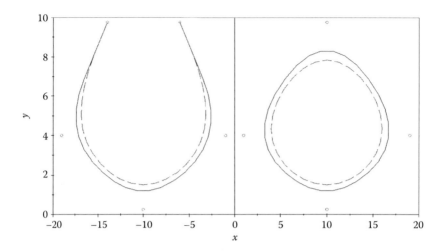

FIGURE 3.10 Plots of quadratic (solid lines) and cubic (dashed lines) B-spline curves with non-monotonic control points. Configuration files **F3_10-1.CON** through **F3_10-4.CON** and **F3_10.CF2**.

Let us consider the function in Equation 1.1 and evaluate numerically its first two derivatives and then plot them on the same graph. Input has been considered the **F1_01. DAT** file from Chapter 1, renamed **F3_11.TXT**. After you run **Util~TXT** with settings from configuration files **F3_11-1.CON** and **F3_11-2.CON**, you will obtain the data files **F3_11-1.D2D** and **F3_11-1.D2D** used to plot the graph in Figure 3.11. Note the very close similarity between the numerically calculated first derivative and the exactly calculated one in Figure 1.3.

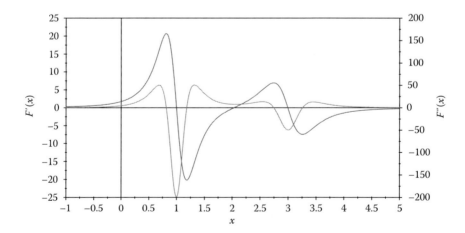

FIGURE 3.11 Plots of the first and second derivative of $F(x)$ in Equation 1.1, evaluated numerically for 400 data points. Configuration files **F3_11-1.CON**, **F3_11-2.CON**, and **F3_11.CF2**.

3.2.5 Angle-Value Rectification

Making a series of angles continuous after being forced within intervals of the form $[-\pi\ldots\pi]$ or $[-\pi/2\ldots3\pi/2]$ by some inverse trigonometric function like **ArcTan** can be occasionally of concern. **Util~TXT** is capable to remedy such defects through the **NghbrAng** function that it calls from unit **LibMath**. **NghbrAng** uses the previous value of an angle series to correct a current value, by adding or subtracting certain number of π values. The program can handle angles expressed both in radians and in degrees. However, only the former case will be exemplified here.

We will first generate a set of angle values that needs to be corrected. A short program named **P3_12.PAS** has been written for this purpose (see Appendix B) that outputs a data file named **F3_12.TXT** with three columns. Column one is an initial angle that increases linearly from -2π to 2π, while the other two columns contain the same angle restricted to $-\pi$ to π and $-\pi/2$ to $3\pi/2$, respectively. The values on columns two and three were obtained by applying the tangent function to the initial angle, followed by the **ArcTan** function (line #**18**) and of the inverse tangent function of two arguments **Atan2** in unit **LibMath** (line #**19**).

Four **F3_12.TXT CON** files have been prepared to generate the **D2D** data files required to plot Figure 3.12. In order to place markers along the jumping portions of the two saw-tooth lines, configuration files **F3_12-1.CON** and **F3_12-2.CON** were formatted to read columns two and three of the **F3_12.TXT** file and generate the new files **F3_12-1.D2D** and **F3_12-2.D2D**. These files include additional nodes placed at a distance of about 0.1 units along the dropping portions of these sawtooth lines. The actual correction of the zigzagging angles on columns two and three of the **F3_12.TXT** file has been done using **CON** files **F3_12-3** and **F3_12-4**, resulting in data files **F3_12-3.D2D** and **F3_12-4.D2D**.

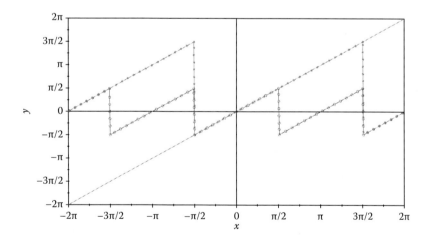

FIGURE 3.12   Plots of the modified angles (the lines with markers) and their corrected version (the line without markers). Configuration files **F3_12-1.CON** through **F3_12-4.CON** and **F3_12. CF2**.

Important: The first numerical row in data file **F3_12.TXT** provides the starting angle for the conversion. These three values ensure that the rectified angles starts from -2π.

3.2.6 Data Decimation

One last example of **Util~TXT** use refers to decimating an input data file, that is, retaining only every **k**th row, with **k** specified on the last line of the **CON** file. The case of decimating an experimentally acquired data file will be discussed, with reference to the phenomena of aliasing and to the importance of properly selecting the sample size in data acquisition (Alciatore and Histand 2007).

ASCII file **F3_13A.DTA** available with the book is organized in five columns, of which columns 3, 4, and 5 were plotted as function of their order (i.e., the column for *x* was assigned to zero in **D_2D** and in the corresponding **CF2** files). Note that two of the three signals plotted have a higher frequency content, and as the data file is decimated (equivalent to reducing the number of samples per second), the appearance of these graphs is altered (see Figure 3.13). The structure of the two **Util~TXT** configuration files (i.e., **F3_13B.CON** and **F3_13C.CON**) used to generate the decimated data files **F3_13B.DTA** and **F3_13C.DTA** with 250 and 125 samples, respectively, can be easily deciphered.

Note that instead of always using as input the original data file the previously generated file can be used as input for the next decimation.

Also note that **Util~TXT** is capable to generate multiple conversions in one run, with parameters read from the same **CON** file. For example, you can concatenate together all the configuration files utilized so far in a single file and then run **Util~TXT** with settings from it. An example of this type will be provided later in this chapter.

3.2.7 **DXF** Output of 2D and 3D Polylines

Util~TXT is capable of generating **DXF** files without actually plotting the respective polylines on the computer screen. You can do this by changing the extension of the output file from **XY** to **DXF**, a case in which the transformed points will be formatted as **AutoCAD R12 DXF** polylines. Likewise, **Util~TXT** can be used to convert sets of *x*, *y*, *z* triplets into **DXF** 3D polylines. See, for example, the **F3_14.CON** file that was used in the process of generating the variable-radius spiral in Figure 3.14 discussed next.

Later, there will be other uses of **Util~TXT** shown, like scaling and translating the vertices of a 2D polyline, obtained by digitizing a raster image using **AutoCAD**.

3.3 **UTIL~DXF** PROGRAM FOR VISUALIZATION OF **R12 DXF** FILES

There are a number of **DXF** viewers available to the interested user, either free, freeware, or open-source programs (e.g., see **eDrawings**). **Util~DXF** supplied with this book, both as executable and as source code, can view 2D and 3D polylines, circles, and arches of circle recorded to a **R12 DXF** file—see the about **Util~DXF** screen. One useful feature of **Util~DXF** is to extract the coordinates of the vertices of a selected polyline or group of polylines and output them to an ASCII file.

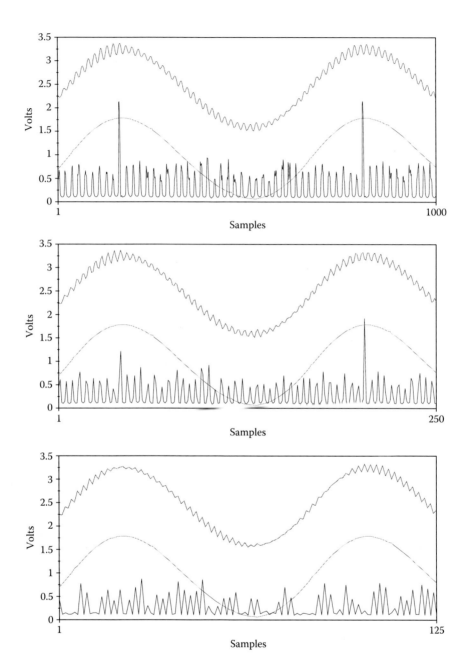

FIGURE 3.13 From top to bottom: plot of initial data with 1000-samples and decimated data with only 250 and 125 samples. Configuration files **F3_13A.CF2**, **F3_13B.CF2**, and **F3_13C.CF2**.

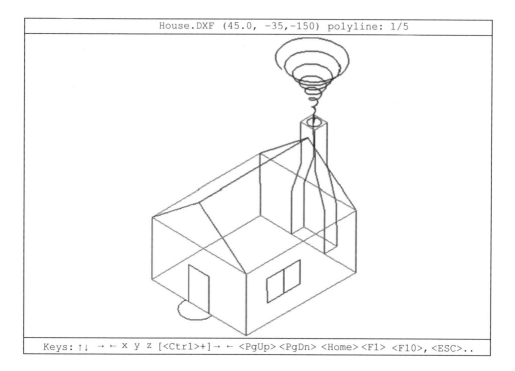

FIGURE 3.14 Screenshot of the **Util_DXF** program when run with settings from the **F3_14.CON** file. The 3D spiral was generated separately, and it has been exported directly to **DXF** using **Util~TXT**.

Important: **Util~DXF** will not represent properly 2D polylines created in planes other than the XY plane of the world coordinate system of **AutoCAD** or a plane parallel to it. Also note that in case of splined polylines or polylines containing arches of a circle, the control polygons will be represented rather than the smooth curve.

3.3.1 Extracting Polyline Vertex Coordinates

Launch **Util~DXF** and open the **House.DXF** file. Note that its polylines are colored in cyan, while the current polyline appears in red (Figure 3.14). From this view window, the available options are *pan* using the arrow keys, *zoom in* and *zoom out* using the <Pg Up> and <Pg Dn> keys, and going *back to the reference view* by pressing <Home>. Holding the <Ctrl> key while pressing <Pg Up>, <Pg Dn>, and the <Home> keys will have slightly different effects that you may want to investigate. If you hold the <Ctrl> key while pressing <←> or <→>, the figure will rotate in 3D about either x-, y-, or z-axis, depending on which of them is active (i.e., the one appearing capitalized at the bottom of the screen). You can change the axis of rotation by pressing the corresponding <X>, <Y>, or <Z> keys. Note that these axes remain aligned with the screen (i.e., x and y will be the horizontal and vertical axes with y positive up, while z will be perpendicular to the screen and oriented away from you).

By pressing the <F1> key, you can skim through the available polylines (note the change of the counter on the top of the screen, which initially read 1/5). <Ctrl> + <F1> will let you type in the polyline number that you want to become current. If you press <F10> or <Ctrl> + <F10>, you will be prompted to edit/confirm the polyline range you want their vertices written to file. With the **House.DXF** file opened, press <F10> and select polyline number 1 (i.e., the 3D helix representing the smoke, which will be discussed in Section 3.6) to have its vertices written to ASCII. This output file will be named by default **Poly0001.XYZ**. If you extract to file other polyline or groups of polylines, the name of the output file will be indexed by one with each new export.

Rename **Poly0001.XYZ** as **F3_15.XYZ** and use **D_2D** to plot z versus x and z versus y of this file. This will result in the side views of the helix as shown in Figure 3.15.

Important: If your **DXF** file contains only 2D polylines, then the extension of the output file will be **XY**. If at least one polyline is elevated above the XY plane or it is a true 3D polyline (like the spiral in the **House.DXF** file), then the extension of the vertex file will be **XYZ**.

Important: The coordinates of the polyline vertices will be expressed relative to the world coordinate system of the original drawing. It is therefore essential to set inside **AutoCAD** the UCS to "world" before exporting your drawing to **R12 DXF**.

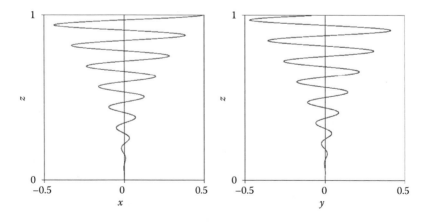

FIGURE 3.15   Side views of polyline number **1** produced with **D_2D** using **XYZ** file output from **Util~DXF**. Configuration files **F3_15A.CF2** and **F3_15B.CF2**.

3.3.2 Raster Curve Digitization Using **Util~DXF** and **Util~TXT**

Another useful application of **Util~DXF** is the possibility of digitizing a curve available as raster image only. While there are computer programs available to do the same thing (like **DigXY** from www.thunderheadeng.com), the method presented here is more accurate because it is done at higher resolution inside **AutoCAD**.

Let us consider the example of digitizing the *stress* versus *elongation* sigmoidal curve of an elastomeric material. Begin by importing the raster image **F3_16.TIF** to **AutoCAD**. Then draw an L-shaped polyline over any two adjacent sides of the plot box such that its three vertices coincide with marked points on the graph (see Figure 3.16). If the available raster image is slightly rotated, as it commonly happens with photocopied documents, use the *align* command in **AutoCAD** to rotate both the picture and the L-shaped polyline and make them parallel with the world coordinate system. Then draw a second polyline, this time overlapping the sigmoidal curve, inserting sufficient number of vertices to capture its shape. If your plot contains multiple curves of the same *x* and *y* categories, simply generate separate polylines for each of these curves.

Delete the picture and then type 'purge' at the **AutoCAD** command line to eliminate any unwanted entities from your drawing. When you are done, type 'dxfout' at the command line and export your drawing under the name **Rubber.DXF**. Make sure you select "AutoCAD 12" as **DXF** output format.

Important: The contour of the raster image will be exported to **R12 DXF** as a three-vertex polyline. Similarly, any block available in the drawing's database will be exported as visible entities to **DXF**, unless you purge your drawing prior to export. For the *purge* command to have the intended effect, you must first explode all blocks of the drawing.

Now open **Rubber.DXF** using **Util~DXF** and extract the vertices of the two polylines to file. This is named automatically **POLY0001.XY**. Usually, the graphic entities of a drawing are written to the **DXF** file in the same order in which they were generated inside **AutoCAD**. This will also be the order in which they will be written to **POLY0001.XY**.

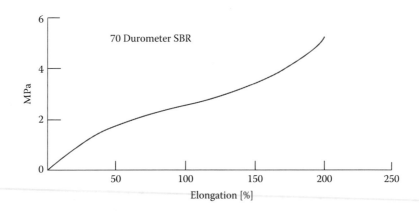

FIGURE 3.16   Stress–strain curve of an elastomeric material (Hertz 1991) available with the book as raster image **F3_16.TIF**.

Before you can use **Util~TXT** to automatically scale and translate the vertices of the main polyline such that their coordinates coincide with the original plot, you must perform the following steps (see Table 3.1): Open the **POLY0001.XY** file using **Notepad**. Make sure that the L-shaped polyline vertices occur before and not after the vertices of the sigmoidal curve. If it is not structured as shown to the left of Table 3.1, you must cut and paste the three lines containing the coordinates of the L-shaped polyline right under the header (it is where **Util~TXT** expects them), and then add a line separator '----------'. Then insert two empty lines under the '**X   Y**' header. On the first of these lines, type the values of x_{min} and y_{min} as they appear on the graph (i.e., **0** and **0**). Similarly, on the second empty line that you have inserted, type x_{max} and y_{max} (i.e., **250** and **6**). Before saving the file as **F3_20.XY**, verify that its top portion looks similar to the right column of Table 3.1. The **{MPa}** comment on line 4 is optional.

Next, you will have to prepare a **CON** file from where **Util~TXT** will take the conversion settings—see the one prepared for this example named **F3_17.CON**. Essentially, the option '**{XY from raster to D_2D format}**' must be set to '**Y**', and all the other transformations must be set to '**N**'. Since the **XY** input file is assumed to have a standard structure, the **{row start}**, **{row finish}**, **{column for X}**, and **{column for Y}** options will all be ignored.

Figure 3.17 is a plot generated using the transformed vertices of the sigmoidal polyline and recorded by **Util~TXT** to the **F3_17.DTA** file.

Important: If you want the units on any of the two axis of your graph changed, simply convert the values of x_{min}, y_{min} or x_{max}, y_{max} on lines three and four of **F3_17.XY** to the new units. In the

TABLE 3.1   Modifications to a Default **XY** File from **Util~DXF** Required for Raster Curve Digitization

POLY0001.XY **(Original File)**		F3_20.XY **(Edited File)**	
Polyline(s) 1 to 2 from RUBBER.DXF		Polyline(s) 1 to 2 from RUBBER.DXF	
X	Y	X	Y
0.8434730	-0.265137	0	0
-0.090741	-0.265137	2.5	6{MPa}
-0.090741	0.2855790	0.8434730	-0.265137
-------------------		-0.090741	-0.265137
-0.090741	-0.265137	-0.090741	0.2855790
-0.077363	-0.248015	-------------------	
-0.055967	-0.224465	-0.090741	-0.265137
-0.034811	-0.202117	-0.077363	-0.248015
-0.005962	-0.174001	-0.055967	-0.224465
0.0200020	-0.151892	-0.034811	-0.202117
0.0457260	-0.132908	-0.005962	-0.174001
0.0657280	-0.119884	0.0200020	-0.151892
0.0907300	-0.105466	0.0457260	-0.132908
0.1224640	-0.089846	0.0657280	-0.119884
0.1486200	-0.079153	0.0907300	-0.105466
0.1789120	-0.068099	0.1224640	-0.089846
0.2123290	-0.056564	0.1486200	-0.079153
0.2390140	-0.047673	0.1789120	-0.068099

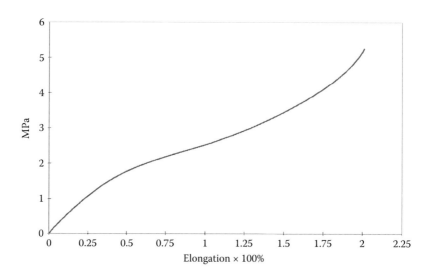

FIGURE 3.17   The vector format of the stress–strain curve in Figure 3.16 generated with the **F3_17. DTA** file output by **Util~TXT** based on the **F3_17.XY** vertex file. Configuration files **F3_17.CON** and **F3_17.D2D**.

example considered, if you want the stress expressed in psi rather than MPa, then on line **4** of the **F3_17.XY** file, you must change the value of y_{max} from **6** to **870.226**.

3.3.3 Transferring Level Curves from **D_3D** to **D_2D**

Another useful application of **Util~DXF** is to transfer level-curve data from **D_3D** to **D_2D** and use it in combined plots, for example, as animation backgrounds. To exemplify, the data file used to plot of the level curves in Figure 2.17 has been copied and renamed **F3_18.D3D**. The *x*- and *y*-axes were swapped so that the graph looks as shown in Figure 3.18a.

Before using **Util~DXF** to convert these curves to a format readable by **D_2D**, perform the following steps: replot function F_3 using **D_3D** with settings from **F3_18A.CF3** and export the level curves as **DXF 1:1** to file **F3_18A.DXF**. Open this file inside **AutoCAD** and insert small closed polylines, about the size of a 0.02 radius circle at the locations and in the layers indicated in the Table 3.2 (see file **F3_18.DWG**). These are the local minima and local maxima of the function $F_3(x, y)$ in Equation 2.3, found numerically as explained in Chapter 4.

Save this drawing to file **F3_18B.DXF** as **AutoCAD** release 12 **DXF** and then use **Util~DXF** to extract all level curves to the vertex file **F3_18B.XY**. When plotting the **F3_18B.XY** data, in order for this new graph to exactly match the original one (see Figure 3.18b), you must set the *x*- and *y*-axis limits inside **D_2D** to the same values as in the initial plot, that is, −1.5 and 2.5 over *x*-axis and −2.5 and 2.5 over *y*-axis.

Important: To preserve the scale coloring information of the level curves, use the <Ctrl> + <F10> option to export their vertices to file rather than option <F10>. **Util~DXF** will use the layer names where these level curves are placed to add color information to the output ASCII file.

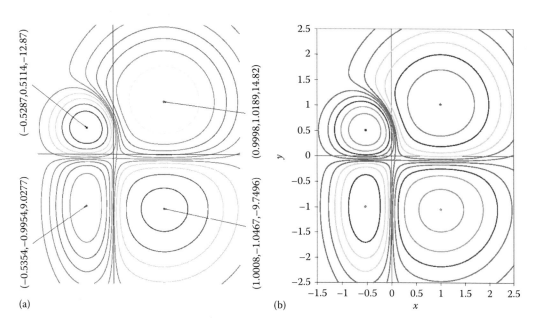

FIGURE 3.18 **DXF 1:1** level-curve plot of function F_3 in Equation 2.3, showing additional editing done using (a) **AutoCAD** and replotted using **D_2D** after conversion to (x, y) format using (b) **Util~DXF**. Configuration files **F3_18A.CF3** and **F3_18B.CF2**.

TABLE 3.2   Local Extrema of Function F_3 in Figure 3.18

x	y	z_{min} or z_{max}	Layer Name
1.0008	-1.0467	-9.7496	C-9_7496
-0.5354	-0.9954	+9.0277	C09_0277

3.4 **Util~PLT** PROGRAM FOR MANIPULATING **PLT** FILES

Instead of directly creating a hardcopy of your **AutoCAD** drawing, it is possible to print it to a file with the extension **PLT**. In this paragraph, it will be explained how to view and manipulate such files using the **Util~PLT** program. The type of **PLT** files **Util~PLT** can read in are those generated with the **HP-GL ADI 4.2** by **Autodesk** #7550 driver, available from the *add printer* menu of **AutoCAD**. Such **PLT** files have a simple structure, consisting essentially of a succession of **PU** (*pen up*), **PD** (*pen down*), and **PA** (*pen absolute*) commands. Of these, the **PA** command requires as integers the x and y coordinates of the point where the pen will go on the surface of the paper, either in the *pen up* or *pen down* mode. In other **PLT** dialects, there are available additional commands for changing color, for drawing arches of circle, etc. **HP-GL ADI 4.2** can generate monochrome plots only, with circles, arches of circle, splined curves, as well as text characters and symbols represented as successions of approximating segments.

The **Util~PLT** program allows drawings with arches of circle, splines, circles, ellipses, texts, etc., to be converted to line segments. Another useful application of **Util~PLT** is to *flatten* 3D models for the purpose of reducing their size on disk or for preventing unauthorized access the original solid model. Likewise, surface plots created with **D_3D** can be "flattened" in the 'hide' mode for the purpose of further editing.

From earlier discussions, it is evident that when plotting a drawing to the **HP-GL ADI 4.2 PLT** file format, any color information will be lost. It is possible however to write each layer (or groups of layers) to separate **PLT** files and then convert them back to **DXF** the **Util~PLT** program. You can then combine these **DXF** files into the same drawings using **AutoCAD** and assign them different colors.

Figure 3.19 is a screenshot of the main view window of the **Util~PLT** program, showing a 3D part originally created with **AutoCAD** (see file **F3_19.DWG** available with the book). To toggle between viewing the part at its normal proportions, or stretching it to fit the view window like in Figure 3.19, press the <F1> key. If you want to copy the screen to **DXF**, press <F10>. You will be prompted to specify the line type, line thickness (defaults are *solid line* and *zero thickness*), and the coincidence and colinearity parameters. Same as in the **D_2D** and **D_3D** programs, these parameters are used when eliminating the polyline vertices that almost coincide or of a vertex that is almost collinear with its two neighbors. Same as the **D_2D** and **D_3D** programs, following a **DXF** export, **Util~PLT** will indicate the limit values of these two parameters at which polyline optimization begins, by eliminating the near coincident and near collinear vertices. When you set the values of these colinearity and coincidence parameters as well as the line thickness, have in mind that a **DXF** copy of the stretched image will fit a box of approximately 640 by 450 units.

2D polylines created inside **AutoCAD** can be digitized by exporting them to **R12 DXF** first, and then to ASCII as (*x*, *y*) pairs using **Util~DXF**. If only scaling and offsetting are required, then the ASCII file that is output by **Util~PLT** simultaneously with the **DXF** file export may be enough. The scaling and offsetting can in this case be done by editing

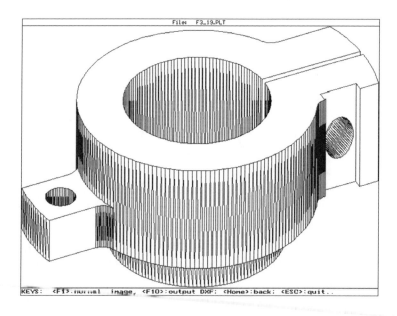

FIGURE 3.19 **Util~PLT** view of the **F3_19.PLT** file in the stretch-to-fit mode. To view it at its original proportions, you must press the <F1> key. See also the **F3_19.DWG** file.

the minimum and maximum limits over x and y from within **Util~PLT**. You can change these limits right after opening the **PLT** file or you can do this later by pressing the <Back> key while in the graphic screen of **Util~PLT**.

3.4.1 Flattening and Retouching Plots Created with **D_2D**

It was pointed in the previous chapter that **AutoCAD** can exhibit defects when it comes to hidden-line elimination defects (e.g., see Figure 2.24). To correct such imperfections, you can print it from **AutoCAD** to **PLT** in the hide mode and then export this **PLT** file to **R12 DXF** using the **Util~PLT** program. This second **DXF** file will be a "flattened" version of the original plot with its hidden lines removed. When all entities have zero elevation, they can be edited much easier using **AutoCAD** (i.e., the *trim* and *extend* commands will work on any entity, because they are now of zero elevation). Figure 3.20 shows the same Figure 2.24, but with the hidden-line defects repaired as explained next.

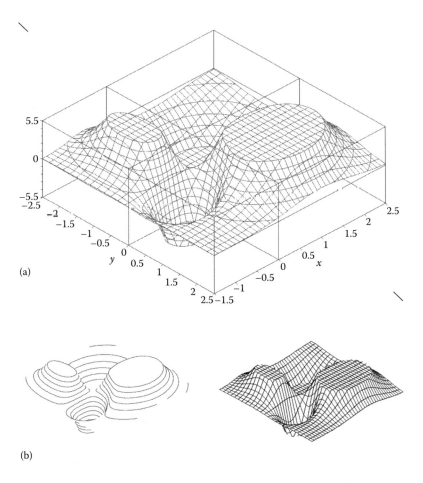

(a)

(b)

FIGURE 3.20   Flattened version of Figure 2.24 (a) obtained by exporting separately to **PLT** the level curves, meshgrid, and bounding box (b—box not shown) and then overlapping them back after being converted to **R12 DXF** using **Util~PLT**. Configuration files **F3_20Z.CF3**, **F3_20XY. CF3**, and **F3_20BOX.CF3**.

The following entities were extracted from **F3_20.DXF** and plotted to files in the *hide* mode: Layers 0, **$_Body**, **$_Top_Bottom**, together with **Border_x**, **Border_y**, **Constant_x**, and **Constant_y** were plotted to file **F3_20XY.PLT**. All level-curve layers (names beginning with **C**), together with layers 0, **$_Body**, and **$_Top_Bottom** were plotted to file **F3_20Z.PLT**. The same 0, **$_Body**, and **$_Top_Bottom** layers, together with layers **_Box**, **_Divisions**, and **_Zero_Lines** were plotted to file **F3_20Box. PLT**. These three **PLT** files were convert back to **R12 DXF** using **Util~PLT** and then were recombined inside **AutoCAD**. In order to have the values along the three axes available as text entities rather than polylines (note that following a **PLT** export, these are no longer editable but rather collections of polylines), the content of layers **0** and **_Text** in the original drawing was copied to the clipboard, then pasted inside the **DWG** file and overlapped with the other components.

The two short oblique lines on the upper-left and bottom-right corners automatically generated by **D_3D** (see Figure 3.20) were used as references when doing the 'move' and 'scale' transformations required to exactly overlap the new graph with the original one from **D_3D**.

Note that the level curves in Figure 3.20 have a smoother appearance than in the original Figure 2.24. This is because the level curves were extracted to **DXF** at a higher resolution (i.e., 101×126 points) using configuration file **F3_20Z.CF3**, while the meshgrid was generated at a lower resolution as set in configuration file **F3_20XY.CF3**.

Important: When plotting a drawing to **PLT**, its dimensions and position relative to a paper reference frame will be different than in the original **AutoCAD** drawing. Since the **PA** command uses integer arguments, some loss of accuracy over the original **DWG** file should be expected.

3.4.2 Alphanumerical Character Discretization

Another potentially useful application of **Util~PLT** is the discretizing of alphanumerical characters (i.e., extracting to file points placed along their contour), as it will be explained next with reference to Figures 3.21 and 3.22. Similarly one can be discretize circles, arches of circle, ellipses, splined polylines etc.

Begin by generating an **AutoCAD** drawing containing the entities that you want discretized. For the time being, just open file **F3_21A.DWG** available with the book using **AutoCAD** (see Figure 3.21) and print it to **F3_21.PLT**. Then open this **PLT** file using **Util~PLT** and export it to **R12 DXF** (make sure it is upstretched). Rename this export file **F3_21B.DXF**. If you view it with either **AutoCAD** or **Util~DXF**, you will notice that the contours of the three characters will appear as polylines. Also note that the border, which originally was a closed polyline consisting of four arches of circle, now appears as a multivertex polyline. If additional editing is necessary, you can always open the **F3_21B.DXF** file with **AutoCAD** and export it back to **R12 DXF** after doing the necessary modifications. To see the limitations of **Util~DXF**, open a **R12 DXF** copy of **F3_21.DWG** generated from inside **AutoCAD** (see the **F3_21A.DXF** file available with the book). Note that the C, A, and D characters will not be visible at all, while the border consisting of a four-arch polyline is represented as a rectangle.

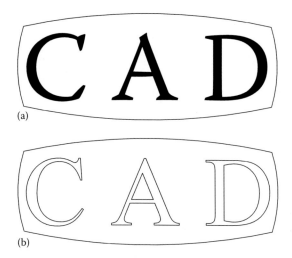

FIGURE 3.21 (a) View of the **F3_21.DWG** drawing as it was generated with **AutoCAD** and (b) view of its **PLT** file (**F3_21.PLT**) converted using **Util~PLT** to **DXF** (the **F3_21B.DXF** file).

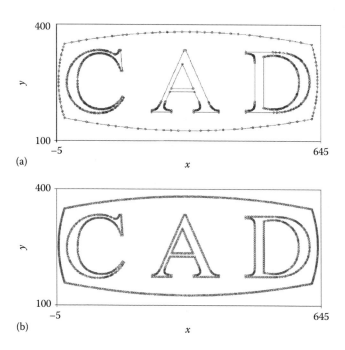

FIGURE 3.22 The drawing in Figure 3.21b exported to a **Poly####.XY** file and plotted using **D_2D** (a) as is and (b) after inserting additional points interpolated linearly using the **Util~TXT** program. Configuration files **F3_22A.CF2**, **F3_22.CON**, and **F3_22B.CF2**.

The vertices of the polylines inside **F3_21B.DXF** will be exported to a **Poly####.XY** file. Then, using the **Util~TXT** program, linearly interpolated points will be added to this **XY** file in a batch conversion operation (i.e., the instructions for several successive transformations will be all read from the same **CON** file).

A plot of the initial vertex file generated with **Util~DXF** (renamed **F3_22A.XY**) is given in Figure 3.22a. Note the six closed polylines that make this new figure ('C' and the border are one single polyline, while 'A' and 'D' consist each of two polylines) and the several rectilinear portions on this plot without markers. We will employ the **Util~TXT** program to add linearly interpolated points to cover these sections such that the distance between every two vertices belonging to the same polyline will not exceed five units of length. This can be done either to one polyline at a time as explained in paragraph 3.2 or by using a **CON** file with multiple records—see the **F3_22.CON** file available with the book. This is essentially a concatenation of six regular **CON** files delimited by a record separator '--------', all six instructing **Util~TXT** to write data to the same output file, that is, **F3_22B.XY**. A plot of file **F3_22B.XY** thus obtained is available in Figure 3.22b.

Important: Once the portion of the input data file between the specified **{row start}** and **{row finish}** is converted, the record separator is then copied to the output file to act as **D_2D** line separator. If this separator read from the **CON** file is an empty line, it will not be copied to the output data file, and as a consequence, the curves will appear connected together in a single polyline.

3.5 **G_3D.LSP** PROGRAM FOR GENERATING 3D CURVES AND SURFACES INSIDE **AutoCAD**

The **G_3D.LSP** program is an **AutoLISP** application that can be used to plot inside **AutoCAD** both meshed surfaces and 3D curves using data read from an ASCII file of extension **G3D**. Such a file should consist of **(x y z)** triplets delimited by spaces (parentheses must be included). When the same data are read using the **D_3D**, **D_2D**, or **Util~TXT** programs, parentheses are optional, but this format should be strictly followed in case of opening them with **G_3D.LSP**.

In this section, we will look at generating 3D curves and 3D meshed surfaces using the **G_3D.LSP** program. The **Util~TXT** program discussed earlier allows you to convert *x*, *y* and *x*, *y*, *z* data sets to 2D and 3D **DXF** polylines. **G_3D.LSP** has similar capabilities but in addition, it can automatically scale the data, so that the given curve will fit a given 3D plot box. Same applies to 3D surfaces, which are true 3D surfaces and can be generated only with the **G_3D AutoLISP** application.

3.5.1 3D Polyline Plotting Using **G_3D.LSP**

Begin by launching **AutoCAD** and start a new drawing. Save this drawing under the name **F3_23.DWG**. Since **G_3D.LSP** is menu driven, a copy of the **DCL_G3D.DCL** dialog definition file must be available in the current directory. At the **AutoCAD** command line, type 'upload' and select **G_3D.LSP**, then type 'g3d' to launch the program and select the desired **G3D** input file. If this input file consists of a regular grid of samples, it can be plotted both as a curve and as a meshed surface; otherwise, it can be plotted as a curve only.

Load as input file **F3_23.G3D** available with the book (see also program **P3_23.PAS** in Appendix B used to generate the **G3D** data file). This will generate the 3D spiral visible in Figure 3.23 having a parabolically increasing radius:

$$
\begin{cases}
x(\theta) = R\left[\dfrac{\theta}{(2\pi \cdot n_c)}\right]^2 \cos(\theta) \\[2em]
y(\theta) = R\left[\dfrac{\theta}{(2\pi \cdot n_c)}\right]^2 \sin(\theta) \\[2em]
z(\theta) = \dfrac{p \cdot \theta}{(2\pi)}
\end{cases}
\tag{3.5}
$$

where
  n_c is the total number of coils
  R is the major radius (the radius of the last coil)
  p is the pitch of the helix (the distance between two successive coils)

G_3D.LSP can plot the helix as is or it can scale it to fit a rectangular box. In the latter case, the x, y, and z limits of the bounding box can be edited by the user, as well as the box width and height, but not its length (its x dimension), which has imposed unit value (see Figure 3.23b).

Important: If you set the box width to zero, you will obtain a projection of the curve (or surface) on the *XOZ* side plane. If you set the box height to zero, the curve will be projected

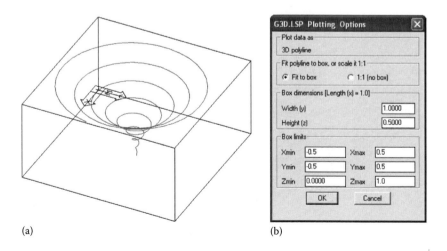

(a)   (b)

FIGURE 3.23   (a) Plot of **F3_23.G3D** data file using **G_3D.LSP** and (b) the corresponding **G_3D. LSP** menu settings.

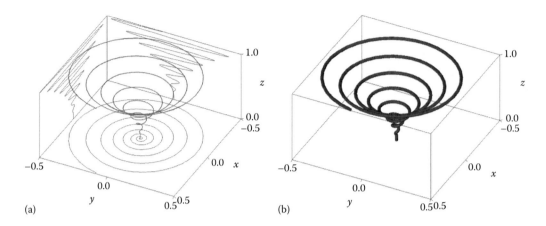

(a)   (b)

FIGURE 3.24   3D curves generated with **G_3D.LSP** and further edited inside **AutoCAD**. (a) Side and bottom projections, and divisions and labels, and (b) a circle extruded along and divisions and labels added. See also the drawing files **F3_24A.DWG** and **F3_24B.DWG** available with the book.

on the bottom plane. If you set both the height and width to zero, the curve will be projected on the *YOZ* back plane, under the forced assumption that **Width (y)** = 1 and **Height (z)** = 0.5 (see Figure 3.24a).

The plot box is oriented such that the **AutoCAD** UCS icon visible in Figure 3.23a is placed at the 'minimum corner' (i.e., at the point of coordinates −0.5, −0.5, 0.0 in case of Figure 3.24) and is oriented in the positive direction of the *x*-, *y*-, and *z*-axes. This is a useful piece of information in case you want to manually add divisions and values along the edges of the bounding box.

Examples of edited 3D curve plots are given in Figure 3.24. Figure 3.24a has been created by running the **G_3D** program four times: (i) with default box-size settings, (ii) with either **Width (y)** or **Height (z)** set to zero, and (iii) with both **Width(y)** and **Height(z)** set to zero. Figure 3.24b is a rendered view of 3D solid obtained by extruding a circle along the 3D spiral originally created with **G_3D.LSP**. In all these cases, before any plot has been generated, the limits along the *x*-, *y*-, and *z*-axes were rounded to the values visible on the menu in Figure 3.23b and on the actual plots in Figure 3.24.

A second example of a 3D curve that will be discussed is that of a toroidal spiral (see Figure 3.25 and the **P3_25.PAS** program in Appendix B used to generate the **G3D** data file to plot it). The parametric equations of this curve are

$$\begin{cases} x(\theta) = \left[r_T + r_S \cos(n\theta) \right] \cdot \cos(\theta) \\ y(\theta) = \left[r_T + r_S \cos(n\theta) \right] \cdot \sin(\theta) \\ z(t) = r_S \sin(n\theta) \end{cases} \tag{3.6}$$

where
  r_T is the middle radius of the torus
  r_S is the radius of the coil
  n is the number of coils

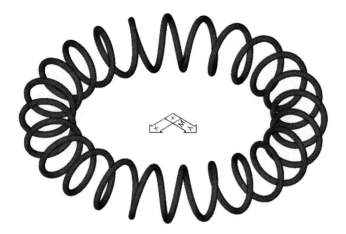

FIGURE 3.25   Plot of the **F3_25.G3D** data file using **G_3D.LSP**. See also the **F3_25.DWG** drawing file.

Figure 3.25 is a rendering of a solid obtained by extruding a circle along the path of Equation 3.6 generated using **G_3D.LSP**. Because **AutoCAD** does not accept closed extrusion paths, the curve had to be broken at one vertex prior to the actual extrusion.

3.5.2 3D Surface Plotting Using **G_3D.LSP**

Similarly to 3D curves, **G_3D.LSP** allows surfaces to be generated directly inside **AutoCAD**. **G_3D** uses the **AutoCAD** *3dmesh* command with vertices read from an ASCII file formatted as $(x_i \; y_j \; z_{ij})$. The file extension should be **G3D**, and the x_i values should be evenly spaced not arbitrarily spaced. The use of **G_3D.LSP** application is facilitated by the ability of **D_3D** to export the current plot to a **G3D** file.

Figure 3.26a is a plot of the *four-hump function* in Equation 2.2, with 41×51 data points read from file **F3_26.G3D** produced using **D_3D**. The view point and limits over the z-axis

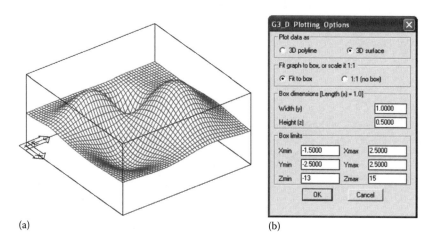

(a)   (b)

FIGURE 3.26   (a) Plot of **F3_26.G3D** data file using **G_3D.LSP** and the (b) corresponding menu settings.

have been rounded from the default values returned by **G_3D.LSP**. Note that shrinking the Zmin and Zmax limits in **G_3D** will not truncate the function surface. It will rather show the 3D surface extending outside the bounding box. Figure 3.27 is a plot of the same double-surface plot in Figure 2.42, this time generated using **G_3D.LSP**. Since **AutoCAD** does not include line and text entities when generating rendered images, the plot in Figure 3.27b is the result of an overlap of three separate screenshots as shown in Figure 3.28 (see also the **F3_27.DWG** drawing file available with the book).

Important: When representing multiple surfaces on the same graph, you must keep the same *x*, *y*, and *z* limits and plot box dimensions for each data file, or otherwise the graph will not depict the true intersection of the two surfaces.

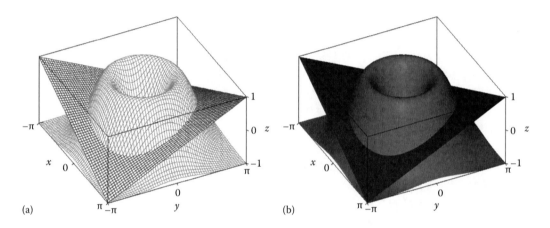

(a)   (b)

FIGURE 3.27 Hidden-line (a) and rendered (b) plots of two intersecting surfaces produced using **G_3D.LSP** with data from files **F3_27-1.G3D** and **F3_27-2.G3D**.

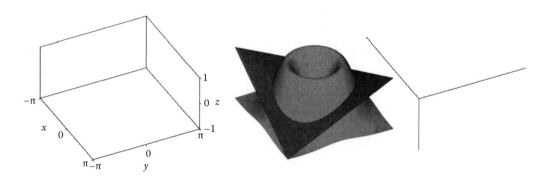

FIGURE 3.28 The three **AutoCAD** screenshots overlapped manually in the order from left to right, required to generate the plot in Figure 3.27b.

3.6 **M_3D.LSP** PROGRAM FOR AUTOMATIC 3D MODEL GENERATION AND ANIMATION INSIDE **AutoCAD**

The **M_3D.LSP** is another useful **AutoLISP** application capable of (i) drawing in specified layers lines, cylinders, cones and cone frustums, spheres, tori, arrows, and cylindrical helixes with specifications read from ASCII file of extension **M3D**, (ii) writing texts, (iii) inserting blocks at positions and with orientations read from the same data file (the blocks must already exist in the database of the current drawing), and (iv) creating animation frames by turning *on* and then back *off* layers **1, 2, 3**, etc., of the current drawing (assuming that these layers already exist) and exporting screenshots of these frames to **BMP** and/ or to **AutoCAD** slide files of extension **SLD**. Layers of names other than **1, 2, 3**, etc., will not be animated and can be used to display immovable background objects. In order to easily animate the **SLD** frame files, **M_3D** will additionally generate a script file (extension **SCR**). When launched with the **AutoCAD** *script* command, this script will load the **SLD** screenshots one by one and display them the amount of **MS_delay** milliseconds, until the user presses the <Esc> key.

3.6.1 Animation of **DXF** Files with Multiple Layers Using **M_3D.LSP**

For a first demonstration of **M_3D.LSP** use, open inside **AutoCAD** file **F3_04003.DXF** generated earlier by the **P3_04.PAS** program. Since the drawing appears flipped compared to the original image, you must mirror everything about the *OX* axis (see Figure 3.29 and the **F3_29.DWG** file). In order not to mirror the text together with the rest of the

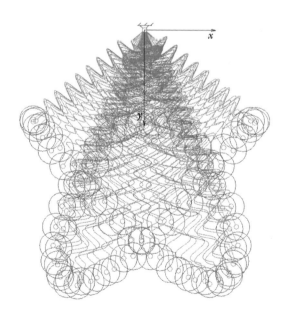

FIGURE 3.29 **AutoCAD** view of the **F3_04003.DXF** file generated with **P3_04.PAS**, after it has been mirrored about the *OX* axis and saved as **F3_29.DWG**. See also animation file **F3_29.GIF**.

drawing, you must first type 'mirrtext' at the command line and change the value of the corresponding system variable from 1 to 0.

Once you mirrored the drawing, load **M_3D.LSP** by typing 'appload' at the command line. After the program is loaded, type 'motion' and confirm the default **SCR** file name (or specify your own). **M_3D.LSP** will turn *off* the existing layers named **1, 2, 3**, etc. (up to at most **999)** and then will turn them back *on* one at a time. Depending on the program settings, each frame will be copied to the hard drive as **BMP** and/or **SLD** file. The name of these **BMP** and **SLD** frame files will be the one you specified after issuing the **M_3D** command *motion*, followed by a three-digit frame number, for example, 001, 002, and 003. To animate the **SLD** files from within **AutoCAD**, type *script* at the command line and open the **SCR** file just created. To stop the animation, press the <Esc> key and then issue the *regen* command to refresh the screen. To create a stand-alone animation, you can assemble the **BMP** frames into a video clip using a moviemaker or create an animated **GIF** as explained in Chapter 1.

Important: The generation of **SLD** animation frames is currently turned *off*. To activate it, open **M_3D.LSP using Notepad** and remove the semicolon in front of the line that reads **(setq SLD_output 1)** located at the end of the file. The animation frame rate can be adjusted by editing the line that reads **(setq MS_delay 10)**. Alternatively, you can edit the script file directly, using the *replace all* function in **Notepad**.

Important: The executable **~Purge.EXE** available with the book (see Chapter 9) allows for rapid deleting all **PCX, BMP, SLD**, and **SCR** files in the current directory. The program will also delete without confirmation all files of extension **BAK** and **OLD**, as well as the **acad. err** and **acadstk.dmp** files in the current directory, if they exist.

3.6.2 3D Model Generation with Data Read from File

The examples that will be discussed next refer to using **M_3D.LSP** for assembling in separate layers 3D objects with specifications read from file, for the purpose of generating animations with these layers as described earlier.

Available with the book, there are six **M3D** files: Data file **F3_30.M3D** intended to work with drawing **F3_30.DWG** and data files **F3_31.M3D, F3_31SW.M3D, F3_31WCS1. M3D, F3_31WCS2.M3D**, and **F3_31UCS.M3D**, all five intended to work with drawing **F3_31.DWG**. These two **DWG** files (see Figures 3.30a and 3.31a) contain the 3D models that form the nonmoving parts of the front of a small wheeled tractor. In association with the aforementioned **M3D** files, they will be used to simulate the motion of the steering linkage of the tractor (see Figures 3.30b, 3.31b, and 3.32 and the companion animated **GIF** files).

Begin by opening file **F3_30.DWG** (Figure 3.30a), then issue the **Auto CAD** *appload* command and load **M_3D.LSP**. Note that in the database of file **F3_30.DWG**, there is already a block named 'wheel', which is a model of one of the wheels of the tractor. Type 'm3d' at the command line and select **F3_30.M3D**. This ASCII file contains the descriptions of the entities that will be assembled to form the front axle with wheels, and steering linkage components, for the tractor being steered lock-to-lock on a flat surface in 14 positions. Once the **M3D** file is uploaded, **AutoCAD** will generate in separate layers these

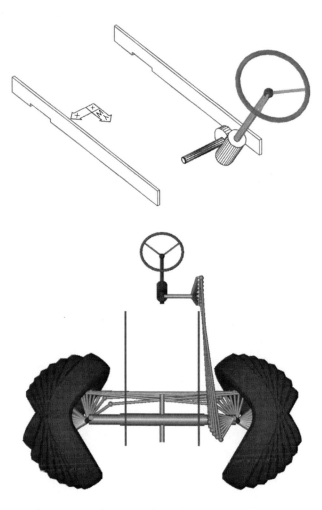

FIGURE 3.30 Horizontal steer simulation with background drawing **F3_30.DWG** shown above and rendered view of the overlapped layers generated with data from **F3_30.M3D** (below). See also animated **GIF** files **F3_30-1.GIF**, **F3_30-2.GIF**, and **F3_30-3.GIF**.

14 positions (Figure 3.30b). Select a suitable viewpoint and amount of zoom (you can also resize the window in which **AutoCAD** runs) and then type 'motion' at the command line to generate the **BMP** frames of the simulation. Animation files **F3_30-1.GIF**, **F3_30-2. GIF**, and **F3_30-3.GIF** have been produced using such **BMP** frames, generated for isometric, top view, and front view points of the tractor model.

Note that the files readable by **M_3D.LSP** can be generated with any computer program or by hand using **Notepad**. To understand how these data files are structured, you can study the **M3D** files available with the book and the **M_3D.LSP** source code where the format and syntax of the acceptable commands are explained. You will learn that spheres are fully defined, by their radii plus, the x, y, and z coordinates of their centers. Lines, cylinders, tori, cones, cone frustums, and arrows are fully positioned and oriented by two 3D points (the amount of rotation about their axes is not relevant for these entities). Angular orientation is not specified in case of cylindrical helixes either, although this may cause occasional loss of realism in certain simulations. **AutoCAD** blocks are the only objects for which you must specify their insertion point, and the coordinates of two additional points, to fully orient them one along the *x*-axis and one along the *y*-axis of the reference frame attached to the respective block.

Important: All 3D points entered in an **M3D** file are assumed specified relative to the world coordinate system (WCS) of the drawing.

Data file **F3_30.M3D** includes the number of notations as follows (see also Figure 3.31): **AB** is the drag link of the control linkage, **A** is the ball joint of the drop arm, **B** the ball joint of the steering arm, and **CD** is the tie rod of the Ackermann linkage. The calculations involved in determining the coordinates of these ball joint centers **A**, **B**, **C**, and **D** can be found in the paper by Simionescu and Talpasanu (2007) listed at the end of the chapter.

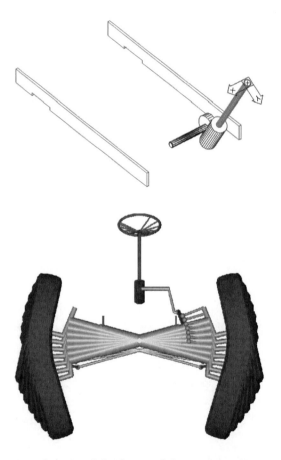

FIGURE 3.31   Bump steer simulation with background drawing **F3_31.DWG** shown above and rendered view of the overlapped layers generated with data from **F3_31.M3D** and **F3_31SW.M3D** (below). See also animation files **F3_31-1.GIF** and **F3_31-2.GIF**.

3.6.3 Automatic Insertion of **AutoCAD** Blocks

One likely criticism of the aforementioned kinematic simulations is that the steering wheel remains immobile as the wheels and the steering linkage move. In the next example, this problem will be addressed, while explaining how to instruct **M_3D** to insert **AutoCAD** blocks at positions and with orientations read from file. This second example is a motion simulation of the cross-coupling between axle oscillation and the steering mechanism known as *bump steer*. In this simulation, the steering knuckles are assumed locked and the front axle is oscillated causing the arm of the steering box (the drop arm) to move off its reference position, which causes the steering wheel to rotate.

We will first edit **F3_30.DWG** and turn the steering wheel into an **AutoCAD** block, which will be inserted in different rotated positions at the end of the steering column. The modified file has been renamed **F3_31.DWG** and is available with the book. Open drawing **F3_30.DWG** and move the UCS at the end of the steering column, such that it is oriented with the *y*-axis in the longitudinal plane of the vehicle, the *x*-axis pointing to the right, and the *z*-axis aligned with the steering column. Note that the steering column is tilted 30° backward—see Figure 3.31a. Also, the steering wheel is placed in a layer called 'Steering_wheel' that you may want to turn *off* as you reposition the UCS. Also notice the small circle at the end of the steering column that will facilitate positioning of the UCS. With the layer 'Steering_wheel' *on*, issue the command *block* and create a new **AutoCAD** block named 'S_wheel'. Specify (0,0,0) as its insertion point and select the steering-wheel rim, spokes, and spherical hub as its constituents. Move the UCS back to the world position and save your drawing as **F3_31.DWG**.

Now upload the **M_3D.LSP** application inside **F3_31.DWG** and type 'm3d' at the command line and select **F3_31.M3D** as input. What you will obtain are seven over-lapped images of the front axle oscillated from −15° to +15° with the steering knuckles locked in the straight ahead position, together with the steering mechanism components. Note the missing steering wheel, which exists however as a block in the drawing's database, same as the block 'wheel'. In order to insert the 'S_wheel' block in its position and rotated due to the *bump steer*, type 'm3d' again and select as input the **F3_31SW. M3D** file. If you issue the *render* command, you should obtain an overlapped image similar to Figure 3.31b.

The steering-wheel angles correlated with the position of the axle as it oscillates are listed in Table 3.3 and have been taken from Simionescu and Talpasanu (2007). When calculating these values, the steering-box reduction ratio was assumed to be 16:1, which means that the drop-arm displacement will be transmitted at the steering wheel amplified 16 times. As the axle oscillates with its steering knuckles locked, the rotation of the steering

TABLE 3.3   Steering-Wheel Angle vs. Axle-Beam Oscillation Angle

Position Number (*i*)	1	2	3	4	5	6	7
Axle-beam angle (ψ)	−15.00°	−10.00°	−5.00°	0.00	5.00°	10.00°	15.00°
Steering-wheel angle (φ)	−45.120°	−35.088°	−19.968°	0.00	24.608°	53.648°	86.992°

wheel can be used as a measure of the cross-coupling between the steering control motion and the axle oscillation, also known as the *bump steer* of the vehicle.

The problem of simulating the steering-wheel motion due to bump steer has been solved using a separately generated file **F3_31SW.M3D** that prescribes the insertion point and orientation of the 'S_wheel' block for the seven positions in Table 3.3. The Pascal program **P3_31.PAS** listed in Appendix B generates these lines and writes them to file **F3_31SW.M3D**. In addition, **P3_31.PAS** outputs file **F3_31UCS.M3D**, which can be used to insert two arrow entities corresponding to the *x*- and *y*-axes of the local reference frame attached to the steering wheel.

In the process of calculating the insertion point and orientation of the block 'S_wheel', and of the end points of the *x* and *y* arrows attached to the steering wheel, program **P3_31.PAS** calls the roto-translation procedure **RT** from unit **LibGe3D.PAS** four times (lines **#44** to **#47**). This is done for the WCS coordinates of the two points attached to the steering wheel of local coordinates (400, 0, 0) and (0, 400, 0)—which are also the ends of the two arrows—to be transformed as follows: (i) one rotation about the *OZ* axis by the *steering-wheel angle* φ (see Table 3.3), (ii) one rotation about the *OX* axis by −30° to account for the backward tilt of the steering column, and (iii) one translation to the point of WCS coordinates (0.000, 971.338, 658.399) where the steering wheel actually attaches to its column.

To exemplify additional capabilities of the **M_3D.LSP AutoLISP** program, two **M3D** command files have been manually generated and are listed on the next page. The **F3_31WCS1.M3D** file includes the command lines to generate the three arrows of the global reference frame *OXY* with the origin at point (0,0,0) located in the middle of the axle beam and extending to points (900,0,0), (0,400,0), and (0,0,500), respectively (see Figure 3.32a).

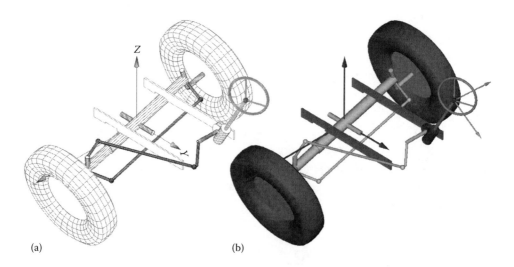

(a)   (b)

FIGURE 3.32 Bump steer simulations with background drawing **F3_31.DWG** and (a) data from **F3_31.M3D**, **F3_31SW.M3D**, and **F3_31WCS1.M3D** and (b) from **F3_31.M3D**, **F3_31SW.M3D**, **F3_31WCS2.M3D**, and **F3_31UCS.M3D**. See also animation files **F3_32a.GIF** and **F3_32b.GIF**.

```
(";")------------------------------------------------------------
(";") M3D command file name:  F3_31WCS1.M3D
(";")
(";") Draw WCS using cones and lines, and label axes as x, y, z
(";")------------------------------------------------------------
("WCS") change layer to "WCS"
(CL "BLUE") change color
(     0.0 0.0 0.0   900.0 0.0 0.0) line
(CO 825.0 0.0 0.0   900.0 0.0 0.0   15.0) cone
(TX "X" 920.0 0.0 0.0   30   0.0) X-axis label
(   0.0   0.0   0.0    0.0 400.0 0.0) line
(CO 0.0 325.0   0.0    0.0 400.0 0.0   15.0) cone
(TX "Y"   0.0 420.0    0.0 30 90) Y-axis label
(   0.0 0.0   0.0    0.0 0.0 500.0) line
(CO 0.0 0.0 425.0    0.0 0.0 500.0   15.0) cone
(TX "Z" 0.0    0.0 520.0 30 0) Z-axis label
(CL "WHITE") back to regular color
```

File **F3_31WCS2.M3D** serves a similar function, that is, to generate the global reference frame *OXY* with the origin in the middle of the axle beam, *OX* to the right, *OY* longitudinally backward, and *OZ* vertically up, but using arrows made of cones and cylinders rather than cones and lines. In this other version, the reference frame will remain visible following the **AutoCAD** *render* command (see Figure 3.32b).

```
(";")------------------------------------------------- ------------------
(";") M3D command file name:   F3_31WCS2.M3D
(";")
(";") Draw the WCS using arrow entities (cones and cylinders)
(";")------------------------------------------------------------
("WCS") change layer to "WCS"
(CL "BLUE") change color
(AR 0.0 0.0 0.0   900.0 0.0 0.0  75.0 5.0) WCS x-axis
(TX "X" 920.0 0.0 0.0   30   0.0) WCS x-axis label
(AR 0.0 0.0 0.0   0.0 400.0 0.0 75.0 5.0)   WCS y-axis
(TX "Y" 0.0 420.0   0.0 30 90) WCS y-axis label
(AR 0.0 0.0 0.0   0.0 0.0 500.0 75.0 5.0) WCS z-axis
(TX "Z" 0.0    0.0 520.0 30 0) WCS Z-axis label
(CL "WHITE") back to regular color
```

That animation file **F3_32B.GIF** is a rendered view of the tractor model, rather than the result of the *hide* command. In this case, its frames have been generated manually by copying to the clipboard the active **AutoCAD** screen using the <Alt> + <Prnt Scrn> keys. The advantage of drawing arrows using cones and slender cylinders rather than cones and plain lines became apparent in this case.

***

The plotting procedures in unit **LibPlots** have been introduced, along with the computer programs **Util~TXT** for manipulating ASCII files and **Util~DXF** and **Util~PLT** for viewing **R12 DXF** files. **AutoLISP** applications **G_3D.LSP** and **M_3D.LSP** allow true 3D entities to be generated with data read from file. **G_3D.LSP** generates 2D and 3D curves and 3D surfaces , either sized 1:1, or scaled to fit a plot box. In turn, **M_3D.LSP** can automatically generate cylinders, cones, spheres, tori, and **AutoCAD** blocks and can also animate them if they are placed in successively numbered layers. Further examples of the use of these programs and procedures are available in the remainder of the book.

REFERENCES AND FURTHER READINGS

Alciatore, D. G. and Histand, M. B. (2007). *Introduction to Mechatronics and Measurement Systems*. New York: McGraw Hill.

Hertz, D. L. Jr. (December 1991). An analysis of rubber under strain from an engineering perspective. *Elastomerics*, 1991. (also available from www.sealseastern.com/pdf/rubberunderstrain.pdf).

Simionescu, P. A. and Talpasanu, I. (2007). Synthesis and analysis of the steering system of an adjustable tread-width four-wheel tractor. *Mechanism and Machine Theory*, 42(5), 526–540.

Zecher, J. E. (1993). *Computer Graphics for CAD/CAM Systems*. New York: Marcel Dekker/Taylor & Francis.

For vibration of single degree of freedom systems, see

Meriam, J. L. and Kraige, L. G. (2006). *Engineering Mechanics: Dynamics*. Hoboken, NJ: Wiley.

Rao, S. S. (2010). *Mechanical Vibrations*. Upper Saddle River, NJ: Prentice Hall.

For numerical differentiation and integration, see

Kreyszig, E. (2006). *Advanced Engineering Mathematics*. New York: Wiley.

Press, W. H., Flannery, B. P., Teukolsky, S. A., and Vetterling, W.T. (1989). *Numerical Recipes in Pascal: The Art of Scientific Computing*. Cambridge, U.K.: Cambridge University Press.

For more information on **AutoCAD** release 12 **DXF** format (i.e., **R12 DXF**), see

http://paulbourke.net/dataformats/dxf/.

www.mediatel.lu/workshop/graphic/3D_fileformat/h_dxf12.html.

For more information on the **HP-GL** (**Hewlett-Packard Graphics Language**) format, see

www.sxlist.com/techref/language/hpgl.htm.

To download a free **DXF** viewer, go to

www.bravaviewer.com/viewers.

www.edrawingsviewer.com.

CHAPTER 4

Root Finding and Minimization or Maximization of Functions

THERE IS A BROAD ARRAY OF ALGORITHMS available for finding roots or extrema of functions, for multicriteria optimization, or for solving sets of equations (either linear or nonlinear). They differ greatly by their ease of implementation, robustness, and speed. Usually, these three characteristics do not go hand in hand, that is, a robust algorithm will be slow, while one that is fast but less robust will require additional preparation effort, like providing information about the derivative, or a good initial guess of the solution to be found. It is considered robust an algorithm that converges even for badly chosen initial conditions or parameter settings, while a fast algorithm will converge after fewer numbers of iterations or function calls—a characteristic desirable particularly if each function evaluation takes significant computational effort or when the algorithm is used in real-time applications. A number of such algorithms will be discussed in this chapter, including a new evolutionary algorithm for exploring the boundary of the feasible space in constrained optimization problems.

4.1 BRENT'S *ZERO* ALGORITHM FOR ROOT FINDING OF NONLINEAR EQUATIONS

A popular algorithm for root finding of functions of one variable that does not require information about their derivatives is the *Zero* algorithm due to Brent (1973). It combines *root bracketing*, *bisection method*, and *inverse quadratic interpolation* to converge within an interval [**a**, **b**] that contains a root of the function. While the details of the algorithm will not be presented here, three different implementations available in unit **LibMath** as procedures **Zero**, **ZeroStart**, and **ZeroGrid** will be discussed, based on the problem of finding the roots of two of the functions considered earlier in Chapter 1, and renamed here

$$F_1(x) = \frac{1}{(x-1)^2 + 0.1} + \frac{1}{(x-3)^2 + 0.2} - 3 \qquad (4.1)$$

and

$$F_2(x) = \frac{x(x^2 - 3)}{x^2 - 4} \qquad (4.2)$$

As Figures 4.1 and 4.2 show, function F_1 has four roots, while F_2 has three roots, approximated with some accuracy by the **D_2D** program and visible on the two graphs. Here, we will investigate the problem of finding better approximations of these roots using Brent's *Zero* algorithm, while in the next paragraph, the minima and maxima of the same two functions will be evaluated numerically with increased precision over the values visible in Figures 4.1 and 4.2.

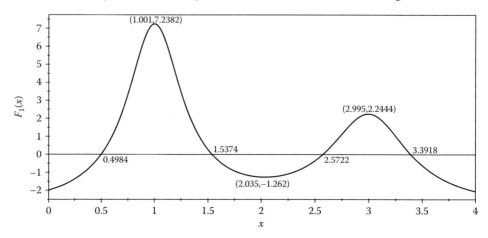

FIGURE 4.1   Plot of the function in Equation 4.1. Configuration file: **F4_01.CF2**.

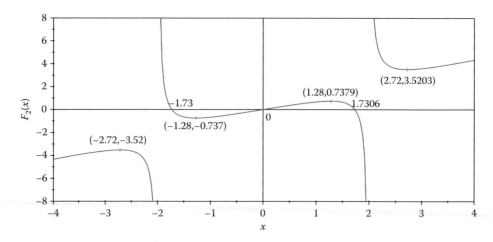

FIGURE 4.2   Plot of the function in Equation 4.2. Configuration file **F4_02.CF2**.

Program **P4_01.PAS** listed in Appendix B calls procedure **Zero** from unit **LibMath** with the following arguments: the function name **F1**, the lower and upper limits of an interval [**a, b**] known to contain a root of the function, and variable **x** where the root will be stored. Note that function **F1** was compiled under the **Force Far Calls** directive by placing its name between switches **{$F+}** and **{$F-}**. The results returned by the program are:

```
x = 4.98525413059701E-0001    F1(x)= 2.62810606610486E-0016
Function calls=11
```

Important: The **Zero** procedure counts the number of function evaluations and stores it in the Interface variable **NrFev0**. If **NrFev0** exceeds **LimFev0** (currently set to 10,000), then the program will terminate without warning. It is therefore advisable to inspect, after the search is over, the value of the function-call counter **NrFev0** as a way of verifying if the algorithm stopped before reaching a root. Evaluating the function at the returned solution **x** (which should be very close to zero) is another way of verifying that the algorithm converged.

The second program named **P4_02.PAS** and listed in Appendix B illustrates the use of a variant of the *Zero* algorithm named **ZeroStart**, which can find the root closest to a given point (or initial guess) that must be assigned to variable **x** prior to calling **ZeroStart**. In addition to the initial guess **x**, the user must specify the size of the constant steps (stored in variable **Step**) that the algorithm will take to inspect the function to the left and to the right of the initial guess. If this step size is too small, there will be too many function evaluations performed in the process of bracketing a root. Conversely, if the step size is too big and the function is multimodal, the nearest root can be missed during the bracketing process.

In the example considered, the root of function F_2 closest to point 1.0 is to be found. Note that the search to the right of the initial guess must go uphill first (i.e., increasing from zero) before it reaches the desired root, that is, **1.732051**. The search to the left of the initial guess results in root 0.0 that will be discarded, however, because it is farther from 1.0 than 1.732051. You can verify that for an initial guess outside the interval [−2, 2] and for the same step size of magnitude 0.1, the **ZeroStart** procedure will not return a valid solution because it is unable to leap over the singular points at −2 and +2. The results output by program **P4_02.PAS** are:

```
x= 1.73205080756888E+0000    F2(x)= 6.01949046163952E-0016
Function calls=37
```

Program **P4_03.PAS** in Appendix B calls procedure **ZeroGrid** that can return up to 52 roots of a given function within the specified interval [**a, b**]. If there are fewer than 52 roots found, the remaining components of vector **X** will be assigned the constant **InfD** equal to 1.0E100 and defined in unit **LibMath**. A grid size specified by the user (set to 25 in the program—see lines #**24** and #**33**) is used to partition the interval [**a, b**] and bracket the zeros of the function over the respective grid. The results are indeed close to the values visible on the graphs in Figures 4.1 and 4.2:

```
x1= 4.98525413059701E-0001    F1(x1)=-1.94289029309402E-0016
x2= 1.53732723677864E+0000    F1(x2)=-7.26632295999785E-0016
x3= 2.57225426015191E+0000    F1(x3)=-5.87203896618149E-0016
x4= 3.39189309000975E+0000    F1(x4)= 9.21571846612679E-0016
x5= 1.00000000000000E+0100    F1(x5)=-3.00000000000000E+0000
Function calls=62
```

```
x1=-1.73205080756888E+0000    F2(x1)=-6.01949046163952E-0016
x2= 0.00000000000000E+0000    F2(x2)=-0.00000000000000E+0000
x3= 1.73205080756888E+0000    F2(x3)= 6.01949046163952E-0016
x4= 1.00000000000000E+0100    F2(x4)= 1.00000000000000E+0100
x5= 1.00000000000000E+0100    F2(x5)= 1.00000000000000E+0100
Function calls=363
```

4.2 BRENT'S METHOD FOR MINIMIZING FUNCTIONS OF ONE VARIABLE

Equally useful to root finding is the problem of determining the minimum and maximum points of functions of one variable. Minimization and maximization in any dimension are related since the minima of $f(x)$ are the maxima of $-f(x)$. Here, we will consider the problem of finding the minimum points of a scalar function of one variable using Brent's method, which is a relatively fast algorithm that does not require derivative information. In searching for a minimum over a given interval, depending on the local behavior of the function, Brent's method switches between a more robust but slow *golden section search* and a faster *parabolic interpolation* minimization (Brent 1973). Three implementations of Brent's method gathered in unit **LibMin1** will be considered, that is, procedures **Brent**, **BrentStart**, and **BrentGrid**, in conjunction with finding the minima and maxima of F_1 and F_2 in Equations 4.1 and 4.2.

Program **P4_04.PAS** in Appendix B calls procedure **Brent** to minimize function F_1 over the interval [1, 3]. It returns the minimum of the function **vF** and its abscissa **x**. Convergence is controlled by constants **Tol** = 10^{-16} and **Eps** = 10^{-16} and by the function-call counter **NrFev1** = 50,000 (i.e., the search will stop if **NrFev1** exceeds **LimFev1**). These are default values set in the implementation section of unit **LibMin1**. With these settings, the results obtained are:

```
x= 2.03510727084125E+0000    F1(x)=-1.26219569494521E+0000
Obj. function calls=41
```

The companion program **P4_05.PAS** in Appendix B employs procedure **BrentStart** that searches around an initial guess **x** provided by the user until a minimum is bracketed, and then procedure **Brent** is called to accurately locate the minimum point. At the beginning, the search inside procedure **BrentStart** moves at a constant step in the descending direction of the function, until either **NrFev1** exceeds half of the maximum allowed number of function calls **LimFev1**, or a minimum is bracketed. Then, the search is performed in the opposite direction starting from the same initial **x** value, until another minimum is bracketed, or the search reaches a point farther away than the previously found minimum.

In both situations, if a minimum is bracketed, **Brent** procedure is called to refine the search. The results returned by program **P4_05.PAS** are:

```
x= 2.03510727426347E+0000    F1(x)=-1.26219569494521E+0000
Obj. function calls=25163
```

Note that in case of function F_1 and initial point at 0.75, the algorithm was able to reach the minimum at 2.035107 that required going uphill first, although the majority of function calls were spent uselessly moving downhill, to the left of the starting point $x = 0.75$.

The third program in the minimization series named **P4_06.PAS** (see Appendix B) uses the **BrentGrid** procedure to isolate all minima within the interval [**a**, **b**]. It is called with functions F_1, F_2 and their negatives so that all their minima and maxima are located. The minimum bracketing scheme implemented in the **BrentGrid** procedure is relatively simple, similar to the way extrema are identified by **D_2D** (see Figures 4.1 and 4.2), that is, the interval [**a**, **b**] is divided into **Npts** equally spaced points and the function value at each of these points is compared with its neighbors. If the middle point is lower than its neighbors, then a minimum is bracketed. Once a minimum is bracketed, **Brent** procedure introduced earlier is called to finish the search. In Press et al. (1989), an adaptive step-size minimum bracketing procedure is described, which is more effective than the one discussed here. The results returned by the program **P4_06.PAS** are:

```
x1= 2.03510727427766E+0000    F1(x1)=-1.26219569494521E+0000
Obj. function calls=57

x1= 1.00113562665303E+0000    F1(x1)= 7.23822399222707E+0000
x2= 2.99520656838562E+0000    F1(x2)= 2.24447267408327E+0000
Obj. function calls=93

x1=-2.00000000000000E+0000    F2(x1)=-1.12589990684263E+0015
x2=-1.27582078565695E+0000    F2(x2)=-7.38017459656381E-0001
x3= 2.00000000000000E+0000    F2(x3)=-1.12589990684262E+0015
x4= 2.71519452607298E+0000    F2(x4)= 3.52034518609217E+0000
Obj. function calls=245

x1=-2.71519452734705E+0000    F2(x1)=-3.52034518609217E+0000
x2=-2.00000000000000E+0000    F2(x2)= 1.12589990684262E+0015
x3= 1.27582078559014E+0000    F2(x3)= 7.38017459656381E-0001
x4= 2.00000000000000E+0000    F2(x4)= 1.12589990684263E+0015
Obj. function calls=243
```

Note that the **BrentGrid** procedure was able to locate all minima and maxima of F_1 and F_2 after a reasonably small number of function calls. These include the $\pm\infty$ asymptotes of F_2 (approximated with values in the range of $\pm 10^{15}$) occurring at points $x = -2$ and $x = 2$.

4.3 NELDER–MEAD ALGORITHM FOR MULTIVARIATE FUNCTION MINIMIZATION

In this and the following sections, the problem of minimizing a scalar function of two or more variables is discussed. This is required in many instances, like curve fitting to data and solving sets of equations (linear or nonlinear). Both unconstrained functions, and the more commonly encountered in practice case where the function to be minimized is subjected to constraints, are considered.

One popular direct search method of function minimization is the *Simplex* method due to Nelder and Mead. It is called direct because the algorithm does not use the derivative of the function. A version of this algorithm is implemented under the name *fminsearch* in **MATLAB** and is also available in Press et al. (1989) as procedure *Amoeba*.

The Nelder–Mead Method uses the concept of a *Simplex*, which is a geometrical figure having $n + 1$ vertices when the search occurs in an n-dimensional space, for example, when minimizing a function of two variables the simplex is a triangle and when the function has three variables the simplex is a tetrahedron. The search begins with an initial nondegenerate simplex that is transformed such that the vertex with the highest function value (the worst vertex) is eliminated. These transformations are (i) *reflection* of the simplex away from the worst vertex; (ii) in case a reflection turned the highest point into a new lowest point, an *expansion* in the same direction will immediately be performed; (iii) *contraction* of the simplex along one direction away from the highest vertex; and (iv) *shrinkage* of the simplex along n directions towards the lowest vertex. Transformations (i), (ii), and (iii) are done relative to the centroid of the n best vertices. For example, the first search step recorded in Figure 4.3a and d is a *reflection*, the first step in Figure 4.3c is a *contraction*, and the first step in Figure 4.3d is a *shrinkage*. See also the animated **GIF**s that accompany these figures for a better understanding of how the simplex morphs as it advances towards a minimum point. The searches recorded in Figure 4.3 have as objective finding the minima and maxima of the *four-hump function* introduced in Chapter 2 are restated here as:

$$F_3(x,y) = 10\left[e^{-(2x+1)^2 - (y+1)^2} - e^{-(x-1)^2 - (y+1)^2} \right]^2 + 15\left[e^{-(x-1)^2 - (y-1)^2} - e^{-(2x+1)^2 - (2y-1)^2} \right]^2 \quad (4.3)$$

Possible stopping criteria of the algorithm are (i) exceeding a certain number of function calls or iterations, (ii) the volume of the simplex becomes too small, or (iii) there is not much difference in the function value between the best and the worst vertices of the simplex.

To illustrate how the Nelder–Mead *Simplex* algorithm works, program **P4_07.PAS** has been written and is available with the book. It minimizes function F_3 in Equation 4.3 using the **NelderMead** procedure called from unit **LibMinN**. As the search progresses, **P4_07.PAS** writes the coordinates of the simplex to an ASCII file as **D_2D** animation frames, with the initial coordinates being randomly generated within the given lower and upper bounds $-1.5 \le x \le 2.5$ and $-2.5 \le y \le 2.5$ (see vector variables **Xmin** and **Xmax**).

Note that there may be instances where the simplex will diverge away from an optimum or may stagnate around a saddle point or in a valley. In such cases, it is useful to restart the search, either with a totally new simplex or with one that has some of the highest vertices replaced with new values, possibly randomly generated.

To provide you with a more refined function-minimization tool, the following features have been included in procedure **NelderMead**: (i) The simplex is forced to remain within

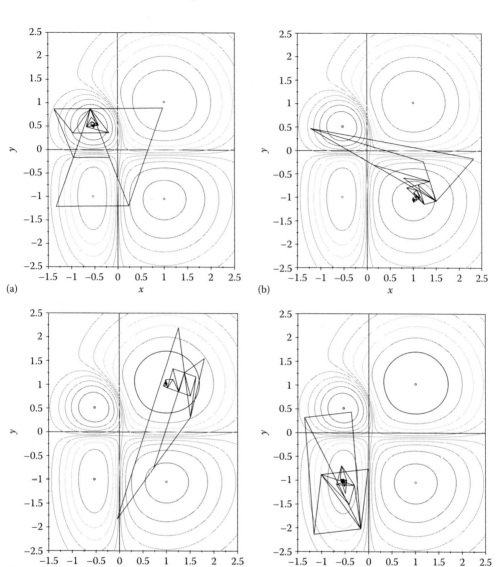

FIGURE 4.3   Histories of Nelder–Mead simplex search with the simplex converging to the two minima (a and b) and the two maxima (c and d) of the function in Equation 4.3. Configuration files **F4_03A.CF2** through **F4_03D.CF2**. See also animations **F4_03a.GIF** through **F4_03d.GIF**.

the interval **XXmin...XXmax**. (ii) The possibility of specifying a starting point that is assigned to one vertex of the simplex, while the coordinates of the remaining vertices are randomly generated. If no initial guess is available, the entire initial simplex will be randomly generated. (iii) The possibility of reading all or part of the vertices of the initial simplex from an ASCII file. If the initial simplex read from file is incomplete, its remaining vertices will be randomly generated. (iv) The possibility of pausing the search by pressing the <Esc> key and allowing the user to inspect the current solution; the user can either accept the current solution or resume the search until one of the programmed stopping conditions is attained.

Program **P4_08.PAS** in Appendix B calls procedure **NelderMead** with the initial simplexes read from ASCII files **P4_08-1.SPX** and **P4_08-2.SPX**. These were chosen such that the search will advance to the local minima (Figure 4.4a) and local maxima (Figure 4.4b) of the function. Note that the search in Figure 4.4a begins with a contraction of the simplex, while the search in Figure 4.4a begins with a shrinkage of the simplex. Regarding the level curves in Figures 4.3 and 4.4, these are read by **D_2D** from ASCII file **F3.XY**, which is a duplicate of the file **F3_18B.XY** in Chapter 3.

The results output by **P4_08.PAS** are (variable **PlsMns** is assigned on line #30 of the program):

```
For PlsMns= +1, the minimum of the function is obtained as
x1 = 1.00075122494239E+0000
x2 =-1.04666091613729E+0000
F_opt=-9.74956322430359E+0000
Obj. function calls=137

For PlsMns= -1, the maximum of the function is obtained as
x1 =-5.35496066732408E-0001
x2 =-9.95424083908040E-0001
F_opt= 9.02774166941257E+0000
Obj. function calls=120
```

In order to locate the global minimum and global maximum of function F_3, a different strategy is proposed in program **P4_09.PAS** (see Appendix B). The search is

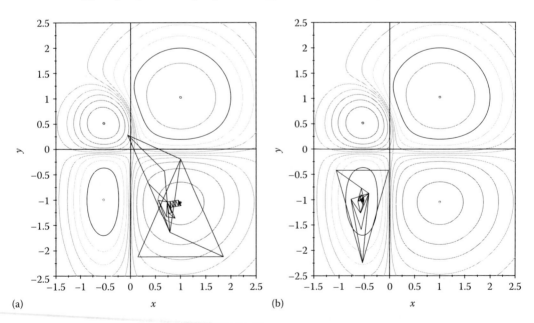

(a) (b)

FIGURE 4.4   Histories of Nelder–Mead *simplex* searches done with program **F4_08.PAS**, converging to the local minima (a) and local maxima (b) of the function. Configuration files: **F4_04A.CF2** and **F4_04B.CF2**. See also animations **F4_04a.GIF** and **F4_04b.GIF**.

repeated several times with new initial simplexes that were randomly generated, and the best solution found among these separate searches is retained and displayed at the end of the run.

The results returned by this new program are (constant **PlsMns** is defined on line **#10** of the program—see Appendix B):

```
For PlsMns= +1, the results are
x1 =-5.28706765640623E-0001
x2 = 5.13746292321373E-0001
F_opt=-1.28850039129209E+0001
Obj. function calls=11913

For PlsMns= -1, the results are
x1 = 1.00024221604051E+0000
x2 = 1.02264234759275E+0000
F_opt= 1.48245012784894E+0001
Obj. function calls=11632
```

On rare occasions, program **P4_09.PAS** may crash if a large value is transmitted to the exponential functions inside **Fn** (i.e., a number too large for Pascal to handle may be generated, and *floating point overflow* will occur—see next section on constraint handling).

Important: By setting the variable **WriteOutN** in procedure **NelderMead** to the logical value TRUE, the user can stop the search and inspect the best solution found so far, with the possibility of retaining this solution and halting the program or resuming the search.

4.4 HANDLING CONSTRAINTS IN OPTIMIZATION PROBLEMS

Most real-world problems require finding minimum or maximum of functions while simultaneously satisfying a number of constraints. The most common of these, called *side constraints*, are boundaries imposed to the design variables, that is, $x_{i\,\min} \le x_i \le x_{i\,\max}$. Additional relationships between some or all variables of the objective function may also be imposed, both as inequalities and as equalities. Because of their more frequent encounter and the more convenient handling, only the case of inequality constraints will be considered here.

Let us assume the problem of minimizing a function introduced earlier (i.e., the *hyperbolic paraboloid* in Figure 2.28):

$$F_4(x,y) = 0.1xy \tag{4.4a}$$

subjected to the following constraint:

$$x^2 + y^2 \le \left[r_T + r_S \cos\left(n \cdot \arctan\left(\frac{x}{y} \right) \right) \right]^2 \tag{4.4b}$$

with $r_T = 1$, $r_S = 0.2$, and $n = 6$. This is equivalent to forcing the search for an optimum point within the closed contour delimited by parametric equations:

$$\begin{cases} x(\theta) = \left[r_T + r_S \cos(n\theta) \right] \cdot \cos(\theta) \\ y(\theta) = \left[r_T + r_S \cos(n\theta) \right] \cdot \sin(\theta) \end{cases} \tag{4.5}$$

where $0 \leq \theta \leq 2\pi$.

Due to the symmetry of both the function and its constraint, there will be two distinct but equal minima (Figure 4.5a). These are located on the boundary between the feasible and infeasible spaces, where the closed curve of Equation 4.5 intersects the second diagonal. Similar observations can be made about the two maxima of the constrained function, which are mirror of the minimum points and are located where the first diagonal intersects the boundary of the feasible space.

The level-curve plot in Figure 4.5a has been generated with **D_3D** using the file **F4_05A. D2D** with 501 × 501 data points, produced by the program **F4_05A.PAS** (see listing in Appendix B).

For solving of the optimization problem defined by Equations 4.4a and b, program **P4_10.PAS** has been written and is listed in Appendix B. To account for the imposed constraint, an easy to implement version of the *penalty function* method was adopted, where function **Fn** returns a very large value (i.e., **InfD** defined in unit **LibMath**) if the constraint is not satisfied. This simple approach was found to work well in many instances, being in addition very convenient to program. In case of functions of two variables, the added benefit of this constraint handling approach is that there is no difference between programming the function for minimum finding purposes, and generating the data for

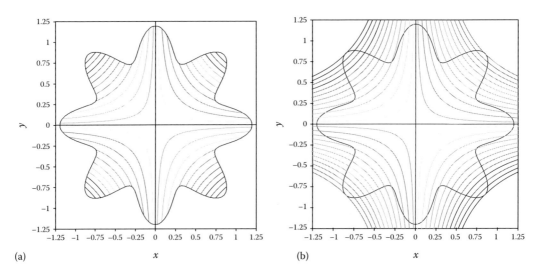

(a)   (b)

FIGURE 4.5 Plot of function 4.4a constrained by inequality 4.4b (a) and how file **F4.XY** has been generated as the overlap and trim inside **AutoCAD** of the function in Equation 4.4a and the parametric curve in Equation 4.5 (b). Configuration files to redo these plots: **F4_05A.CF3** and **F4_05B.CF3** and **F4_05B.CF2**.

plotting it using **D_3D** (see program **F4_05A.PAS** in Appendix B). Remember however that this form of the penalty method is not suitable for handling equality constraints.

Note on line #**9** of program **P4_10.PAS** the definition of constant **PlsMns** that can be assigned either value +1 in case minimization is performed or –1 in case of maximization. This constant multiplies the feasible values of objective function **Fn** (see line #**27**).

Also note on line #**31** the use of variable **WriteOutN** defined in **Interface** section of unit **LibMinN**, which is assigned the logical value FALSE. This will turn off the search status, which means that it will not be possible to pause the search and inspect the best solution found so far by the procedure **NelderMead**.

Because of the mentioned symmetry of both the function and its constraint, identifying any of the minimum or maximum points of the function allows the other extrema to be verified through exact calculations. Their values are as follows:

```
x = 0.84852813;  y =-0.84852813;  F_opt=-0.72;
x =-0.84852813;  y = 0.84852813;  F_opt=-0.72;
x = 0.84852813;  y =-0.84852813;  F_opt= 0.72;
x =-0.84852813;  y =-0.84852813;  F_opt= 0.72;
```

The search history in Figure 4.6 and the accompanying animated **GIF** file **F4_06.GIF** have been generated with program **P4_11.PAS** (listing not included, but available with the book), which has a structure similar to **P4_07.PAS** used to produce Figures 4.3 and 4.4.

To reduce the size of the data file used by **D_2D** to plot the background when animating Figure 4.6 and to increase the resolution at which the boundary of the feasible space is plotted, the following procedure has been implemented. Firstly, a low-resolution level-curve plot of function F_4 has been generated using **D_3D** and then was exported to **DXF**. The boundary of the feasible space was created separately by plotting the parametric curve in Equation 4.5 using the data file output by program **F4_05B. PAS** (see Appendix B) and in turn was exported to **DXF**. The two **DXF** files were opened

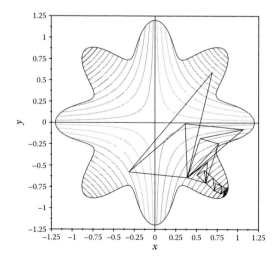

FIGURE 4.6 History of Nelder–Mead *simplex* searches according to program **P4_11.PAS**. Configuration file **F4_06.CF2**. See also animated **GIF** file **F4_06.GIF**.

in separate **DWG** files and then were overlapped as visible in Figure 4.5b. Finally, the level curves were trimmed to look as shown in Figure 4.5a. The trimmed level curves of the constrained function F_4 were exported from **AutoCAD** to **R12 DXF**. Using the **Util~DXF** program, the vertices of these polylines were finally exported to ASCII file **F4.XY**. This smaller ASCII file was ultimately used to plot the background curves in Figure 4.6 that show the Nelder–Mead *simplex* search histories.

A second constrained optimization problem considered is that of a speed reducer design, translated by Li and Papalambros (1985) into minimizing the following objective function:

$$F_5(x_1 \ldots x_7) = 0.7845 x_1 r_2^2 \left(3.3333 r_2^2 + 14.9334 x_3 - 43.0934\right)$$
$$-1.508 x_1 \left(x_2^2 + x_7^2\right) + 7.477 x_1 \left(x_6^3 + x_7^3\right) + 0.7854 \left(x_4 x_6^2 + x_5 x_7^2\right) \quad (4.6)$$

subjected to side constraints:

$$
\begin{array}{ll}
2.6 \le x_1 \le 3.6 & 0.7 \le x_2 \le 0.8 \\
17 \le x_3 \le 28 & 7.3 \le x_4 \le 8.3 \\
7.3 \le x_5 \le 8.3 & 2.9 \le x_6 \le 3.9 \\
5.0 \le x_7 \le 5.5
\end{array}
\qquad (4.7)
$$

Program **P4_12.PAS** listed in Appendix B implements a multistart solution method to this optimization problem, which uses procedure **NelderMead** called from unit **LibMinN**. Note that the initial guess is updated after each trial by adding a random perturbation to the previously calculated optimum (see lines **#52** to **#54**) and by truncating out of the decimals of the previous search result **XX** (line **#59**—note the use of procedures **MyVal** and **MySt** called from unit **LibInOut**). Truncating out of decimals also helps with reporting the search results, because fewer number of significant digits are retained by the user. Also note the use of the **BackUpFile** command on line **#67** that changes the extension of ASCII file **Results** from **TXT** to **OLD**, so that the results obtained previously are not immediately lost.

After several runs of program **P4_12.PAS**, one of the best solutions found was

```
Obj. function calls=6008446
F_opt= 2.35245309676356E+0003
x1= 2.60000000000000E+0000
x2= 7.00000000000000E-0001
x3= 1.70000000000000E+0001
x4= 7.30020000000000E+0000
x5= 7.30020000000000E+0000
x6= 2.90000000000000E+0000
x7= 5.00000000000000E+0000
```

As the above results indicate, the global minimum of F_5 occurs for the lowest possible values of x_1 through x_7 and it equals **2.35244784872076E+3**. This suggests that the optimum is bounded, that is, all constraints are active at the minimum point.

4.5 EVOLUTIONARY ALGORITHM FOR BOUNDED-OPTIMUM SEARCH

Optimization algorithms, like that of Nelder and Mead discussed earlier, do not always return the global minima, due to either the multimodal or noisy behavior of the objective functions, or the form of its constraints. Such problems can be better handled using *evolutionary algorithms*, which employ mechanisms inspired by biological evolution, that is, *reproduction, mutation, recombination, selection*, and *survival of the fittest* applied to a population of solution. The individuals in this population are points in the design space that are evolved using the mechanisms listed earlier, such that the function value (also known as fitness) at these points is improved. There is a wealth of literature, including numerous online resources, which those less familiar to the subject of *evolutionary computation* may want to consult.

Here, a novel two-population *evolutionary algorithm* will be presented, which has the ability to explore the boundary between the feasible and the infeasible spaces of objective function. In many practical problems, like it was the case of the speed reducer of Li and Papalambros (1985) considered earlier, the optimum is located right on the boundary of the feasible space. A number of implementations of this new *female–male evolutionary algorithm* as it was called are discussed in Simionescu et al. (2006). These implementations will be abbreviated F–M(μF, μM), where μF is the size of the female population (the feasible individuals) and μM is the size of the male population (the infeasible individuals).

The main steps of a generic *female–male evolutionary algorithm* are as follows:

Step 1: Generate an initial female population of μF individuals and an initial male population of μM individuals as uniform random points within the extended intervals:

$$\left[x_{i\ min} - k_{ext} \cdot \left(x_{i\ max} - x_{i\ min} \right), \ x_{i\ max} + k_{ext} \cdot \left(x_{i\ max} - x_{i\ min} \right) \right] \tag{4.8}$$

with $1 \leq i \leq n$ and n as the number of variables of the objective function. In this equation, coefficient k_{ext} with values greater-equal zero, assigned by the user, controls the amount with which the infeasible space is expanded, so that an initial male population can be created and evolved. This is particularly important when only side constraints are imposed in an optimization problem. If additional constraints are present, the infeasible region may be sufficiently large and coefficient k_{ext} can be set to a smaller value, including zero. Evidently, when the objective function is evaluated, the side constraints are verified as they were posed in the original problem.

Step 2: Rank females based on their fitness using complete or partial sorting, or just identify the best-fit female (the α-female).

Step 3: Mutate females by replacing a fraction **RepF** of the lowest ranked females with randomly generated new ones.

Step 4: Mutate males by replacing a fraction **RepM** of their population with randomly generated new males.

Step 5: (Crossover) Form female–male pairs by assigning one male to each female based on their closeness in the n-dimensional Euclidean space. Begin with the α-female and continue in a rank-decreasing order until all available males are assigned to a female. In other implementations called *polygamous-males algorithms*, males are permitted to recombine with more than one female. Also possible is to do *multifemale crossover*, which can be *unrestricted* (i.e., a male can recombine with any number of females in one generation), *restricted* (when the number of crossovers a male can perform is limited to a fraction of the total female population), or *monogamous* (a male cannot recombine during the same generation more than once). After female–male pairs are formed, offspring are generated using midpoint or random crossover. Offspring can be females (if they result inside the feasible space) or can be males (if they result outside the feasible space).

Step 6: (Selection) The selection step is performed concomitant with offspring generation as follows: if the child results outside the feasible space, he replaces his father unconditionally; if the child is a female, she replaces her mother either unconditionally or only if there is an improvement in fitness.

Stopping criteria: Steps 2 through 6 are repeated until an imposed condition is satisfied, that is, either exceeding a maximum number of function calls or generations, attaining an imposed threshold fitness, or recording the same α–female over a given number of generations.

Program **P4_13.PAS** (source code available with the book) is an implementation of a *monogamous* version of the algorithm, with µF females and µM males, or F–M(µF, µM) in short. The F–M(µF, µM) algorithm minimizes function F_4 in Equation 4.4a subjected to constraints 4.4b. If you set the variable **NrTrials** equal to one, the program also writes to ASCII file **F4-FmM.POP** the female and male individuals, together with the marker type, color information, and animation-frame separators.

Using **D_2D** with ASCII files **F4.XY** and **F4-FmM.POP** as inputs (the former to plot as background the level curves in Figure 4.5a, and the latter to animate the female and male populations as they evolve), Figure 4.7 and animation file **F4_7-1.GIF** have been generated. **F4_7-1.GIF** uses the **PCF** frames exactly as they were generated by **D_2D**, while the frames in the companion file **F4_7-2.GIF** are the **BMP** screenshots output using **M_3D.LSP** based on the **DXF** file generated by **D_2D**, with each frame written to a separate layer—see also the **F4_7.DWG** drawing file available with the book.

The **F4-FmM.REZ** ASCII file produced by setting **NrTrials** to 1000 contains the results of 1000 search trials and was used to generate the plots in Figure 4.8. For the expansion coefficient k_{ext} = 0.04, population sizes µF = 4 and µM = 8, replacement rates **RepF** = 0.15 and **RepM** = 0.25, and threshold value −0.071, the success rate was around 94%. The success rate was measured as the number of solutions below the threshold value for a given maximum number of function calls.

The search report appended to the **F4-FmM.REZ** file included the following additional information:

```
Best = -0.0719999990
Worst= -0.0661771431
Avg. = -0.0717428861
Total function calls =   1002324
Average function calls =    1002
```

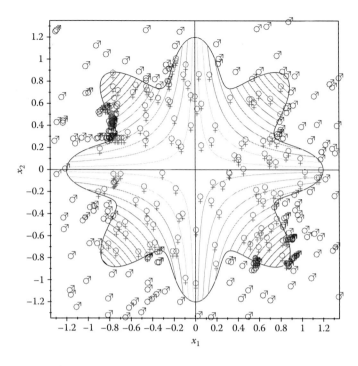

FIGURE 4.7 Overlap of the female and male populations in a sample run of **P4_13.PAS**. Configuration files **F4_07-1.CF3** and **F4_07-2.CF3**. See also animation files **F4_07-1.GIF** and **F4_07-2.GIF**.

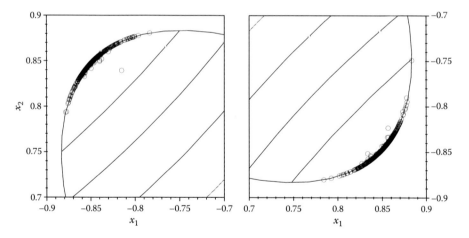

FIGURE 4.8 Overlap of the 1000 search results read from **F4-FmM.REZ**, detailed around the two minima in Figure 4.5a. Configuration files **F4_08A.CF3** and **F4_08B.CF3**.

As you experiment with **P4_13.PAS**, you will realize that the success rates differ depending on the sizes of the female and male populations, extension coefficient k_{ext}, crossover settings, and degree of sorting of the female population (complete, partial, or just α-female identification). You may want to try different female–male crossover schemes or multiparent recombinations, as well as to optimize functions of more than two variables, like F_5 in Equations 4.6 and 4.7.

4.6 MULTICRITERIA OPTIMIZATION PROBLEMS

The examples considered so far dealt with minimizing only one function at a time. There are practical problems where two or more objective functions must be minimized and/or maximized simultaneously in the presence of constraints. Such problems are called *multicriteria* or *multiple objective optimization problems*. Because the imposed objectives are in most cases conflicting, in a multicriteria problem, there is not one single solution, but rather a family of solutions called *Pareto set* or *Pareto frontier*. Simply put, a point in the design space is considered a Pareto solution to the problem (i.e., belongs to the Pareto frontier) if no single criterion (i.e., single objective function) can be improved without worsening at least one other criterion. In the remainder of this section, two bicriterion optimization problems in two variables will be considered, and some basic concepts related to multiobjective optimization will be discussed.

4.6.1 Cantilever Beam Design Example

Figure 4.9 shows a cantilever beam loaded with a down force F = 15,000 N applied at the free end. The beam is hallow and of imposed length L = 1000 mm. The outside diameters of the two sections have fixed values, that is, D_1 = 100 mm and D_2 = 80 mm. The material of the beam is assumed to have an elastic modulus E = 206·10³ N/mm and yield strength σ_Y = 220 N/mm.

The two variables allowed in this design are the length of the thinner section x_1 and the internal diameter of the beam x_2. The design problem is to simultaneously minimize the total volume of the beam $f_1(x_1, x_2)$ and the deflection at its free end $f_2(x_1, x_2)$:

$$f_1\left(x_1,\ x_2\right) = \frac{\pi}{4}\left[(L-x_1)\cdot(D_1^2-x_2^2)+x_1\cdot(D_2^2-x_2^2)\right] \tag{4.9}$$

$$f_2\left(x_1,x_2\right) = \frac{64F}{3\pi E}\left[\frac{L^3-x_1^3}{D_1^4-x_2^4}+\frac{x_1^3}{D_2^4-x_2^4}\right] \tag{4.10}$$

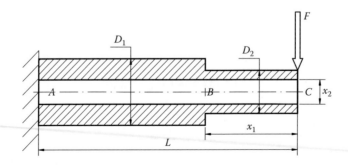

FIGURE 4.9 Staggered cantilevered beam in the bicriterion optimization problem.

Additionally, (i) the maximum bending stress at cross sections A and B should remain below the yield strength σ_Y of the material, which translates into the following inequality constraints (the stress raiser effect at cross section B due to the change in the diameter is ignored):

$$\sigma_{A\,max} = \frac{F \cdot L}{\pi(D_1^4 - x_2^4)/(32D_1)} \leq \sigma_Y \tag{4.11}$$

$$\sigma_{B\,max} = \frac{F \cdot x_1}{\pi(D_2^4 - x_2^4)/(32D_2)} \leq \sigma_Y \tag{4.12}$$

Further constraints imposed to this problem are (ii) length x_1 should be positive and less equal than L and (iii) the inner diameter x_2 of the beam should range between 40 and 75 mm:

$$x_1 \geq 0.0 \quad \text{and} \quad x_1 \leq L$$
$$x_2 \geq 40 \quad \text{and} \quad x_2 \leq 75 \tag{4.13}$$

4.6.2 Design Space and Performance Space Plots

Insight into a given multicriteria optimization problem can be gained by inspecting its *feasible space* and its *performance space* prior to optimization. The feasible space (also known as *space of the design variables* or *design space*) consists of the points that satisfy all the constraints. The *performance space* is a mapping of the feasible space points into the space of the objective functions. For problems of two variables, plotting the feasible space can be relatively easily done. Likewise, plotting the performance space in a bicriterion problem is also possible.

For more than two criteria or design variables, plotting the feasible space and performance space requires employing some *dimension reduction* method as discussed in several references listed at the end of this chapter.

Program **P4_14.PAS** in Appendix B has been written to generate the data required to represent graphically using **D_2D**, the design space and performance space of the cantilever beam problem introduced earlier. The design space $[x_{1min}...x_{1max}] \times [x_{2min}...x_{2max}]$ is divided into a **nx1** × **nx2** grid, and then the constraints and the two objective functions are evaluated at these nodes. If at a given grid point all constraints are satisfied, then the corresponding (x_1, x_2) pair is written to **F4_10A.D2D** file, while the corresponding $f_1(x_1, x_2)$ and $f_2(x_1, x_2)$ values are written to **F4_10B.D2D** file. The plots generated using these data files are visible in Figure 4.10a and b. The left boundary of the performance space in Figure 4.10a is exactly the Pareto frontier of the optimization problem. Since no information was retained about how a point (f_1, f_2) on the Pareto frontier is associated with a point (x_1, x_2) in the feasible space, the optimization problem is not yet solved.

Note in Figure 4.10a that to its far right the Pareto front exhibits a short horizontal section. The points along this horizontal section are called *week Pareto solutions* because criterion f_1 can be further improved while criterion f_2 remains unchanged.

Alternative to evaluating the design space over a regular grid when producing the data points for plotting the feasible and performance spaces, random points can be generated

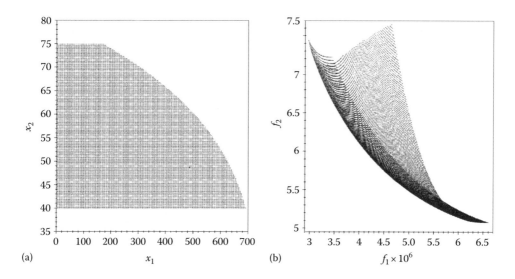

FIGURE 4.10 Plot of the feasible space (a) and performance space (b) of the cantilevered beam problem. Configuration files **F4_10A.CF2** and **F4_10B.CF2**.

within the same $[x_{min}...x_{max}]$ intervals. This second approach was implemented in program **P4_15.PAS** listed in Appendix B, which generates files **F4_12A.D2D** and **F4_12B.D2D**, used in plotting the feasible space and performance space of a second bicriterion optimization problem, that is, that of minimizing

$$f_1(x_1,\ x_2) = 0.4x_1 + x_2$$
$$f_2(x_1,\ x_2) = 1 + x_1^2 - x_2 + 0.2\cos(4.75x_2) \tag{4.14}$$

while satisfying the inequality

$$0.8x_1^2 + x_2^2 \le 1.0 \tag{4.15}$$

As expected, the design space is an ellipse centered at origin and of semiaxes 1.118 and 1.0 (see Figure 4.11a). Since function $f_2(x_1, x_2)$ is multimodal and therefore nonconvex, the problem itself is nonconvex and so is its performance space as Figure 4.11a shows. The fact that the performance space is nonconvex will render deterministic multicriteria optimization methods unable to identify the full range of the Pareto frontier (Osyczka 2002; Deb 2009).

4.6.3 Pareto Front Search

There are a number of algorithms available in literature for solving multicriteria optimization problems. Most of these work by combining the individual objective functions into a single function called *preference function*, which is minimized using known methods. Examples of preference functions are *weighted sum of the objectives*, *normed weighted sum of the objectives*, and *mini–max methods*.

Metaheuristics (like *evolutionary algorithms* and *simulated annealing*) permit obtaining a multitude of Pareto solutions to the problem in one run. These algorithms can also cope better with nonconvex problems. By contrast, in weighted sum of the objectives methods,

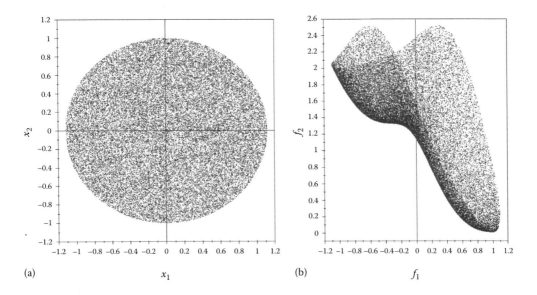

FIGURE 4.11   Plot of the design space (a) and performance space (b) of problem in Equations 4.14 and 4.15. Configuration files **F4_11A.CF2** and **F4_11B.CF2**.

the search must be repeated for several combinations of weighting coefficients, each run generating a separate point on the Pareto frontier. Moreover, for nonconvex problems like the one in Equations 4.14 and 4.15, only the convex regions of the Pareto frontier can be located by such an algorithm (Osyczka 2002).

Program **P4_16.PAS** listed in Appendix B finds the convex portions of the Pareto frontier of the second optimization problem earlier. It implements a *normed weighted sum of the objectives* with a preference function of the form

$$f_{12}(x_1,\ x_2) = w_1\,\frac{f_1(x_1,\ x_2) - f_{1\min}}{f_{1\max} - f_{1\min}} + w_2\,\frac{f_2(x_1,\ x_2) - f_{2\min}}{f_{2\max} - f_{2\min}} \tag{4.16}$$

In this equation, $f_{1\min}, f_{1\max}, f_{2\min}$, and $f_{2\max}$ are the global minima and maxima of functions $f_1(x_1, x_2)$ and $f_2(x_1, x_2)$, respectively, evaluated separately in the presence of constraint (4.15), while w_1 and w_2 are weighting coefficients where $w_1 + w_2 = 1$. These upper and lower limits of functions f_1 and f_2 are evaluated in program **P4_16.PAS** by executing the code between lines #35 and #52, while the Pareto frontier points are searched for and written to file by executing lines #54 to #76.

A plot of the Pareto frontier points overlapped with the feasible and performance spaces of this second bicriterion optimization problem is available in Figure 4.12. Note that indeed the nonconvex portion of the Pareto frontier has not been mapped in these diagrams. Similar plots of the Pareto frontier points overlapped with the feasible and performance spaces of the cantilever beam design problem are given in Figure 4.13. The Pareto front points were in this case determined using program **P4_17.PAS** listed in Appendix B.

★★★

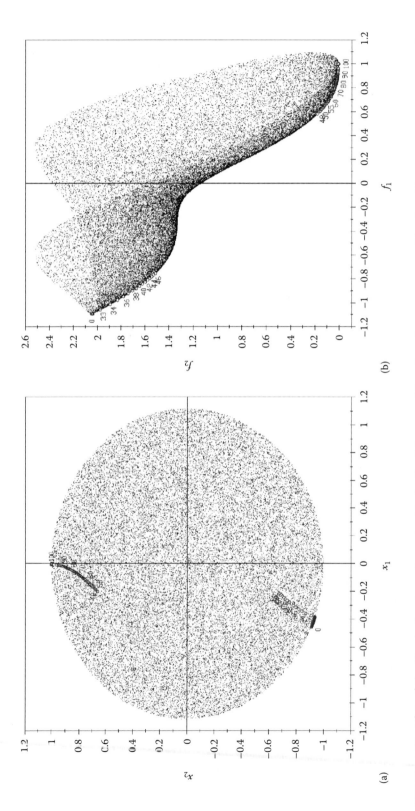

FIGURE 4.12 Plot of the Pareto front overlapped with the feasible space (a) and performance space (b) of problem in Equations 4.14 and 4.15. Configuration files **F4_11A.CF2**, **F4_12A.CF2**, **F4_11B.CF2**, and **F4_12B.CF2**.

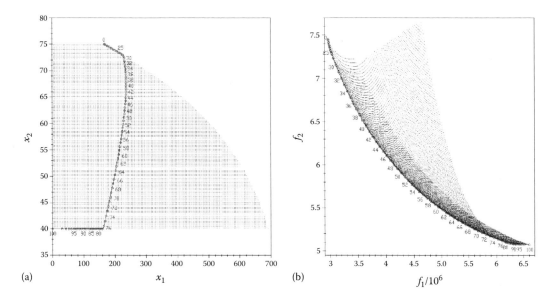

(a)   (b)

FIGURE 4.13   Plot of the Pareto front overlapped with the feasible space (a) and performance space (b) of the cantilevered beam optimization problem. Configuration files **F4_10A.CF2**, **F4_13A. CF2**, **F4_10B.CF2**, and **F4_13B.CF2**.

A number of procedures for finding the zeros and minimum or maximum of functions of one variable have been presented, all based on algorithms originally developed by Brent. For minimizing functions of more than one variable, Nelder–Mead algorithm and the corresponding procedure **NelderMead** were discussed. A two-population *evolutionary algorithm* capable of exploring the boundary between the feasible and infeasible regions of the design space has been further presented, together with illustrative animation graphs. At the end of the chapter, the *normed weighted sum of the objectives* method of bicriterion optimization problem solving, and two techniques for plotting design space and performance space in multicriteria optimization problems have been presented.

REFERENCES AND FURTHER READINGS

Brent, R. P. (1973). *Algorithms for Minimization without Derivatives*. Upper Saddle River, NJ: Prentice Hall.

Deb, K. (2009). *Multi-Objective Optimization Using Evolutionary Algorithms*. Hoboken, NJ: John Wiley & Sons.

Li, H. L. and Papalambros, P. (1985). A production system for use of global optimization knowledge. *ASME Journal of Mechanisms Transmission and Automation in Design*, 107(2), 277–284.

Osyczka, A. (2002). *Evolutionary Algorithms for Single and Multicriteria Optimization*. Heidelberg, Germany: Physica-Verlag/Springer.

Press, W. H., Flannery, B. P., Teukolsky, S. A., and Vetterling, W. T. (1989). *Numerical Recipes in Pascal: The Art of Scientific Computing*. Cambridge, U.K.: Cambridge University Press.

Simionescu, P. A., Dozier, G. V., and Wainwright, R. L. A two-population evolutionary algorithm for constrained optimization problems. *IEEE World Congress on Computational Intelligence*, Vancouver, British Columbia, Canada, July 16–21, 2006, pp. 1647–1653.

Singer, S. and Nelder, J. (2009). Nelder-Mead algorithm. *Scholarpedia*, 4(7): 2928.

For multicriteria optimization including visualization of Pareto frontiers see:

Agrawal, G., Lewis, K. E., Chugh, K., Huang, C.-H., Parashar, S., and Bloebaum, C. L. (2004). Intuitive visualization of Pareto frontier for multi-objective optimization in n-dimensional performance space. *Proceedings of the 10th AIAA/ISSMO Multidisciplinary Analysis and Optimization Conference*, AIAA-2004, Albany, New York, pp. 4434–4445.

Spears, W. M., De Jong, K. A., Bäck, T., Fogel, D. B., and de Garis, H. (1993). An overview of evolutionary computation. In *Machine Learning: ECML-93, Lecture Notes in Computer Science* (Brazdil, P.B. ed.), Berlin, Germany: Springer, Vol. 667, pp. 442–459.

For more information on evolutionary computation, in addition to the books by Osyczka and Deb listed earlier, see also

Eiben, A. E. and Smith, J. E. (2010). *Introduction to Evolutionary Computing*. Berlin, Germany: Springer.

Haupt, R. L. and Haupt, S. E. (2004). *Practical Genetic Algorithms*. Hoboken, NJ: John Wiley & Sons.

CHAPTER 5

Procedures for Motion Simulation of Planar Mechanical Systems

IN THIS CHAPTER, A number of procedures gathered in units **LibMecIn** and **LibMec2D** available with the book are presented as a preamble to Chapter 6 where the kinematic analysis of planar linkage mechanisms using *Assur groups* is discussed. Some of these procedures are also used in the synthesis and analysis of cam-follower mechanisms, the subject of Chapter 7. They allow for automatic graphical representation and animation of rotary and linear motors, of offset points and complex shapes attached to mobile links, and for visualizing the velocity and acceleration vectors and of loci of moving points. Procedure **Spring** used in the elastic pendulum example in Chapter 3 is also available from unit **LibMec2D**. At the end of the chapter, two approaches to generating mechanical system simulations accompanied by dynamic plots with scan lines and scan points will be discussed, one using the procedures in unit **LibPlots** and the other using the **D_2D** program.

5.1 SAMPLE PROGRAM USING THE **LibMec2D** UNIT AND PROCEDURES **Locus** AND **CometLocus**

P5_01.PAS, listed in Appendix B, although not of a mechanical system simulation type, has the main elements of an animation program that uses **LibMec2D** procedures. It was used to produce Figure 5.1 and the companion animated **GIF** files **F5_01a.GIF** and **F5_01b.GIF** available with the book. The program calculates the x and y coordinates of a number of discrete points belonging to an array of **n** *Archimedean spirals* (see Equation 1.11), and outputs each animation frames to layers **1**, **2**, **3**, etc., of file **F5_01.DXF**. Additionally, it writes the **n** spirals as polylines to the background layers **p1** to **p8** of the same **F5_01. DXF** file. Remember that layers with names integer positive numbers are interpreted by

Archimedean spirals

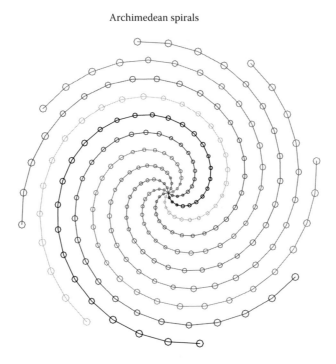

FIGURE 5.1 Accumulated frames of a polar array of *Archimedean spirals* with variable size markers output by program **P5_01.PAS**. See also animation file **F5_01a.GIF** and **F5_01b.GIF**.

M_3D.LSP as animation frames. Any other layer including **0**, **Ground**, and locus layers will remain *on* and thus the entities drawn to these layers will appear as background images.

Procedure **OpenMecGraph** on line #**17** of program **P5_01.PAS** launches the graphic system and establishes the limits of the workspace. Procedure **InitDXFfile** (line #**18**) opens for writing file **F5_01.DXF** and copies to it the content of the **DXF** header file **DXF. HED**. Procedure **SetTitle** on line #**19** sets the simulation title that is written to layer **Ground** of the output **DXF** file once, and to the computer screen every time procedure **NewFrame** is called (line #**24**). In addition, procedure **NewFrame** holds the current frame *on* for 500 ms and then refreshes the screen. It also indexes the current **DXF** layer number.

To simulate the circles with progressively increasing diameter as they move along the **n** spirals, procedure **SetJointSize** (line #**25**) is called after each animation frame with its argument defined as function of the frame number. In any other simulation program, **SetJointSize** should be called only once, somewhere at the beginning of the program. If the default size (i.e., 4) and appearance (i.e., full view) of the joints and motors are satisfactory, then there is no need to call **SetJointSize** at all. If **SetJointSize** is called with a negative argument, then the motors and joints in that simulation will be represented in a simplified manner and without hiding the overlapped joints (see Figure 5.2). Note that the points generated by calling procedure **PutPoint** and the ground points drawn by procedure **PutGPoint** are one unit smaller than the argument of the **SetJointSize**.

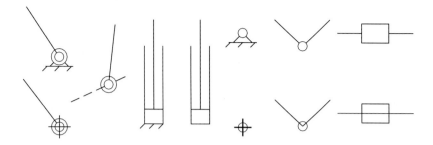

FIGURE 5.2 Rotary and linear motors and turning and sliding joints used by procedures **LibMecIn**, **LibAssur**, and **LibMec2D**. Their full view or simplified representation and relative size are controlled by calling procedure **SetJointSize**.

Procedure **Locus** on line **#31** of program **P5_01.PAS** is responsible for writing to separate temporary files of double (extension **$2D**) the *x* and *y* coordinates of each curve. The color of these curves is defined by the equation **iC MOD 15 + 1**, where **iC** is the curve number. Also note that the name of the layers where the polylines are written begins with letter "p", so they remain *on* all the time in an animation done using the **M_3D.LSP** **AutoLisp** application. If procedure **CometLocus** is used instead, when animated with **M_3D.LSP**, the Archimedean spirals will appear growing as the eight points move outwards. No temporary **$2D** file will be generated this time.

Animation continues until the user presses the <Esc> key. If line **#35** is replaced with line **#36**, then the animation is repeated until the global variable **MecOut** becomes FALSE (which is caused by calling procedure **CloseMecDXF** on line **#22**) and the user presses <Esc>.

5.2 JOINTS AND ACTUATORS AVAILABLE FOR MECHANICAL SYSTEM SIMULATION

Figure 5.2 is a summary of actuators and joints useful in the simulation of planar mechanical systems as they are output by the procedures in units **LibAssur** and **LibMecIn**. The actuators can be of rotational type (i.e., powered cranks) and of linear type (i.e., hydraulic or pneumatic cylinders, screw jacks, solenoids, membrane actuators) and can be either attached to the ground or to another moving link. The RTRTR and RTRR powered dyads discussed in detail in Chapter 6 also utilize linear motors that are represented graphically in the same manner.

The position, velocity, and acceleration equations of the rotary and linear motors in Figure 5.2 have been programmed in a number of procedures available from unit **LibMecIn**. In the remainder of this chapter the kinematic equations of these motors and their computer implementations, that is, procedures **Crank**, **gCrank**, **Slider**, and **gSlider**, will be discussed. Also discussed in this chapter are Pascal procedures: **Ang3PVA**, **Ang4PVA**, **Base**, **gShape**, **LabelJoint**, **Link**, **ntAccel**, **Offset**, **OffsetV**, **PutAng**, **PutDist**, **PutGPoint**, **PutGText**, **PutPoint**, **PutRefSystem**, **PutText**, **PutVector**, **Shape**, **VarDist**; these are useful in the simulation and analysis of planar mechanical systems.

Important: In case there is interest only in the position results or only in the position and velocity results, procedures **Crank**, **gCrank**, **Slider**, **gSlider**, **Offset**, **OffsetV**, **Ang3PVA**, and **Ang4PVA** can be called with their velocity and/or acceleration input and output variables set to **InfD** (a constant defined in unit **LibMath** and equal to 10^{100}). A generic variable named _ (i.e., the underscore symbol) declared in the interface section of unit **LibMath** and set equal to **InfD** should be used for this purpose.

5.2.1 Kinematic Analysis of Input Rotational Members

A turning link, named *crank* when it rotates continuously and in the same direction or *rocker* when it oscillates back and forth, is the most common input element in mechanism kinematics. Figure 5.3 shows two instances of such a link, where joint A can be attached either to a mobile element (Figure 5.3a) or to the ground (Figure 5.3b).

The general case where the crank is jointed to a mobile member (i.e., the velocity and acceleration of point A in Figure 5.3a are nonzero) will be considered first. The kinematic equations for the case where the crank is pin jointed to the ground (Figure 5.3b) can be easily derived by setting the velocity and acceleration of joint center A to zero.

At any instant of time in a simulation, the following parameters are assumed known:

- The coordinates xP and yP relative to the fixed reference frame OXY of a point P of the moving member to which the crank is attached.

- The projections $\dot{x}P$ and $\dot{y}P$ of the velocity of P onto the fixed reference frame.

- The projections $\ddot{x}P$ and $\ddot{y}P$ of the acceleration of P onto the fixed reference frame.

- The coordinates xA and yA of the joint center A relative to the fixed reference frame.

- The projections $\dot{x}A$ and $\dot{y}A$ of the velocity of point A onto the fixed reference frame.

- The projections $\ddot{x}A$ and $\ddot{y}A$ of the acceleration of point A onto the fixed reference frame.

- The crank length AB.

- The angle φ between an extension of the reference line PA and the link AB.

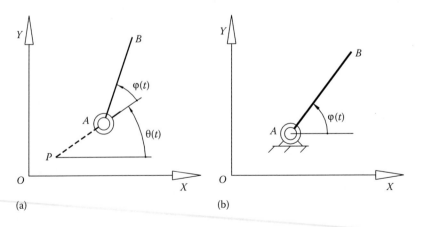

(a) (b)

FIGURE 5.3 Schematic for calculating the displacement, velocity, and acceleration of a point B of a rotational element AB when it is jointed (a) to a mobile member PA and (b) to the ground.

- The angular velocity $\dot{\varphi}$ of the crank relative to link *PA*.

- The angular acceleration $\ddot{\varphi}$ of the crank relative to link *PA*.

For these given inputs, we will calculate the following parameters:

- The coordinates *xB* and *yB* of point *B* relative to the fixed reference frame.

- The projections $\dot{x}B$ and $\dot{y}B$ of the velocity of *B* onto the axes of the fixed reference frame.

- The projections $\ddot{x}B$ and $\ddot{y}B$ of the acceleration of *B* onto the axes of the fixed reference frame.

By projecting the vector equation **OB** = **OA** + **AB** onto the axes of the *OXY* reference frame, coordinates *xB* and *yB* result as follows:

$$\begin{cases} xB = xA + AB\cos(j+\theta) \\ yB = yA + AB\sin(j+\theta) \end{cases} \tag{5.1}$$

where θ is the angle measured between *OX* and vector **PA** (see Figure 5.3a); θ can be easily calculated using the known coordinates of points *A* and *P*.

Differentiating Equation 5.1 with respect to time yields the projections of the linear velocity of point *B* onto *OX* and *OY*:

$$\begin{cases} \dot{x}B = \dot{x}A - (\varphi + \dot{\theta}) \cdot AB \cdot \sin(\varphi + \theta) \\ \dot{y}B = \dot{y}A + (\dot{\varphi} + \dot{\theta}) \cdot AB \cdot \cos(\varphi + \theta) \end{cases} \tag{5.2}$$

which, by using the results in Equation 5.1, further writes

$$\begin{cases} \dot{x}B = \dot{x}A - (\dot{\varphi} + \dot{\theta})(yB - yA) \\ \dot{y}B = \dot{y}A + (\dot{\varphi} + \dot{\theta})(xB - xA) \end{cases} \tag{5.3}$$

The components of the linear acceleration of *B* are obtained by differentiating Equation 5.3:

$$\begin{cases} \ddot{x}B = \ddot{x}A - (\ddot{\varphi} + \ddot{\theta})(yB - yA) - (\dot{\varphi} + \dot{\theta})(\dot{y}B - \dot{y}A) \\ \ddot{y}B = \ddot{y}A + (\ddot{\varphi} + \ddot{\theta})(xB - xA) + (\dot{\varphi} + \dot{\theta})(\dot{x}B - \dot{x}A) \end{cases} \tag{5.4}$$

Angle θ and its time derivatives $\dot{\theta}$ and $\ddot{\theta}$ occurring in these equations can be calculated using the position, velocity, and acceleration components of points *P* and *A*, as it will be explained later in this chapter when procedures **AngPVA** and **VarDist** are introduced.

In case of a rotational member jointed to the ground as shown in Figure 5.3b, angle θ and its first and second derivatives $\dot{\theta}$ and $\ddot{\theta}$ become zero. Therefore, Equation 5.1 simplified to

$$\begin{cases} xB = xA + AB\cos\varphi \\ yB = yA + AB\sin\varphi \end{cases} \tag{5.5}$$

while Equations 5.4 and 5.5 become

$$\begin{cases} \dot{x}B = \dot{x}A - \dot{\varphi} \cdot (yB - yA) \\ \dot{y}B = \dot{y}A + \dot{\varphi} \cdot (xB - xA) \end{cases} \tag{5.6}$$

and finally

$$\begin{cases} \ddot{x}B = \ddot{x}A - \ddot{\varphi} \cdot (yB - yA) - \dot{\varphi} \cdot (\dot{y}B - \dot{y}A) \\ \ddot{y}B = \ddot{y}A + \ddot{\varphi} \cdot (xB - xA) + \dot{\varphi} \cdot (\dot{x}B - \dot{x}A) \end{cases} \tag{5.7}$$

5.2.2 Procedures **Crank** and **gCrank**

Equations 5.1, 5.3, and 5.4 have been programmed in procedure **Crank** part of the unit **LibMecIn**. The procedure calculates the position, velocity, and acceleration of point B of a crank AB that rotates relative to a mobile element PA. If the graphic system is on, the procedure also draws in color **Color** (less if **Color** equals zero or the **BGI** constant **Black**) a line connecting A and B, and by calling procedure **Motor** from unit **LibMec2D**, it draws at point A a moving rotary-motor symbol. If **Color** is a negative number, then only the motor symbol will be drawn in color **-Color**. Angle φ and its time derivatives $\dot{\varphi}$ and $\ddot{\varphi}$ are measured counterclockwise from an extension of line PA shown by procedure **Crank** as a short segment drawn on the side of the motor opposite to point P (see Figure 5.3a).

Procedure **Crank** has the following heading:

```
procedure Crank(Color: Integer; xP,yP, vxP,vyP, axP,ayP, xA,yA,
vxA,vyA, axA,ayA, Phi, dPhi, ddPhi, AB: double; var xB,yB,
vxB,vyB, axB,ayB: double);
```

The correspondence between the formal parameters of the procedures and the notations used in Equations 5.1 through 5.4 and in Figure 5.3a is as follows:

Input parameters of procedure **Crank**:

-16...16	xP	yP	$\dot{x}P$	$\dot{y}P$	$\ddot{x}P$	$\ddot{y}P$	xA	yA	$\dot{x}A$	$\dot{y}A$	$\ddot{x}A$	$\ddot{y}A$
Color	xP	yP	vxP	vyP	axP	ayP	xA	yA	vxA	vyA	axA	ayA

φ	$\dot{\varphi}$	$\ddot{\varphi}$	AB
Phi	dPhi	ddPhi	AB

Output parameters of procedure **Crank**:

xB	yB	$\dot{x}B$	$\dot{y}B$	$\ddot{x}B$	$\ddot{y}B$
xB	yB	vxB	vyB	axB	ayB

The companion procedure **gCrank** with the heading

```
procedure gCrank(Color: Integer; xA,yA, Phi, dPhi, ddPhi, AB:
double; var xB,yB, vxB,vyB, axB,ayB: double);
```

calculates the position, velocity, and acceleration of point B of a crank AB for the case where joint A is connected to the ground (Figure 5.3b). The correspondence between procedure's formal parameters and the notations used in Equations 5.5 through 5.7 and Figure 5.3b is summarized in the following, where angle φ and its derivatives $\dot{\varphi}$ and $\ddot{\varphi}$ are measured counterclockwise from the OX axis.

Input parameters of procedure **gCrank**:

-16...16	xA	yA	φ	$\dot{\varphi}$	$\ddot{\varphi}$	AB
Color	xA	yA	Phi	dPhi	ddPhi	AB

Output parameters of procedure **gCrank**:

xB	yB	$\dot{x}B$	$\dot{y}B$	$\ddot{x}B$	$\ddot{y}B$
xB	yB	vxB	vyB	axB	ayB

If the graphic system is on, procedure **gCrank** draws in color **Color** (less if **Color** is zero) a line connecting points A and B, and by calling procedure **gMotor** from unit **LibMecGr**, it draws at point A a grounded rotary-motor symbol. If parameter **Color** is a negative number, then only the motor symbol will be drawn in color **-Color**.

To exemplify the use of procedures **Crank** and **gCrank**, program **P5_04.PAS** (see Appendix B) has been written. The program animates a ground crank of lengths AB in series with a second crank of length BC and also plots the locus of the end point C of the second crank, which is an epicycloid of equation

$$\begin{cases} x(t) = AB\cos(0.25\pi + 2\pi \cdot t) + BC\cos(-8\pi \cdot t) \\ y(t) = AB\sin(0.25\pi + 2\pi \cdot t) + BC\sin(-8\pi \cdot t) \end{cases} \tag{5.8}$$

The relative size of the joints, motors, and actuators in an animation was set by calling procedure **SetJointSize** on line #**15**. Sample screenshot of the simulations generated by program **P5_04.PAS** are available in Figure 5.4, done for **SetJointSize** called with both a negative and a positive argument. Corresponding to these figures are animation files **F5_04a.GIF** and **F5_04b.GIF** produced using the **M_3D.LSP** application. When generating the frames of animated **GIF** file **F5_04b.GIF**, procedure **Locus** on line #**31** has been

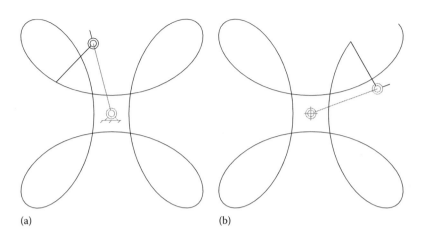

FIGURE 5.4   Epicycloid generated with procedures **gCrank** and **Crank** arranged in series shown in (a) full view and (b) simplified joint. See also animated **GIF** files **F5_04a.GIF** and **F5_04b.GIF**.

replaced with **CometLocus**, which has the effect of showing in **AutoCAD** the locus of point *C* as it progresses, same as on the computer screen during the first run. A third animated **GIF** named **F5_04C.GIF** has also been generated to illustrate the effect of calling procedures **gCrank** with its parameter **Color** set to a negative value.

Note that the open-loop mechanism in Figure 5.4 can be assumed to be a simple serial manipulator of the SCARA type (Craig 2004), and the program **P5_04.PAS** actually solves the direct kinematics problem of this manipulator. See also Chapter 9 where the subject of SCARA robot kinematics is discussed in more detail.

5.2.3 Kinematic Analysis of Input Translational Members

Translational input members (also called linear actuators or linear motors) come in a variety of configurations. Since the hydraulic or pneumatic cylinders are the most common embodiment of a linear motor, a generic representation as shown in Figure 5.5 will be assumed.

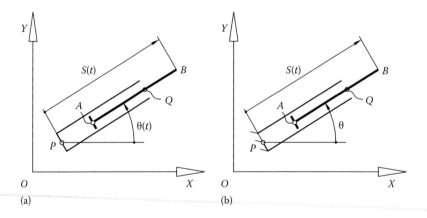

FIGURE 5.5   Schematic of a linear motor attached via its connecting points *P* and *Q* (a) to a mobile member and (b) to the ground.

The cylinder that guides to the inside the piston can be attached to a mobile element or can be connected to the ground.

Referring to Figure 5.5a, the analysis will be performed for the situation where the velocity and acceleration of two points A and Q of the cylinder are nonzero (i.e., the linear motor is mounted on a mobile element), while the kinematic equations of the linear motor attached to the ground will be derived as a particular case.

At any instant of time of the simulation, the following parameters are assumed known:

- The coordinates xP, yP and xQ, yQ relative to the fixed reference frame OXY of two points located on the cylinder's axis.

- The projections $\dot{x}P$ and $\dot{y}P$ of the velocity of point P onto the fixed reference frame.

- The projections $\ddot{x}P$ and $\ddot{y}P$ of the accelerations of point P onto the fixed reference frame.

- The projections $\dot{x}Q$ and $\dot{y}Q$ of the velocity of point Q onto the fixed reference frame.

- The projections $\ddot{x}Q$ and $\ddot{y}Q$ of the accelerations of point Q onto the fixed reference frame.

- The piston displacement s and its time derivatives $\dot{s}$ and $\ddot{s}$ (all measured relative to the member to which it is attached and assumed positive when oriented as shown in Figure 5.5).

- The piston length AB.

Given these parameters, it is required to determine the following variables:

- The coordinates xB and yB of point B relative to the fixed reference frame OXY.

- The projections $\dot{x}B$ and $\dot{y}B$ of the velocity of B onto the axes of the fixed reference frame.

- The projections $\ddot{x}B$ and $\ddot{y}B$ of the acceleration of B onto the axes of the fixed reference frame.

- The coordinates xA and yA of point A of the piston relative to the fixed reference frame.

- The projections $\dot{x}A$ and $\dot{y}A$ of the velocity of A onto the axes of the fixed reference frame.

- The projections $\ddot{x}A$ and $\ddot{y}A$ of the acceleration of A onto the axes of the fixed reference frame.

By projecting the vector equation $\mathbf{OB} = \mathbf{OP} + \mathbf{PB}$ on the OXY reference frame, the coordinates of point B are obtained as follows:

$$\begin{cases} xB = xP + s \cdot \cos\theta \\ yB = yP + s \cdot \sin\theta \end{cases} \tag{5.9}$$

where the angle θ between vector $\mathbf{PB}$ and axis OX (Figure 5.5a) can be easily calculated.

Differentiating Equation 5.9 once with respect to time, the components of the linear velocity of point B are obtained as

$$
\begin{cases}
\dot{x}B = \dot{x}P + \dot{s} \cdot \cos\theta - \dot{\theta} \cdot s \cdot \sin\theta \\
\dot{y}B = \dot{y}P + \dot{s} \cdot \sin\theta + \dot{\theta} \cdot s \cdot \cos\theta
\end{cases}
\tag{5.10}
$$

and by combining in the position Equation 5.9, they further become

$$
\begin{cases}
\dot{x}B = \dot{x}P + \dot{s} \cdot \cos\theta - \dot{\theta} \cdot (yB - yP) \\
\dot{y}B = \dot{y}P + \dot{s} \cdot \sin\theta + \dot{\theta} \cdot (xB - xP)
\end{cases}
\tag{5.11}
$$

The components of acceleration of point B are determined by differentiating Equations 5.11:

$$
\begin{cases}
\ddot{x}B = \ddot{x}P + \ddot{s} \cdot \cos\theta - \dot{\theta} \cdot \dot{s} \cdot \sin\theta - \ddot{\theta} \cdot (yB - yA) - \dot{\theta} \cdot (\dot{y}B - \dot{y}A) \\
\ddot{y}B = \ddot{y}P + \ddot{s} \cdot \sin\theta + \dot{\theta} \cdot \dot{s} \cdot \cos\theta + \ddot{\theta} \cdot (xB - xA) + \dot{\theta} \cdot (\dot{x}B - \dot{x}A)
\end{cases}
\tag{5.12}
$$

Angle θ and its time derivatives $\dot{\theta}$ and $\ddot{\theta}$ can be calculated using the known coordinates xP, yP, xQ, and yQ and their time derivatives as explained in Section 5.3.

The coordinates xA and yA of point A of the piston are obtained by projecting vector equation **OA** = **OP** + **PA** onto the axes of the fixed reference frame OXY:

$$
\begin{cases}
xA = xP + (s - AB) \cdot \cos\theta \\
yA = yP + (s - AB) \cdot \sin\theta
\end{cases}
\tag{5.13}
$$

The components of the linear velocity of point A are obtained through differentiation as

$$
\begin{cases}
\dot{x}A = \dot{x}P + \dot{s} \cdot \cos\theta - \dot{\theta} \cdot (s - AB) \cdot \sin\theta \\
\dot{y}A = \dot{y}P + \dot{s} \cdot \sin\theta + \dot{\theta} \cdot (s - AB) \cdot \cos\theta
\end{cases}
\tag{5.14}
$$

Using the results in Equation 5.13, these two equations become

$$
\begin{cases}
\dot{x}A = \dot{x}P + \dot{s} \cdot \cos\theta - \dot{\theta} \cdot (yA - yP) \\
\dot{y}A = \dot{y}P + \dot{s} \cdot \sin\theta + \dot{\theta} \cdot (xA - xP)
\end{cases}
\tag{5.15}
$$

The components of the linear acceleration of point A are obtained by differentiating Equation 5.15:

$$\begin{cases} \ddot{x}A = \ddot{x}P + \ddot{s} \cdot \cos\theta - \dot{\theta} \cdot \dot{s} \cdot \sin\theta - \ddot{\theta} \cdot (yA - yP) - \dot{\theta} \cdot (\dot{y}A - \dot{y}P) \\ \ddot{y}A = \ddot{y}P + \ddot{s} \cdot \sin\theta + \dot{\theta} \cdot \dot{s} \cdot \cos\theta + \ddot{\theta} \cdot (xA - xP) + \dot{\theta} \cdot (\dot{x}A - \dot{x}P) \end{cases} \quad (5.16)$$

If the linear motor is mounted to the ground as shown in Figure 5.5b, then $\dot{\theta}$ and $\ddot{\theta}$ will be both zero. The coordinates of points B and A can be calculated using Equations 5.9 and 5.13 given earlier , while the scalar components of their velocities and accelerations are the following:

$$\begin{cases} \dot{x}A = \dot{x}B = \dot{s} \cdot \cos\theta \\ \dot{y}A = \dot{y}B = \dot{s} \cdot \sin\theta \end{cases} \quad (5.17)$$

and

$$\begin{cases} \ddot{x}A = \ddot{x}B = \ddot{s} \cdot \cos\theta \\ \ddot{y}A = \ddot{y}B = \ddot{s} \cdot \sin\theta \end{cases} \quad (5.18)$$

5.2.4 Procedures `Slider` and `gSlider`

Equations 5.9, 5.11 through 5.13, 5.15, and 5.16 have been programmed inside procedure **Slider** that calculates the position, velocity, and acceleration of points A and B of the a linear actuator, when its cylinder is attached to a mobile element at P and Q. The positions, velocities, and accelerations of points P and Q must be provided as inputs, together with the displacement s of the piston and its first and second time derivatives $\dot{s}$ and $\ddot{s}$. The heading of procedure **Slider** is

```
procedure Slider(Color: Integer; xP,yP,vxP,vyP,axP,ayP, xQ,yQ,
vxQ,vyQ, axQ,ayQ, AB, s, ds, dds: double; var xB,yB, vxB,vyB,
axB,ayB, xA,yA, vxA,vyA, axA,ayA: double);
```

and the correspondence between its formal parameters and the notations used earlier are listed next:

Input parameters of procedure **Slider**:

-16...16	xP	yP	$\dot{x}P$	$\dot{y}P$	$\ddot{x}P$	$\ddot{y}P$	xQ	yQ	$\dot{x}Q$	$\dot{y}Q$	$\ddot{x}Q$	$\ddot{y}Q$
Color	xP	yP	vxP	vyP	axP	ayP	xQ	yQ	vxQ	vyQ	axQ	ayQ

AB	s	$\dot{s}$	$\ddot{s}$
AB	s	ds	dds

Output parameters of procedure **Slider**:

xB	yB	$\dot{x}B$	$\dot{y}B$	$\ddot{x}B$	$\ddot{y}B$	xA	yA	$\dot{x}A$	$\dot{y}A$	$\ddot{x}A$	$\ddot{y}A$
xB	yB	vxB	vyB	axB	ayB	xA	yA	vxA	vyA	axA	ayA

If the graphic system is on, procedure **Slider** draws in color **Color** (less if it is equal to zero or the **BGI** constant **Black**) the piston and the cylinder, similar to Figure 5.5a. If either distance AB or distance PQ is less than five times the joint size, then the procedure will draw a slider block at B and its sliding axis PQ. If **Color** is a negative number, then only the slider block will be drawn without its axis.

The companion procedure **gSlider** calculates the position, velocity, and acceleration of points A and B of the piston for the case when the cylinder is fixed to the ground. The procedure implements Equations 5.9, 5.10, 5.17, and 5.18 and has the following heading:

```
procedure gSlider(Color:Word; xP,yP,xQ,yQ, PQ, s,ds,dds:double;
var xB,yB, vxB,vyB, axB,ayB, xA,yA, vxA,vyA, axA,ayA:double);
```

while and the correspondence between its formal parameters and the notations used earlier is as follows:

Input parameters of procedure **gSlider**:

0...16	xP	yP	xQ	yQ	PQ	s	$\dot{s}$	$\ddot{s}$
Color	xP	yP	xQ	yQ	PQ	s	ds	Dds

Output parameters of procedure **gSlider**:

xB	yB	$\dot{x}B$	$\dot{y}B$	$\ddot{x}B$	$\ddot{y}B$	xA	yA	$\dot{x}A$	$\dot{y}A$	$\ddot{x}A$	$\ddot{y}A$
xB	yB	vxB	vyB	axB	ayB	xA	yA	vxA	vyA	axA	ayA

If the graphic system is on, procedure **gSlider** draws in color **Color** (less if it equals zero or the **BGI** constant **Black**) the cylinder connected to the ground and its piston similar to Figure 5.5b. Similarly to procedure **Slider**, if distance AB or distance PQ is less than five times the joint size, the procedure will draw a slider block at B and its sliding axis PQ. If points P and Q coincide, then the procedure assumes the sliding axis to be horizontal.

For both procedures, the piston and cylinder diameter or block size can be controlled by calling procedure **SetJointSize** as shown in the sample program **P5_06A.PAS** (see Appendix B) and in program **P5_06B.PAS** available with the book. Both programs use procedures **gSlider** and **Slider** to simulate the motion of two perpendicular linear motors connected in series (see Figure 5.6), where the piston of the second motor traces a Lissajous curve of equation

$$\begin{cases} x(t) = 30 \ \sin\left(2\pi \times t - \dfrac{\pi}{4}\right) \\ y(t) = 25 \ \sin\left(4\pi \times t + \dfrac{\pi}{4}\right) \end{cases}$$

(5.19)

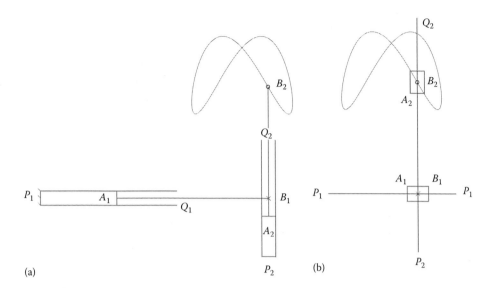

(a)   (b)

FIGURE 5.6   Two sliders in series tracing a Lissajous curve generated using programs **P5_06A.PAS** (a) and **P5_06B.PAS** (b). See also animation files **F5_06a.GIF** and **F5_06b.GIF**.

Sample outputs generated by these two programs are given in Figure 5.6 and in the animated **GIF** files **F5_06a.GIF**, **F5_06b.GIF**, and **F5_06c.GIF**. Program **P5_06B.PAS** illustrates the case when the two linear motors are drawn as slider blocks. By calling the procedure **Slider** in this second program with its color parameter set to **Cyan**, the simulation changes as shown in animation file **F5_06c.GIF**, that is, the sliding axis will not be drawn.

Note the use in procedures **LabelJoint** and **PutPoint** of the underscore character to specify subscripts (lines #**41** to #**49**). The same subscript labeling is available in procedures **PutAng**, **PutDist**, and **PutRefSystem**, and it is done internally by calling procedure **PD_text** from unit **LibDXF**.

You may want to experiment with circular frequencies other than 2π and 4π and phase angles other than $\pm\pi/4$ in Equation 5.19 and observe their effect upon the appearance of the locus of point B_2. Pen plotters and computer numerically controlled machines (CNC milling machines, torch, or plasma cutters) operate on the principle illustrated by program **P5_06A.PAS**. Planar Cartesian coordinate robots, also known as linear robots, have similar configurations (Craig 2004).

5.3 POSITION, VELOCITY, AND ACCELERATION OF POINTS AND MOVING LINKS

In the kinematic simulation of mechanical systems, it is frequently required to determine the angular position, velocity, and acceleration of a moving link for which the scalar coordinates of two points are known, together with their first and second time derivatives, or to determine the position, velocity, and acceleration of a point connected to a moving body of known motion.

Let us assume a rigid link defined by points A and B in planar motion. The case where the coordinates of a point P attached to this link are specified relative to a local reference frame will be discussed in more detail. Figure 5.7a shows such an arrangement, where local

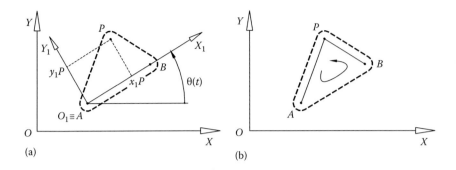

FIGURE 5.7   Schematic for calculating the position, velocity, and acceleration of a point P attached to a moving link AB knowing the local coordinates (a) x_1P and y_1P or distances AP and BP and the orientation of the (b) APB loop.

reference frame $O_1X_1Y_1$ has its O_1X_1 axis oriented from A to B and its origin is coincident with point A, and where offset point P is specified using local coordinates (x_1P, y_1P).

If lengths AP and BP are specified instead, the location of point P can be determined as the intersection of two circles centered at A and B and of radii AP and BP. In order to distinguish between the two intersection points of the two circles, the orientation of the triangular loop APB has to be additionally specified. This second approach is conveniently solved using procedure **Int2CirPVA** discussed in Chapter 6 and will not be detailed here beyond its computer implementation in procedure **OffsetV**.

At any instant of time, the followings parameters are assumed known:

- The coordinates xA and yA of point A relative to the fixed reference frame OXY.

- The projections $\dot{x}A$ and $\dot{y}A$ of the velocity of point A onto the fixed reference frame.

- The projections $\ddot{x}A$ and $\ddot{y}A$ of the acceleration of point A onto the fixed reference frame.

- The coordinates xB and yB of point B relative to the fixed reference frame OXY.

- The projections $\dot{x}B$ and $\dot{y}B$ of the velocity of point B onto the fixed reference frame.

- The projections $\ddot{x}B$ and $\ddot{y}B$ of the acceleration of point B onto the fixed reference frame.

- The local coordinates x_1P and y_1P of a point of interest P attached to the moving link relative to the moving reference frame $O_1X_1Y_1$ (Figure 5.7a) or distances AP and BP together with the orientation of triangular loop ABP (Figure 5.7b).

The unknown kinematic parameters are

- The coordinates xP and yP of point P relative to the fixed reference frame OXY.

- The projections $\dot{x}P$ and $\dot{y}P$ of the linear velocity of point P onto the fixed reference frame.

- The projections $\ddot{x}P$ and $\ddot{y}P$ of the acceleration of point P onto the fixed reference frame.

For the case illustrated in Figure 5.7a, we project vector equation **OP** = **OA** + **AP** onto the axes of the fixed reference frame and obtain:

$$\begin{cases} xP = xA + x_1P \cdot \cos\theta + y_1P \cdot \cos\left(\theta + \dfrac{\pi}{2}\right) = xA + x_1P \cdot \cos\theta - y_1P \cdot \sin\theta \\[3mm] yP = yA + x_1P \cdot \sin\theta + y_1P \cdot \sin\left(\theta + \dfrac{\pi}{2}\right) = yA + x_1P \cdot \sin\theta + y_1P \cdot \cos\theta \end{cases} \tag{5.20}$$

where angle θ is measured between axis OX and vector **AB** (see Figure 5.7a) and is given by the formula

$$\theta = \tan^{-1}\left[\frac{yB - yA}{xB - xA}\right]. \tag{5.21}$$

Note that the same two Equations 5.20 can be obtained by applying a rotation by angle θ to point P of coordinates x_1P and y_1P, followed by a translation from $(0, 0)$ to (xA, yA).

Differentiating Equation 5.20 once with respect to time, the projections of the velocity of point P onto the axes of the fixed reference frame are obtained:

$$\begin{cases} \dot{x}P = \dot{x}A - \dot{\theta} \cdot (x_1P \cdot \sin\theta + y_1P \cdot \cos\theta) \\[2mm] \dot{y}P = \dot{y}A + \dot{\theta} \cdot (x_1P \cdot \cos\theta - y_1P \cdot \sin\theta) \end{cases} \tag{5.22}$$

equivalent to

$$\begin{cases} \dot{x}P = \dot{x}A - \dot{\theta} \cdot (yP - yA) \\[2mm] \dot{y}P = \dot{y}A + \dot{\theta} \cdot (xP - xA) \end{cases} \tag{5.23}$$

The angular velocity $\dot{\theta}$ of the moving member AB can be determined by writing equations similar to (5.23) for point B instead of P, the velocity of which is known:

$$\begin{cases} \dot{x}B = \dot{x}A - \dot{\theta} \cdot (yB - yA) \\[2mm] \dot{y}B = \dot{y}A + \dot{\theta} \cdot (xB - xA) \end{cases} \tag{5.24}$$

yielding the following two equivalent equations:

$$\dot{\theta} = -\frac{\dot{x}B - \dot{x}A}{yB - yA} \quad \text{or} \quad \dot{\theta} = \frac{\dot{y}B - \dot{y}A}{xB - xA} \tag{5.25}$$

The x and y components of the acceleration of point P were obtained by differentiating Equations 5.23, that is,

$$\begin{cases} \ddot{x}P = \ddot{x}A - \ddot{\theta}\cdot(x_1P\cdot\sin\theta + y_1P\cdot\cos\theta) - \dot{\theta}^2\cdot(x_1P\cdot\cos\theta - y_1P\cdot\sin\theta) \\ \ddot{y}P = \ddot{y}A + \ddot{\theta}\cdot(x_1P\cdot\cos\theta - y_1P\cdot\sin\theta) - \dot{\theta}^2\cdot(x_1P\cdot\sin\theta + y_1P\cdot\cos\theta) \end{cases} \tag{5.26}$$

equivalent to

$$\begin{cases} \ddot{x}P = \ddot{x}A - \ddot{\theta}\cdot(yP - yA) - \dot{\theta}^2\cdot(xP - xA) \\ \ddot{y}P = \ddot{y}A + \ddot{\theta}\cdot(xP - xA) - \dot{\theta}^2\cdot(yP - yA) \end{cases} \tag{5.27}$$

Note that Equations 5.23 and 5.27 are the scalar form of Euler's equation for the velocity and acceleration of a rigid body in 2D motion (Goldstein et al. 2001).

The angular acceleration of the AB member is determined by extracting $\ddot{\theta}$ from Equations 5.26 for the particular case of point P coinciding with point B:

$$\ddot{\theta} = -\frac{\ddot{x}B - \ddot{x}A + \dot{\theta}^2(xB - xA)}{yB - yA} \quad \text{or} \quad \ddot{\theta} = \frac{\ddot{y}B - \ddot{y}A + \dot{\theta}^2(yB - yA)}{xB - xA}. \tag{5.28}$$

In order to avoid a possible division by zero, depending on the value of denominators $(yB - yA)$ and $(xB - xA)$, either the first or the second of Equations 5.25 and 5.28 should be used when calculating $\dot{\theta}$ and $\ddot{\theta}$.

5.3.1 Procedures **Offset** and **OffsetV**

Equations 5.20, 5.23, and 5.27 have been implemented in procedure **Offset0** part of unit **LibMec2D**, which calculates the position, velocity, and acceleration of a point P attached to a mobile link, giving its relative coordinates x_1P and y_1P. Procedure **Offset0** is not visible outside unit **LibMec2D**, but it is used by procedure **Offset**. The companion procedure **OffsetV** is based on the procedure **Int2CirPVA** in unit **LimMec2D** and uses distances AP and BP to point P and the orientation of the triangular loop APB as inputs (see Figure 5.7b). Both procedures have graphic output capabilities and are easily interchangeable. Their headings are as follows:

```
procedure Offset(Color:Integer; Style:char; xA,yA, vxA,vyA,
axA,ayA, xB,yB, vxB,vyB, axB,ayB, x1P,y1P:double; var xP,yP,
vxP,vyP, axP,ayP:double);
```

and

```
procedure OffsetV(Color:Integer; Style:char; xA,yA, vxA,vyA,
axA,ayA, xB,yB, vxB,vyB, axB,ayB, AP,BP, APB:double; var xP,yP,
vxP,vyP, axP,ayP:double);
```

The correspondence between the formal parameters of these two procedures and the notations used in Equations 5.20 through 5.28 and in Figure 5.4 are as follows:

Input parameters of procedure **Offset**:

−16..16	T, I, /, \, V, A	xA	yA	$\dot{x}A$	$\dot{y}A$	$\ddot{x}A$	$\ddot{y}A$	xB	yB	$\dot{x}B$	$\dot{y}B$	$\ddot{x}B$	$\ddot{y}B$	x_1P	y_1P
Color	Style	xA	yA	vxA	vyA	axA	ayA	xB	yB	vxB	vyB	axB	ayB	x1P	y1P

Input parameters of procedure **OfsetV**:

−16..16	T, I, /, \, V, A	xA	yA	$\dot{x}A$	$\dot{y}A$	$\ddot{x}A$	$\ddot{y}A$	xB	yB	$\dot{x}B$	$\dot{y}B$	$\ddot{x}B$	$\ddot{y}B$	AP	BP	ABP
Color	Style	xA	yA	vxA	vyA	axA	ayA	xB	yB	vxB	vyB	axB	ayB	AP	BP	ABP

Output parameters of procedures **Ofset** and **OfsetV**:

xP	yP	$\dot{x}P$	$\dot{y}P$	$\ddot{x}P$	$\ddot{y}P$
xP	yP	vxP	vyP	axP	ayP

If the graphic system is on, apart from returning the coordinates of point P and of their first and second time derivatives, these procedures plot on the computer screen and to the current **DXF** file additional graphic entities that help locating point P (see Figure 5.8 and programs **P5_08A.PAS** and **P5_08B.PAS** in Appendix B). If parameter **Style** equals "/" or "\", the procedures will draw a line connecting points A and P or points B and P, respectively. If parameter **Style** equals "**I**" or "**T**", a line from point P perpendicular to AB will be drawn, while for **Style** equals "**T**", a line connecting points A and B will be additionally drawn. If **Style** equals "**V**", then polyline APB will be drawn, while if **Style** equals "**A**", then the complete triangle APB will be drawn. If parameter **Color** is positive and procedure **SetJointSize** is called with a positive argument (i.e., the joints are set to full view), then triangle ABP will be filled with color. Otherwise, a transparent triangle

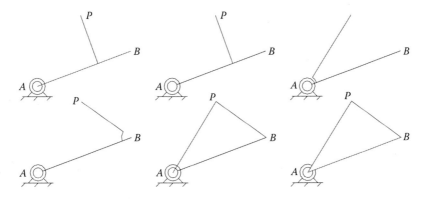

FIGURE 5.8   Various representations of an offset point P attached to a crank done by procedures **Offset** and **OffsetV**. See programs **P5_08A.PAS** and **P5_08A.PAS** and animation files **F5_08a.GIF** and **F5_08b.GIF**.

ABP will be drawn. If the parameter `Color` is set equal to 0, or `Style` is assigned the blank character, then there will be no line or triangle drawn.

5.3.2 Procedures `AngPVA`, `Ang3PVA`, and `Ang4PVA`

Equations 5.21, 5.25, and 5.28 can be employed to calculate the angular position, angular velocity, and angular acceleration of a rigid link for which the position, velocity, and acceleration of two points *A* and *B* attached to it are known. These equations were implemented in procedure `AngPVA` with the heading

```
procedure AngPVA(xA,yA, vxA,vyA, axA,ayA, xB,yB, vxB,vyB,
axB,ayB:double; var Theta, dTheta, ddTheta:double);
```

The correspondence between the formal parameters of the procedure and the notations used earlier are summarized in the following tables:

Input parameters of procedure `AngPVA`:

xA	yA	$\dot{x}A$	$\dot{y}A$	$\ddot{x}A$	$\ddot{y}A$	xB	yB	$\dot{x}B$	$\dot{y}B$	$\ddot{x}B$	$\ddot{y}B$
xA	yA	vxA	vyA	axA	ayA	xB	yB	vxB	vyB	axB	ayB

Output parameters of procedure `AngPVA`:

θ	$\dot{\theta}$	$\ddot{\theta}$
Theta	dTheta	ddTheta

When calling procedure `AngPVA`, variable `Theta` must carry a meaningful value, that is, either zero when `AngPVA` is called for the first time or the previous value of `Theta`. This is required to ensure the continuity of the returned angle, done by calling procedure `NghbrAng` described in Chapter 3. Procedure `AngPVA` is also used by procedures `Crank`, `Slider`, and `Offset` and by procedures `Ang3PVA` and `Ang4PVA`. Procedure `Ang3PVA` returns the angle defined by points *A*, *B*, and *C* (i.e., by vectors **BA** and **BC**) and its first and second time derivatives, while procedure `Ang4PVA` calculates the angle between vectors **AB** and **CD** and the first and second time derivatives of this angle. These procedures have the following heading:

```
Ang3PVA(xA,yA, vxA,vyA, axA,ayA, xB,yB, vxB,vyB, axB,ayB, xC,yC,
vxC,vyC, axC,ayC:double; var Theta, dTheta, ddTheta:double);
```

```
Ang4PVA(xA,yA, vxA,vyA, axA,ayA, xB,yB, vxB,vyB, axB,ayB, xC,yC,
vxC,vyC, axC,ayC, xD,yD, vxD,vyD, axD,ayD:double;  var Theta,
dTheta, ddTheta:double);
```

The user must provide the *x* and *y* coordinates of points *A*, *B*, *C* or *A*, *B*, *C*, and *D* together with the scalar components of their velocities and accelerations. The same requirement about angle `Theta` carrying an initial meaningful value or the previous value of the angle applies to procedures `Ang3PVA` and `Ang4PVA` as well.

Important: The procedures in unit **LibMec2D** assume that all angles are in radians (rad), that angular velocities are in rad/s, and that angular accelerations are in rad/s².

Important: In procedures **Offset**, **OffsetV**, **AngPVA**, **Ang3PVA**, and **Ang4PVA**, it is essential that distances *AB* and *BC* or *AB* and *CD* remain constant. Otherwise, the time derivatives returned by these procedures will not be correct. The cases where these distance do not remain constant are addressed in the next section.

5.4 POSITION, VELOCITY, AND ACCELERATION IN RELATIVE MOTION: PROCEDURE **VarDist**

A more general case than the one discussed earlier is that where the distance between points *A* and *B* does not remain constant (see Figure 5.9). Given coordinates *xA*, *yA*, and *xB*, *yB* of these two points relative to the *OXY* frame, and the *OX* and *OY* projections of their velocities and accelerations (i.e., $\dot{x}A$, $\dot{y}A$, $\dot{x}B$, $\dot{y}B$, $\ddot{x}A$, $\ddot{y}A$, $\ddot{x}B$, and $\ddot{y}B$), we want to find the distance *r* between these points, the angle θ formed by line *AB* with the *OX* axis. Also of interest are the first and second time derivatives of the variable distance *r* and of angle θ, that is, $\dot{r}$, $\ddot{r}$, $\dot{\theta}$, and $\ddot{\theta}$, respectively.

Distance *r* can be calculated with the known formula:

$$r = \sqrt{(xB - xA)^2 + (yB - yA)^2} \tag{5.29}$$

To determine the velocity and acceleration components, we project the vector equation **AB** = **OB** − **OA** onto the axes of the fixed reference frame *OXY* and obtain

$$\begin{cases} r \cdot \cos\theta = xB - xA \\ r \cdot \sin\theta = yB - yA \end{cases} \tag{5.30}$$

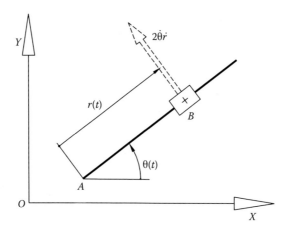

FIGURE 5.9 Schematic for calculating the variable distance *r* and angle θ determined by moving points *A* and *B*, and of the time derivatives $\dot{r}$, $\ddot{r}$, $\dot{\theta}$, and $\ddot{\theta}$. Also shown in dashed line is the Coriolis acceleration vector of the slider moving relative to its guide.

where angle θ can be calculated with Equation 5.21. Differentiating Equation 5.30 once with respect to time yields a set of two linear equations in the unknowns $\dot{r}$ and $\dot{\theta}$:

$$\begin{cases} \dot{r}\cdot\cos\theta - \dot{\theta}\cdot r\cdot\sin\theta = \dot{x}B - \dot{x}A \\ \dot{r}\cdot\sin\theta + \dot{\theta}\cdot r\cdot\cos\theta = \dot{y}B - \dot{y}A \end{cases} \tag{5.31}$$

which combined with Equation 5.30 become

$$\begin{cases} \dot{r}\cdot\cos\theta - \dot{\theta}\cdot(yB - yA) = \dot{x}B - \dot{x}A \\ \dot{r}\cdot\sin\theta + \dot{\theta}\cdot(xB - xA) = \dot{y}B - \dot{y}A \end{cases} \tag{5.32}$$

Accelerations are obtained by differentiating Equation 5.32

$$\begin{cases} \ddot{r}\cdot\cos\theta - \dot{\theta}\cdot\dot{r}\cdot\sin\theta - \ddot{\theta}\cdot(yB - yA) - \dot{\theta}\cdot(\dot{y}B - \dot{y}A) = \ddot{x}B - \ddot{x}A \\ \ddot{r}\cdot\sin\theta + \dot{\theta}\cdot\dot{r}\cdot\cos\theta + \ddot{\theta}\cdot(xB - xA) + \dot{\theta}\cdot(\dot{x}B - \dot{x}A) = \ddot{y}B - \ddot{y}A \end{cases} \tag{5.33}$$

These are equivalent to the following set of two linear equations in the unknowns $\ddot{s}$ and $\ddot{\theta}$:

$$\begin{cases} \ddot{r}\cdot\cos\theta - \ddot{\theta}\cdot(yB - yA) = \ddot{x}B - \ddot{x}A + \dot{\theta}\cdot\dot{r}\cdot\sin\theta + \dot{\theta}\cdot(\dot{y}B - \dot{y}A) \\ \ddot{r}\cdot\sin\theta + \ddot{\theta}\cdot(xB - xA) = \ddot{y}B - \ddot{y}A - \dot{\theta}\cdot\dot{r}\cdot\cos\theta - \dot{\theta}\cdot(\dot{x}B - \dot{x}A) \end{cases} \tag{5.34}$$

The systems of two linear equations (5.32) and (5.34) can be easily solved using Cramer's rule or the inverse matrix method, and together with Equations 5.29 and 5.21 have been implemented in procedure **VarDist** with the heading:

```
VarDist(xA,yA, vxA,vyA, axA,ayA, xB,yB, vxB,vyB, axB,ayB:double;
var r, dr, ddr, Theta, dTheta, ddTheta:double);
```

The correspondence between the formal parameters of the procedure and the notations used earlier is as follows:

Input parameters of procedure **VarDist**:

xA	yA	$\dot{x}A$	$\dot{y}A$	$\ddot{x}A$	$\ddot{y}A$	xB	yB	$\dot{x}B$	$\dot{y}B$	$\ddot{x}B$	$\ddot{y}B$
xA	yA	vxA	vyA	axA	ayA	xB	yB	vxB	vyB	axB	ayB

Output parameters of procedure **VarDist**:

r	$\dot{r}$	$\ddot{r}$	θ	$\dot{\theta}$	$\ddot{\theta}$
r	dr	ddr	Theta	dTheta	ddTheta

In order to ensure the continuity of angle **Theta** returned by **VarDist**, same as for procedure **AngPVA**, variable **Theta** must be assigned a seed value, that is, zero, when the procedure is first called or the previously calculated value of angle **Theta**.

Important: If points A and B overlap and distance r becomes zero, then angle θ cannot be evaluated. Additionally, because of the way the distance between points A and B is calculated, no distinction can be made between negative and positive r values. Therefore, point B should always remain on the same side of point A (Figure 5.9).

5.5 CORIOLIS ACCELERATION EXAMPLE: PROCEDURE **PutVector**

Program **P5_10.PAS**, listed in Appendix B, exemplifies the use of procedure **VarDist** and of procedure **PutVector**. It also provides an example of writing data to an output ASCII file. The program simulates the motion of a slider block B moving along a guide QQ' that is perpendicular to the end of a rocker OP (see Figure 5.10). Both the rocker and the slider are driven back and forth sinusoidally (see lines #35 through #40 of the program). Note that the displacement of point B of the slider is measured from point P, and it can be both positive and negative. Procedure **VarDist** returns the distance from point Q' to point B and the first and second time derivatives of this distance. It also returns the angular position, velocity, and acceleration of the slider block, which coincide with those of the crank. Before writing them to ASCII files **P5_10A.TXT** and **P5_10B.TXT**, the slider displacement and its angular position were offset by the amounts –Q'Q/2 and –π/2, respectively. This way, they can be easily compared with the inputs applied to the crank and to the slider. Additionally, the angle values were converted from radians to degrees.

By inspecting the content of output files **P5_10A.TXT** and **P5_10B.TXT**, it can be seen that identical values were recorded for both the angular and linear inputs of the crank and of the slider, which confirms the correctness of the equations programmed inside procedure **VarDist**.

In addition to the ASCII output, program **P5_10.PAS** also represents graphically the mechanism in **nPoz** discrete positions, together with the locus of point B (Figure 5.10).

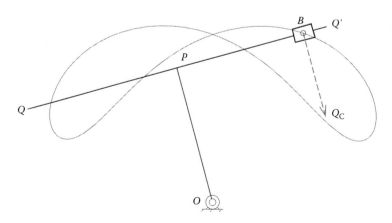

FIGURE 5.10   One of the frames generated by program **P5_10.PAS** showing the locus of point B on the slider and its Coriolis acceleration vector. See also animation file **F5_10.GIF**.

By calling procedure **PutVector** (line #**58**), the Coriolis acceleration vector of the slider moving along guide QQ' is also plotted with the simulation (see Figure 5.5 and Meriam and Kraige 2012). The x and y components of this acceleration vector are calculated prior to calling procedure **PutVector** on lines #**56** and #**57** of the program.

Because the angular position, velocity, and acceleration of the slider are identical with those of the entire T-shaped guide, procedure **AngPVA** applied to points P and Q or to points Q' and Q can be used instead of procedure **VarDist**.

5.6 MODEL VALIDATION: PROCEDURE **ntAccel**

One way of checking the validity of the results obtained using the procedures in unit **LibMec2D** is to compare them with results known to be correct. Such verifications should be done to ensure that the input motors of a mechanism are assigned consistent motions, for example, that their velocities and accelerations are indeed the first and second time derivatives of their displacements. Similar verifications are also proper when modifying an existing kinematic procedure or when developing a new one.

If kinematic data calculated with a concomitant method are not available for comparison, alternative techniques can be applied. In order to verify that the position results are correct, you can open the **DXF** frames of the simulation inside **AutoCAD** and check that the lengths and angles of links known to be rigid remain constant throughout the motion cycle of the mechanism. Using the inquiry procedures **PutDist** and **PutAng** discussed in Section 5.7, the same can be verified directly from within the simulation program.

Once position results are known to be correct, velocities and accelerations can be evaluated by applying finite difference formulae to the displacement data and then compared with the results returned by the program or by the procedure under scrutiny. To illustrate this concept, the aforementioned program **P5_10.PAS** has been duplicated as **P5_11.PAS** and further modified so that the coordinates of point B and the scalar components of its velocity and acceleration ($\dot{x}B$, $\dot{y}B$, $\ddot{x}B$, and $\ddot{y}B$) are evaluated and output to ASCII file **F5_11.TXT** (Figure 5.11). The time **t** values were

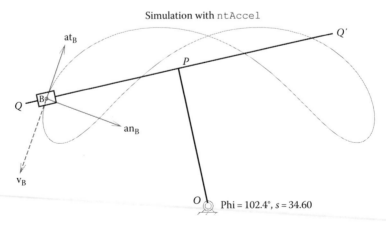

FIGURE 5.11 Simulation done with program **P5_11.PAS** showing the velocity and the normal and tangential acceleration of point B on the slider. See also animation file **F5_11.GIF**.

also recorded to this ASCII file (see lines #**58** and #**59** of the source code **P5_11.PAS** given in Appendix B).

ASCII file **F5_11.TXT** was then opened inside **Excel** (see file **F5_11.XLS** available with the book), and the first derivatives with respect to time of the x and y coordinates of point B were evaluated using finite differences (see Equation B.24). Using these newly calculated velocities noted **vxB*** and **vyB*** in Figure 5.12, approximations of the x and y acceleration components of point B were also generated using finite differences. By plotting the exact (**axB** and **ayB**) and approximate (**axB*** and **ayB***) accelerations of point B on the same graph, almost overlapping lines were obtained (Figure 5.12), thus validating the results output by program **P5_11.PAS**.

Another way of verifying the correctness of a kinematic simulation (although more of a qualitative nature) is to observe the velocity and acceleration vectors of one or more points of interest of the mechanism. It is known that the velocity vector v should remain tangent to the path of the point, while its acceleration vector a should always be oriented towards the inside of the path (Figures 5.13). If this is not happening, then calculation or computer implementation errors are to be expected.

Before drawing the normal a_n and tangential a_t acceleration vectors using procedure **PutVector**, program **P5_11.PAS** calls procedure **ntAccel** with inputs

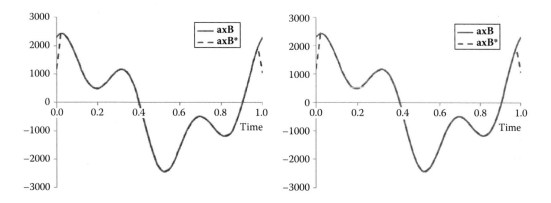

FIGURE 5.12 Comparison between the x and y components of the acceleration of point B output by program **P5_10.PAS** and the same components calculated using finite differences.

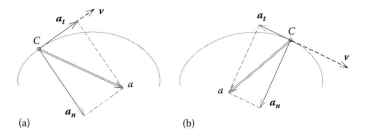

FIGURE 5.13 Velocity vector v and the normal and tangential acceleration vectors of point C on an (a) accelerating and (b) decelerating section of its path.

the x and y components of the velocity and acceleration of point B (see line **#54**). Procedure **ntAccel** calculates the x and y components of the $\boldsymbol{a}_n$ and $\boldsymbol{a}_t$ vectors using the following equations:

$$\boldsymbol{a}_t = \left(\boldsymbol{a} \cdot \boldsymbol{v}\right) \frac{\boldsymbol{v}}{|\boldsymbol{v}|^2} = \left(\frac{a_x v_x^2 + a_y v_x v_y}{v_x^2 + v_y^2}, \ \frac{a_x v_x v_y + a_y v_y^2}{v_x^2 + v_y^2} \right) \tag{5.35}$$

$$\boldsymbol{a}_n = \boldsymbol{a} - \boldsymbol{a}_t = \left(\frac{a_x v_y^2 - a_y v_x v_y}{v_x^2 + v_y^2}, \ \frac{a_y v_x^2 - a_x v_x v_y}{v_x^2 + v_y^2} \right) \tag{5.36}$$

As mentioned earlier, ensuring that the simulation is correct is always of concern. Animation of the mechanism provides a first good indication that its links assemble as intended. Labeling joints and placing stationary markers at different locations can be additionally helpful in this respect. In program **P5_11.PAS**, procedures **PutGPoint** (line **#41**) and **PutPoint** (line **#51**) draw on the screen and to the output **DXF** file a point of selected type, and also label this point. Characters available to control the type of point generated by procedures **PutGPoint** and **PutPoint** are: "." for one pixel, "**x**", "**X**", "**o**", "**0**" and "**O**" for × and š points of two or three sizes respectively. Also available as control characters are "**^**" and "**v**" for a grounded pin joint normal or reversed orientation. In turn, procedure **LabelJoint** (lines **#42**, **#45** and **#48**) allows moving point labels to be aligned with a specified direction.

Important: Procedures **PutPoint** and **LabelJoint** write the label (procedure **PutPoint** also draws the point of specified type) in the current layer of the **DXF** file output by the program. Procedure **PutGPoint** draws the point and writes its label to the **Ground** layer. These two procedures can be also used to display one-time information about the mechanism (procedure **PutGPoint**) or some variable parameter (procedure **PutPoint**) as it has been done on lines **#32** and **#57** of program **P5_11.PAS**. Procedures **SetTitle**, **PutGText,** and **PutText** available from unit **LibMec2D** are however better suited for such purposes.

If procedure **CloseMechGraph** is called with its argument set to TRUE (see line **#37** of program **P5_01.PAS** and line **#62** of program **P5_11.PAS**), then the temporary files of extension **$2D** used to record the loci are not deleted. Instead, their extension is changed to **D2D** so that they can be represented graphically using the **D_2D** program. Figure 5.14a is a plot of the loci of points *P1* to *P8* saved to file by program **P5_01.PAS**, while Figure 5.14b shows overlapped the loci of point *B* of the mechanism in Figure 5.11 generated for several crank OP length values. Note that the default names of the **D2D** files have been changed to **F15_14-1.D2D**, **F15_14-2.D2D**, etc. as they were generated by program **P5_11.PAS**. Also note that the color information is recorded to these loci files and can be interpreted by the **D_2D** program as explained in Chapter 1.

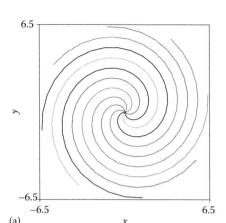

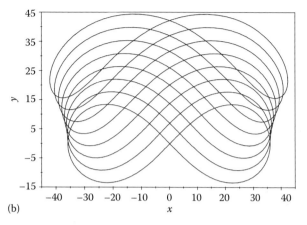

(a)

(b)

FIGURE 5.14 Plot of the eight spiral loci in Figure 5.1 (a) and of the overlapped loci of point *B* of the mechanism in Figure 5.10 for link length *OP* equal to 35, 30, 25, 20, 15, 10, 4 and 1E-6, i.e., near zero (b). Configuration files **F5_14a.CF2** and **F5_14b.CF2**.

5.7 WORKSPACE LIMITS AND INQUIRY PROCEDURES **PutDist** AND **PutAng**

Program **P5_11.PAS** in Appendix B shows how interface variables **XminWS**, **XmaxWS**, **YminWS**, and **YmaxWS** defined in unit **LibMec2D** can be used to best set up the limits of the view window. The first simulation cycle is performed without visualizing the mechanism, only to gather the workspace limits of its members. After this first cycle, procedure **OpenMechGraph** is called (see lines #**26** to #**29** of program **P5_11.PAS**). Alternatively, interface variables **XminWS**, **XmaxWS**, **YminWS**, and **YmaxWS** can be printed at the end of the run, so that the limits of the workspace can be manually adjusted for later runs (see lines #**63** and #**64** of the same program).

Either for verification purpose or to present the results of a simulation, it is possible to write data to file for inspection or to display it in tabular or graphical form. Plotting kinematic parameters as 2D line graphs together with the simulation is also possible, as explained in a separate section later in this chapter, but requires additional programming effort. It is easier to output the values of interest directly on the computer screen as it has been done in program **P5_11.PAS** using **PutGPoint** and **PutPoint** (lines #**32** and #**57**). More specialized procedures are available, that is, **PutGText** for static text (like the title of the simulation, although the use of procedure **SetTitle** is recommended) and **PutText** for text that changes content or location during the simulation. Program **P5_15A.PAS** in Appendix B and the companion program **P5_15B.PAS** (listing not included) exemplify the use of these text output procedures and that of the inquiry procedures **PutDist** and **PutAng**, all four available from unit **LibMec2D**. The distance and angle inquiry procedures **PutDist** and **PutAng** have the following headings:

```
PutDist(Color:Word; xA,yA,xB,yB:double; ExtL:double; Dim:string);

PutAng(Color:Word; x1,y1,x0,y0,x2,y2:double; ExtL:double;
Dim:string);
```

They display in color **Color** the distance from point (**xA,yA**) to point (**xB,yB**), or the angle at (**x0,y0**) formed with additional points (**x1,y1**) and (**x0,y0**), respectively. If points (**x1,y1**) and (**x0,y0**) coincide, then the angle displayed by **PutAng** will be measured from a line parallel to the *OX* axis. The first two and the last two characters of parameter **Dim** can be set to either "|<", "|", "<" or to ">|", "|", ">" respectively, to control the insertion of the extension lines and arrow heads of the dimension line or dimension arc. If the remainder of the characters in the string **Dim** are empty spaces, then the angle (in degrees) or distance will be calculated using the available point coordinates, and will be displayed on the screen. By default, the number of digits used to display these angles or distances is four, but it can be increased by calling procedures **PutDist** and **PutAng** with the **Dim** parameter set equal to five or more consecutive spaces (flanked or not by combinations of "|", "<", ">", or "|" characters). If parameter **Dim** transmitted to these procedures is other than an empty string or consecutive blank spaces, the actual **Dim** value will be displayed (less the control characters, if provided).

Program **P5_15.PAS** in Appendix B (which is a modification of earlier program **P5_04.PAS**—see also Figure 5.4) illustrates the use of procedures **PutDist** and **PutAng**. The program simulates two cranks jointed in series as shown in Figure 5.15a. Using procedure **PutText**, the input values of the two crank angles are displayed on top of the screen, together with the distance measured between ground joint *A* and endpoint *C* (see lines #**38** and #**39**). Similarly, the title of the simulation is displayed by calling procedure **PutGText** on line #**27**. Note the use with these procedures of the generic variable " _ " to designate the *x* and *y* coordinates of the left corner and top of the screen, and of the separator "**n**" to break the text in multiple lines. Also note on line #**35** of the program how procedure **PutDist** was called with the extension line length **ExtLLgt** set equal to either +8 or −8, depending on the orientation of the vector loop *ABC*. By doing so, the dimension line does not intersect the two cranks as they rotate during the simulation.

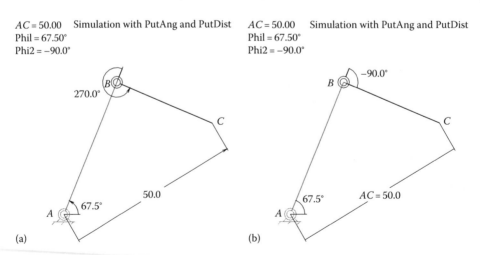

FIGURE 5.15   Simulation of two cranks jointed in series that are independently driven, produced with programs (a) **P5_15a.PAS** and (b) **P5_15b.PAS**. See also animation files **P5_15a.GIF** and **P5_15b.GIF**.

Because the second crank angle **Phi2** is always negative, there is a mismatch between the value displayed by procedure **PutAng** and the value printed on the top of the screen by procedure **PutText** (see Figure 5.15a). This is because procedure **PutAng** always measures angles in the positive direction. One way of displaying negative angles in a simulation is to transmit the angle value to the procedure via parameter **Dim**, as it has been done in program **P5_15B.PAS** is used to produce Figure 5.15b. Program **P5_15OLD.PAS** (see also animation file **P5_15OLD.GIF**—both available with the book) are additional examples of procedures **PutAng** and **PutDist** use.

5.8 ADDING COMPLEX SHAPES TO SIMULATIONS: PROCEDURES **Base**, **Link**, **gShape**, AND **Shape**

In order to add realism to a simulation, or to check for possible interferences between moving bodies or between them and other surroundings objects, it is helpful to include shapes in a simulation. Distinction is made between shapes attached to the ground, which do not change location and are written only once to the **DXF** file, and shapes attached to moving links, which change their position and orientation and must be written to separate **DXF** layers. Procedures **Base**, **Link**, **gShape**, and **Shape** available from unit **LibMech2D** serve such purposes. The first two of these procedures have the following syntaxes:

```
Base(Color, xA,yA,xB,yB, w, rA,rB);

Link(Color, xA,yA,xB,yB, w, rA,rB);
```

They allow rectangular shapes of color **Color** (filleted or chamfered at the corners) to be aligned with points (**xA,yA**) and (**xB,yB**). The width of the rectangle is specified through the parameter **w**, while **rA** and **rB** are the fillet radii of the corners adjacent to end **A** and end **B**, respectively. If either **rA** or **rB** is a negative number, then chamfering rather than filleting at the respective corners of the rectangle is performed instead. Program **P5_16A.PAS** in Appendix B exemplifies the use of these two procedures to animate a rectangular crank that rotates about a base—see Figure 5.16a and animation files **F5_16a.GIF**, as well as **F5_16a-1.GIF** and **F5_16a-2.GIF**. The frames in these last two files have been obtained by setting the parameter **Col** on line #**15** of the program to **-2** and **0**, respectively.

Note in program **P5_16A.PAS** the use of procedures **gShape** and **Shape** to plot a stationary circle of radius 1.6 representing the driving shaft of the crank (line #**31**) and a circle of radius 0.8 centered at point (**xA,yA**) of the crank (line #**32**). However, the full merit of procedures **gShape** and **Shape** is that they allow complex shapes to be read from file and be placed to the ground or attached them to moving links. The headings of these two procedures are

```
gShape(FxyName, Color, xA,yA);

Shape(FxyName, Color, xA,yA, xB,yB);
```

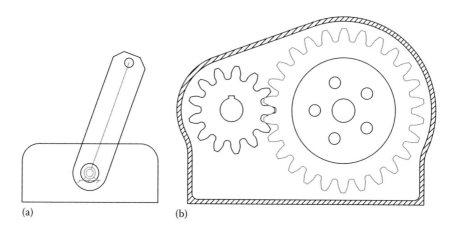

(a)   (b)

FIGURE 5.16   Kinematic simulation of a crank rotating about a (a) base and of a (b) one-stage gear reducer created using procedures **Base**, **Link**, **gShape**, and **Shape**. See also animation files **F5_16a.GIF** and **F5_16b.GIF**.

where **FxyName** is the name of the ASCII file from where the x and y coordinates of the polylines forming the shape are read. The polylines read from file will be plotted on the screen in color **Color** and will have their origin translated to the point of coordinates **xA** and **yA**. Procedure **Shape** requires one additional distinct point (**xB**, **yB**) that serves as a point along the x-axis of shape.

If parameter **FxyName** transmits to the procedure a number rather than a file name, then a circle centered at (**xA**, **yA**) and of radius **FxyName** will be plotted on the screen and to the current **DXF** file. In case of procedure **Shape** only, if **FxyName** equals the empty string or the name of an inexistent file, then a circle centered at (**xA**, **yA**) and passing through point (**xB**, **yB**) will be drawn instead.

Program **P5_16B.PAS**, listed in Appendix B, is a second example of a kinematic simulation that uses complex shapes in the form of two gears attached to two synchronously rotating cranks. The result of the simulation is visible in Figure 5.16b and in the animation file **F5_16b.GIF** available with the book. Note that the pinion is provided with a center hole and a keyway also read from the **Pinion.XY** file. The driven gear also includes a center hole, a rim circle, and **nh** peripheral holes. All these circles are drawn by separately calling procedure **Shape** with no file name as argument (see lines #**40**, #**41** and #**44** of the program).

The shapes supplied as ASCII files to procedures **Shape** and **gShape** were recorded as x and y coordinates of polyline vertices. Multiple polylines can be written to the same file using "-----" separators or a pair of **InfD** values, and their color can be changed from file as discussed in Chapter 1. A convenient way to generate complex shapes is to draw them in **AutoCAD**, export them to **R12 DXF**, and then use the **UTIL~DXF** program to write the x and y coordinates of selected polylines to ASCII files. If any of these shapes include arches of circles, full circles, or splined polylines, then such entities must be discretized by plotting them to a **PLT** file first (see Chapter 3), then using program **UTIL~PLT**, the shapes from **PLT** are then converted to **DXF**, so they can be opened into **AutoCAD** to be scaled and translated back to their original size and location. Only now are they ready for **DXF** export and vertex extraction to ASCII file using the **UTIL~DXF** program. The same steps can be applied to decimate the number of vertices of involute gears generated with **Gears. LSP** for the purpose of shortening the refreshing time in an animation.

5.9 SIMULATIONS ACCOMPANIED BY PLOTS WITH SCAN LINES AND SCAN POINTS

Program **P3_04.PAS** introduced in Chapter 3 was a first example of a simulation accompanied by dynamic plots with scan lines and scan points and **PCX** output. Note that the scan lines and scan points generated by calling procedures **PlotScanLine** and **PlotScanPoint** are drawn on the screen only. In this section, the example of a rotating vector (a phasor) accompanied by a plot of the projection of the tip of the vector on the vertical axis will be discussed (Figure 5.17). Here (see program **P5_17A.PAS** in Appendix B), both **PCX** and multilayer **DXF** output are possible. An alternative approach discussed with reference to program **P5_17B.PAS** is one where a multilayer **DXF** file of a simulation is combined inside **AutoCAD** with the **DXF** export of a comet plot generated using the **D_2D** program.

Compared to the example considered in Chapter 3, program **P5_17A.PAS** performs the kinematic calculations inside the main simulation loop, rather than only once before the animation begins. Such a strategy is better suited to simulations that employ the procedures in units **LibMecIn** and **LibAssur** discussed in Chapter 6. To animate scan lines and scan points inside **AutoCAD** using the **M_3D.LSP** application, procedures **DXFScanLine** and **DXFScanPoint** are called from unit **LibMech2D** in the process of generating the **DXF** file output (see lines #**70** and #**71**). Adding the scan lines and scan points to the **DXF** file actually occurs when calling procedure **PlotCurve** on line #**73**.

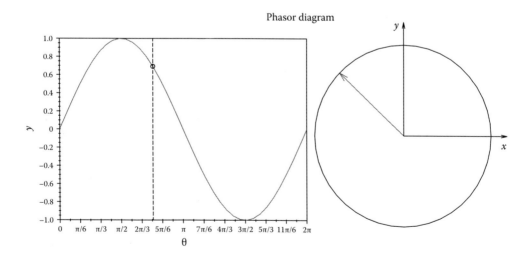

FIGURE 5.17 Kinematic simulation of a phasor accompanied by a dynamic plot of the projection of the end of the vector onto the y-axis. See also animation files **F5_17-PCX.GIF**, **F5_17-DXF.GIF**, and **F5_17-D_2D.GIF**.

The screen version of the same scan lines are turned *on* when lines #49 and #50 are executed and are turned back *off* when lines #55 and #56 are executed.

One full animation cycle occurs inside the *repeat–until* loop (lines #35 and #79), where the successive positions of the phasor are calculated and displayed (see also lines #36 to #58). Inside this same *for* loop, vectors **_yA** and **_Theta** required to plot the graph to the left in Figure 5.17 are also generated. After the first kinematic calculations are completed and only if **FirstTime** is TRUE, the graph of **_yA** vs. **_Theta** is generated (see lines #60 to #76).

The **DXF** files **F5_17-1.DXF** (the phasor) and **F5_17-2.DXF** (the graph with scan line and scan point) that were juxtaposed inside **AutoCAD** and served to generate the animated **GIF** file **F5_17a-DXF.GIF** occur during the first simulation cycle. During the second simulation cycle, the program generates **PCX** copies of the entire screen. These **PCX** frames were then assembled in the animated **GIF** file **F5_17a-PCX.GIF**.

Comparable results can be alternatively obtained by combining inside **AutoCAD** a multilayer **DXF** copy of the animated phasor and a **DXF** file export of the phasor projection vs. phasor angle done using **D_2D**. The program that produces both the phasor animation and the data file for **D_2D** plotting (i.e., **F5_17B.D2D**) is listed in Appendix B.

Because **D_2D** cannot generate an overlap of a scan line and scan point, **DXF** exports of the two type of comet plots with nonaccumulating frames had to be generated separately—see also configuration files **F5_17B-1.CF2** and **F5_17-B 2.CF2**. These were then assembled inside **AutoCAD** and the result visible in the animation file **F5_17B.GIF** obtained.

Note that the phasor length *OA* was set equal to half of the plot box height (line #11) so that no scaling is required when the **DXF** export of the vector simulation and of the animated plot are combined inside **AutoCAD**. However, before writing it to the data file **F5_17B.D2D**, the *y*-axis projection of the phasor is normalized (see line #31).

***

The procedures in unit **LibMec2D** discussed in this chapter allow the simulation of rotary and linear motors and actuators (procedures **Crank**, **gCrank**, **Slider**, and **gSlider**) and of the motion of points attached to moving links (**Offset** and **OffsetV**). Also available from unit **LibMec2D** are procedures **AngPVA**, **Ang3PVA**, and **Ang4PVA**, useful for calculating the position, velocity, and acceleration of moving links, and procedure **VarDist**, which allows the calculation of the variable distance between two moving points and its first and second time derivatives. Inquiry procedures **PutAng** and **PutDist** can be used to monitor the change of angles and distance of interest. For adding complex shapes to a simulation in the form of polylines read from files, procedures **Base**, **Link**, **gShape**, and **Shape** are provided in unit **LibMec2D**. Vectors can be represented as arrows using procedure **PutVector**. At the end, two approaches to producing simulations accompanied by animated graph with scan lines and scan points were given.

REFERENCES AND FURTHER READINGS

Craig, J. J. (2004). *Introduction to Robotics: Mechanics and Control.* Upper Saddle River, NJ: Prentice Hall.

Giancoli, D. C. (2008). *Physics for Scientists & Engineers with Modern Physics.* Upper Saddle River, NJ: Prentice Hall.

Goldstein, H., Poole Jr., C. P., and Safko, J. L. (2001). *Classical Mechanics.* Boston, MA: Addison-Wesley.

Meriam, J. L. and Kraige, L. G. (2012). *Engineering Mechanics: Dynamics.* Hoboken, NJ: John Wiley & Sons.

CHAPTER 6

Kinematic Analysis of Planar Linkage Mechanisms Using Assur Groups

THIS CHAPTER IS DEVOTED TO THE KINEMATIC ANALYSIS of planar mechanisms that employ turning and sliding joints only, also known as *linkage mechanisms* or *linkages* in short. Numerous such mechanisms can be analyzed by decomposing them into input link(s), plus subassemblies of links and joints that stand alone have zero degrees of freedom (DOFs). These subassemblies are known as *Assur groups*, named after the Russian engineer L. V. Assur who discovered them at the turn of the twentieth century. When such a zero DOF subassembly consists of two links and three joints, known as *dyad*, the corresponding kinematic equations can be solved analytically rather than numerically, and therefore allow for very fast computer implementations. The kinematic equations of all known dyads are derived in this chapter. They were also programmed in a number of Pascal procedures gathered in unit **LibAssur** available with the book. By calling these procedures in the same order in which the actual linkage mechanism has been formed, starting with the input member(s), the position, velocity, and acceleration of any moving link or point of the mechanism can be calculated, while supplementary, the whole mechanism can be animated over a given motion range.

6.1 ASSUR GROUP–BASED KINEMATIC ANALYSIS OF LINKAGE MECHANISMS

It is assumed that the reader has some knowledge of mechanism kinematics, including link and joint identification and mobility calculation. If this knowledge is limited, then a review of the relevant sections from any of the textbooks on Mechanism Theory listed at the end of the chapter is recommended.

Given a planar mechanism with n total number of links (including the fixed link), j_1 total number of joints with one DOF, and j_2 total number of joints with two DOFs, the mobility of the mechanism is given by the following formula:

$$m = 3(n-1) - 2j_1 - j_2 \tag{6.1}$$

Equation 6.1, known as the *Gruebler–Kutzbach criterion*, essentially indicates that in order for all the links of the mechanism to have a determinate motion, the mechanism must have m independent inputs. These inputs can be in the form of powered joints or of links driven by external forces or moments.

For the needle drive mechanism of a sewing machine in Figure 6.1, the mobility equation writes

$$m = 3(6-1) - 2 \cdot 7 = 1 \tag{6.2}$$

Note that at B there is a turning pair (a pin joint) overlapped with a prismatic pair (a sliding joint), and both must be accounted for when evaluating the total number of single *DOF* joints j_1. Topologically, the mechanisms in Figure 6.1 are formed by amplifying a crank OA with an RRT dyad (the two Rs stand for the two rotational joints and T stands for the translational or sliding joint) and with an *RRR* dyad with three rotational joints R. The way these

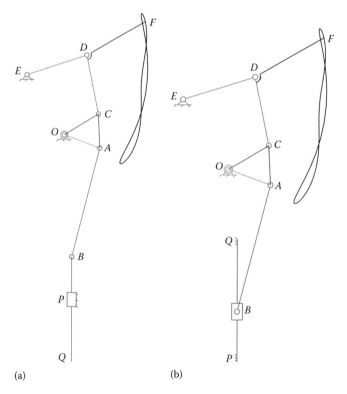

(a)   (b)

FIGURE 6.1 Mechanisms of a sewing machine simulated with program **P6_01.PAS**, which employs a crank OA with an offset point C, an RRR dyad with a coupler point F, and an RRT dyad. Mechanism (a) uses an RR_T isomer, and mechanism (b) uses an RRT_ isomer of the RRT dyad. See also animation file **F6_01.GIF**.

entities are assembled will become apparent after viewing the animation file **F6_01.GIF** and from studying the simulation programs **P6_01.PAS** listed in Appendix B.

Based on the name of the output **DXF** file being either **F6_01A** or **F6_01B** (see line #**14**), the program calls from unit **LibAssur** procedure **RRT_** or procedure **RR_T** to model the RRT dyad. Figure 6.1 and animation file **F6_01.GIF** show the differences between the ways the needle slider is represented by these two procedures. These two embodiments of the RRT dyad will be called *isomers*. The animated **GIF** file **F6_01.GIF** available with the book has been produced inside **AutoCAD** using the **M_3D.LSP** application, by combining together the corresponding **DXF** files generated by the **P6_01.PAS** program.

Other than the rotary and linear motors discussed in Chapter 5, actuators like those shown in Figure 6.2 can be used as inputs in the construction of linkage mechanisms. Of these, the RTRR actuator (Figure 6.2b) is more widely used in practice, while the RTRTR actuator (Figure 6.2a) occurs in rope shovels and some parallel robots (see also Chapter 9).

A summary of all known dyads and of their possible isomers is given in Figure 6.3, of which the RRR dyad and the RR_T and RRT_ isomers of the RRT dyad have already been mentioned with reference to program **P6_01.PAS**. Figure 6.3 shows these dyads and their isomers in their most general as well as simplified configurations. Representative linkage mechanisms that can be modeled using the respective dyads are also given on the last row in Figure 6.3. Note that no distinction has been made between the TRR and RRT dyads, and the TTR and RTT dyads. This is because the kinematic equations are independent of the direction in which motion is transmitted between their links. Also notice that a TTT dyad has not been included in this classification since by itself it has a stand-alone mobility of one rather than zero.

In the remainder of this chapter, the kinematic equations of the actuators in Figure 6.2 and of the *Assur groups* in Figure 6.3 will be derived. These equations have been programmed in a number of Pascal procedures gathered in unit **LibMecIn** (i.e., procedures **RTRTR, RTRTRc, RTRR,** and **RTRRc**) and in unit **LibAssur** (i.e., procedures **RRR, RRRc, RR_T, RRT_, RT_R, T_R_T, _TRT_, T_RT_, R_T_T, RT__T, R_TT_,** and **RT_T_**).

Important: If there is interest only in the position results or only in the position and velocity results, these procedures can be called with their velocity or velocity and acceleration parameters set equal to constant **InfD** defined in unit **LibMath**. Moreover, the names

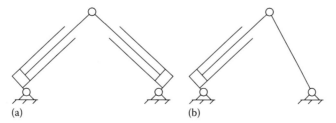

FIGURE 6.2   Double oscillating-slide actuator RTRTR (a) and single oscillating-slide actuator RTRR (b), available as procedures in unit **LibMecIn**. They can be pin-jointed to the ground (as shown) and jointed to the same moving link or to two separate moving links.

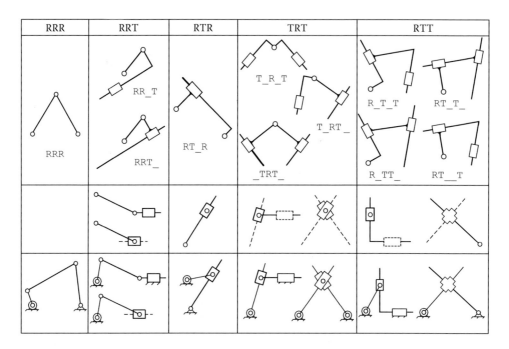

RRR	RRT	RTR	TRT	RTT	

FIGURE 6.3   Isomers of the five known dyads available as procedures in unit **LibAssur**, their simplified embodiments with overlap joints (third row), and a few representative applications (fourth row).

assigned to these output velocity and/or acceleration variables do not have to be distinct. The generic variable _ defined in the interface section of unit **LimMath**, which is preassigned the value **InfD**, should be used according to the aforementioned convention (see program **P6_01.PAS** in Appendix B).

6.2 INTERSECTION BETWEEN TWO CIRCLES: PROCEDURE **INT2CIR**

The position analysis of the RTRTR and RTRR oscillating-slide actuators and that of the RRR dyad can be reduced to finding the coordinates of the intersection points between two circles centered at A and B and of radii r_1 and r_2 as shown in Figure 6.4. The (x, y) coordinates of these intersection points C_1 and C_2 must simultaneously satisfy the following equations:

$$\begin{cases} (x - xA)^2 + (y - yA)^2 = r_1^2 \\ (x - xB)^2 + (y - yB)^2 = r_2^2 \end{cases} \tag{6.3}$$

which after expanding the squared binomials become

$$\begin{cases} x^2 - 2xA \cdot x + xA^2 + y^2 - 2yA \cdot y + yA^2 = r_1^2 \\ x^2 - 2xB \cdot x + xB^2 + y^2 - 2yB \cdot y + yB^2 = r_2^2 \end{cases} \tag{6.4}$$

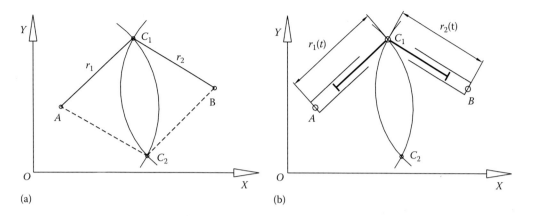

FIGURE 6.4 Schematic for calculating the intersection points between two circles (a) and for calculating the velocity and acceleration of point C, when A and B are moving and r_1 and r_2 change with time (b).

Subtracting the first equation from the second one yields

$$2(xA - xB) \cdot x + 2(yA - yB) \cdot y = r_2^2 - r_1^2 + xA^2 - xB^2 + yA^2 - yB^2 \tag{6.5}$$

which allow unknown coordinates x and y to be explicited one with respect to the other, that is,

$$x = -\frac{yA - yB}{xA - xB} \cdot y + \frac{r_2^2 - r_1^2 + xA^2 - xB^2 + yA^2 - yB^2}{2(xA - xB)} \tag{6.6a}$$

$$y = -\frac{xA - xB}{yA - yB} \cdot x + \frac{r_2^2 - r_1^2 + xA^2 - xB^2 + yA^2 - yB^2}{2(yB - yA)} \tag{6.6b}$$

For convenience, we introduce the following notations in Equation 6.6:

$$\begin{aligned} x &= a_1 \cdot y + b_1 \\ y &= a_2 \cdot x + b_2 \end{aligned} \tag{6.7}$$

where coefficients a_1, b_1, a_2, and b_2 can be easily identified by matching terms. We then substitute Equation 6.7 back into Equation 6.4 and obtain

$$\begin{cases} (a_1 \cdot y + b_1)^2 - 2xA \cdot (a_1 \cdot y + b_1) + xA^2 + y^2 - 2yA \cdot y + yA^2 = r_1^2 \\ x^2 - 2xB \cdot x + xB^2 + (a_2 \cdot x + b_2)^2 - 2yB \cdot (a_2 \cdot x + b_2) + yB^2 = r_2^2 \end{cases} \tag{6.8}$$

After squaring the binomials and rearranging terms, we further get

$$\begin{cases} (a_1^2 + 1) \cdot y^2 + 2(a_1 \cdot b_1 - xA \cdot a_1 - yA) \cdot y + xA^2 - 2xA \cdot b_1 + b_1^2 + yA^2 - r_1^2 = 0 \\ (a_2^2 + 1) \cdot x^2 + 2(a_2 \cdot b_2 - yB \cdot a_2 - xB) \cdot x + yB^2 - 2yB \cdot b_2 + b_2^2 + xB^2 - r_2^2 = 0 \end{cases} \tag{6.9}$$

These are two independent quadratic equations of solutions:

$$y = \frac{\left(-a_1 \cdot b_1 + xA \cdot a_1 + yA \pm \sqrt{\Delta_1}\right)}{\left(a_1^2 + 1\right)} \tag{6.10a}$$

$$x = \frac{\left(-a_2 \cdot b_2 + yB \cdot a_2 + xB \pm \sqrt{\Delta_2}\right)}{\left(a_2^2 + 1\right)} \tag{6.10b}$$

where

$$\Delta_1 = (a_1 \cdot b_1 - xA \cdot a_1 - yA)^2 - (a_1^2 + 1)(xA^2 - 2xA \cdot b_1 + b_1^2 + yA^2 - r_1^2)$$
$$\Delta_2 = (a_2 \cdot b_2 - yB \cdot a_2 - xB)^2 - (a_2^2 + 1)(yB^2 - 2yB \cdot b_2 + b_2^2 + xB^2 - r_2^2) \tag{6.11}$$

These equations were implemented in procedure **x2Circles** available from unit **LibGe2D**, which is in turn called by procedure **Int2Cir** in unit **LibMec2D**. This latter procedure has the heading

```
procedure Int2Cir(xA,yA, xB,yB, r1,r2:double; LftRgt:shortint;
var xC,yC, Delta:double);
```

The correspondence between the formal parameters of this procedure and the notations used in Equations 6.3 through 6.11 and in Figure 6.4 is summarized in the following tables:

Input parameters of procedure **Int2Cir**:

xA	yA	r_1	r_2	± 1
xA	yA	r1	r2	LftRgt

Output parameters of procedure **Int2Cir**:

x	y	Δ_1 if $|xA-xB|>|yA-yB|$ or Δ_2 if $|xA-xB|<|yA-yB|$
xC	yC	Delta

Note that for certain relative positions of points A and B, Equations 6.6 can result in divisions by zero. To avoid this, inside procedure **x2Circles**, denominators $(xA-xB)$ and $(yA-yB)$ are evaluated first, and depending on the magnitude of their absolute values, either Equations 6.6a and 6.10a or Equations 6.6b and 6.10b are employed. Consequently, variable **Delta** returned by procedure **Int2Cir** may exhibit occasional first- and higher-order discontinuities, as discussed in more detail in Section 6.6.

The double sign $\pm$ in Equation 6.10 denotes the two possible intersection configurations shown in Figure 6.4, resulting in point C_1 or point C_2. To resolve this ambiguity, **Int2Cir** checks the orientation of the triangular loops AC_1B and AC_2B by evaluating the cross product $\mathbf{AC} \times \mathbf{AB}$ using procedure **S123** from unit **LibGe2D**. Of the two variants, the x and y pair for which the sign of the cross product $\mathbf{AC} \times \mathbf{AB}$ is equal to the input variable **LftRgt** will be returned as solution.

6.3 VELOCITY AND ACCELERATION OF THE INTERSECTION POINTS BETWEEN TWO CIRCLES: PROCEDURE **Int2CirPVA**

For added generality, we now assume that points A and B move with known velocities and accelerations. We also assume that radii r_1 and r_2 of the two intersecting circles do not remain constant, but rather vary smoothly (i.e., time derivatives $\dot{r}_1, \dot{r}_2, \ddot{r}_1$ and $\ddot{r}_2$ exist and are continuous functions) with time. The velocities $\dot{x}$ and $\dot{y}$ and accelerations $\ddot{x}$ and $\ddot{y}$ of intersection points C_1 and C_2 can be determined through differentiation, yielding sets of two linear equations that are very easy to solve. For scalar velocities, we differentiate once with respect to time Equation 6.3 and obtain:

$$\begin{cases} 2(x-xA)\cdot(\dot{x}-\dot{x}A)+2(y-yA)\cdot(\dot{y}-\dot{y}A)=2r_1\cdot\dot{r}_1 \\ 2(x-xB)\cdot(\dot{x}-\dot{x}B)+2(y-yB)\cdot(\dot{y}-\dot{y}B)=2r_2\cdot\dot{r}_2 \end{cases} \quad (6.12)$$

After rearranging terms, these two equations become:

$$\begin{cases} (x-xA)\cdot\dot{x}+(y-yA)\cdot\dot{y}=(x-xA)\cdot\dot{x}A+(y-yA)\cdot\dot{y}A+r_1\cdot\dot{r}_1 \\ (x-xB)\cdot\dot{x}+(y-yB)\cdot\dot{y}=(x-xB)\cdot\dot{x}B+(y-yB)\cdot\dot{y}B+r_2\cdot\dot{r}_2 \end{cases} \quad (6.13)$$

The acceleration equations are obtained by differentiating Equations 6.13:

$$\begin{cases} (\dot{x}-\dot{x}A)\cdot\dot{x}+(x-xA)\cdot\ddot{x}+(\dot{y}-\dot{y}A)\cdot\dot{y}+(y-yA)\cdot\ddot{y}= \\ \quad (\dot{x}-\dot{x}A)\cdot\dot{x}A+(x-xA)\cdot\ddot{x}A+(\dot{y}-\dot{y}A)\cdot\dot{y}A+(y-yA)\cdot\ddot{y}A+\dot{r}_1^2+r_1\cdot\ddot{r}_1 \\ (\dot{x}-\dot{x}B)\cdot\dot{x}+(x-xB)\cdot\ddot{x}+(\dot{y}-\dot{y}B)\cdot\dot{y}+(y-yB)\cdot\ddot{y}= \\ \quad (\dot{x}-\dot{x}B)\cdot\dot{x}B+(x-xB)\cdot\ddot{x}B+(\dot{y}-\dot{y}B)\cdot\dot{y}B+(y-yB)\cdot\ddot{y}B+\dot{r}_2^2+r_2\cdot\ddot{r}_2 \end{cases} \quad (6.14)$$

equivalent to

$$\begin{cases} (x-xA)\cdot\ddot{x}+(y-yA)\cdot\ddot{y}=(x-xA)\cdot\ddot{x}A+(y-yA)\cdot\ddot{y}A- \\ \quad (\dot{x}-\dot{x}A)^2-(\dot{y}-\dot{y}A)^2+\dot{r}_1^2+r_1\cdot\ddot{r}_1 \\ (x-xB)\cdot\ddot{x}+(y-yB)\cdot\ddot{y}=(x-xB)\cdot\ddot{x}B+(y-yB)\cdot\ddot{y}B- \\ \quad (\dot{x}-\dot{x}B)^2-(\dot{y}-\dot{y}B)^2+\dot{r}_2^2+r_2\cdot\ddot{r}_2 \end{cases} \quad (6.15)$$

Equations 6.13 through 6.15 have been implemented in procedure `Int2CirPVA` part of unit `LibMec2D`, which returns the position (by calling procedure `Int2Cir`), velocity, and acceleration of the desired intersection point C between the two circles of moving centers A and B and of variable radii r_1 and r_2. The heading of procedure `Int2CirPVA` is

```
procedure Int2CirPVA(xA,yA,vxA,vyA,axA,ayA, xB,yB,vxB,vyB,axB,ayB,
r1,dr1,ddr1, r2,dr2,ddr2:double; LftRgt:shortint; var xC,yC,vxC,
vyC,axC,ayC, Delta:double);
```

The correspondence between its formal parameters and the notations used in these equations and in Figure 6.4 is summarized in the following two tables:

Input parameters of procedure `Int2CirPVA`:

xA	yA	$\dot{x}A$	$\dot{y}A$	$\ddot{x}A$	$\ddot{y}A$	xB	yB	$\dot{x}B$	$\dot{y}B$	$\ddot{x}B$	$\ddot{y}B$	AB
xA	yA	vxA	vyA	axA	ayA	xC	yC	vxC	vyC	axC	ayC	AB

r_1	$\dot{r}_1$	$\ddot{r}_1$	r_2	$\dot{r}_2$	$\ddot{r}_2$	± 1
r1	dr1	ddr1	r2	dr2	ddr2	LftRgt

Output parameters of procedure `Int2CirPVA`:

xP	yP	$\dot{x}P$	$\dot{y}P$	$\ddot{x}P$	$\ddot{y}P$	Δ
xP	yP	vxP	vyP	axP	ayP	Delta

Input parameter `LftRgt` must be assigned either +1 or −1 depending on the desired orientation of the *ACB* loop. For a counterclockwise or right-hand orientation (i.e., as you walk around the considered loop, your right hand should always point toward the outside of the loop), parameter `LftRgt` must be set to +1 or to constant `Right`. For a clockwise or left-hand orientation, parameter `LftRgt` must be set equal to −1 or `Left`, where constants `Left` and `Right` are predefined in the interface section of unit `LibMec2D`.

6.4 KINEMATICS OF THE RTRTR DOUBLE LINEAR INPUT ACTUATOR: PROCEDURE `RTRTRc`

The RTRTR double linear input actuator (Figure 6.5) has some practical applications in rope shovels, as well as in robotics and some automatic machinery. Its active elements are the two linear motors, represented in Figure 6.5a as cylinder–piston pairs. Potential joints A and B can be connected to separate moving links and to the same moving link or can be connected to the ground.

In this section, the kinematic equations of the RTRTR kinematic chain will be derived as intersections between two circles (this is known as the *constraint equation* approach) for

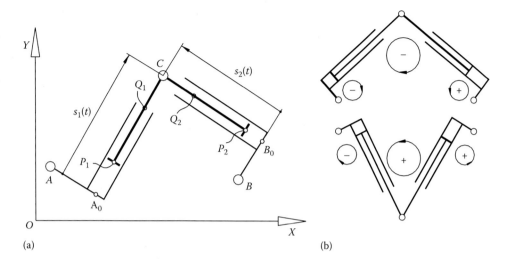

(a)   (b)

FIGURE 6.5 Notations used in the RTRT double linear input actuator kinematics (a) and the sign conventions for eccentricities A_0A and B_0B (the smaller oriented circle) and orientation of the ACB loop (the larger oriented circle) (b). Note that in figure (a), both A_0A and B_0B are positive.

the general case where potential joints A and B are attached to separate moving links. The situations where one or both of these joints are attached to the ground can be obtained as particular cases where the velocities and accelerations of point A or point B are zero.

At any instant of time, the following parameters are assumed given:

- Coordinates xA and yA of joint center A relative to the fixed reference frame OXY.

- Projections $\dot{x}A$ and $\dot{y}A$ of the velocity of A onto the axes of the fixed reference frame.

- Projections $\ddot{x}A$ and $\ddot{y}A$ of the accelerations of A onto the axes of the fixed reference frame.

- Coordinates xB and yB of joint B relative to the fixed reference frame.

- Projections $\dot{x}B$ and $\dot{y}B$ of the velocity of B onto the axes of the fixed reference frame.

- Projections $\ddot{x}B$ and $\ddot{y}B$ of the accelerations of B onto the axes of the fixed reference frame.

- Displacements s_1 and s_2 of the two pistons and their time derivatives $\dot{s}_1$, $\dot{s}_2$, $\ddot{s}_1$, and $\ddot{s}_2$ (considered positive when oriented such that the actuator expands—see Figure 6.5a).

- Lengths of the two cylinders A_0Q_1 and B_0Q_2.

- Lengths of the two pistons P_1C and P_2C.

- Eccentricities of the two cylinders A_0A and B_0B. These can be either positive or negative according to the orientation of the triangular loops AA_0C and BB_0C (see Figure 6.5b).

- Orientation of the ACB loop (see Figure 6.12b).

With these given, it is now possible to determine the following unknown parameters:

- Coordinates xC and yC of joint center C relative to the fixed reference frame.
- Projections $\dot{x}C$ and $\dot{y}C$ of the velocity of C onto the axes of the fixed reference frame.
- Projections $\ddot{x}C$ and $\ddot{y}C$ of the acceleration of C onto the axes of the fixed reference frame.
- Coordinates xP_1 and yP_1 of joint center P_1 relative to the fixed reference frame.
- Coordinates xQ_1 and yQ_1 of joint center Q_1 relative to the fixed reference frame.
- Coordinates xP_2 and yP_2 of joint center P_2 relative to the fixed reference frame.
- Coordinates xQ_2 and yQ_2 of joint center Q_2 relative to the fixed reference frame.

The coordinates of point C and its velocity and acceleration components can be found by calling procedure **Int2CirPVA** with r_1 and r_2 and their time derivatives assigned as follows:

$$r_1^2 = AC^2 = A_0 A^2 + s_1^2 \qquad\qquad r_2^2 = BC^2 = B_0 B^2 + s_2^2$$
$$\dot{r}_1 = s_1 \cdot \dot{s}_1 / r_1 \qquad\qquad \dot{r}_2 = s_2 \cdot \dot{s}_2 / r_2 \qquad\qquad (6.16)$$
$$\ddot{r}_1 = \frac{(\dot{s}_1^2 + s_1 \cdot \ddot{s}_1 - \dot{r}_1^2)}{r_1} \qquad\qquad \ddot{r}_2 = \frac{(\dot{s}_2^2 + s_2 \cdot \ddot{s}_2 - \dot{r}_2^2)}{r_2}$$

The coordinates of points A_0 and B_0 can be found by solving the following two pairs of constraint equations, similar to 6.3:

$$\begin{cases} (xA_0 - xA)^2 + (yA_0 - yA)^2 = A_0 A^2 \\ (xA_0 - xC)^2 + (yA_0 - yC)^2 = s_1^2 \end{cases} \qquad\qquad (6.17)$$

and

$$\begin{cases} (xB_0 - xB)^2 + (yB_0 - yB)^2 = B_0 B^2 \\ (xB_0 - xC)^2 + (yB_0 - yC)^2 = s_2^2 \end{cases} \qquad\qquad (6.18)$$

by simply calling procedure **Int2Cir**. In turn, the coordinates of points P_1 and Q_1 collinear with points A_0 and C can be calculated with

$$\frac{xC - xP_1}{P_1 C} = \frac{xC - xA_0}{s_1} \qquad\qquad \frac{yC - yP_1}{P_1 C} = \frac{yC - yA_0}{s_1}$$
$$\frac{xQ_1 - xA_0}{A_0 Q_1} = \frac{xC - xA_0}{s_1} \qquad\qquad \frac{yQ_1 - yA_0}{A_0 Q_1} = \frac{yC - yA_0}{s_1} \qquad (6.19)$$

Likewise, the coordinates of points P_2 and Q_2 collinear with points B_0 and C result from equations:

$$\frac{xC - xP_2}{P_2C} = \frac{xC - xB_0}{s_2} \qquad \frac{yC - yP_2}{P_2C} = \frac{yC - yB_0}{s_2}$$
$$\frac{xQ_2 - xB_0}{B_0Q_2} = \frac{xC - xB_0}{s_2} \qquad \frac{yQ_2 - yB_0}{B_0Q_2} = \frac{yC - yB_0}{s_2} \qquad (6.20)$$

Coordinates of points P_1, Q_1, P_2, and Q_2 are needed by procedure **RTRTRc** to represent graphically the two cylinders and their pistons.

Using the equations derived earlier, procedure **RTRTRc** in unit **LibMecIn** calculates the position, velocity, and acceleration of the center of pin joint C, for the case where potential joints A and B are attached to mobile elements. If the graphic system is on, the procedure also draws a schematic of the mechanism in a manner similar to Figure 6.5b. The heading of procedure **RTRTRc** is

```
procedure RTRTRc(Color:Word; xA,yA, vxA,vyA, axA,ayA, xB,yB, vxB,
vyB, axB, ayB, A0A, A0Q1, P1C, BB0, B0Q2, P2C, s1,ds1,dds1,
s2,ds2,dds2:double; LftRgt:shortint; var xC, yC, vxC, vyC, axC,
ayC, Delta:double);
```

The correspondence between the formal parameters and the notations used in Figure 6.5 and the related equations is summarized next:

Input parameters of procedure **RTRTRc**:

0...16	xA	yA	$\dot{x}A$	$\dot{y}A$	$\ddot{x}A$	$\ddot{y}A$	xB	yB	$\dot{x}B$	$\dot{y}B$	$\ddot{x}B$	$\ddot{y}B$
Color	xA	yA	vxA	vyA	axA	ayA	xB	yB	vxB	vyB	axB	ayB

A_0A	A_0Q_1	P_1C	B_0B	B_0Q_2	P_2C	s_1	$\dot{s}_1$	$\ddot{s}_1$	s_2	$\dot{s}_2$	$\ddot{s}_2$	± 1
A0A	A0Q1	P1C	B0B	B0Q2	P2C	s1	ds1	dds1	s2	ds2	dds2	LftRgt

The possible values of the input parameter **LftRgt** are −1 and +1. If the mechanism must have right-hand assembly configuration, **LftRgt** should be set equal to +1 or to constant **Right**, while for a left-hand assembly, configuration **LftRgt** must be set equal to −1 or to constant **Left** (Figure 6.5b).

Output parameters of procedure **RTRTRc** are as follows:

xC	yC	$\dot{x}C$	$\dot{y}C$	$\ddot{x}C$	$\ddot{y}C$	Δ_1 or Δ_2
xC	yC	vxC	vyC	axC	ayC	Delta

Of these, variable **Delta** returns the value of either discriminant Δ_1 or Δ_2 in Equations 6.11, as selected by procedure **Int2CirPVA**. The value of variable **Delta** can be used to assess efficiency with which the motion is transmitted throughout the mechanism or, if the mechanism does not assemble, it can be used to estimate how far from an assembly configuration the mechanism actually is (see Section 6.6).

If the graphic system is on, procedure **RTRTRc** will additionally draw in color **Color** (less if **Color** equals zero or the **BGI** constant **Black**) the pistons and their assembled cylinder as shown in Figures 6.6 and 6.7. If either joint *A*, joint *B*, or both are connected to the ground (i.e., their velocities and accelerations are zero), the respective joint is no longer represented as a circle, but rather using the grounded pin joint symbol. The pin joint at *C* is represented as a circle of radius **JtSz**, that is, an interface variable defined in unit **LibMec2D** that can be set by calling the procedure **SetJointSize**.

Figures 6.6 and 6.7 were output using programs **P6_06.PAS** (see Appendix B) and **P6_07.PAS** (source code not included). They simulate the motion of an RTRTR kinematic

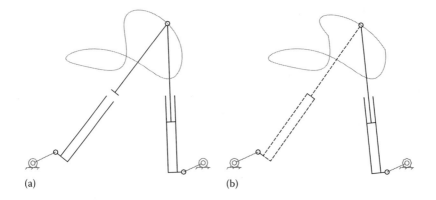

(a)   (b)

FIGURE 6.6   Simulations of an RTRTR actuator attached to two rotating cranks done with program **P6_06.PAS**. See also animation files **F6_06a.GIF** and **F6_06b.GIF**. Figure (a) has been obtained by setting variable **BumpPiston** to FALSE and figure (b) by setting variable **BumpPiston** to TRUE.

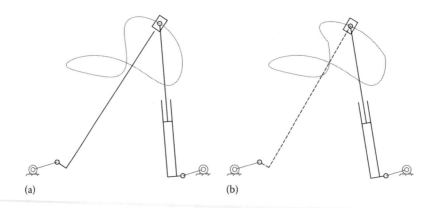

(a)   (b)

FIGURE 6.7   Simulations of an RTRTR actuator attached to two rotating cranks done with program **P6_07.PAS**. See also animation files **F6_07a.GIF** and **F6_07b.GIF**. Figure (a) corresponds to variable **BumpPiston** to FALSE and figure (b) to **BumpPiston** to TRUE.

chain having joints A and B driven by two rockers. The input angles φ_1 and φ_2 of these rockers were defined as harmonic functions of time according to equations

$$\varphi_1(t) = \frac{\pi}{3}\sin(2\pi \cdot t) \quad \text{and} \quad \varphi_2(t) = \pi + \frac{\pi}{3}\sin(2\pi \cdot t) \tag{6.21}$$

The pistons of the two actuators are also harmonically driven, that is,

$$s_1(t) = 0.65 + 0.15\cos(2\pi \cdot t) \quad \text{and} \quad s_2(t) = 0.6 + 0.12\cos(4\pi \cdot t) \tag{6.22}$$

If variable **BumpPiston** in unit **LibMec2D** is set to TRUE (see line #**17** of program **P6_06.PAS**), then procedure **RTRTRc** constrains its two pistons to remain inside their cylinders. Additionally, if the piston rod of any of the two actuator is shorter than its cylinder (i.e., $P_1C < A_0Q_1$ or $P_2C < B_0Q_2$), then joint C will not be allowed to slide inside the respective cylinder. If any of these two limit situations occur, then procedure **RTRTRc** will perform the kinematic calculations with the velocity and acceleration of the respective piston set to zero. Additionally, during the animation of the mechanism, if the linear motor becomes locked, it will be represented graphically in dashed line.

Figure 6.6 shows the screenshots of program **P6_06.PAS** generated for the case where the two pistons are constrained to remain inside their cylinders (Figure 6.6a) and for the case when they are not (Figure 6.6b), that is, **BumpPiston** equals TRUE and **BumpPiston** equals FALSE, respectively.

The companion program **P6_07.PAS** available with the book shows how procedure **RTRTRc** represents graphically the mechanism when length P_1C or P_2C is shorter than five times the current joint size as set by calling procedure **SetJointSize**. Figure 6.7 is a two-screenshot output by **P6_07.PAS** for the case where the left piston has a zero length rod, that is, $P_1C = 0$. Note that the cylinder is now represented as an L-shaped guide with a slide moving along it. By setting the **BumpPiston** parameter to TRUE, the slide block (now centered at C) will not be allowed to move outside its guide, that is, P_1 will be constrained to remain between points A_0 and Q_1 (Figure 6.5). Same as before, in its limit position, the velocity and acceleration of the slide will be forced to $\dot{s}_1 = 0$ and $\ddot{s}_1 = 0$.

Important: If procedure **RTRTRc** is called with the displacement of any of the two pistons having negative values, then the respective displacement will be automatically set to zero (as well as their time derivatives $\dot{s}_1$ and $\ddot{s}_1$ or $\dot{s}_2$ and $\ddot{s}_2$), irrespective of the **BumpPiston** setting.

Program **P6_07.PAS** was modified to drive the RTRTR actuator using longer cranks ($O_1A = 0.45$ and $O_2B = 0.35$) that oscillate at twice the initial amplitude according to the equations

$$\varphi_1(t) = \pi \cdot \sin(2\pi \cdot t) \quad \text{and} \quad \varphi_2(t) = \pi + \pi \cdot \sin(2\pi \cdot t) \tag{6.23}$$

With these modifications, the program was renamed **P6_08.PAS** and is available with the book. Note the use in this program of the *getter* procedures **GetA0**, **GetB0**, **GetP1**, **GetP2**, etc., which return the coordinates of points A_0, B_0, P_1, P_2, Q_1, and Q_2 of the RTRTR

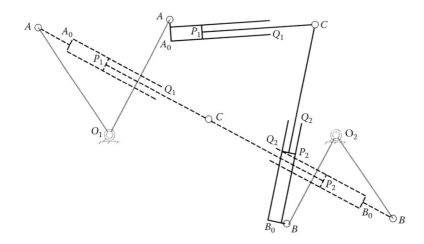

FIGURE 6.8 Two overlapped frames of an RTRTR actuator driven by two rotating cranks generated using program **P6_08.PAS**. When the mechanism cannot be assembled, its linear motors are represented in stretched and dashed lines. See also animation file **F6_08.GIF**.

actuator, not supplied by **RTRTRc**. Two overlapped positions of the mechanism generated with **P6_08.PAS** are shown in Figure 6.8, of which the one in dashed line is outside the assembly range of the mechanism. When the mechanism does not assemble, it is represented in a stretched and dashed line. The stretching of the cylinder eccentricities is due to the calculations being performed inside procedure **RTRTRc** by forcing **Delta** to zero from its negative value.

6.5 KINEMATICS OF THE RTRTR DOUBLE LINEAR INPUT ACTUATOR USING A VECTOR EQUATION APPROACH: PROCEDURE **RTRTR**

An alternative method of solving the position, velocity, and accelerations of the RTRTR double oscillating-slide actuator is the *vector-loop method*. This section discusses this second approach with reference to Figure 6.9.

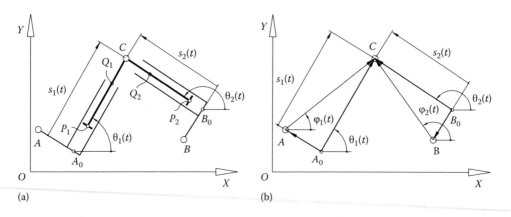

FIGURE 6.9 Oscillating-slide actuator notations (a) and vector assignment to its links (b). The sign of eccentricities A_0A and B_0B and orientation of ACB loop follow the same convention in Figure 6.5.

Begin by projecting on the axes of the OXY reference frame vector equation:

$$\mathbf{AC} - \mathbf{BC} - \mathbf{AB} = 0 \tag{6.24}$$

which yields the following pair of scalar equations:

$$\begin{cases} AC \cdot \cos(\varphi_1) - (x_B - x_A) = BC \cdot \cos(\varphi_2) \\ AC \cdot \sin(\varphi_1) - (y_B - y_A) = BC \cdot \sin(\varphi_2) \end{cases} \tag{6.25}$$

We square these equations

$$\begin{cases} AC^2 \cdot \cos^2(\varphi_1) - 2AC \cdot (x_B - x_A) \cdot \cos(\varphi_1) + (x_B - x_A)^2 = BC^2 \cdot \cos^2(\varphi_2) \\ AC^2 \cdot \sin^2(\varphi_1) - 2AC \cdot (y_B - y_A) \cdot \sin(\varphi_1) + (y_B - y_A)^2 = BC^2 \cdot \sin^2(\varphi_2) \end{cases} \tag{6.26}$$

and after adding them and rearranging terms, we obtain:

$$2AC \cdot (x_B - x_A) \cdot \cos(\varphi_1) + 2AC \cdot (y_B - y_A) \cdot \sin(\varphi_1) = AC^2 - BC^2 + (x_B - x_A)^2 + (y_B - y_A)^2 \tag{6.27}$$

Equation 6.27 is of the form $a_1 \cdot \cos(\varphi_1) + b_1 \cdot \sin(\varphi_1) = c_1$ with solutions

$$\varphi_1 = Atan2(b_1, a_1) \pm Atan2\left(\sqrt{\Delta}, c_1\right) \quad \text{where } \Delta = a_1^2 + b_1^2 - c_1^2 \tag{6.28}$$

where $Atan2(Dy, Dx) = \tan^{-1}(Dy/Dx)$ is the inverse tangent function of two arguments that uses the signs of Dx and Dy to determine the quadrant of the resultant angle (see function **Atan2** in unit **LibMath**).

A similar procedure applied to Equations 6.25 formatted as:

$$\begin{cases} BC \cdot \cos(\varphi_2) + (x_B - x_A) = AC \cdot \cos(\varphi_1) \\ BC \cdot \sin(\varphi_2) + (y_B - y_A) = AC \cdot \sin(\varphi_1) \end{cases} \tag{6.29}$$

yields an equation of the form $a_2 \cdot \cos(\varphi_2) + b_2 \cdot \sin(\varphi_2) = c_2$ with

$$a_2 = 2BC \cdot (x_B - x_A)$$

$$b_2 = 2BC \cdot (y_B - y_A) \tag{6.30}$$

$$c_2 = AC^2 - BC^2 - (x_B - x_A)^2 - (y_B - y_A)^2$$

The solutions of this trigonometric equation are:

$$\varphi_2 = Atan2(b_2, a_2) \pm Atan2\left(\sqrt{\Delta}, c_2\right) \quad \text{where } \Delta = a_2^2 + b_2^2 - c_2^2 \tag{6.31}$$

Note that discriminants Δ in Equations 6.28 and 6.31 are the same and equal to

$$\Delta = 4AC^2 \cdot BC^2 - \left[AC^2 + BC^2 - (x_B - x_A)^2 - (y_B - y_A)^2\right]^2 \tag{6.32}$$

The coordinates of joint center C that are of main interest result from projecting on the x- and y-axes of the fixed reference frame of the following vector equation:

$$\mathbf{OC} = \mathbf{OA} - \mathbf{A_0A} + \mathbf{A_0C} \tag{6.33}$$

which yields

$$\begin{cases} x_C = x_A + A_0A \cdot \sin(\theta_1) + s_1 \cdot \cos(\theta_1) \\ y_C = y_A - A_0A \cdot \cos(\theta_1) + s_1 \cdot \sin(\theta_1) \end{cases} \tag{6.34}$$

Unknown angles θ_1 and θ_2 occurring in Equation 6.34 can be obtained using the following two vector equations:

$$\begin{aligned} \mathbf{A_0A} + \mathbf{AC} - \mathbf{A_0C} &= 0 \\ \mathbf{B_0B} + \mathbf{BC} - \mathbf{B_0C} &= 0 \end{aligned} \tag{6.35}$$

The first of these vector equations projects on the x- and y-axes as

$$\begin{cases} A_0A \cdot \cos\left(\dfrac{\pi}{2+\theta_1}\right) + AC \cdot \cos(\varphi_1) - s_1 \cdot \cos(\theta_1) = 0 \\ A_0A \cdot \sin\left(\dfrac{\pi}{2+\theta_1}\right) + AC \cdot \sin(\varphi_1) - s_1 \cdot \sin(\theta_1) = 0 \end{cases} \tag{6.36}$$

and can be further written as

$$\begin{cases} s_1 \cdot \cos(\theta_1) + A_0A \cdot \sin(\theta_1) = AC \cdot \cos(\varphi_1) \\ -A_0A \cdot \sin(\theta_1) + s_1 \cdot \cos(\theta_1) = AC \cdot \sin(\varphi_1) \end{cases} \tag{6.37}$$

Similarly, the second of vector Equation 6.35 yields

$$\begin{cases} s_2 \cdot \cos(\theta_2) + B_0 B \cdot \sin(\theta_2) = BC \cdot \cos(\varphi_2) \\ -B_0 B \cdot \cos(\theta_2) + s_1 \cdot \sin(\theta_2) = BC \cdot \sin(\varphi_2) \end{cases} \tag{6.38}$$

Equations 6.37 and 6.38 are sets of two linear equations in the unknowns $\cos(\theta_1)$, $\sin(\theta_1)$, and $\cos(\theta_2)$, $\sin(\theta_2)$ that are very easy to solve.

When plotting the mechanism in a simulation, use is made, in addition to the coordinates of joint center C, of the coordinates of points A_0, B_0, P_1, P_2, Q_1, and Q_2. The coordinates of points A_0 and B_0 result from projecting on the axes of the OXY reference frame the following vector equations:

$$\mathbf{OA_0} = \mathbf{OA} - \mathbf{A_0A} \quad \text{and} \quad \mathbf{OB_0} = \mathbf{OB} - \mathbf{B_0B} \tag{6.39}$$

while the coordinates of points P_1, Q_1, P_2, and Q_2 result from vector equations

$$\begin{aligned} \mathbf{OP_1} &= \mathbf{OA_0} + \mathbf{A_0C} - \mathbf{P_1C} \quad \text{and} \quad \mathbf{OQ_1} = \mathbf{OA_0} + \mathbf{A_0Q_1} \\ \mathbf{OP_2} &= \mathbf{OB_0} + \mathbf{B_0C} - \mathbf{P_2C} \quad \text{and} \quad \mathbf{OQ_2} = \mathbf{OB_0} + \mathbf{B_0Q_2} \end{aligned} \tag{6.40}$$

Of further interest are the scalar components of the velocity of point C noted $\dot{x}_C$ and $\dot{y}_C$, obtainable by differentiating Equation 6.34 with respect to time, that is,

$$\begin{cases} \dot{x}_C = \dot{x}_A + A_0A \cdot \cos(\theta_1) \cdot \dot{\theta}_1 + \dot{s}_1 \cdot \cos(\theta_1) - s_1 \cdot \sin(\theta_1) \cdot \dot{\theta}_1 \\ \dot{y}_C = \dot{y}_A + A_0A \cdot \sin(\theta_1) \cdot \dot{\theta}_1 + \dot{s}_1 \cdot \sin(\theta_1) + s_1 \cdot \cos(\theta_1) \cdot \dot{\theta}_1 \end{cases} \tag{6.41}$$

The unknown angular velocities $\dot{\theta}_1$ and $\dot{\theta}_2$ in Equations 6.41 can be calculated using vector equation

$$-\mathbf{A_0A} + \mathbf{A_0C} - \mathbf{B_0C} + \mathbf{B_0B} = \mathbf{AB} \tag{6.42}$$

which projects on the x- and y-axes of the OXY reference frame as

$$\begin{cases} -A_0A \cdot \cos\left(\theta_1 + \dfrac{\pi}{2}\right) + s_1 \cdot \cos(\theta_1) - s_2 \cdot \cos(\theta_1) + B_0B \cdot \cos\left(\theta_2 + \dfrac{\pi}{2}\right) = x_B - x_A \\ -A_0A \cdot \sin\left(\theta_1 + \dfrac{\pi}{2}\right) + s_1 \cdot \sin(\theta_1) - s_2 \cdot \sin(\theta_1) + B_0B \cdot \sin\left(\theta_2 + \dfrac{\pi}{2}\right) = y_B - y_A \end{cases} \tag{6.43}$$

equivalent to

$$\begin{cases} A_0A \cdot \sin(\theta_1) + s_1 \cdot \cos(\theta_1) - s_2 \cdot \cos(\theta_2) - B_0B \cdot \sin(\theta_2) = x_B - x_A \\ -A_0A \cdot \cos(\theta_1) + s_1 \cdot \sin(\theta_1) - s_2 \cdot \sin(\theta_2) + B_0B \cdot \cos(\theta_2) = y_B - y_A \end{cases} \tag{6.44}$$

By differentiating Equation 6.44 with respect to time, we get

$$\begin{cases} A_0A \cdot \cos(\theta_1) \cdot \dot{\theta}_1 + \dot{s}_1 \cdot \cos(\theta_1) - s_1 \cdot \sin(\theta_1) \cdot \dot{\theta}_1 - \dot{s}_2 \cdot \cos(\theta_2) + s_2 \cdot \sin(\theta_2) \cdot \dot{\theta}_2 \\ \quad - B_0B \cdot \cos(\theta_2) \cdot \dot{\theta}_2 = \dot{x}_B - \dot{x}_A \\ A_0A \cdot \sin(\theta_1) \cdot \dot{\theta}_1 + \dot{s}_1 \cdot \sin(\theta_1) + s_1 \cdot \cos(\theta_1) \cdot \dot{\theta}_1 - \dot{s}_2 \cdot \sin(\theta_2) - s_2 \cdot \cos(\theta_2) \cdot \dot{\theta}_2 \\ \quad - B_0B \cdot \sin(\theta_2) \cdot \dot{\theta}_2 = \dot{y}_B - \dot{y}_A \end{cases} \tag{6.45}$$

After collecting terms, we obtain the sought-after set of two linear equations in the unknowns $\dot{\theta}_1$ and $\dot{\theta}_2$, that is,

$$\begin{cases} \left[A_0A \cdot \cos(\theta_1) - s_1 \cdot \sin(\theta_1) \right] \cdot \dot{\theta}_1 + \left[s_2 \cdot \sin(\theta_2) - B_0B \cdot \cos(\theta_2) \right] \cdot \dot{\theta}_2 = \\ \quad \dot{x}_B - \dot{x}_A - \dot{s}_1 \cdot \cos(\theta_1) + \dot{s}_2 \cdot \cos(\theta_2) \\ \left[A_0A \cdot \sin(\theta_1) + s_1 \cdot \cos(\theta_1) \right] \cdot \dot{\theta}_1 - \left[s_2 \cdot \cos(\theta_2) + B_0B \cdot \sin(\theta_2) \right] \cdot \dot{\theta}_2 = \\ \quad \dot{y}_B - \dot{y}_A - \dot{s}_1 \cdot \sin(\theta_1) + \dot{s}_2 \cdot \sin(\theta_2) \end{cases} \tag{6.46}$$

Finally, the scalar components of the acceleration of point C result from differentiating Equation 6.41 with respect to time:

$$\begin{cases} \ddot{x}_C = \ddot{x}_A - A_0A \cdot \sin(\theta_1) \cdot \dot{\theta}_1^2 + A_0A \cdot \cos(\theta_1) \cdot \ddot{\theta}_1 + \ddot{s}_1 \cdot \cos(\theta_1) - \dot{s}_1 \cdot \sin(\theta_1) \cdot \dot{\theta}_1 \\ \quad - \dot{s}_1 \cdot \sin(\theta_1) \cdot \dot{\theta}_1 - s_1 \cdot \cos(\theta_1) \cdot \dot{\theta}_1^2 - s_1 \cdot \sin(\theta_1) \cdot \dot{\theta}_1^2 \\ \ddot{y}_C = \ddot{y}_A + A_0A \cdot \cos(\theta_1) \cdot \dot{\theta}_1^2 + A_0A \cdot \sin(\theta_1) \cdot \ddot{\theta}_1 + \ddot{s}_1 \cdot \sin(\theta_1) + \dot{s}_1 \cdot \cos(\theta_1) \cdot \dot{\theta}_1 \\ \quad + \dot{s}_1 \cdot \cos(\theta_1) \cdot \dot{\theta}_1 - s_1 \cdot \sin(\theta_1) \cdot \dot{\theta}_1^2 + s_1 \cdot \cos(\theta_1) \cdot \ddot{\theta}_1 \end{cases} \tag{6.47}$$

The unknown angular accelerations $\ddot{\theta}_1$ and $\ddot{\theta}_2$ are solutions to the following linear equations:

$$
\begin{cases}
\left[A_0 A \cdot \cos(\theta_1) - s_1 \cdot \sin(\theta_1) \right] \cdot \ddot{\theta}_1 + \left[s_2 \cdot \sin(\theta_2) - B_0 B \cdot \cos(\theta_2) \right] \cdot \ddot{\theta}_2 = \\[4pt]
\quad - \left[-A_0 A \cdot \sin(\theta_1) \cdot \dot{\theta}_1 - \dot{s}_1 \cdot \sin(\theta_1) - s_1 \cdot \cos(\theta_1) \cdot \dot{\theta}_1 \right] \cdot \dot{\theta}_1 \\[4pt]
\quad - \left[\dot{s}_2 \cdot \sin(\theta_2) + s_2 \cdot \cos(\theta_2) \cdot \dot{\theta}_2 + B_0 B \cdot \sin(\theta_2) \cdot \dot{\theta}_2 \right] \cdot \dot{\theta}_2 \\[4pt]
\quad + \ddot{x}_B - \ddot{x}_A - \dot{s}_1 \cdot \cos(\theta_1) + \dot{s}_1 \cdot \sin(\theta_1) \cdot \dot{\theta}_1 + \dot{s}_2 \cdot \cos(\theta_2) - \dot{s}_2 \cdot \sin(\theta_2) \cdot \dot{\theta}_2 \\[6pt]
\left[A_0 A \cdot \sin(\theta_1) + s_1 \cdot \cos(\theta_1) \right] \cdot \ddot{\theta}_1 - \left[s_2 \cdot \cos(\theta_2) + B_0 B \cdot \sin(\theta_2) \right] \cdot \ddot{\theta}_2 = \\[4pt]
\quad - \left[A_0 A \cdot \cos(\theta_1) \cdot \dot{\theta}_1 + \dot{s}_1 \cdot \cos(\theta_1) - s_1 \cdot \sin(\theta_1) \cdot \dot{\theta}_1 \right] \cdot \dot{\theta}_1 \\[4pt]
\quad + \left[\dot{s}_2 \cdot \cos(\theta_2) - s_2 \cdot \sin(\theta_2) \cdot \dot{\theta}_2 + B_0 B \cdot \cos(\theta_2) \cdot \dot{\theta}_2 \right] \cdot \dot{\theta}_2 \\[4pt]
\quad + \ddot{y}_B - \ddot{y}_A - \dot{s}_1 \cdot \sin(\theta_1) - \dot{s}_1 \cdot \cos(\theta_1) \cdot \dot{\theta}_1 + \dot{s}_2 \cdot \sin(\theta_2) + \dot{s}_2 \cdot \cos(\theta_2) \cdot \dot{\theta}_2
\end{cases}
\tag{6.48}
$$

obtained by differentiating Equations 6.46 with respect to time.

Procedure **RTRTR** in unit **LibMechIn** implements these newly derived equations to solve the position, velocity, and acceleration of the RTRTR kinematic chain. Procedure **RTRTR** is fully interchangeable with procedure **RTRTRc**. Parameter **Delta** calculated with Equation 6.32 will be different however from the one returned by procedure **RTRTRc**. Also different is the way in which the mechanism is represented by **RTRTR** in the positions where it cannot be assembled and when calculations are continued by assuming **Delta** being equal to zero inside **RTRTR**. This results in the alignment of joints A, C, and B, but not together with the sliding joints of the two linear motors as procedure **RTRTRc** does (see Figure 6.10).

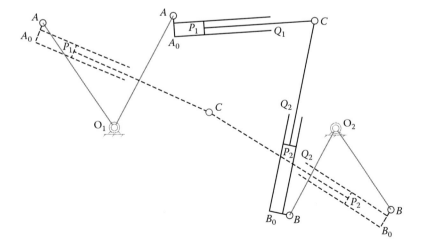

FIGURE 6.10 The same mechanism in Figure 6.8 generated using the program **P6_08.PAS** with procedure **RTRTRc** substituted with **RTRTR**. See also animation file **F6_10.GIF**.

6.6 MOTION TRANSMISSION CHARACTERISTICS OF RTRTR-BASED MECHANISMS

One of the main concerns when designing a linkage mechanism is its capability to transmit motion efficiently without overloading or jamming its joints due to the large reaction forces that may be generated. Without performing any force analysis, it is possible to estimate the motion transmission characteristics of a linkage using kinematics only. In case of the four bar and slider–crank mechanisms, the concept of *transmission angle* has been introduced for this purpose, while for cams and gear mechanisms, the *pressure angle* is utilized instead (see Chapters 7 and 8).

It will be shown that parameters **Delta** returned by procedures **RTRTRc** and **RTRTR** can be used as a measure of how close an RTRTR kinematic chain gets to a position in which it cannot be assembled. This position can be a *branching configuration* (i.e., one where the orientation of the *ACB* loop can toggle from a left-hand orientation to a right-hand orientation) or a *jamming configuration*.

Depending on the magnitude of denominators $(xA–xB)$ and $(yA–yB)$, procedure **Int2Cir** employs either Δ_1 or Δ_2 in Equations 6.11 to calculate the coordinates of point *C*, and the chosen discriminant is then assigned by procedure **RTRTRc** to variable **Delta**. Procedure **RTRTR** always assigns to variable **Delta** the value of the discriminant Δ in Equation 6.32.

To verify if there is any similarity between the discriminants that procedures **RTRTRc** and **RTRTR** operate with, program **P6_11.PAS** has been written and its listing is inserted in Appendix B. The program runs in parallel two RTRTR actuators driven by the same two cranks and having the same input functions applied to their linear motors. For every position of the simulation, the program writes to ASCII file **F6_11.TXT** the time **t**, the values of parameters **Delta** returned by both procedure **RTRTRc** and procedure **RTRTR**, and the value of the angle formed by lines *AC* and *BC* of the RTRTR loop. It also records the parameter

$$k_{ACB} = \frac{AB}{(AC + BC)} \tag{6.49}$$

Named the *triangular ratio,*

where

    AB is the distance between joints *A* and *B*
    AC is the distance between joints *A* and *C*
    BC is the distance between joints *A* and *B*

As the plots in Figure 6.11 show, there is a good degree of correlation between parameters **Delta**, the angle $<ACB$ formed by the two linear motors, and the triangular ratio k_{ACB}. It means that any of them can be used as a measure of the motion transmission characteristics of the mechanism. Note that if the mechanism does not assemble, the difference between the two **Delta** values becomes more prominent, with the one returned by

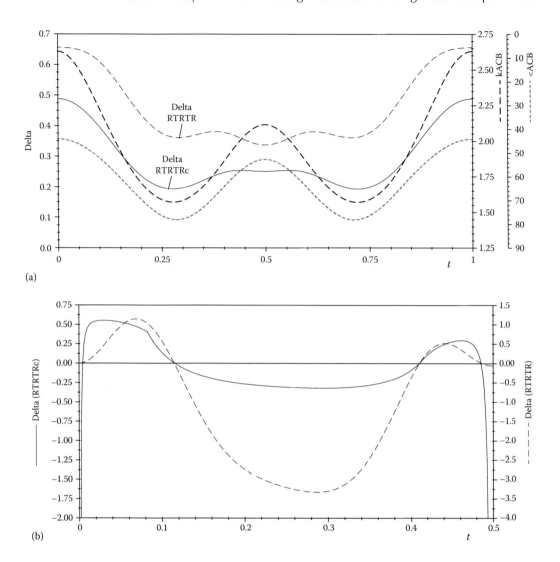

(a)

(b)

FIGURE 6.11 Plot of performance parameters returned by procedures **RTRTRc** and **RTRTR** when the mechanism assembles (a) (see also animation file **F6_11a.GIF**) and of parameters **Delta** only when the mechanism does not always assemble (b) (see also animation file **F6_11b.GIF**). Configuration files **F6_11A.CF2** and **F6_11B.CF2**.

procedure **RTRTTc** having a bigger value range. Additionally, **Delta** returned by procedure **RTRTRc** is nonsmooth, a consequence of switching between Δ_1 and Δ_2.

Animation files **F6_11a.GIF** and **F6_11b.GIF** available with the book were generated for two different sets of actuator data (see lines #**14** to #**18** of program **P6_11.PAS**). When the two mechanisms do not assemble, the differences between how procedures **RTRTRc** and **RTRTR** represent graphically the respective actuators become immediately evident (see also animation file **F6_11b.GIF**).

Useful for setting up the mechanism simulation properly is the use of procedure **SizeLinMotor** (see lines #**43**, #**44**, #**47**, and #**48** of the same program). Provided that $\max(s_1)/\min(s_1) < 2$ and $\max(s_2)/\min(s_2) < 2$, after running a complete set of linear motor displacements s_1 and s_2 through procedure **SizeLinMotor**, the values of variables **P1C**, **P2C**, **A0Q1**, and **B0Q2** will be modified such that the pistons of the two linear motors will remain inside their cylinders. Note that procedure **SizeLinMotor** must be first called with the piston-displacement argument set to variable _, which equals **InfD** (see lines #**43** and #**44**). This will reset the piston and cylinder length values so they can be updated as the simulation progresses.

Alternatively, procedure **SizeLinMotor** can be called inside the main animation loop (see program **P6_11BIS.PAS** available with the book), a case in which the continuous updating of the piston and cylinder lengths is visible during the first animation cycle of the mechanism. After that, the optimum values of variables **P1C**, **P2C**, **A0Q1**, and **B0Q2** can be printed on the screen and used for manually assigning the corresponding piston and cylinder lengths.

6.7 KINEMATIC ANALYSIS OF THE RTRR OSCILLATING-SLIDE ACTUATOR USING EQUATIONS OF CONSTRAINT: PROCEDURE **RTRRc**

A common inversion of the slider–crank mechanism is the RTRR oscillating-slide actuator (Figure 6.12). It has numerous applications in earth moving and agricultural equipment, landing gears and flight control surfaces of aircrafts, dump trucks, industrial automation, etc. The input element of an RTRR actuator is a linear motor, and its potential joints (noted A and B in Figure 6.12a) are in most cases connected to the same movable element or to the ground. Note that, with very few exceptions, in order to

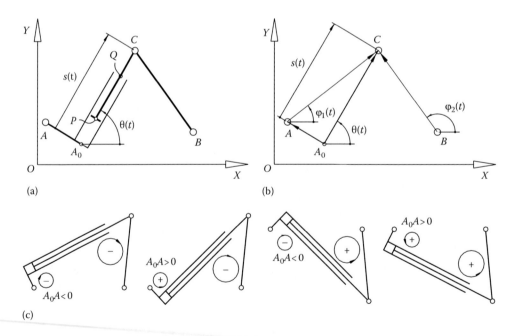

FIGURE 6.12   Notations used in solving the kinematics of the oscillating-slide actuator (a), vector assignment to its links (b), and sign convention of piston eccentricity A_0A (the smaller oriented circle) and orientation of the ACB loop (the larger oriented circle) (c).

minimize the transverse reaction forces between the piston and the cylinder, eccentricity A_0A is set equal to zero.

Following the same approach applied to the RTRTR kinematic chain, the analysis of the RTRR oscillating-slide actuator will be performed for the general case where the velocity and acceleration of joints A and B are nonzero, and the extension of its linear motor varies continuously with time. The following parameters are assumed known at any given moment of a simulation:

- Coordinates xA and yA of point A relative to the fixed reference frame OXY.

- Projections $\dot{x}A$ and $\dot{y}A$ of the velocity of A onto the axes of the fixed reference frame.

- Projections $\ddot{x}A$ and $\ddot{y}A$ of the accelerations of point A onto the fixed reference frame.

- Coordinates xB and yB of point B relative to the fixed reference frame.

- Projections $\dot{x}B$ and $\dot{y}B$ of the velocity of point B onto the fixed reference frame.

- Projections $\ddot{x}B$ and $\ddot{y}B$ of the accelerations of point B onto the fixed reference frame.

- Piston displacement s and its time derivatives $\dot{s}$ and $\ddot{s}$ relative to the cylinder, considered positive when oriented such that the actuator expands.

- Cylinder length A_0Q.

- Piston length PC.

- Rocker length BC.

- Piston eccentricity A_0A.

- Orientation of the ACB triangular loop.

We are interested in determining the following unknown parameters:

- The coordinates xC and yC of joint center C relative to the fixed reference frame OXY.

- The projections $\dot{x}C$ and $\dot{y}C$ of the velocity of point C onto the axes of the fixed reference frame.

- The projections $\ddot{x}C$ and $\ddot{y}C$ of the acceleration of C onto the axes of the fixed reference frame.

- The coordinates xP and yP of point P relative to the fixed reference frame.

- The coordinates xQ and yQ of point Q relative to the fixed reference frame.

The RTRR actuator can be viewed as a particular embodiment of the RTRTR double linear input actuator discussed earlier, where one of the motors does not change length.

The position, velocity, and acceleration components of joint center C can be calculated by calling procedure **Int2CirPVA** with r_1 and r_2 and their time derivatives assigned as follows:

$$r_1^2 = AC^2 = A_0A^2 + s_1^2 \qquad r_2 = BC$$

$$\dot{r}_1 = \frac{s_1 \cdot \dot{s}_1}{r_1} \qquad \dot{r}_2 = 0$$

$$\ddot{r}_1 = \frac{\left(\dot{s}_1^2 + s_1 \cdot \ddot{s}_1 - \dot{r}_1^2\right)}{r_1} \qquad \ddot{r}_2 = 0 \qquad (6.50)$$

The coordinates of joint center A_0 result from solving Equations 6.17 using procedure **Int2Cir**, while the coordinates of points P and Q result from solving Equations 6.19. The coordinates of these points are required to represent graphically the linear motor as the mechanism moves.

Procedure **RTRRc** calculates, using a constraint equation approach, the position, velocity, and acceleration of pin joint center C and displacements, velocity, and accelerations to joints A and B for given inputs s, $\dot{s}$, and $\ddot{s}$ of the linear motor. If the graphic system is on, procedure **RTRRc** also draws in color **Color** (less if it is assigned the **BGI** constant **Black** or the value zero) the piston, its cylinder, and the rocker BC, in a manner similar to Figure 6.12. The heading of procedure **RTRRc** is as follows:

```
procedure RTRRc(Color:Word; xA,yA, vxA,vyA, axA,ayA, xB,yB,
vxB,vyB, axB,ayB, A0A, A0Q, PC, s,ds,dds:double;  LftRgt:shortint;
var xC,yC, vxC,vyC, axC,ayC, Delta:double);
```

The correspondence between the formal parameters and the notations used earlier is summarized in the following tables:

Input parameters of procedure **RTRRc**:

0...16	xA	yA	$\dot{x}A$	$\dot{y}A$	$\ddot{x}A$	$\ddot{y}A$	xB	yB	$\dot{x}B$	$\dot{y}B$	$\ddot{x}B$	$\ddot{y}B$
Color	xA	yA	vxA	vyA	axA	ayA	xB	yB	vxB	vyB	axB	ayB

A_0A	A_0Q	PC	BC	s	$\dot{s}$	$\ddot{s}$	± 1
A0A	A0Q	PC	BC	s	ds1	dds1	LftRgt

Output parameters of procedure **RTRRc**:

xC	yC	$\dot{x}C$	$\dot{y}C$	$\ddot{x}C$	$\ddot{y}C$	Δ
xC	yC	vxC	vyC	axC	ayC	Delta

LftRgt controls the orientation of the mechanism, that is, for a right-hand assembly configuration, it must be set to +1 or to constant **Right**, while for a left-hand assembly configuration, it must be set to −1 or to constant **Left** (see Figure 6.12c). If the scalar velocities and accelerations of either joint A or B are zero, then the respective pivot joint is assumed connected to the ground and will be represented graphically accordingly, by calling procedure **gPivotJoint** in unit **LibMec2D**. Output variable **Delta** is identical to the variable with the same name from procedure **Int2CirPVA** that is called inside **RTRRc**.

6.8 KINEMATIC ANALYSIS OF THE RTRR OSCILLATING-SLIDE ACTUATOR USING A VECTOR-LOOP APPROACH: PROCEDURE **RTRR**

The kinematics of the RTRR actuator can also be solved using a vector-loop method. We recognize first that angles φ_1 and φ_2 are given by the same Equations 6.28 and 6.31 derived earlier for the case of the RTRTR kinematic chain.

Likewise, the coordinates of joint center C are

$$\begin{cases} x_C = x_A + A_0A \cdot \sin(\theta) + s \cdot \cos(\theta) \\ y_C = y_A - A_0A \cdot \cos(\theta) + s \cdot \sin(\theta) \end{cases} \tag{6.51}$$

where unknown angle θ results from solving linear equations

$$\begin{cases} s \cdot \cos(\theta) + A_0A \cdot \sin(\theta) = AC \cdot \cos(\varphi_1) \\ -A_0A \cdot \sin(\theta) + s \cdot \cos(\theta) = AC \cdot \sin(\varphi_1) \end{cases} \tag{6.52}$$

in the unknowns $\cos(\theta)$ and $\sin(\theta)$.

In order to represent graphically the mechanism, the x and y coordinates of points A_0, P, and Q must be calculated first. These result from projecting on the axes of the OXY frame the following vector equations:

$$\mathbf{OA_0} = \mathbf{OA} - \mathbf{A_0A}$$
$$\mathbf{OP} = \mathbf{OA_0} + \mathbf{A_0C} - \mathbf{PC} \tag{6.53}$$
$$\mathbf{OQ} = \mathbf{OA_0} + \mathbf{A_0Q}$$

Regarding the scalar components of the velocity of point C, these result from differentiating Equations 6.51 with respect to time:

$$\begin{cases} \dot{x}_C = \dot{x}_A + A_0A \cdot \cos(\theta) \cdot \dot{\theta} + \dot{s} \cdot \cos(\theta) - s \cdot \sin(\theta) \cdot \dot{\theta} \\ \dot{y}_C = \dot{y}_A + A_0A \cdot \sin(\theta) \cdot \dot{\theta} + \dot{s} \cdot \sin(\theta) + s \cdot \cos(\theta) \cdot \dot{\theta} \end{cases} \tag{6.54}$$

To calculate the unknown angular velocity $\dot{\theta}$, we begin with vector equation

$$-\mathbf{A}_0\mathbf{A} + \mathbf{A}_0\mathbf{C} - \mathbf{BC} = \mathbf{AB} \tag{6.55}$$

which projects on the axes of the fixed reference frame as

$$\begin{cases} -A_0A \cdot \cos\left(\dfrac{\pi}{2} + \theta\right) + A_0C \cdot \cos(\theta) - BC \cdot \cos(\varphi_2) = x_B - x_A \\[3mm] -A_0A \cdot \sin\left(\dfrac{\pi}{2} + \theta\right) + A_0C \cdot \sin(\theta) - BC \cdot \sin(\varphi_2) = y_B - y_A \end{cases} \tag{6.56}$$

Equations 6.56 simplify to

$$\begin{cases} A_0A \cdot \sin(\theta) + s \cdot \cos(\theta) - BC \cdot \cos(\varphi_2) = x_B - x_A \\[2mm] -A_0A \cdot \cos(\theta) + s \cdot \sin(\theta) - BC \cdot \sin(\varphi_2) = y_B - y_A \end{cases} \tag{6.57}$$

and by differentiating them with respect to time, we further get

$$\begin{cases} A_0A \cdot \cos(\theta) \cdot \dot{\theta} + \dot{s} \cdot \cos(\theta) - s \cdot \sin(\theta) \cdot \dot{\theta} + BC \cdot \sin(\varphi_2) \cdot \dot{\varphi}_2 = \dot{x}_B - \dot{x}_A \\[2mm] A_0A \cdot \sin(\theta) \cdot \dot{\theta} + \dot{s} \cdot \sin(\theta) + s \cdot \cos(\theta) \cdot \dot{\theta} - BC \cdot \cos(\varphi_2) \cdot \dot{\varphi}_2 = \dot{y}_B - \dot{y}_A \end{cases} \tag{6.58}$$

After collecting terms, the following set of linear equations in the unknowns $\dot{\theta}$ and $\dot{\varphi}_2$ is obtained:

$$\begin{cases} \left[A_0A \cdot \cos(\theta) - s \cdot \sin(\theta) \right] \cdot \dot{\theta} + BC \cdot \sin(\varphi_2) \cdot \dot{\varphi}_2 = \dot{x}_B - \dot{x}_A - \dot{s} \cdot \cos(\theta) \\[2mm] \left[A_0A \cdot \sin(\theta) + s \cdot \cos(\theta) \right] \cdot \dot{\theta} - BC \cdot \cos(\varphi_2) \cdot \dot{\varphi}_2 = \dot{y}_B - \dot{y}_A - \dot{s} \cdot \sin(\theta) \end{cases} \tag{6.59}$$

The scalar components of the acceleration of point C result from differentiating Equations 6.54:

$$\begin{cases} \ddot{x}_C = \ddot{x}_A - A_0A \cdot \sin(\theta) \cdot \dot{\theta}^2 + A_0A \cdot \cos(\theta) \cdot \ddot{\theta} + \ddot{s} \cdot \cos(\theta) - \dot{s} \cdot \sin(\theta) \cdot \dot{\theta} \\[2mm] \qquad - \dot{s} \cdot \sin(\theta) \cdot \dot{\theta} - s \cdot \cos(\theta) \cdot \dot{\theta}^2 - s \cdot \sin(\theta) \cdot \ddot{\theta} \\[2mm] \ddot{y}_C = \ddot{y}_A + A_0A \cdot \cos(\theta) \cdot \dot{\theta}^2 + A_0A \cdot \sin(\theta) \cdot \ddot{\theta} + \ddot{s} \cdot \sin(\theta) + \dot{s} \cdot \cos(\theta) \cdot \dot{\theta} \\[2mm] \qquad + \dot{s} \cdot \cos(\theta) \cdot \dot{\theta} - s \cdot \sin(\theta) \cdot \dot{\theta}^2 + s \cdot \cos(\theta) \cdot \ddot{\theta} \end{cases} \tag{6.60}$$

where unknown angular acceleration $\ddot{\theta}$ and $\ddot{\varphi}_2$ are solutions of equations

$$
\begin{cases}
\left[A_0A \cdot \cos(\theta) - s \cdot \sin(\theta)\right] \cdot \ddot{\theta} + BC \cdot \sin(\varphi_2) \cdot \ddot{\varphi}_2 = \\
\quad -\left[-A_0A \cdot \sin(\theta) \cdot \dot{\theta} - \dot{s} \cdot \sin(\theta) - s \cdot \cos(\theta) \cdot \dot{\theta}\right] \cdot \dot{\theta} - BC \cdot \cos(\varphi_2) \cdot \dot{\varphi}_2^2 \\
\quad + \ddot{x}_B - \ddot{x}_A - \ddot{s} \cdot \cos(\theta) + \dot{s} \cdot \sin(\theta) \cdot \dot{\theta} \\
\left[A_0A \cdot \sin(\theta) + s \cdot \cos(\theta)\right] \cdot \ddot{\theta} - BC \cdot \cos(\varphi_2) \cdot \ddot{\varphi}_2 = \\
\quad -\left[A_0A \cdot \cos(\theta) \cdot \dot{\theta} + \dot{s} \cdot \cos(\theta) - s \cdot \sin(\theta) \cdot \dot{\theta}\right] \cdot \dot{\theta} - BC \cdot \sin(\varphi_2) \cdot \dot{\varphi}_2^2 \\
\quad + \ddot{y}_B - \ddot{y}_A - \ddot{s} \cdot \sin(\theta) - \dot{s} \cdot \cos(\theta) \cdot \dot{\theta}
\end{cases}
\tag{6.61}
$$

obtained by differentiating with respect to time Equation 6.59. Both Equations 6.59 and 6.61 are very easy to solve using Cramer's rule or the inverse matrix method.

Procedure **RTRR** that implements this vector-loop approach to perform the kinematic simulation of the RTRR actuator has the same heading and is totally interchangeable with procedure **RTRRc**. The differences between the two procedures are the way the mechanism is represented in the positions in which it cannot be assembled, and the value of the discriminant returned by variable **Delta**. It is to be expected that variables **Delta** of these two procedures will exhibit characteristics comparable to **Delta** returned by **RTRTRc** and **RTRTR** discussed earlier with reference to Figure 6.11.

Programs **P6_13A.PAS** and **P6_13B.PAS** available with the book are applications of procedures **RTRRc** and **RTRR** (see Figure 6.13). Program **P6_13A.PAS** animates

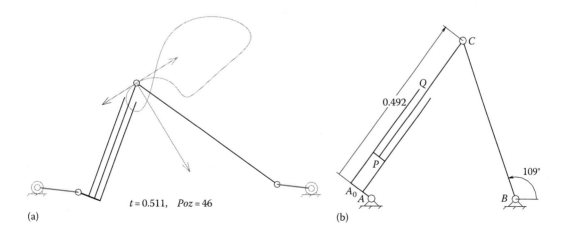

(a) $t = 0.511, \quad Poz = 46$

(b)

FIGURE 6.13   Oscillating-slide actuator with harmonic linear motor input with potential joints A and B driven by two rockers (a) and connected directly to the ground (b) generated with programs **P6_13A.PAS** and **P6_13B.PAS**. See also animation files **F6_13a.GIF** and **F6_13b.GIF**.

simultaneously (i.e., overlapped) these two kinematic chains driven by the same cranks. It also plots separately their velocity and acceleration vectors of point C. Notice that there is no visible difference between the **RTRRc** and **RTRR** output, unless the linear motor range is extended to reveal the different way in which the RTRR kinematic chain is represented when assembly is not permitted. Both programs use the **SizeLinMotor** procedure to adjust the piston rod and cylinder lengths such that piston P always remains between points A_0 and Q (see Figure 6.12a). In addition, program **P6_13B.PAS** implements an option where the minimum clearances between the piston and the two cylinder ends can be specified—see the use of variables **A0_Pmin** and **Q_Pmin**. Also notice, in the same program **P6_13B.PAS**, the use of procedures **GetA0**, **GetP**, and **GetQ** that return the coordinates of points A_0, P, and Q of the actuator.

6.9 KINEMATIC ANALYSIS OF THE RRR DYAD: PROCEDURES RRRc AND RRR

The RRR dyad is one of the most commonly encountered *Assur groups*. In this paragraph, its position, velocity, and acceleration equations will be derived using both equations of constrain and the vector-loop method.

For both approaches, the following parameters are assumed known at any given time (Figure 6.14):

- Coordinates xA, yA and xB, yB of potential joints A and B relative to the OXY fixed reference frame.

- Projections $\dot{x}A$ and $\dot{y}A$ of the velocity of joint center A onto the fixed reference frame.

- Projections $\ddot{x}A$ and $\ddot{y}A$ of the accelerations of A onto the fixed reference frame.

- Projections $\dot{x}B$ and $\dot{y}B$ of the velocity of joint center B onto the fixed reference frame.

- Projections $\ddot{x}Q$ and $\ddot{y}Q$ of the accelerations of point B onto the fixed reference frame.

- Lengths AC and BC of the two links of the dyad.

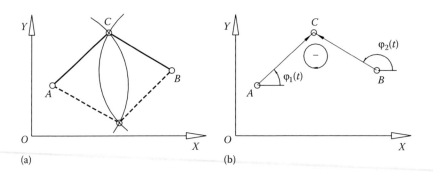

(a) (b)

FIGURE 6.14 Notations used in solving the kinematics of the RRR dyad (a) and vector and angle assignment to its links (b).

The purpose of the analysis is to determine

- Coordinates xC and yC of joint center C relative to the OXY reference frame.

- Projections $\dot{x}C$ and $\dot{y}C$ of the velocity of C onto the axes of the fixed reference frame.

- Projections $\ddot{x}C$ and $\ddot{y}C$ of the acceleration of C onto the axes of the fixed reference frame.

The coordinates of point C can be calculated using constraint Equations 6.6 and 6.10 with $r_1 = AC$ and $r_2 = BC$. The scalar components of the velocity and acceleration of point C can be calculated using Equations 6.13 and 6.15 with $\dot{r}_1 = \dot{r}_2 = \ddot{r}_1 = \ddot{r}_2 = 0$. Procedure **RRRc** in unit **LibAssur** calls procedure **Int2CirPVA** rather than implementing these equations new. Its heading is as follows:

```
procedure RRRc(Color:Word; xA,yA, vxA,vyA, axA,ayA, xB,yB,
vxB,vyB, axB,ayB, AC, BC:double;  LftRgt:shortint; var xC,yC,
vxC,vyC, axC,ayC, Delta:double);
```

The correspondence between the formal parameters and the notations used earlier is: Input parameters of procedure **RRRc**:

0...16	xA	yA	$\dot{x}A$	$\dot{y}A$	$\ddot{x}A$	$\ddot{y}A$	xB	yB	$\dot{x}B$	$\dot{y}B$	$\ddot{x}B$	$\ddot{y}B$
Color	xA	yA	vxA	vyA	axA	ayA	xB	yB	vxB	vyB	axB	ayB

AC	BC	x	y	$\dot{x}$	$\dot{y}$	$\ddot{x}$	$\ddot{y}$	± 1
AC	BC	xC	yC	dxC	dyC	ddxC	ddyC	LftRgt

Output parameters of procedure **RRRc**:

xC	yC	$\dot{x}C$	$\dot{y}C$	$\ddot{x}C$	$\ddot{y}C$	Δ
xC	yC	vxC	vyC	axC	ayC	Delta

Variable **LftRgt** specifies the orientation of the triangular loop ACB as discussed earlier. For a negative orientation of the dyad, shown in solid lines in Figure 6.14a, variable **LftRgt** should be assigned the value −1 or constant **Left**. The positive orientation of the RRR dyad, enforced by setting input variable **LftRgt** to +1 or constant **Right**, is shown by the dashed lines in Figure 6.14a.

In addition to returning the output parameters listed earlier, procedure **RRRc** also draws in color **Color** (less if **Color** equals zero or **Black** or if the graphic system is off) two lines connecting points A and C and points B and C. It also represents joint C as a circle of radius **JtSz**. If the velocities and accelerations of joints A and/or B are zero, procedure **RRRc** draws the respective joint using a grounded pin joint symbol.

For a vector-loop kinematic analysis of the RRR dyad, Equation 6.24 as well as Equations 6.28 and 6.31 can be used to calculate angles φ_1 and φ_2 (Figure 6.14). Once angles φ_1 and φ_2 are determined, the coordinates of joint C can be obtained starting from the following vector equation:

$$\mathbf{AC} = \mathbf{OA} + \mathbf{AC} = 0 \tag{6.62}$$

which projects on the axes of the OXY reference frame as

$$\begin{cases} x_C = x_A + AC \cdot \cos(\varphi_1) \\ y_C = y_A + AC \cdot \sin(\varphi_1) \end{cases} \tag{6.63}$$

The scalar components of the velocity of joint center C result from differentiation Equations 6.63 with respect to time:

$$\begin{cases} \dot{x}_C = \dot{x}_A - AC \cdot \sin(\varphi_1) \cdot \dot{\varphi}_1 \\ \dot{y}_C = \dot{y}_A + AC \cdot \cos(\varphi_1) \cdot \dot{\varphi}_1 \end{cases} \tag{6.64}$$

By differentiating Equations 6.64, the x and y components of the acceleration of point C are obtained:

$$\begin{cases} \ddot{x}_C = \ddot{x}_A - AC \cdot \cos(\varphi_1) \cdot \dot{\varphi}_1^2 - AC \cdot \sin(\varphi_1) \cdot \ddot{\varphi}_1 \\ \ddot{y}_C = \ddot{y}_A - AC \cdot \sin(\varphi_1) \cdot \dot{\varphi}_1^2 + AC \cdot \cos(\varphi_1) \cdot \ddot{\varphi}_1 \end{cases} \tag{6.65}$$

To solve for the unknown angular velocity and acceleration $\dot{\varphi}_1$ and $\ddot{\varphi}_1$, we resort to Equations 6.25 rearranged as

$$\begin{cases} AC \cdot \cos(\varphi_1) - BC \cdot \cos(\varphi_2) = x_B - x_A \\ AC \cdot \sin(\varphi_1) - BC \cdot \sin(\varphi_2) = y_B - y_A \end{cases} \tag{6.66}$$

Differentiating them once with respect to time yields a set of two linear equations in the unknowns $\dot{\varphi}_1$ and $\dot{\varphi}_2$:

$$\begin{cases} -AC \cdot \sin(\varphi_1) \cdot \dot{\varphi}_1 + BC \cdot \sin(\varphi_2) \cdot \dot{\varphi}_2 = \dot{x}_B - \dot{x}_A \\ AC \cdot \cos(\varphi_1) \cdot \dot{\varphi}_1 - BC \cdot \cos(\varphi_2) \cdot \dot{\varphi}_2 = \dot{y}_B - \dot{y}_A \end{cases} \tag{6.67}$$

The second derivatives of the same angles φ_1 and φ_2 result from solving equations:

$$\begin{cases} -AC\cdot\sin(\varphi_1)\cdot\ddot{\varphi}_1 + BC\cdot\sin(\varphi_2)\cdot\ddot{\varphi}_2 = \ddot{x}_B - \ddot{x}_A + AC\cdot\cos(\varphi_1)\cdot\dot{\varphi}_1^2 - BC\cdot\cos(\varphi_2)\cdot\dot{\varphi}_2^2 \\ AC\cdot\cos(\varphi_1)\cdot\ddot{\varphi}_1 - BC\cdot\cos(\varphi_2)\cdot\ddot{\varphi}_2 = \ddot{y}_B - \ddot{y}_A + AC\cdot\sin(\varphi_1)\cdot\dot{\varphi}_2^2 - BC\cdot\sin(\varphi_2)\cdot\dot{\varphi}_2^2 \end{cases} \quad (6.68)$$

Note that the vector method also yields the angular velocities and accelerations of the two links of the dyad directly, which can be of interest in some analyses.

Procedure **RRR** in unit **LibAssur** calculates the position, velocity, and acceleration of the RRR dyad using Equations 6.62 through 6.68, and it is interchangeable with procedure **RRRc**. Its variable **Delta** calculated using Equation 6.32 will evidently be different than the one returned by procedure **RRRc**. If the dyad cannot be assembled, both procedure **RRRc** and **RRR** will represent links AC and BC with joints A, C, and B collinear and in dashed lines.

To verify the correctness of the output by procedures **RRRc** and **RRR**, program **P6_15.PAS** has been written and is available with the book. It simulates the motion of an RTRTR kinematic chain having its linear motors locked (modeled using procedure **RTRTRc**) driven by two cranks, overlapped with an RRR dyad (modeled using procedure **RRRc** or **RRR**) driven by the same two cranks as shown in Figure 6.15. The separate graphing of the velocity and acceleration vectors of joint C renders these vectors indistinguishable over the entire motion cycle of the mechanism, a confirmation that procedures **RRRc** and **RRR** produce correct results. The needle drive mechanism simulation program **P6_01.PAS** introduced earlier is another example of procedure **RRR** and **RRRc** use.

FIGURE 6.15   Five-bar linkage simulation generated with program **P6_15.PAS** that calls procedure **RTRTR** with its linear motors locked, overlapped with procedures **RRRc** and **RRR**. See also animation file **F6_15.GIF**.

6.10 KINEMATIC ANALYSIS OF THE RRT DYAD USING A VECTOR-LOOP APPROACH

In this section, the position, velocity, and acceleration problem of the RR_T and RRT_ isomers of the RRT dyad is solved using a vector-loop approach. The kinematic equations were derived for two distinct slider configurations (see Figure 6.3) and were implemented in procedures **RRT_** and **RR_T**. In case of isomer RR_T, the potential joint is a sleeve, while in case of isomer RRT_, the translating potential joint is a rod that can be fixed or can perform some kind of controlled motion.

6.10.1 RR_T Dyadic Isomer: Procedure **RR_T**

The RRT_ isomer of the RRT dyad depicted in Figure 6.16 will be analyzed first, with the following parameters assumed specified at any instant of time:

- Length AC of the connecting rod.

- Slider eccentricity BC perpendicular to BQ (can be either positive or negative).

- Length of slider rod BQ (can be either positive or negative).

- Coordinates xA, yA of potential joint A relative to the fixed reference frame OXY.

- Angle θ of the slider axis measured counterclockwise from a parallel to the OX axis.

- Projections $\dot{x}A$ and $\dot{y}A$ of the velocity of joint center A onto the fixed reference frame.

- Projections $\ddot{x}A$ and $\ddot{y}A$ of the accelerations of A onto the fixed reference frame.

- Projections $\dot{x}P$ and $\dot{y}P$ of the velocity of joint center P onto the fixed reference frame.

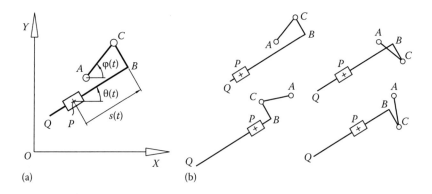

(a)   (b)

FIGURE 6.16 Kinematic diagrams of the RR_T isomer of the RRT dyad with potential joints A and P (as shown, length BQ is negative) (a) and its assembly configurations based on the sign of eccentricity BC and the double sign in Equation 6.74, which may result in a longer or shorter displacement s (b).

- Projections $\ddot{x}P$ and $\ddot{y}P$ of the accelerations of point P onto the fixed reference frame.

- First time derivative $\dot{\theta}$ of angle θ (the angular velocity of the slider).

- Second time derivative $\ddot{\theta}$ of angle θ (the angular acceleration of the slider).

The purpose of this kinematic analysis is to determine

- The coordinates xC and yC of joint center C relative to the OXY reference frame.

- Projections $\dot{x}C$ and $\dot{y}C$ of the velocity of C onto the axes of the fixed reference frame.

- Projections $\ddot{x}C$ and $\ddot{y}C$ of the acceleration of C onto the axes of the fixed reference frame.

- The coordinates xB and yB of point B relative to the OXY reference frame.

- Projections $\dot{x}B$ and $\dot{y}B$ of the velocity of B onto the axes of the fixed reference frame.

- Projections $\ddot{x}B$ and $\ddot{y}B$ of the acceleration of B onto the axes of the fixed reference frame.

- The coordinates xQ and yQ of point Q relative to the OXY reference frame.

- Projections $\dot{x}Q$ and $\dot{y}Q$ of the velocity of Q onto the axes of the fixed reference frame.

- Projections $\ddot{x}Q$ and $\ddot{y}Q$ of the acceleration of Q onto the axes of the fixed reference frame.

We begin with the following vector equation:

$$\mathbf{OA} + \mathbf{AC} = \mathbf{OP} + \mathbf{PB} + \mathbf{BC} \tag{6.69}$$

and project it on the axes of the OXY reference frame:

$$\begin{cases} x_A + AC \cdot \cos(\varphi) = x_P - s \cdot \cos(\theta) + BC \cdot \cos\left(\theta + \frac{\pi}{2}\right) \\ y_A + AC \cdot \sin(\varphi) = y_P - s \cdot \sin(\theta) + BC \cdot \sin\left(\theta + \frac{\pi}{2}\right) \end{cases} \tag{6.70}$$

which are equivalent to

$$\begin{cases} AC \cdot \cos(\varphi) = -s \cdot \cos(\theta) - BC \cdot \sin(\theta) + x_P - x_A \\ AC \cdot \sin(\varphi) = -s \cdot \sin(\theta) + BC \cdot \cos(\theta) + y_P - y_A \end{cases} \tag{6.71}$$

We then square Equations 6.71 and obtain

$$
\begin{cases}
AC^2 \cdot \cos^2(\varphi) = s^2 \cdot \cos^2(\theta) + BC^2 \cdot \sin^2(\theta) + (x_P - x_A)^2 + 2s \cdot BC \cdot \sin(\theta) \cdot \cos(\theta) \\
\quad -2s(x_P - x_A)\cos(\theta) - 2BC \cdot (x_P - x_A) \cdot \sin(\theta) \\
AC^2 \cdot \sin^2(\varphi) = s^2 \cdot \sin^2(\theta) + BC^2 \cdot \cos^2(\theta) + (y_P - y_A)^2 - 2s \cdot BC \cdot \sin(\theta) \cdot \cos(\theta) \\
\quad -2s(y_P - y_A)\cos(\theta) + 2BC \cdot (y_P - y_A) \cdot \cos(\theta)
\end{cases}
\tag{6.72}
$$

After adding them and rearranging terms, we obtain the following quadratic equation in the unknown s:

$$
s^2 - 2\big[(x_P - x_A)\cos(\theta) + (y_P - y_A)\sin(\theta)\big] \cdot s + 2BC \cdot \big[(y_P - y_A)\cos(\theta) - (x_P - x_A)\sin(\theta)\big]
$$
$$
+ (x_P - x_A)^2 + (y_P - y_A)^2 - AC^2 + BC^2 = 0
\tag{6.73}
$$

with solutions

$$
s = \big[(x_P - x_A)\cos(\theta) + (y_P - y_A)\sin(\theta)\big] \pm \sqrt{\Delta}
\tag{6.74}
$$

where

$$
\Delta = \big[(x_P - x_A)\cos(\theta) + (y_P - y_A)\sin(\theta)\big]^2
$$
$$
-2BC \cdot \big[(y_P - y_A)\cos(\theta) - (x_P - x_A)\sin(\theta)\big]
$$
$$
-(x_P - x_A)^2 - (y_P - y_A)^2 + AC^2 - BC^2
\tag{6.75}
$$

Note that there are four assembly configurations of the RR_T isomer, corresponding to the sign of the eccentricity BC and the choice of the double sign in Equation 6.74.

The coordinates of point B and Q, of interest when plotting the mechanism or when the dyad is amplified with additional *Assur groups*, are

$$
\begin{cases}
x_B = x_P - s \cdot \cos(\theta) \\
y_B = y_P - s \cdot \sin(\theta)
\end{cases}
\tag{6.76}
$$

and

$$
\begin{cases}
x_Q = x_B - BQ \cdot \cos(\theta) \\
y_Q = y_B - BQ \cdot \sin(\theta)
\end{cases}
\tag{6.77}
$$

The coordinates of point C are obtained by projecting vector equation

$$\mathbf{OC} = \mathbf{OP} + \mathbf{PB} + \mathbf{BC} \tag{6.78}$$

on the axes of the OXY frame:

$$\begin{cases} x_C = x_P - s \cdot \cos(\theta) - BC \cdot \sin(\theta) = x_B - BC \cdot \sin(\theta) \\ y_C = y_P - s \cdot \sin(\theta) + BC \cdot \cos(\theta) = y_B + BC \cdot \cos(\theta) \end{cases} \tag{6.79}$$

Once coordinates of joint center C become available, the trigonometric functions of angle φ can be straightforwardly evaluated, that is,

$$\cos(\varphi) = \frac{(x_C - x_A)}{AC} \quad \text{and} \quad \sin(\varphi) = \frac{(y_C - y_A)}{AC} \tag{6.80}$$

By differentiating Equation 6.71 with respect to time, a set of two linear equations in the unknowns $\dot{\varphi}$ and $\dot{s}$ is obtained:

$$\begin{cases} AC \cdot \sin(\varphi) \cdot \dot{\varphi} - \cos(\theta) \cdot \dot{s} = \dot{x}_P - \dot{x}_A - [BC \cdot \cos(\theta) - s \cdot \sin(\theta)] \cdot \dot{\theta} \\ -AC \cdot \cos(\varphi) \cdot \dot{\varphi} - \sin(\theta) \cdot \dot{s} = \dot{y}_P - \dot{y}_A - [BC \cdot \sin(\theta) + s \cdot \cos(\theta)] \cdot \dot{\theta} \end{cases} \tag{6.81}$$

which are very easy to solve.

The components of the velocity of point B and Q are obtained by differentiating Equations 6.76 and 6.77, that is,

$$\begin{cases} \dot{x}_B = \dot{x}_P - \dot{s} \cdot \cos(\theta) + s \cdot \sin(\theta) \cdot \dot{\theta} \\ \dot{y}_B = \dot{y}_P - \dot{s} \cdot \sin(\theta) - s \cdot \cos(\theta) \cdot \dot{\theta} \end{cases} \tag{6.82}$$

and

$$\begin{cases} \dot{x}_Q = \dot{x}_B + BQ \cdot \sin(\theta) \cdot \dot{\theta} \\ \dot{y}_Q = \dot{y}_B - BQ \cdot \cos(\theta) \cdot \dot{\theta} \end{cases} \tag{6.83}$$

while those of point C are obtained by differentiating Equations 6.79 in their second form:

$$\begin{cases} \dot{x}_C = \dot{x}_B - BC \cdot \cos(\theta) \cdot \dot{\theta} \\ \dot{y}_C = \dot{y}_B - BC \cdot \sin(\theta) \cdot \dot{\theta} \end{cases} \tag{6.84}$$

Accelerations $\ddot{\varphi}$ and $\ddot{s}$ are obtained by differentiating Equations 6.81 which yield another set of two linear equations:

$$\begin{cases} +AC\cdot\sin(\varphi)\cdot\ddot{\varphi}-\cos(\theta)\cdot\ddot{s} = \ddot{x}_P - \ddot{x}_A + \left[BC\cdot\sin(\theta)-s\cdot\cos(\theta)\right]\cdot\dot{\theta}^2 \\[4pt] \quad -\left[BC\cdot\cos(\theta)-s\cdot\sin(\theta)\right]\cdot\ddot{\theta}+2\dot{s}\cdot\sin(\theta)\cdot\dot{\theta}+AC\cdot\cos(\varphi)\cdot\dot{\varphi}^2 \\[4pt] AC\cdot\cos(\varphi)\cdot\ddot{\varphi}+\sin(\theta)\cdot\ddot{s} = \ddot{y}_P - \ddot{y}_A - \left[BC\cdot\cos(\theta)-s\cdot\sin(\theta)\right]\cdot\dot{\theta}^2 \\[4pt] \quad -\left[BC\cdot\sin(\theta)+s\cdot\cos(\theta)\right]\cdot\ddot{\theta}-2\dot{s}\cdot\cos(\theta)\cdot\dot{\theta}+AC\cdot\sin(\varphi)\cdot\dot{\varphi}^2 \end{cases} \tag{6.85}$$

The x and y components of the acceleration of points B, Q, and C are obtained by differentiating Equations 6.82, 6.83, and 6.84, that is,

$$\begin{cases} \ddot{x}_B = \ddot{x}_P - \ddot{s}\cdot\cos(\theta)+2\dot{s}\cdot\sin(\theta)\cdot\dot{\theta}+s\cdot\cos(\theta)\cdot\dot{\theta}^2+s\cdot\sin(\theta)\cdot\ddot{\theta} \\[4pt] \ddot{y}_B = \ddot{y}_P - \ddot{s}\cdot\sin(\theta)-2\dot{s}\cdot\cos(\theta)\cdot\dot{\theta}+s\cdot\sin(\theta)\cdot\dot{\theta}^2-s\cdot\cos(\theta)\cdot\ddot{\theta} \end{cases} \tag{6.86}$$

$$\begin{cases} \ddot{x}_Q = \ddot{x}_B + BQ\cdot\cos(\theta)\cdot\dot{\theta}^2+BQ\cdot\sin(\theta)\cdot\ddot{\theta} \\[4pt] \ddot{y}_Q = \ddot{y}_B + BQ\cdot\sin(\theta)\cdot\dot{\theta}^2-BQ\cdot\cos(\theta)\cdot\ddot{\theta} \end{cases} \tag{6.87}$$

$$\begin{cases} \ddot{x}_C = \ddot{x}_B + BC\cdot\sin(\theta)\cdot\dot{\theta}^2-BC\cdot\cos(\theta)\cdot\ddot{\theta} \\[4pt] \ddot{y}_C = \ddot{y}_B - BC\cdot\cos(\theta)\cdot\dot{\theta}^2-BC\cdot\sin(\theta)\cdot\ddot{\theta} \end{cases} \tag{6.88}$$

Procedure **RR_T** in unit **LibAssur** that implements the equations derived earlier has the following heading:

```
procedure RR_T(Color:Word; xA,yA, vxA,vyA, axA,ayA, xP,yP,
vxP,vyP, axP,ayP, Theta,dTheta,ddTheta, AC,BC,BQ:double;
PlsMns:shortint; var xB,yB, vxB,vyB, axB,ayB, xC,yC, vxC,vyC,
axC,ayC, xQ,yQ, vxQ,vyQ, axQ,ayQ, Delta:double);
```

The correspondence between the formal parameters of procedure **RRT_** and the notations used in the earlier equations is summarized in the following tables:

Input parameters of procedure **RR_T**:

0...16	x_A	y_A	$\dot{x}_A$	$\dot{y}_A$	$\ddot{x}_A$	$\ddot{y}_A$	x_P	y_P	$\dot{x}_P$	$\dot{y}_P$	$\ddot{x}_P$	$\ddot{y}_P$
Color	xA	yA	vxA	vyA	axA	ayA	xP	yP	vxP	vyP	axP	ayP

θ		$\dot{\theta}$		$\ddot{\theta}$		AC		BC		BQ		± 1
Theta		dTheta		ddTheta		AC		BC		BQ		PlsMns

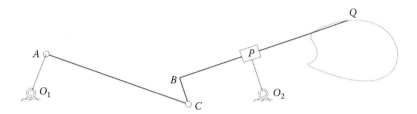

FIGURE 6.17　A two-DOF mechanism consisting of a crank-driven RR_T dyad with the slider block mounted at the end of rocker O_2P that oscillates according to equation $\varphi_2 = \pi/2 + \pi/9 \cdot \sin(2\pi t)$. See also the animation file **F6_17.GIF** available with the book.

Output parameters of procedure **RR_T**:

xB	$\dot{x}B$	$\dot{y}B$	$\ddot{x}B$	$\ddot{y}B$	xC	yC	$\dot{x}C$	$\dot{y}C$	$\ddot{x}C$	$\ddot{y}C$
xB	vxB	vyB	axB	ayB	xC	yC	vxC	vyC	axC	ayC

xQ		yQ		$\dot{x}Q$		$\dot{y}Q$		$\ddot{x}Q$		$\ddot{y}Q$		Δ
xQ		yQ		vxQ		vyQ		axQ		ayQ		Delta

Input parameter **PlsMns** specifies the nature of the double sign in Equation 6.74, and it should be assigned either the value −1 or +1.

If of interest, the displacement s of the slider block relative to its guide and the first and second time derivatives $\dot{s}$ and $\ddot{s}$ can be easily determined by calling procedure **VarDist** from unit **LibMec2D**.

The simulation of the sample mechanism in Figure 6.17 has been done with program **P6_17.PAS**, listed in Appendix B. Since there was no interest in the velocity and acceleration output by procedures **gCrank** and **RR_T**, the generic variable _ defined in unit **LibMec2D** that is preassigned the value **InfD** has been used in a number of places (both as input and as output). Same was applied in place of a variable **Delta** returned by procedure **RRT_**.

6.10.2　RRT_ Dyadic Isomer: Procedure **RRT_**

The kinematic analysis problem considered previously was restated for the case of the RRT dyad configured as shown in Figure 6.18, that is, the RR_T isomer. The following parameters are assumed known at any instant of time of a simulation:

- Length AC of the connecting rod.

- Slider eccentricity BC assumed perpendicular to PQ, which can be either positive or negative.

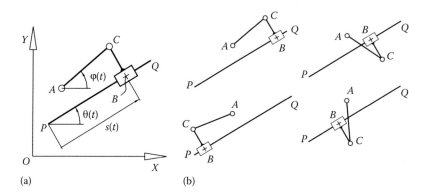

FIGURE 6.18 The RRT_ isomer of the RRT dyad with a potential sliding rod PQ (a) and its four possible assembly configurations (b).

- Coordinates xA, yA of potential pin joint A relative to the fixed reference frame OXY.

- Projections $\dot{x}A$ and $\dot{y}A$ of the velocity of joint center A onto the fixed reference frame.

- Projections $\ddot{x}A$ and $\ddot{y}A$ of the accelerations of A onto the fixed reference frame.

- Coordinates xP, yP and xQ, yQ relative to the fixed reference frame of points P and Q of the slider guide.

- Projections $\dot{x}P$ and $\dot{y}P$ of the velocity of joint center P onto the fixed reference frame.

- Projections $\ddot{x}P$ and $\ddot{y}P$ of the accelerations of point P onto the fixed reference frame.

- Projections $\dot{x}Q$ and $\dot{y}Q$ of the velocity of point Q onto the fixed reference frame.

- Projections $\ddot{x}Q$ and $\ddot{y}Q$ of the accelerations of point Q onto the fixed reference frame.

The objective of the analysis is to determine the following unknown parameters:

- Coordinates xB and yB of sliding joint center B relative to the OXY reference frame.

- Projections $\dot{x}B$ and $\dot{y}B$ of the velocity of point B onto the axes of the fixed reference frame.

- Projections $\ddot{x}B$ and $\ddot{y}B$ of the acceleration of point B onto the axes of the fixed reference frame.

- Coordinates xC and yC of joint center C relative to the OXY reference frame.

- Projections $\dot{x}C$ and $\dot{y}C$ of the velocity of C onto the axes of the fixed reference frame.

- Projections $\ddot{x}C$ and $\ddot{y}C$ of the acceleration of C onto the axes of the fixed reference frame.

Equations 6.69 through 6.88 are also applicable in the kinematic analysis of the RRT dyad in Figure 6.18, with the exception of Equations 6.77, 6.83, and 6.87, which should be replaced with the following six equations:

$$\begin{cases} x_Q = x_P + PQ \cdot \cos(\theta) \\ y_Q = y_P + PQ \cdot \sin(\theta) \end{cases} \tag{6.89}$$

$$\begin{cases} \dot{x}_Q = \dot{x}_P - PQ \cdot \sin(\theta) \cdot \dot{\theta} \\ \dot{y}_Q = \dot{y}_P + PQ \cdot \cos(\theta) \cdot \dot{\theta} \end{cases} \tag{6.90}$$

and

$$\begin{cases} \ddot{x}_Q = \ddot{x}_P - PQ \cdot \cos(\theta) \cdot \dot{\theta}^2 - PQ \cdot \sin(\theta) \cdot \ddot{\theta} \\ \ddot{y}_Q = \ddot{y}_P - PQ \cdot \sin(\theta) \cdot \dot{\theta}^2 + PQ \cdot \cos(\theta) \cdot \ddot{\theta} \end{cases} \tag{6.91}$$

where

$$PQ = \sqrt{(x_Q - x_P)^2 + (y_Q - y_P)^2} \tag{6.92}$$

Procedure **RRT_** in unit **LibAssur** that performs the kinematic analysis of the RR_T dyadic isomer has the following heading:

```
RRT_(Color, xA,yA,vxA,vyA,axA,ayA, xP,yP,vxP,vyP,axP,ayP,
xQ,yQ,vxQ,vyQ,axQ,ayQ, AC,BC, PlsMns, xB,yB,vxB,vyB,axB,ayB,
xC,yC,vxC,vyC,axC,ayC, Delta)
```

The correspondence between the formal parameters and the notations used in these equations and in Figure 6.18 is summarized in the following tables:

Input parameters of procedure **RRT_**:

−16...16	xA	yA	$\dot{x}A$	$\dot{y}A$	$\ddot{x}A$	$\ddot{y}A$	xP	yP	$\dot{x}P$	$\dot{y}P$	$\ddot{x}P$	$\ddot{y}P$
Color	xA	yA	vxA	vyA	axA	ayA	xP	yP	vxP	vyP	axP	ayP

$\dot{x}Q$	$\dot{y}Q$	$\ddot{x}Q$	$\ddot{y}Q$	AC	BC	±1
vxQ	vyQ	axQ	ayQ	AC	BC	PlsMns

Output parameters of procedure **RRT_**:

xB	$\dot{x}B$	$\dot{y}B$	$\ddot{x}B$	$\ddot{y}B$	xC	yC	$\dot{x}C$	$\dot{y}C$	$\ddot{x}C$	$\ddot{y}C$	Δ
xB	vxB	vyB	axB	ayB	xC	yC	vxC	vyC	axC	ayC	Delta

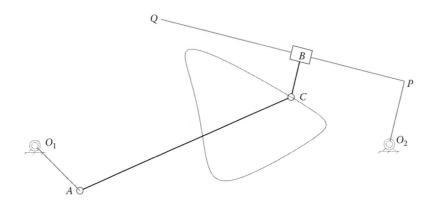

FIGURE 6.19 The same mechanism in Figure 6.17 simulated using the **RRT_** procedure. See also animation file **F6_19.GIF** available with the book.

The input parameter **Color** specifies the color in which the dyad will be plotted on the screen and to the **DXF** file. In comparison with the previous procedures, parameter **Color** can be assigned either a positive or a negative value. If **Color** is negative, then line *PQ* representing the slider guide will not be plotted.

Same as in case of procedure **RR_T**, parameter **PlsMns** controls the double sign in Equation 6.74 and should be set equal to either –1 or +1, depending on the desired closure of the dyad (left hand or right hand).

Angle θ formed by guide *PQ* with the horizontal axis and its first and second derivatives $\dot{\theta}$ and $\ddot{\theta}$ (which are not readily available as before) can be calculated by calling procedures **AngPVA** from unit **LibMec2D**. Inputs to procedure **AngPVA** will be the *x* and *y* coordinates of points *P* and *Q* and their first and second time derivatives.

Computer program **P6_19.PAS** available with the book repeats the simulation done with **P6_17.PAS** this time using procedure **RRT_** instead of **RR_T**. A screenshot generated with this new program is given in Figure 6.19, while **F6_19.GIF** provides a full cycle simulation of the mechanism.

6.11 KINEMATIC ANALYSIS OF THE RTR DYAD USING A VECTOR-LOOP APPROACH: PROCEDURE **RT_R**

Here, the position, velocity, and acceleration problem of the RTR dyad will be solved using a vector-loop approach. Figure 6.20a shows a generalized RTR dyad with both potential joints *A* and *B* offset from the axis of the sliding rod. Note that RT_R and R_TR are not distinct isomers, and therefore the kinematic equations remain the same. With the notations in Figure 6.20, the following parameters are assumed known at any instant of time of the kinematic analysis:

- Eccentricity *AC* assumed perpendicular to *PQ* (can be either positive or negative).

- Eccentricity *BP* assumed perpendicular to *PQ* (can be either positive or negative).

- Slider guide rod length *PQ* (can be either positive or negative).

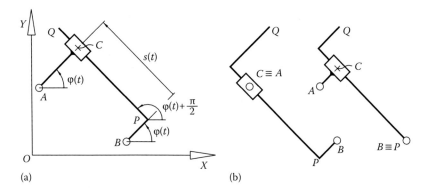

(a)   (b)

FIGURE 6.20   Notations used in solving the kinematics of the RTR dyad (a) and equivalent configurations from the perspective of point Q motion (b).

- Coordinates xA, yA of potential joint A relative to the fixed reference frame OXY.

- Projections $\dot{x}A$ and $\dot{y}A$ of the velocity of joint center A onto the fixed reference frame.

- Projections $\ddot{x}A$ and $\ddot{y}A$ of the accelerations of A onto the fixed reference frame.

The purpose of the analysis is to determine

- Displacement s of the slider measured as shown and its first and second derivatives $\dot{s}$ and $\ddot{s}$, respectively.

- Coordinates xC and yC of joint center C relative to the OXY reference frame.

- Projections $\dot{x}C$ and $\dot{y}C$ of the velocity of C onto the axes of the fixed reference frame.

- Projections $\ddot{x}C$ and $\ddot{y}C$ of the acceleration of C onto the axes of the fixed reference frame.

- Coordinates xP, yP and xQ, yQ of the respective points on the slider guide relative to the fixed reference frame.

- Projections $\dot{x}P$ and $\dot{y}P$ of the velocity of joint center P onto the fixed reference frame.

- Projections $\ddot{x}P$ and $\ddot{y}P$ of the accelerations of point P onto the fixed reference frame.

- Projections $\dot{x}Q$ and $\dot{y}Q$ of the velocity of point Q onto the fixed reference frame.

- Projections $\ddot{x}Q$ and $\ddot{y}Q$ of the accelerations of point Q onto the fixed reference frame.

Note that if eccentricities AC and BP are modified equal amounts, slider displacement s and its time derivatives will remain the same. Consequently, point Q attached offset to the sliding rod performs the same motion that can be obtained with either AC or BP being set equal to zero (see also Figure 6.20b).

We begin the analysis by writing the following vector equation:

$$\mathbf{OA} + \mathbf{AC} = \mathbf{OB} + \mathbf{BP} + \mathbf{PC} \tag{6.93}$$

and project it on the axes of the fixed reference frame as

$$\begin{cases} x_A + AC \cdot \cos(\varphi) = x_B + BP \cdot \cos(\varphi) + s \cdot \cos\left(\varphi + \dfrac{\pi}{2}\right) \\[4mm] y_A + AC \cdot \sin(\varphi) = y_B + BP \cdot \sin(\varphi) + s \cdot \sin\left(\varphi + \dfrac{\pi}{2}\right) \end{cases} \tag{6.94}$$

equivalent to

$$\begin{cases} (AC - BP) \cdot \cos(\varphi) + s \cdot \sin(\varphi) = x_B - x_A \\[2mm] (AC - BP) \cdot \sin(\varphi) - s \cdot \cos(\varphi) = y_B - y_A \end{cases} \tag{6.95}$$

We then square these last two equations:

$$\begin{cases} (AC - BP)^2 \cdot \cos^2(\varphi) + s^2 \cdot \sin^2(\varphi) + 2s \cdot (AC - BP) \cdot \cos(\varphi) \cdot \sin(\varphi) = (x_B - x_A)^2 \\[2mm] (AC - BP)^2 \cdot \sin^2(\varphi) + s^2 \cdot \cos^2(\varphi) - 2s \cdot (AC - BP) \cdot \cos(\varphi)\sin(\varphi) = (y_B - y_A)^2 \end{cases} \tag{6.96}$$

and after adding them and rearranging terms, we obtain

$$(AC - BP)^2 + s^2 = (x_B - x_A)^2 + (y_B - y_A)^2 \tag{6.97}$$

The unknown slider displacement s will therefore result as

$$s = \pm\sqrt{(x_B - x_A)^2 + (y_B - y_A)^2 - (AC - BP)^2} \tag{6.98}$$

In order to calculate the angle φ, we resort to Equation 6.95 and multiply the first one by s and the second one by $(AC–BP)$:

$$\begin{cases} s \cdot (AC - BP) \cdot \cos(\varphi) + s^2 \cdot \sin(\varphi) = s \cdot (x_B - x_A) \\[2mm] (AC - BP)^2 \cdot \sin(\varphi) - s \cdot (AC - BP)\cos(\varphi) = (AC - BP) \cdot (y_B - y_A) \end{cases} \tag{6.99}$$

and then add them together to obtain

$$\sin(\varphi) = \frac{s \cdot (x_B - x_A) + (AC - BP) \cdot (y_B - y_A)}{(x_B - x_A)^2 + (y_B - y_A)^2} \tag{6.100}$$

We repeat the procedure and multiply the first of Equations 6.95 by $(AC - BP)$ and the second one by $-s$:

$$\begin{cases} (AC - BP)^2 \cdot \cos(\varphi) + s \cdot (AC - BP) \sin(\varphi) = (AC - BP) \cdot (x_B - x_A) \\ -s \cdot (AC - BP) \cdot \sin(\varphi) + s^2 \cdot \cos(\varphi) = -s \cdot (y_B - y_A) \end{cases} \tag{6.101}$$

After adding the two equations together, we get

$$\cos(\varphi) = \frac{(AC - BP) \cdot (x_B - x_A) - s \cdot (y_B - y_A)}{(x_B - x_A)^2 + (y_B - y_A)^2} \tag{6.102}$$

The actual angle φ can be calculated using equation

$$\varphi = \arctan \left[\frac{s \cdot (x_B - x_A) + (AC - BP) \cdot (y_B - y_A)}{(AC - BP) \cdot (x_B - x_A) - s \cdot (y_B - y_A)} \right] \tag{6.103}$$

Using the sin and cos function of angle φ in Equations 6.100 and 6.102, the coordinates of points C, P, and Q result as follows:

$$\begin{cases} x_C = x_A + AC \cdot \cos(\varphi) \\ y_C = y_A + AC \cdot \sin(\varphi) \end{cases} \tag{6.104}$$

$$\begin{cases} x_P = x_B + BP \cdot \cos(\varphi) \\ y_P = y_B + BP \cdot \sin(\varphi) \end{cases} \tag{6.105}$$

$$\begin{cases} x_Q = x_P - PQ \cdot \sin(\varphi) \\ y_Q = y_P + PQ \cdot \cos(\varphi) \end{cases} \tag{6.106}$$

Unknown angular velocities $\dot{s}$ and $\dot{\varphi}$ are determined by differentiating Equations 6.95:

$$\begin{cases} -(AC - BP) \cdot \sin(\varphi) \cdot \dot{\varphi} + \dot{s} \cdot \sin(\varphi) + s \cdot \cos(\varphi) \cdot \dot{\varphi} = \dot{x}_B - \dot{x}_A \\ (AC - BP) \cdot \cos(\varphi) \cdot \dot{\varphi} - \dot{s} \cdot \cos(\varphi) + s \cdot \sin(\varphi) \cdot \dot{\varphi} = \dot{y}_B - \dot{y}_A \end{cases} \tag{6.107}$$

which after collecting terms yield a set of two equations in the unknowns $\dot{s}$ and $\dot{\varphi}$:

$$\begin{cases} \left[s\cdot\cos(\varphi)-(AC-BP)\cdot\sin(\varphi) \right]\cdot\dot{\varphi}+\dot{s}\cdot\sin(\varphi)=\dot{x}_B-\dot{x}_A \\ \left[s\cdot\sin(\varphi)+(AC-BP)\cdot\cos(\varphi) \right]\cdot\dot{\varphi}-\dot{s}\cdot\cos(\varphi)=\dot{y}_B-\dot{y}_A \end{cases} \tag{6.108}$$

To determine the second time derivative of s and φ, we differentiate Equations 6.108 and get:

$$\begin{cases} \left[s\cdot\cos(\varphi)-(AC-BP)\cdot\sin(\varphi) \right]\cdot\ddot{\varphi}+\ddot{s}\cdot\sin(\varphi)=\ddot{x}_B-\ddot{x}_A \\ \quad -\left[\dot{s}\cdot\cos(\varphi)-s\cdot\sin(\varphi)\cdot\dot{\varphi}-(AC-BP)\cdot\cos(\varphi)\cdot\dot{\varphi} \right]\cdot\dot{\varphi}-\dot{s}\cdot\cos(\varphi)\cdot\dot{\varphi} \\ \left[s\cdot\sin(\varphi)+(AC-BP)\cdot\cos(\varphi) \right]\cdot\ddot{\varphi}-\ddot{s}\cdot\cos(\varphi)=\ddot{y}_B-\ddot{y}_A \\ \quad -\left[\dot{s}\cdot\sin(\varphi)+s\cdot\cos(\varphi)\cdot\dot{\varphi}-(AC-BP)\cdot\sin(\varphi)\cdot\dot{\varphi} \right]\cdot\dot{\varphi}-\dot{s}\cdot\sin(\varphi)\cdot\dot{\varphi} \end{cases} \tag{6.109}$$

and finally we obtain a set of two linear equations in the unknowns $\ddot{s}$ and $\ddot{\varphi}$ that can be easily solved by eliminating one of the variables, or using Cramer's rule:

$$\begin{cases} \left[s\cdot\cos(\varphi)-(AC-BP)\cdot\sin(\varphi) \right]\cdot\ddot{\varphi}+\ddot{s}\cdot\sin(\varphi)=\ddot{x}_B-\ddot{x}_A \\ \quad -2\dot{s}\cdot\cos(\varphi)\cdot\dot{\varphi}+s\cdot\sin(\varphi)\cdot\dot{\varphi}^2+(AC-BP)\cdot\cos(\varphi)\cdot\dot{\varphi}^2 \\ \left[s\cdot\sin(\varphi)+(AC-BP)\cdot\cos(\varphi) \right]\cdot\ddot{\varphi}-\ddot{s}\cdot\cos(\varphi)=\ddot{y}_B-\ddot{y}_A \\ \quad -2\dot{s}\cdot\sin(\varphi)\cdot\dot{\varphi}-s\cdot\cos(\varphi)\cdot\dot{\varphi}^2+(AC-BP)\cdot\sin(\varphi)\cdot\dot{\varphi}^2 \end{cases} \tag{6.110}$$

The scalar components of the velocities and accelerations of points P, C, and Q are obtained by differentiating Equations 6.104, 6.105, and 6.106 as follows:

$$\begin{cases} \dot{x}_C=\dot{x}_A-AC\cdot\sin(\varphi)\cdot\dot{\varphi} \\ \dot{y}_C=\dot{y}_A+AC\cdot\cos(\varphi)\cdot\dot{\varphi} \end{cases} \tag{6.111}$$

$$\begin{cases} \dot{x}_P=\dot{x}_B-BP\cdot\sin(\varphi)\cdot\dot{\varphi} \\ \dot{y}_P=\dot{y}_B+BP\cdot\cos(\varphi)\cdot\dot{\varphi} \end{cases} \tag{6.112}$$

$$\begin{cases} \dot{x}_Q=\dot{x}_P-PQ\cdot\cos(\varphi)\cdot\dot{\varphi} \\ \dot{y}_Q=\dot{y}_P-PQ\cdot\sin(\varphi)\cdot\dot{\varphi} \end{cases} \tag{6.113}$$

Differentiating one more time the same equations, we obtain the following acceleration equations:

$$\begin{cases} \ddot{x}_C = \ddot{x}_A - AC \cdot [\cos(\varphi) \cdot \dot{\varphi}^2 + \sin(\varphi) \cdot \ddot{\varphi}] \\ \ddot{y}_C = \ddot{y}_A - AC \cdot [\sin(\varphi) \cdot \dot{\varphi}^2 - \cos(\varphi) \cdot \ddot{\varphi}] \end{cases} \tag{6.114}$$

$$\begin{cases} \ddot{x}_P = x_B - BP \cdot [\cos(\varphi) \cdot \dot{\varphi}^2 + \sin(\varphi) \cdot \ddot{\varphi}] \\ \ddot{y}_P = y_B - BP \cdot [\sin(\varphi) \cdot \dot{\varphi}^2 - \cos(\varphi) \cdot \ddot{\varphi}] \end{cases} \tag{6.115}$$

$$\begin{cases} \ddot{x}_Q = x_P + PQ \cdot [\sin(\varphi) \cdot \dot{\varphi}^2 - \cos(\varphi) \cdot \ddot{\varphi}] \\ \ddot{y}_Q = y_P - PQ \cdot [\cos(\varphi) \cdot \dot{\varphi}^2 + \sin(\varphi) \cdot \ddot{\varphi}] \end{cases} \tag{6.116}$$

Procedure **RT_R** in unit **LibAssur** implements the equations discussed earlier and has the following heading:

```
procedure RT_R(Color:Word; xA,yA, vxA,vyA, axA,ayA, xB,yB,
vxB,vyB, axB,ayB, AC,BP,PQ:double; PlsMns:shortint; var xP,yP,
vxP,vyP, axP,ayP, xC,yC, vxC,vyC, axC,ayC, xQ,yQ, vxQ,vyQ,
axQ,ayQ, Delta:double);
```

The correspondence between the formal parameters and the notations used in Equations 6.93 through 6.110 and in Figure 6.20 is summarized in the following two tables:
Input parameters of procedure **RT_R**:

0...16	xA	yA	$\dot{x}A$	$\dot{y}A$	$\ddot{x}A$	$\ddot{y}A$	xB	yB	$\dot{x}B$	$\dot{y}B$	$\ddot{x}B$	$\ddot{y}B$	AC	BC	BQ
Color	xA	yA	vxA	vyA	axA	ayA	xB	yB	vxB	vyB	axB	ayB	AC	BC	BQ

Output parameters of procedure **RT_R**:

xP	yP	$\dot{x}P$	$\dot{y}P$	$\ddot{x}P$	$\ddot{y}P$	xC	yC	$\dot{x}C$	$\dot{y}C$	$\ddot{x}C$	$\ddot{y}C$
xP	yP	vxP	vyP	axP	ayP	xC	yC	vxC	vyC	axC	ayC

xQ	yQ	$\dot{x}Q$	$\dot{y}Q$	$\ddot{x}Q$	$\ddot{y}Q$	Δ
xQ	yQ	vxQ	vyQ	axQ	ayQ	Delta

Note that the displacement s of the slider block relative to its guide and the first and second time derivatives $\dot{s}$ and $\ddot{s}$ are not returned by the procedure. If of interest, they can be easily determined by calling procedure **VarDist** from unit **LibMec2D**.

The simulation of the mechanism in Figure 6.21 has been done using program **P6_21.PAS**, the listing of which is available in Appendix B. The program calls procedure **RT_R** with

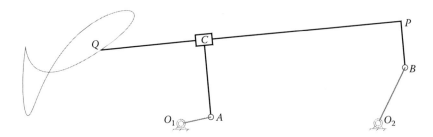

FIGURE 6.21   Simulation of a two-DOF mechanism consisting of an RTR dyad, driven by a crank and a rocker. See also animation file **F6_21.GIF**.

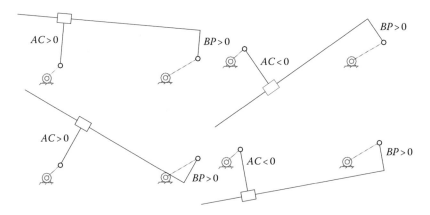

FIGURE 6.22   The four possible configurations of the mechanism in Figure 6.21 based on the sign of the eccentricities of the potential A and B of the RTR dyad. See also animation **F6_22.GIF**.

some of its output variables assigned the generic variable _. To ease identification, all joints are labeled in Figure 6.21. Also shown in the figure is the locus of point Q.

The companion Figure 6.22 and animation file **F6_22.GIF** illustrate the four possible arrangements that can be obtained by alternating the signs of the eccentricities AC and BP of the RTR dyad. Note that of these four mechanisms, two have full cycle mobility, while the other two experience locking positions.

6.12 KINEMATIC ANALYSIS OF THE TRT DYAD USING A VECTOR-LOOP APPROACH

The TRT and RTT dyads have fewer applications than the ones analyzed so far. On the other hand, both have increased number of isomers (i.e., T_R_T, _TRT_, T_RT_ and R_T_T, RT_T_, R_TT_, RT__T, respectively), due to the presence of two prismatic joints that can be configured either with the sliding block first followed by its conjugate sliding rod, or vice versa.

In this section, a vector-based kinematic analysis of the three possible isomers of the TRT dyad will be discussed. The corresponding kinematic equations have been implemented in procedures **T_R_T**, **_TRT_**, and **T_RT_** part of unit **LibAssur**. Examples of mechanism simulations done with these procedures are also provided.

6.12.1 T_R_T Dyadic Isomer: Procedure **T_R_T**

In a kinematic simulation of the TRT dyad with two potential sliding blocks, noted T_R_T (Figure 6.23), the following parameters are assumed known at any instant of time:

- Coordinates xA, yA, of potential turning joint A relative to the fixed reference frame OXY.

- Coordinates xB, yB, of point B measured relative to the fixed reference frame.

- Projections $\dot{x}A$ and $\dot{y}A$ of the velocity of joint center A onto the fixed reference frame.

- Projections $\ddot{x}A$ and $\ddot{y}A$ of the accelerations of A onto the fixed reference frame.

- Projections $\dot{x}B$ and $\dot{y}B$ of the velocity of joint center B onto the fixed reference frame.

- Projections $\ddot{x}B$ and $\ddot{y}B$ of the accelerations of point B onto the fixed reference frame.

- Orientation angle θ_1 of slider block A and its first and second time derivatives $\dot{\theta}_1$ and $\ddot{\theta}_1$, respectively.

- Orientation angle θ_2 of slider block B and its first and second time derivatives $\dot{\theta}_2$ and $\ddot{\theta}_2$, respectively.

- Lengths P_1Q_1 and Q_1C of the L-shaped link supporting slider block A (both can be either positive or negative).

- Lengths P_2Q_2 and Q_2C of the L-shaped link supporting slider block B (either positive or negative).

These being given will allow us to calculate the following dependent parameters:

- Slider displacements s_1 and s_2 measured as shown and their time derivatives $\dot{s}_1, \dot{s}_2, \ddot{s}_1$, and $\ddot{s}_2$.

- Coordinates xC and yC of pin joint center C relative to the OXY reference frame.

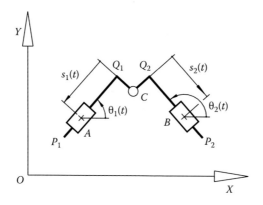

FIGURE 6.23 Notations used in the kinematic analysis of the T_R_T isomer of the TRT dyad.

- Projections $\dot{x}C$ and $\dot{y}C$ of the velocity of point C onto the axes of the fixed reference frame.

- Projections $\ddot{x}C$ and $\ddot{y}C$ of the acceleration of point C onto the axes of the fixed reference frame.

- Coordinates xP_1 and yP_1 of point P_1 relative to the OXY reference frame.

- Projections $\dot{x}P_1$ and $\dot{y}P_1$ of the velocity of point P_1 onto the axes of the fixed reference frame.

- Projections $\ddot{x}P_1$ and $\ddot{y}P_1$ of the acceleration of point P_1 onto the axes of the fixed reference frame.

- Coordinates xQ_1 and yQ_1 of point Q_1 relative to the OXY reference frame.

- Projections $\dot{x}Q_1$ and $\dot{y}Q_1$ of the velocity of point Q_1 onto the axes of the fixed reference frame.

- Projections $\ddot{x}Q_1$ and $\ddot{y}Q_1$ of the acceleration of point Q_1 onto the axes of the fixed reference frame.

- Coordinates xP_2 and yP_2 of point P_2 relative to the OXY reference frame.

- Projections $\dot{x}P_2$ and $\dot{y}P_2$ of the velocity of point P_2 onto the axes of the fixed reference frame.

- Projections $\ddot{x}P_2$ and $\ddot{y}P_2$ of the acceleration of point P_2 onto the axes of the fixed reference frame.

- Coordinates xQ_2 and yQ_2 of point Q_2 relative to the OXY reference frame.

- Projections $\dot{x}Q_2$ and $\dot{y}Q_2$ of the velocity of point Q_2 onto the axes of the fixed reference frame.

- Projections $\ddot{x}Q_2$ and $\ddot{y}Q_2$ of the acceleration of point Q_2 onto the axes of the fixed reference frame.

We begin by writing the following vector-loop equation:

$$\mathbf{OA} + \mathbf{AQ_1} + \mathbf{Q_1C} = \mathbf{OB} + \mathbf{BQ_2} + \mathbf{Q_2C} \tag{6.117}$$

which projects on the x- and y-axes of the fixed reference frame as

$$\begin{cases} x_A - s_1 \cdot \cos(\theta_1) + Q_1C \cdot \cos\left(\theta_1 - \dfrac{\pi}{2}\right) = x_B - s_2 \cdot \cos(\theta_2) + Q_2C \cdot \cos\left(\theta_2 - \dfrac{\pi}{2}\right) \\ y_A - s_1 \cdot \sin(\theta_1) + Q_1C \cdot \sin\left(\theta_1 - \dfrac{\pi}{2}\right) = y_B - s_2 \cdot \sin(\theta_2) + Q_2C \cdot \sin\left(\theta_2 - \dfrac{\pi}{2}\right) \end{cases} \tag{6.118}$$

We rearrange them as a set of two linear equations in the unknowns s_1 and s_2 that is easy to solve:

$$\begin{cases} s_1 \cdot \cos(\theta_1) - s_2 \cdot \cos(\theta_2) = x_A - x_B + Q_1 C \cdot \sin(\theta_1) + Q_2 C \cdot \sin(\theta_2) \\ s_1 \cdot \sin(\theta_1) - s_2 \cdot \sin(\theta_2) = y_A - y_B - Q_1 C \cdot \cos(\theta_1) - Q_2 C \cdot \cos(\theta_2) \end{cases} \tag{6.119}$$

Note that Equations 6.119 have unique solutions only if

$$\sin(\theta_1) \cdot \cos(\theta_2) - \cos(\theta_1) \cdot \sin(\theta_2) \neq 0 \quad \text{or} \quad \theta_1 \neq \theta_2 \tag{6.120}$$

Once slider displacements s_1 and s_2 become known, the coordinates of points C, P_1, P_2, Q_1, and Q_2 required to represent graphically the T_R_T dyadic isomer can be calculated using the following equations:

$$\begin{cases} x_{Q_1} = x_A - s_1 \cdot \cos(\theta_1) \\ y_{Q_1} = y_A - s_1 \cdot \sin(\theta_1) \end{cases} \tag{6.121}$$

$$\begin{cases} x_{P_1} = x_{Q_1} - P_1 Q_1 \cdot \cos(\theta_1) \\ y_{P_1} = y_{Q_1} - P_1 Q_1 \cdot \sin(\theta_1) \end{cases} \tag{6.122}$$

$$\begin{cases} x_{Q_2} = x_B - s_2 \cdot \cos(\theta_2) \\ y_{Q_2} = y_B - s_2 \cdot \sin(\theta_2) \end{cases} \tag{6.123}$$

$$\begin{cases} x_{P_2} = x_{Q_2} - P_2 Q_2 \cdot \cos(\theta_2) \\ y_{P_2} = y_{Q_2} - P_2 Q_2 \cdot \sin(\theta_2) \end{cases} \tag{6.124}$$

The coordinates of joint C can be calculated using either of the following two equations:

$$\begin{cases} x_C = x_A - s_1 \cdot \cos(\theta_1) + Q_1 C \cdot \sin(\theta_1) = x_{Q_1} + Q_1 C \cdot \sin(\theta_1) \\ y_C = y_A - s_1 \cdot \sin(\theta_1) - Q_1 C \cdot \cos(\theta_1) = y_{Q_1} - Q_1 C \cdot \cos(\theta_1) \end{cases} \tag{6.125a}$$

or

$$\begin{cases} x_C = x_B - s_2 \cdot \cos(\theta_2) - Q_2 C \cdot \sin(\theta_2) = x_{Q_2} - Q_2 C \cdot \sin(\theta_2) \\ y_C = y_B - s_2 \cdot \sin(\theta_2) + Q_2 C \cdot \cos(\theta_2) = y_{Q_2} + Q_2 C \cdot \cos(\theta_2) \end{cases} \tag{6.125b}$$

To solve the velocity problem, we differentiate with respect to time Equation 6.119 and obtain a set of two linear equations in the unknowns $\dot{s}_1$ and $\dot{s}_2$, that is:

$$
\begin{cases}
\dot{s}_1 \cdot \cos(\theta_1) - \dot{s}_2 \cdot \cos(\theta_2) = \dot{x}_A - \dot{x}_B + \dot{\theta}_1 \cdot s_1 \cdot \sin(\theta_1) - \dot{\theta}_2 \cdot s_2 \cdot \sin(\theta_2) \\
\quad + \dot{\theta}_1 \cdot Q_1 C \cdot \cos(\theta_1) + \dot{\theta}_2 \cdot Q_2 C \cdot \cos(\theta_2) \\
\dot{s}_1 \cdot \sin(\theta_1) - \dot{s}_2 \cdot \sin(\theta_2) = \dot{y}_A - \dot{y}_B - \dot{\theta}_1 \cdot s_1 \cdot \cos(\theta_1) + \dot{\theta}_2 \cdot s_2 \cdot \cos(\theta_2) \\
\quad + \dot{\theta}_1 \cdot Q_1 C \cdot \sin(\theta_1) + \dot{\theta}_2 \cdot Q_2 C \cdot \sin(\theta_2)
\end{cases}
\tag{6.126}
$$

By applying Equations 6.121, 6.123, and 6.125, these last two equations simplify to

$$
\begin{cases}
\dot{s}_1 \cdot \cos(\theta_1) - \dot{s}_2 \cdot \cos(\theta_2) = \dot{x}_A - \dot{x}_B + \dot{\theta}_1 \cdot (y_A - y_C) - \dot{\theta}_2 \cdot (y_B - y_C) \\
\dot{s}_1 \cdot \sin(\theta_1) - \dot{s}_2 \cdot \sin(\theta_2) = \dot{y}_A - \dot{y}_B - \dot{\theta}_1 \cdot (x_A - x_C) + \dot{\theta}_2 \cdot (x_B - x_C)
\end{cases}
\tag{6.127}
$$

The x and y components of the velocities of points C, P_1, P_2, Q_1, and Q_2 result through differentiation with respect to time of Equations 6.121 through 6.125:

$$
\begin{cases}
\dot{x}_{Q_1} = \dot{x}_A - \dot{s}_1 \cdot \cos(\theta_1) + \dot{\theta}_1 \cdot s_1 \cdot \sin(\theta_1) \\
\dot{y}_{Q_1} = \dot{y}_A - \dot{s}_1 \cdot \sin(\theta_1) - \dot{\theta}_1 \cdot s_1 \cdot \cos(\theta_1)
\end{cases}
\tag{6.128}
$$

$$
\begin{cases}
\dot{x}_{P_1} = \dot{x}_{Q_1} + \dot{\theta}_1 \cdot P_1 Q_1 \cdot \sin(\theta_1) \\
\dot{y}_{P_1} = \dot{y}_{Q_1} - \dot{\theta}_1 \cdot P_1 Q_1 \cdot \cos(\theta_1)
\end{cases}
\tag{6.129}
$$

$$
\begin{cases}
\dot{x}_{Q_2} = \dot{x}_B - \dot{s}_2 \cdot \cos(\theta_2) + \dot{\theta}_2 \cdot s_2 \cdot \sin(\theta_2) \\
\dot{y}_{Q_2} = \dot{y}_B - \dot{s}_2 \cdot \sin(\theta_2) - \dot{\theta}_2 \cdot s_2 \cdot \cos(\theta_2)
\end{cases}
\tag{6.130}
$$

$$
\begin{cases}
\dot{x}_{P_2} = \dot{x}_{Q_2} + \dot{\theta}_2 \cdot P_2 Q_2 \cdot \sin(\theta_2) \\
\dot{y}_{P_2} = \dot{y}_{Q_2} - \dot{\theta}_2 \cdot P_2 Q_2 \cdot \cos(\theta_2)
\end{cases}
\tag{6.131}
$$

and for joint C,

$$\begin{cases} \dot{x}_C = \dot{x}_{Q_1} + \dot{\theta}_1 \cdot Q_1 C \cdot \cos(\theta_1) \\ \dot{y}_C = \dot{y}_{Q_1} + \dot{\theta}_1 \cdot Q_1 C \cdot \sin(\theta_1) \end{cases} \tag{6.132a}$$

or

$$\begin{cases} \dot{x}_C = \dot{x}_{Q_2} - \dot{\theta}_1 \cdot Q_2 C \cdot \cos(\theta_2) \\ \dot{y}_C = \dot{y}_{Q_2} - \dot{\theta}_1 \cdot Q_2 C \cdot \sin(\theta_2) \end{cases} \tag{6.132b}$$

To solve the acceleration problem, we first differentiate Equations 6.127, which yield the following set of two linear equations in the unknowns $\ddot{s}_1$ and $\ddot{s}_2$:

$$\begin{cases} \ddot{s}_1 \cdot \cos(\theta_1) - \ddot{s}_2 \cdot \cos(\theta_2) = \ddot{x}_A - \ddot{x}_B + \ddot{\theta}_1 \cdot (y_A - y_C) + \dot{\theta}_1 \cdot (\dot{y}_A - \dot{y}_C) \\ \quad - \ddot{\theta}_2 \cdot (y_B - y_C) - \dot{\theta}_2 \cdot (\dot{y}_B - \dot{y}_C) + \dot{\theta}_1 \cdot \dot{s}_1 \cdot \sin(\theta_1) - \dot{\theta}_2 \cdot \dot{s}_2 \cdot \sin(\theta_2) \\ \ddot{s}_1 \cdot \sin(\theta_1) - \ddot{s}_2 \cdot \sin(\theta_2) = \ddot{y}_A - \ddot{y}_B - \ddot{\theta}_1 \cdot (x_A - x_C) + \dot{\theta}_1 \cdot (\dot{x}_A - \dot{x}_C) \\ \quad + \ddot{\theta}_2 \cdot (x_B - x_C) + \dot{\theta}_2 \cdot (\dot{x}_B - \dot{x}_C) - \dot{\theta}_1 \cdot \dot{s}_1 \cdot \cos(\theta_1) + \dot{\theta}_2 \cdot \dot{s}_2 \cdot \cos(\theta_2) \end{cases} \tag{6.133}$$

In turn, the components of the accelerations of points C, P_1, P_2, Q_1, and Q_2 result by differentiating with respect to time Equations 6.128 through 6.132, that is,

$$\begin{cases} \ddot{x}_{Q_1} = \ddot{x}_A - \ddot{s}_1 \cdot \cos(\theta_1) + 2\dot{\theta}_1 \cdot \dot{s}_1 \cdot \sin(\theta_1) + \ddot{\theta}_1 \cdot s_1 \cdot \sin(\theta_1) + \dot{\theta}_1^2 \cdot s_1 \cdot \cos(\theta_1) \\ \ddot{y}_{Q_1} = \ddot{y}_A - \ddot{s}_1 \cdot \sin(\theta_1) - 2\dot{\theta}_1 \cdot \dot{s}_1 \cdot \cos(\theta_1) - \ddot{\theta}_1 \cdot s_1 \cdot \cos(\theta_1) + \dot{\theta}_1^2 \cdot s_1 \cdot \sin(\theta_1) \end{cases} \tag{6.134}$$

$$\begin{cases} \ddot{x}_{P_1} = \ddot{x}_{Q_1} + \ddot{\theta}_1 \cdot P_1 Q_1 \cdot \sin(\theta_1) + \dot{\theta}_1^2 \cdot P_1 Q_1 \cdot \cos(\theta_1) \\ \ddot{y}_{P_1} = \ddot{y}_{Q_1} - \ddot{\theta}_1 \cdot P_1 Q_1 \cdot \cos(\theta_1) + \dot{\theta}_1^2 \cdot P_1 Q_1 \cdot \sin(\theta_1) \end{cases} \tag{6.135}$$

$$\begin{cases} \ddot{x}_{Q_2} = \ddot{x}_B - \ddot{s}_2 \cdot \cos(\theta_2) + 2\dot{\theta}_2 \cdot \dot{s}_2 \cdot \sin(\theta_2) + \ddot{\theta}_2 \cdot s_2 \cdot \sin(\theta_2) + \dot{\theta}_2^2 \cdot s_2 \cdot \cos(\theta_2) \\ \ddot{y}_{Q_2} = \ddot{y}_B - \ddot{s}_2 \cdot \sin(\theta_2) - 2\dot{\theta}_2 \cdot \dot{s}_2 \cdot \cos(\theta_2) - \ddot{\theta}_2 \cdot s_2 \cdot \cos(\theta_2) + \dot{\theta}_2^2 \cdot s_2 \cdot \sin(\theta_2) \end{cases} \tag{6.136}$$

$$\begin{cases} \ddot{x}_{P_2} = \ddot{x}_{Q_2} + \ddot{\theta}_2 \cdot P_2 Q_2 \cdot \sin(\theta_2) + \dot{\theta}_2^2 \cdot P_2 Q_2 \cdot \cos(\theta_2) \\ \ddot{y}_{P_2} = \ddot{y}_{Q_2} - \ddot{\theta}_2 \cdot P_2 Q_2 \cdot \cos(\theta_2) + \dot{\theta}_2^2 \cdot P_2 Q_2 \cdot \sin(\theta_2) \end{cases} \tag{6.137}$$

and the same for joint C,

$$\begin{cases} \ddot{x}_C = \ddot{x}_{Q_1} + \ddot{\theta}_1 \cdot Q_1 C \cdot \cos(\theta_1) - \dot{\theta}_1^2 \cdot Q_1 C \cdot \sin(\theta_1) \\ \ddot{y}_C = \ddot{y}_{Q_1} + \ddot{\theta}_1 \cdot Q_1 C \cdot \sin(\theta_1) + \dot{\theta}_1^2 \cdot Q_1 C \cdot \cos(\theta_1) \end{cases} \qquad (6.138a)$$

or

$$\begin{cases} \ddot{x}_C = \ddot{x}_{Q_2} - \ddot{\theta}_2 \cdot Q_2 C \cdot \cos(\theta_2) + \dot{\theta}_2^2 \cdot Q_2 C \cdot \sin(\theta_2) \\ \ddot{y}_C = \ddot{y}_{Q_2} - \ddot{\theta}_2 \cdot Q_2 C \cdot \sin(\theta_2) - \dot{\theta}_2^2 \cdot Q_2 C \cdot \cos(\theta_2) \end{cases} \qquad (6.138b)$$

The equations derived earlier have been implemented in procedure **T_R_T** part of unit **LibAssur**. The heading of this procedure is

```
T_R_T(Color:Word; xA,yA, vxA,vyA, axA,ayA, xB,yB, vxB,vyB,
axB,ayB, Theta1,dTheta1,ddTheta1, Theta2,dTheta2,ddTheta2, P1Q1,
Q1C, P2Q2, Q2C:double; var xP1,yP1, vxP1,vyP1, axP1,ayP1, xQ1,yQ1,
vxQ1,vyQ1, axQ1,ayQ1, xP2,yP2, vxP2,vyP2, axP2,ayP2, xQ2,yQ2,
vxQ2,vyQ2, axQ2,ayQ2, xC,yC, vxC,vyC, axC,ayC: double;
var OK:Boolean);
```

The correspondence between the formal parameters and the notations used in Equations 6.118 through 6.138 and in Figure 6.23 is summarized in the following tables:

Input parameters of procedure **T_R_T**:

0...16	xA	yA	$\dot{x}A$	$\dot{y}A$	$\ddot{x}A$	$\ddot{y}A$	xB	yB	$\dot{x}B$	$\dot{y}B$	$\ddot{x}B$	$\ddot{y}B$
Color	xA	yA	vxA	vyA	axA	ayA	xB	yB	vxB	vyB	axB	ayB

θ_1	$\dot{\theta}_1$	$\ddot{\theta}_1$	θ_2	$\dot{\theta}_2$	$\ddot{\theta}_2$	$P_1 Q_1$	$Q_1 C$	$P_2 Q_2$	$Q_2 C$
Theta1	dTheta1	ddTheta1	Theta2	dTheta2	ddTheta2	P1Q1	Q1C	P2Q2	Q2C

Output parameters of procedure **T_R_T**:

xP_1	yP_1	$\dot{x}P_1$	$\dot{y}P_1$	$\ddot{x}P_1$	$\ddot{y}P_1$	xQ_1	yQ_1	$\dot{x}Q_1$	$\dot{y}Q_1$	$\ddot{x}Q_1$	$\ddot{y}Q_1$
xP1	yP1	vxP1	vyP1	axP1	ayP1	xQ1	yQ1	vxQ1	vyQ1	axQ1	ayQ1

xP_2	yP_2	$\dot{x}P_2$	$\dot{y}P_2$	$\ddot{x}P_2$	$\ddot{y}P_2$	xQ_2	yQ_2	$\dot{x}Q_2$	$\dot{y}Q_2$	$\ddot{x}Q_2$	$\ddot{y}Q_2$	$\theta_1 \neq \theta_2$
xP2	yP2	vxP2	vyP2	axP2	ayP2	xQ2	yQ2	vxQ2	vyQ2	axQ2	ayQ2	OK

Note that the relative displacements s_1 and s_2 of slider blocks A and B and their first and second time derivatives are not returned by the procedure because they can be easily calculated using procedure **VarDist**.

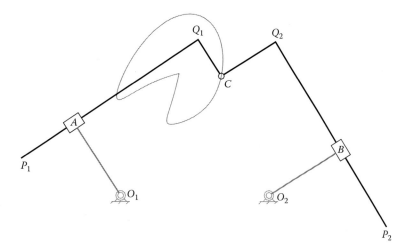

FIGURE 6.24   A two-DOF mechanism consisting of a T_R_T dyadic isomer with two potential sliding blocks driven by two rockers. See also animated file **F6_24.GIF**, the frames of which have been generated using program **P6_24.PAS**.

Program **P6_24.PAS** listed in Appendix B is a sample mechanism simulation (Figure 6.24) that employs procedure **T_R_T**. For simplicity, no velocity and acceleration values are transmitted to procedure **T_R_T**, and therefore no velocity and acceleration results are returned by the program.

6.12.2 _TRT_ Dyadic Isomer: Procedure _**TRT**_

In the kinematic simulation of the _TRT_ isomer of the TRT dyad with two potential sliding rods (Figure 6.25), the following parameters are assumed known at any given time:

- Coordinates xP_1, yP_1 and xQ_1, yQ_1 relative to the fixed reference frame of the ends of slider rod P_1Q_1.

- Projections $\dot{x}P_1$ and $\dot{y}P_1$ of the velocity of point P_1 onto the fixed reference frame.

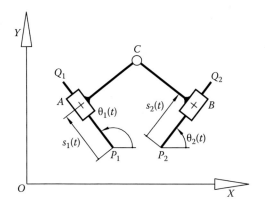

FIGURE 6.25   Notations used in the kinematics of the _TRT_ dyadic isomer.

- Projections $\ddot{x}P_1$ and $\ddot{y}P_1$ of the accelerations of point P_1 onto the fixed reference frame.

- Projections $\dot{x}Q_1$ and $\dot{y}Q_1$ of the velocity of point Q_1 onto the fixed reference frame.

- Projections $\ddot{x}Q_1$ and $\ddot{y}Q_1$ of the accelerations of point Q_1 onto the fixed reference frame.

- Coordinates xP_2, yP_2 and xQ_2, yQ_2 relative to the fixed reference frame of the ends of slider rod P_2Q_2.

- Projections $\dot{x}P_2$ and $\dot{y}P_2$ of the velocity of point P_2 onto the fixed reference frame.

- Projections $\ddot{x}P_2$ and $\ddot{y}P_2$ of the accelerations of point P_2 onto the fixed reference frame.

- Projections $\dot{x}Q_2$ and $\dot{y}Q_2$ of the velocity of joint center Q_2 onto the fixed reference frame.

- Projections $\ddot{x}Q_2$ and $\ddot{y}Q_2$ of the accelerations of point Q_2 onto the fixed reference frame.

- Offset AC between sliding block A and pin joint C (can be either positive or negative).

- Offset BC between sliding block B and pin joint C (can be either positive or negative).

The purpose of the analysis is to determine

- Displacements s_1 and s_2 of slider blocks A and B measured as shown and their first and second time derivatives $\dot{s}_1$, $\dot{s}_2$, $\ddot{s}_1$, and $\ddot{s}_2$.

- The coordinates xC and yC of pin joint center C relative to the OXY reference frame.

- The projections $\dot{x}C$ and $\dot{y}C$ of the velocity of C onto the axes of the fixed reference frame.

- The projections $\ddot{x}C$ and $\ddot{y}C$ of the acceleration of C onto the axes of the fixed reference frame.

Same as before, we begin by writing the vector-loop equation of the dyad, that is,

$$\mathbf{OP_1} + \mathbf{P_1A} + \mathbf{AC} = \mathbf{OP_2} + \mathbf{P_2B} + \mathbf{BC} \tag{6.139}$$

which projects on the x- and y-axes of the OXY reference frame as

$$\begin{cases} x_{P_1} + s_1 \cdot \cos(\theta_1) + AC \cdot \cos\left(\theta_1 - \dfrac{\pi}{2}\right) = x_{P_2} + s_2 \cdot \cos(\theta_2) + BC \cdot \cos\left(\theta_2 - \dfrac{\pi}{2}\right) \\[3mm] y_{P_1} + s_1 \cdot \sin(\theta_1) + AC \cdot \sin\left(\theta_1 - \dfrac{\pi}{2}\right) = y_{P_2} + s_2 \cdot \sin(\theta_2) + BC \cdot \sin\left(\theta_2 - \dfrac{\pi}{2}\right) \end{cases} \tag{6.140}$$

We rearrange them as a set of two linear equations in the unknowns s_1 and s_2 that is easy to solve:

$$\begin{cases} s_1 \cdot \cos(\theta_1) - s_2 \cdot \cos(\theta_2) = x_{P_2} - x_{P_1} - AC \cdot \sin(\theta_1) - BC \cdot \sin(\theta_2) \\ s_1 \cdot \sin(\theta_1) - s_2 \cdot \sin(\theta_2) = y_{P_2} - y_{P_1} + AC \cdot \cos(\theta_1) + BC \cdot \cos(\theta_2) \end{cases} \tag{6.141}$$

For these two equations to have distinct solutions, the same condition (6.120) must hold true and should be verified before further kinematic calculations are performed.

Once slider displacements s_1 and s_2 become available, the coordinates of points A, B can be calculated with the following equations:

$$\begin{cases} x_A = x_{P_1} + s_1 \cdot \cos(\theta_1) \\ y_A = y_{P_1} + s_1 \cdot \sin(\theta_1) \end{cases} \tag{6.142}$$

$$\begin{cases} x_B = x_{P_2} + s_2 \cdot \cos(\theta_2) \\ y_B = y_{P_2} + s_2 \cdot \sin(\theta_2) \end{cases} \tag{6.143}$$

The coordinates of pin joint center C can be calculated with either equation:

$$\begin{cases} x_C = x_A + AC \cdot \cos\left(\theta_1 - \dfrac{\pi}{2}\right) = x_A + AC \cdot \sin(\theta_1) \\ y_C = y_A + AC \cdot \sin\left(\theta_1 - \dfrac{\pi}{2}\right) = y_A - AC \cdot \cos(\theta_1) \end{cases} \tag{6.144a}$$

or equation

$$\begin{cases} x_C = x_B + BC \cdot \cos\left(\theta_2 - \dfrac{\pi}{2}\right) = x_B - BC \cdot \sin(\theta_2) \\ y_C = y_B + BC \cdot \sin\left(\theta_2 - \dfrac{\pi}{2}\right) = y_B + BC \cdot \cos(\theta_2) \end{cases} \tag{6.144b}$$

The velocity problem requires solving the following equations:

$$
\begin{cases}
\dot{s}_1 \cdot \cos(\theta_1) - \dot{s}_2 \cdot \cos(\theta_2) = \\
\quad \dot{x}_{P_2} - \dot{x}_{P_1} - \dot{\theta}_1 \cdot AC \cdot \cos(\theta_1) - \dot{\theta}_2 \cdot BC \cdot \cos(\theta_2) + \dot{\theta}_1 \cdot s_1 \cdot \sin(\theta_1) - \dot{\theta}_2 \cdot s_2 \cdot \sin(\theta_2) \\
\dot{s}_1 \cdot \sin(\theta_1) - \dot{s}_2 \cdot \sin(\theta_2) = \\
\quad \dot{y}_{P_2} - \dot{y}_{P_1} - \dot{\theta}_1 \cdot AC \cdot \sin(\theta_1) - \dot{\theta}_2 \cdot BC \cdot \sin(\theta_2) - \dot{\theta}_1 \cdot s_1 \cdot \cos(\theta_1) + \dot{\theta}_2 \cdot s_2 \cdot \cos(\theta_2)
\end{cases}
\tag{6.145}
$$

in the unknown relative displacements $\dot{s}_1$ and $\dot{s}_2$ of the slider. These equations were obtained by differentiating with respect to time Equations 6.141. By further applying Equations 6.142, 6.143, and 6.144b, we further get

$$
\begin{cases}
\dot{s}_1 \cdot \cos(\theta_1) - \dot{s}_2 \cdot \cos(\theta_2) = \dot{x}_{P_2} - \dot{x}_{P_1} + \dot{\theta}_1 \cdot (y_C - y_{P_1}) - \dot{\theta}_2 \cdot (y_C - y_{P_2}) \\
\dot{s}_1 \cdot \sin(\theta_1) - \dot{s}_2 \cdot \sin(\theta_2) = \dot{y}_{P_2} - \dot{y}_{P_1} - \dot{\theta}_1 \cdot (x_C - x_{P_1}) + \dot{\theta}_2 \cdot (x_C - x_{P_2})
\end{cases}
\tag{6.146}
$$

In turn, the scalar velocities of points A, B, and C are obtained through differentiation with respect to time of Equations 6.142 through 6.144. These are as follows:

$$
\begin{cases}
\dot{x}_A = \dot{x}_{P_1} + \dot{s}_1 \cdot \cos(\theta_1) - \dot{\theta}_1 \cdot s_1 \cdot \sin(\theta_1) \\
\dot{y}_A = \dot{y}_{P_1} + \dot{s}_1 \cdot \sin(\theta_1) + \dot{\theta}_1 \cdot s_1 \cdot \cos(\theta_1)
\end{cases}
\tag{6.147}
$$

$$
\begin{cases}
\dot{x}_B = \dot{x}_{P_2} + \dot{s}_2 \cdot \cos(\theta_2) - \dot{\theta}_2 \cdot s_2 \cdot \sin(\theta_2) \\
\dot{y}_B = \dot{y}_{P_2} + \dot{s}_2 \cdot \sin(\theta_2) + \dot{\theta}_2 \cdot s_2 \cdot \cos(\theta_2)
\end{cases}
\tag{6.148}
$$

and

$$
\begin{cases}
\dot{x}_C = \dot{x}_A + \dot{\theta}_1 \cdot AC \cdot \cos(\theta_1) \\
\dot{y}_C = \dot{y}_A + \dot{\theta}_1 \cdot AC \cdot \sin(\theta_1)
\end{cases}
\tag{6.149a}
$$

or

$$
\begin{cases}
\dot{x}_C = \dot{x}_B - \dot{\theta}_2 \cdot BC \cdot \cos(\theta_2) \\
\dot{y}_C = \dot{y}_B - \dot{\theta}_2 \cdot BC \cdot \sin(\theta_2)
\end{cases}
\tag{6.149b}
$$

To solve the acceleration problem, we first differentiate Equation 6.146, which yields the following set of two linear equations in the unknowns $\ddot{s}_1$ and $\ddot{s}_2$:

$$
\begin{cases}
\ddot{s}_1 \cdot \cos(\theta_1) - \ddot{s}_2 \cdot \cos(\theta_2) = \ddot{x}_{P_2} - \ddot{x}_{P_1} + \ddot{\theta}_1 \cdot (y_C - y_{P_1}) - \ddot{\theta}_2 \cdot (y_C - y_{P_2}) \\
\quad + \dot{\theta}_1 \cdot (\dot{y}_C - \dot{y}_{P_1}) - \dot{\theta}_2 \cdot (\dot{y}_C - \dot{y}_{P_2}) + \dot{\theta}_1 \cdot \dot{s}_1 \cdot \sin(\theta_1) - \dot{\theta}_2 \cdot \dot{s}_2 \cdot \sin(\theta_2) \\
\ddot{s}_1 \cdot \sin(\theta_1) - \ddot{s}_2 \cdot \sin(\theta_2) = \ddot{y}_{P_2} - \ddot{y}_{P_1} - \ddot{\theta}_1 \cdot (x_C - x_{P_1}) + \ddot{\theta}_2 \cdot (x_C - x_{P_2}) \\
\quad - \dot{\theta}_1 \cdot (\dot{x}_C - \dot{x}_{P_1}) + \dot{\theta}_2 \cdot (\dot{x}_C - \dot{x}_{P_2}) - \dot{\theta}_1 \cdot \dot{s}_1 \cdot \cos(\theta_1) + \dot{\theta}_2 \cdot \dot{s}_2 \cdot \sin(\theta_2)
\end{cases}
\tag{6.150}
$$

The accelerations of points A, B, and C result from the differentiation with respect to time Equations 6.147 through 6.149b:

$$
\begin{cases}
\ddot{x}_A = \ddot{x}_{P1} + \ddot{s}_1 \cdot \cos(\theta_1) - 2\dot{\theta}_1 \cdot \dot{s}_1 \cdot \sin(\theta_1) - \ddot{\theta}_1 \cdot s_1 \cdot \sin(\theta_1) - \dot{\theta}_1^2 \cdot s_1 \cdot \cos(\theta_1) \\
\ddot{y}_A = \ddot{y}_{P1} + \ddot{s}_1 \cdot \sin(\theta_1) + 2\dot{\theta}_1 \cdot \dot{s}_1 \cdot \cos(\theta_1) + \ddot{\theta}_1 \cdot s_1 \cdot \cos(\theta_1) - \dot{\theta}_1^2 \cdot s_1 \cdot \sin(\theta_1)
\end{cases}
\tag{6.151}
$$

$$
\begin{cases}
\ddot{x}_B = \ddot{x}_{P_2} + \ddot{s}_2 \cdot \cos(\theta_2) - 2\dot{\theta}_2 \cdot \dot{s}_2 \cdot \sin(\theta_2) - \ddot{\theta}_2 \cdot s_2 \cdot \sin(\theta_2) - \dot{\theta}_2^2 \cdot s_2 \cdot \cos(\theta_2) \\
\ddot{y}_B = \ddot{y}_{P_2} + \ddot{s}_2 \cdot \sin(\theta_2) + 2\dot{\theta}_2 \cdot \dot{s}_2 \cdot \cos(\theta_2) + \ddot{\theta}_2 \cdot s_2 \cdot \cos(\theta_2) - \dot{\theta}_2^2 \cdot s_2 \cdot \sin(\theta_2)
\end{cases}
\tag{6.152}
$$

and

$$
\begin{cases}
\ddot{x}_C = \ddot{x}_A + \ddot{\theta}_1 \cdot AC \cdot \cos(\theta_1) - \dot{\theta}_1^2 \cdot AC \cdot \sin(\theta_1) \\
\ddot{y}_C = \ddot{y}_A + \ddot{\theta}_1 \cdot AC \cdot \sin(\theta_1) + \dot{\theta}_1^2 \cdot AC \cdot \cos(\theta_1)
\end{cases}
\tag{6.153a}
$$

or

$$
\begin{cases}
\ddot{x}_C = \ddot{x}_B - \ddot{\theta}_2 \cdot BC \cdot \cos(\theta_2) + \dot{\theta}_2^2 \cdot BC \cdot \sin(\theta_2) \\
\ddot{y}_C = \ddot{y}_B - \ddot{\theta}_2 \cdot BC \cdot \sin(\theta_2) - \dot{\theta}_2^2 \cdot BC \cdot \cos(\theta_2)
\end{cases}
\tag{6.153b}
$$

Note that angles θ_1 and θ_2 occurring in these equations and their time derivatives $\dot{\theta}_1$, $\dot{\theta}_2$, $\ddot{\theta}_1$, and $\ddot{\theta}_2$ must be evaluated first. This can be done conveniently by calling procedure **AngPVA** with its arguments equal to the x and y coordinates of points P_1, Q_1, P_2, and Q_2 and to the first and second time derivatives of these coordinates.

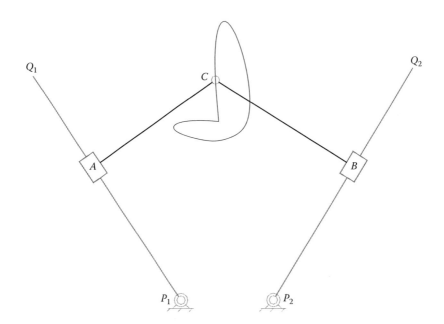

FIGURE 6.26 A two-DOF mechanism consisting of a _TRT_ dyadic isomer driven by two rockers. See also animation file **F6_26.GIF**, the frames of which have been generated using program **P6_26.PAS**.

These equations have been programmed in procedure **_TRT_** part of unit **LibAssur** with the following heading:

```
_TRT_(Color:Word; xP1,yP1, vxP1,vyP1, axP1,ayP1, xQ1,yQ1,
vxQ1,vyQ1, axQ1,ayQ1, xP2,yP2, vxP2,vyP2, axP2,ayP2, xQ2,yQ2,
vxQ2,vyQ2, axQ2,ayQ2, AC,BC:double; var xA,yA, vxA,vyA, axA,ayA,
xB,yB, vxB,vyB, axB,ayB, xC,yC, vxC,vyC, axC,ayC: double;
var OK:Boolean);
```

The correspondence between the formal parameters and the notations used in Equations 6.140 through 6.153 and in Figure 6.25 is summarized in the following tables:

Input parameters of procedure **_TRT_**:

0...16	xP_1	yP_1	$\dot{x}P_1$	$\dot{y}P_1$	$\ddot{x}P_1$	$\ddot{y}P_1$	xQ_1	yQ_1	$\dot{x}Q_1$	$\dot{y}Q_1$	$\ddot{x}Q_1$	$\ddot{y}Q_1$
Color	xP1	yP1	vxP1	vyP1	axP1	ayP1	xQ1	yQ1	vxQ1	vyQ1	axQ1	ayQ1

xP_2	yP_2	$\dot{x}P_2$	$\dot{y}P_2$	$\ddot{x}P_2$	$\ddot{y}P_2$	xQ_2	yQ_2	$\dot{x}Q_2$	$\dot{y}Q_2$	$\ddot{x}Q_2$	$\ddot{y}Q_2$	AC	BC
xP2	yP2	vxP2	vyP2	axP2	ayP2	xQ2	yQ2	vxQ2	vyQ2	axQ2	ayQ2	AC	BC

Output parameters of procedure **_TRT_**:

xA	yA	$\dot{x}A$	$\dot{y}A$	$\ddot{x}A$	$\ddot{y}A$	xB	yB	$\dot{x}B$	$\dot{y}B$	$\ddot{x}B$	$\ddot{y}B$
xA	yA	vxA	vyA	axA	ayA	xB	yB	vxB	vyB	axB	ayB

xC	yC	$\dot{x}C$	$\dot{y}C$	$\ddot{x}C$	$\ddot{y}C$	$\theta_1 \neq \theta_2$
xC	yC	vxC	vyC	axC	ayC	OK

Displacements s_1 and s_2 of rods P_1Q_1 and P_2Q_2 relative to their sliders and their time derivatives $\dot{s}_1, \dot{s}_2, \ddot{s}_1,$ and $\ddot{s}_2$ are not returned by procedure **_TRT_**. However, they can be calculated using procedure **VarDist** with inputs of the position velocity and accelerations of points P_1 and A and P_2 and B, respectively.

Program **P6_26.PAS** (see Appendix B) uses the **_TRT_** procedure to simulate the motion of a _TRT_ dyadic isomer driven by two rockers (see Figure 6.26). Again, for simplicity, all the input and output velocity and acceleration values were assigned the generic variable _. The coordinates of point C returned by **_TRT_** were then used to plot its path by calling procedure **Locus** from unit **LibMec2D** (see Figure 6.26).

6.12.3 T_RT_ Dyadic Isomer: Procedure **T_RT_**

The third possible isomer of the TRT dyad is shown in Figure 6.27. T_RT_ has one of its potential joints shaped as a sliding block and the other potential joint shaped as a sliding rod. T_RT_ can be assumed to be a combination of the previously discussed isomers of the same dyad, also reflected in the similarity of their kinematic equations.

With the notations in Figure 6.27, the following parameters are assumed known at any given time of a simulation:

- Coordinates xA, yA of sliding block A relative to the fixed reference frame OXY.

- Coordinates xP_2, yP_2 and xQ_2, yQ_2 of the ends of the potential sliding rod P_2Q_2 measured relative to the fixed reference frame.

- Orientation angle θ_1 of the slider block A and its first and second time derivatives $\dot{\theta}_1$ and $\ddot{\theta}_1$.

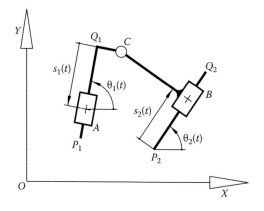

FIGURE 6.27 Notations used in solving the kinematics of the T_RT_ isomer of the TRT dyad with one potential sliding block and one potential sliding rod.

- Projections $\dot{x}A$ and $\dot{y}A$ of the velocity of joint center A onto the fixed reference frame.

- Projections $\ddot{x}A$ and $\ddot{y}A$ of the accelerations of A onto the fixed reference frame.

- Projections $\dot{x}P_2$ and $\dot{y}P_2$ of the velocity of point P_2 onto the fixed reference frame.

- Projections $\ddot{x}P_2$ and $\ddot{y}P_2$ of the accelerations of point P_2 onto the fixed reference frame.

- Projections $\dot{x}Q_2$ and $\dot{y}Q_2$ of the velocity of point Q_2 onto the fixed reference frame.

- Projections $\ddot{x}Q_2$ and $\ddot{y}Q_2$ of the accelerations of point Q_2 onto the fixed reference frame.

- Lengths P_1Q_1 and Q_1C of the L-shaped rod supporting slider block A (can be either positive or negative).

- Length BC of pin joint C offset (can be either positive or negative).

The purpose of the analysis is to determine

- Displacements s_1 and s_2 of slider blocks A and B measured as shown and their first and second time derivatives $\dot{s}_1$, $\dot{s}_2$, $\ddot{s}_1$, and $\ddot{s}_2$.

- Projections xC and yC of joint center C onto the OXY reference frame and their first and second time derivatives $\dot{x}C$, $\dot{y}C$, $\ddot{x}C$, and $\ddot{y}C$.

- Coordinates xP_1, yP_1 and xQ_1, yQ_1 of points P_1 and Q_1 of the L-shaped link, measured relative to the fixed reference frame.

- Projections $\dot{x}P_1$ and $\dot{y}P_1$ of the velocity of point P_1 onto the fixed reference frame.

- Projections $\ddot{x}P_1$ and $\ddot{y}P_1$ of the accelerations of point P_1 onto the fixed reference frame.

- Projections $\dot{x}Q_1$ and $\dot{y}Q_1$ of the velocity of point Q_1 onto the fixed reference frame.

- Projections $\ddot{x}Q_1$ and $\ddot{y}Q_1$ of the accelerations of point Q_1 onto the fixed reference frame.

The vector-loop equation of the dyadic isomer

$$\mathbf{OA} + \mathbf{AQ}_1 + \mathbf{Q}_1\mathbf{C} = \mathbf{OP}_2 + \mathbf{P}_2\mathbf{B} + \mathbf{BC} \tag{6.154}$$

yields the following two scalar equations:

$$
\begin{cases}
x_A - s_1 \cdot \cos(\theta_1) + Q_1C \cdot \cos\left(\theta_1 - \dfrac{\pi}{2}\right) = x_{P_2} + s_2 \cdot \cos(\theta_2) + BC \cdot \cos\left(\theta_2 - \dfrac{\pi}{2}\right) \\[4mm]
y_A - s_1 \cdot \sin(\theta_1) + Q_1C \cdot \sin\left(\theta_1 - \dfrac{\pi}{2}\right) = y_{P_2} + s_2 \cdot \sin(\theta_2) + BC \cdot \sin\left(\theta_2 - \dfrac{\pi}{2}\right)
\end{cases}
\tag{6.155}
$$

After rearranging terms, a set of two linear equations in the unknowns s_1 and s_2 is obtained:

$$\begin{cases} s_1 \cdot \cos(\theta_1) + s_2 \cdot \cos(\theta_2) = x_A - x_{P_2} + Q_1C \cdot \sin(\theta_1) + BC \cdot \sin(\theta_2) \\ s_1 \cdot \sin(\theta_1) + s_2 \cdot \sin(\theta_2) = y_A - y_{P_2} - Q_1C \cdot \cos(\theta_1) - BC \cdot \cos(\theta_2) \end{cases} \quad (6.156)$$

These two linear equations have distinct solutions only if the earlier condition (6.120) is satisfied.

Once slider displacements s_1 and s_2 become known, the coordinates of points Q_1, P_1, and B can be calculated using Equations 6.121, 6.122, and 6.143. In turn, the coordinates of pin joint center C can be calculated using either Equation 6.125a or 6.144b.

To solve the velocity problem, we differentiate Equation 6.156 with respect to time:

$$\begin{cases} \dot{s}_1 \cdot \cos(\theta_1) + \dot{s}_2 \cdot \cos(\theta_2) = \dot{x}_A - \dot{x}_{P_2} \\ \quad + \dot{\theta}_1 \cdot Q_1C \cdot \cos(\theta_1) + \dot{\theta}_2 \cdot BC \cdot \cos(\theta_2) + \dot{\theta}_1 \cdot s_1 \cdot \sin(\theta_1) + \dot{\theta}_2 \cdot s_2 \cdot \sin(\theta_2) \\ \dot{s}_1 \cdot \sin(\theta_1) + \dot{s}_2 \cdot \sin(\theta_2) = \dot{y}_A - \dot{y}_{P_2} \\ \quad + \dot{\theta}_1 \cdot Q_1C \cdot \sin(\theta_1) + \dot{\theta}_2 \cdot BC \cdot \sin(\theta_2) - \dot{\theta}_1 \cdot s_1 \cdot \cos(\theta_1) - \dot{\theta}_2 \cdot s_2 \cdot \cos(\theta_2) \end{cases} \quad (6.157)$$

By applying the position results, this set of two linear equations in the unknowns $\dot{s}_1$ and $\dot{s}_2$ becomes:

$$\begin{cases} \dot{s}_1 \cdot \cos(\theta_1) + \dot{s}_2 \cdot \cos(\theta_2) = \dot{x}_A - \dot{x}_{P_2} - \dot{\theta}_1 \cdot (y_C - y_A) + \dot{\theta}_2 \cdot (y_C - y_{P_2}) \\ \dot{s}_1 \cdot \sin(\theta_1) + \dot{s}_2 \cdot \sin(\theta_2) = \dot{y}_A - \dot{y}_{P_2} + \dot{\theta}_1 \cdot (x_C - x_A) - \dot{\theta}_2 \cdot (x_C - x_{P_2}) \end{cases} \quad (6.158)$$

The x and y components of the velocities of points Q_1, P_1, and B can be calculated using Equations 6.128, 6.129, and 6.148, while those of point C using either Equation 6.132a or 6.153b.

To solve the acceleration problem, we differentiate Equation 6.158 with respect to time and obtain the following two linear equations in the unknowns $\ddot{s}_1$ and $\ddot{s}_2$:

$$\begin{cases} \ddot{s}_1 \cdot \cos(\theta_1) + \ddot{s}_2 \cdot \cos(\theta_2) = \ddot{x}_A - \ddot{x}_{P_2} - \ddot{\theta}_1 \cdot (y_C - y_A) + \ddot{\theta}_2 \cdot (y_C - y_{P_2}) \\ \quad - \dot{\theta}_1 \cdot (\dot{y}_C - \dot{y}_A) + \dot{\theta}_2 \cdot (\dot{y}_C - \dot{y}_{P_2}) + \dot{\theta}_1 \cdot \dot{s}_1 \cdot \sin(\theta_1) + \dot{\theta}_2 \cdot \dot{s}_2 \cdot \sin(\theta_2) \\ \ddot{s}_1 \cdot \sin(\theta_1) + \ddot{s}_2 \cdot \sin(\theta_2) = \ddot{y}_A - \ddot{y}_{P_2} + \ddot{\theta}_1 \cdot (x_C - x_A) - \ddot{\theta}_2 \cdot (x_C - x_{P_2}) \\ \quad + \dot{\theta}_1 \cdot (\dot{x}_C - \dot{x}_A) - \dot{\theta}_2 \cdot (\dot{x}_C - \dot{x}_{P_2}) - \dot{\theta}_1 \cdot \dot{s}_1 \cdot \cos(\theta_1) - \dot{\theta}_2 \cdot \dot{s}_2 \cdot \sin(\theta_2) \end{cases} \quad (6.159)$$

The accelerations of points Q_1, P_1, and B can be calculated using Equations 6.134, 6.135, and 6.152, while those of point C using either Equation 6.138a or 6.153b.

Note that angle θ_2 together with its time derivatives $\dot{\theta}_2$ and $\ddot{\theta}_2$ must be determined before moving forward with the velocity and acceleration problem. This can be done by calling procedure **AngPVA** with arguments set equal to the x and y coordinates of points P_2 and Q_2 and to their respective first and second derivatives with respect to time.

These kinematic equations have been implemented in procedure **T_RT_** part of unit **LibAssur**. The procedure has the following heading:

```
T_RT_(Color:Word; xA,yA, vxA,vyA, axA,ayA,
Theta1,dTheta1,ddTheta1, xP2,yP2, vxP2,vyP2, axP2,ayP2, xQ2,yQ2,
vxQ2,vyQ2, axQ2,ayQ2, P1Q1, Q1C, BC:double; var xP1,yP1,
vxP1,vyP1, axP1,ayP1, xQ1,yQ1, vxQ1,vyQ1, axQ1,ayQ1, xB,yB,
vxB,vyB, axB,ayB, xC,yC, vxC,vyC, axC,ayC:double; var OK:Boolean);
```

The correspondence between the formal parameters and the notations used in Figure 6.27 and the corresponding kinematic equations is summarized in the following tables:

Input parameters of procedure **T_RT_**:

0...16	xA	yA	$\dot{x}A$	$\dot{y}A$	$\ddot{x}A$	$\ddot{y}A$	θ_1	$\dot{\theta}_1$	$\ddot{\theta}_1$	xP_2	yP_2
Color	xA	yA	vxA	vyA	axA	ayA	Theta1	dTheta1	ddTheta1	xP2	yP2

$\dot{x}P_2$	$\dot{y}P_2$	$\ddot{x}P_2$	$\ddot{y}P_2$	xQ_2	yQ_2	$\dot{x}Q_2$	$\dot{y}Q_2$	$\ddot{x}Q_2$	$\ddot{y}Q_2$	P_1Q_1	Q_1C	BC
vxP2	vyP2	axP2	ayP2	xQ2	yQ2	vxQ2	vyQ2	axQ2	ayQ2	P1Q1	Q1C	BC

Output parameters of procedure **T_RT_**:

xP_1	yP_1	$\dot{x}P_1$	$\dot{y}P_1$	$\ddot{x}P_1$	$\ddot{y}P_1$	xQ_1	yQ_1	$\dot{x}Q_1$	$\dot{y}Q_1$	$\ddot{x}Q_1$	$\ddot{y}Q_1$
xP1	yP1	vxP1	vyP1	axP1	ayP1	xQ1	yQ1	vxQ1	vyQ1	axQ1	ayQ1

xB	yB	$\dot{x}B$	$\dot{y}B$	$\ddot{x}B$	$\ddot{y}B$	xC	yC	$\dot{x}C$	$\dot{y}C$	$\ddot{x}C$	$\ddot{y}C$	$\theta_1 \neq \theta_2$
xB	yB	vxB	vyB	axB	ayB	xC	yC	vxC	vyC	axC	ayC	OK

Same as before, displacements s_1 and s_2 of slider blocks A and B and their time derivatives $\dot{s}_1, \dot{s}_2, \ddot{s}_1,$ and $\ddot{s}_2$ are not returned by procedure **T_RT_**. If of interest, they can be calculated with the help of procedure **VarDist**.

Program **P6_28.PAS** listed in Appendix B calls the **T_RT_** procedure to simulate a TRT dyad of the T_RT_ type driven by two rockers (see Figure 6.28). The program labels all joints and plots the locus of the center of joint C. All velocity and acceleration parameters are ignored by assigning them the generic variable _ defined in the interface section of unit **LibMath**.

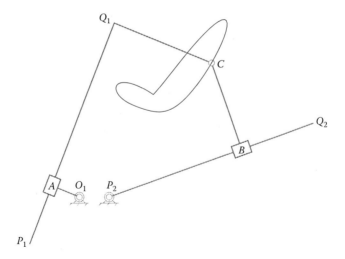

FIGURE 6.28   A two-DOF mechanism consisting of a T_RT_ dyadic isomer driven by two rockers simulated using program **P6_28.PAS**. See also animated file **F6_28.GIF**.

6.13 KINEMATIC ANALYSIS OF THE RTT DYAD USING A VECTOR-LOOP APPROACH

The RTT dyad is the last *Assur group* with two links and three joints. Because of its two back-to-back sliding joints, it has four possible isomers (Figure 6.3). The kinematic equations of these four isomers are discussed in the remaining of this chapter and implemented in procedures **R_T_T, R_TT_, RT__T**, and **RT_T_**. While the kinematic equations required to calculate the displacements, velocities, and accelerations of the two slider blocks differ among these four isomers, some of the remaining kinematic equations are coincident, thus simplifying the analytical derivations.

6.13.1 R_T_T Dyadic Isomer: Procedure **R_T_T**

The R_T_T isomer of the RTT dyad has the translating potential joint consisting of a rod *PQ* that moves to the inside of a sleeve *B* (see Figure 6.29). When a kinematic analysis is performed, the following parameters are assumed given:

- Coordinates xA, yA, of potential turning joint A relative to the fixed reference frame OXY.

- Coordinates xB, yB, of point B measured relative to the fixed reference frame.

- Projections $\dot{x}A$ and $\dot{y}A$ of the velocity of joint center A onto the fixed reference frame.

- Projections $\ddot{x}A$ and $\ddot{y}A$ of the accelerations of A onto the fixed reference frame.

- Projections $\dot{x}B$ and $\dot{y}B$ of the velocity of point B onto the fixed reference frame.

- Projections $\ddot{x}B$ and $\ddot{y}B$ of the accelerations of point B onto the fixed reference frame.

- Orientation angle φ of slider B and its first and second time derivatives $\dot{\varphi}$ and $\ddot{\varphi}$, respectively.

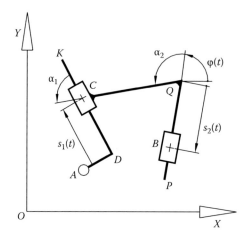

FIGURE 6.29   Notations used in solving the kinematics of the R_T_T dyadic isomer.

- Lengths AD and DK of the two sides of the L-shaped link supporting slider C (can be either positive or negative).

- Lengths PC and BQ of the two sections of the link connecting slider blocks C and B (can be either positive or negative).

- The values of the constant angles α_1 and α_2 measured as shown in Figure 6.29.

The purpose of the analysis is to determine the following unknown kinematic parameters:

- Relative displacements s_1 and s_2 of sliders B and C measured as shown and their first and second time derivatives $\dot{s}_1$, $\dot{s}_2$, $\ddot{s}_1$, and $\ddot{s}_2$.

- The coordinates xC and yC of joint center C relative to the OXY reference frame.

- The projections $\dot{x}C$ and $\dot{y}C$ of the velocity of C onto the axes of the fixed reference frame.

- The projections $\ddot{x}C$ and $\ddot{y}C$ of the acceleration of C onto the axes of the fixed reference frame.

- The coordinates xP and yP of point P relative to the OXY reference frame.

- The projections $\dot{x}P$ and $\dot{y}P$ of the velocity of P onto the axes of the fixed reference frame.

- The projections $\ddot{x}P$ and $\ddot{y}P$ of the acceleration of P onto the axes of the fixed reference frame.

- The coordinates xQ and yQ of point Q relative to the OXY reference frame.

- The projections $\dot{x}Q$ and $\dot{y}Q$ of the velocity of Q onto the axes of the fixed reference frame.

- The projections $\ddot{x}Q$ and $\ddot{y}Q$ of the acceleration of Q onto the axes of the fixed reference frame.

We begin by writing the following vector equation:

$$\mathbf{OA} + \mathbf{AD} + \mathbf{DC} = \mathbf{OB} + \mathbf{BQ} + \mathbf{QC} \tag{6.160}$$

which projects on the x- and y-axes of the fixed reference frame as

$$\begin{cases} x_A + AD\cos\left(\varphi + \alpha_2 - \alpha_1 - \dfrac{\pi}{2}\right) + s_1 \cos(\varphi + \alpha_2 - \alpha_1) = \\[2mm] \quad x_B + s_2 \cos(\varphi) + QC \cdot \cos(\varphi + \alpha_2) \\[4mm] y_A + AD\sin\left(\varphi + \alpha_2 - \alpha_1 - \dfrac{\pi}{2}\right) + s_1 \sin(\varphi + \alpha_2 - \alpha_1) = \\[2mm] \quad y_B + s_2 \sin(\varphi) + QC \cdot \sin(\varphi + \alpha_2) \end{cases} \tag{6.161}$$

We rearrange them as a set of two linear equations in the unknowns s_1 and s_2 that are easy to solve:

$$\begin{cases} s_1 \cos(\varphi + \alpha_2 - \alpha_1) - s_2 \cos(\varphi) = x_B - x_A - AD\sin(\varphi + \alpha_2 - \alpha_1) + QC\cos(\varphi + \alpha_2) \\[2mm] s_1 \sin(\varphi + \alpha_2 - \alpha_1) - s_2 \sin(\varphi) = y_B - y_A + AD\cos(\varphi + \alpha_2 - \alpha_1) + QC\sin(\varphi + \alpha_2) \end{cases} \tag{6.162}$$

Equations 6.162 have solutions for any x_A, y_A, x_B, y_B and angle φ values, provided that the following inequality holds:

$$\sin(\varphi + \alpha_2 - \alpha_1) \cdot \cos(\varphi) - \cos(\varphi + \alpha_2 - \alpha_1) \cdot \sin(\varphi) \neq 0 \quad \text{equivalent to } \alpha_1 \neq \alpha_2 \tag{6.163}$$

Once slider displacements s_1 and s_2 are determined, the coordinates of points $D, K, Q, P,$ and C required to represent graphically the dyad can be calculated with the following equations:

$$\begin{cases} x_D = x_A + AD\cos\left(\varphi + \alpha_2 - \alpha_1 - \dfrac{\pi}{2}\right) = x_A + AD\sin(\varphi + \alpha_2 - \alpha_1) \\[2mm] y_D = y_A + AD\sin\left(\varphi + \alpha_2 - \alpha_1 - \dfrac{\pi}{2}\right) = y_A - AD\cos(\varphi + \alpha_2 - \alpha_1) \end{cases} \tag{6.164}$$

$$\begin{cases} x_K = x_D + DK\cos(\varphi + \alpha_2 - \alpha_1) \\[2mm] y_K = y_D + DK\sin(\varphi + \alpha_2 - \alpha_1) \end{cases} \tag{6.165}$$

$$\begin{cases} x_Q = x_B + s_2 \cos(\varphi) \\[2mm] y_Q = y_B + s_2 \sin(\varphi) \end{cases} \tag{6.166}$$

$$\begin{cases} x_P = x_Q - PQ\cos(\varphi) \\ y_P = y_Q - PQ\sin(\varphi) \end{cases} \tag{6.167}$$

$$\begin{cases} x_C = x_D + s_1\cos(\varphi + \alpha_2 - \alpha_1) \\ y_C = y_D + s_1\sin(\varphi + \alpha_2 - \alpha_1) \end{cases} \tag{6.168a}$$

or

$$\begin{cases} x_C = x_Q + QC\cos(\varphi + \alpha_2) \\ y_C = y_Q + QC\sin(\varphi + \alpha_2) \end{cases} \tag{6.168b}$$

To solve the velocity problem, we differentiate with respect to time Equations 6.162, resulting in a set of two linear equations in the unknowns $\dot{s}_1$ and $\dot{s}_2$:

$$\begin{cases} \dot{s}_1\cos(\varphi + \alpha_2 - \alpha_1) - \dot{s}_2\cos(\varphi) = \dot{x}_B - \dot{x}_A \\ \quad + \dot{\varphi}\big[s_1\sin(\varphi + \alpha_2 - \alpha_1) - s_2\sin(\varphi) - AD\cdot\cos(\varphi + \alpha_2 - \alpha_1) - QC\sin(\varphi + \alpha_2)\big] \\ \dot{s}_1\sin(\varphi + \alpha_2 - \alpha_1) - \dot{s}_2\sin(\varphi) = \dot{y}_B - \dot{y}_A \\ \quad - \dot{\varphi}\big[s_1\cos(\varphi + \alpha_2 - \alpha_1) - s_2\cos(\varphi) + AD\sin(\varphi + \alpha_2 - \alpha_1) - QC\cos(\varphi + \alpha_2)\big] \end{cases} \tag{6.169}$$

and by applying Equations 6.162, they can be further written as

$$\begin{cases} \dot{s}_1\cos(\varphi + \alpha_2 - \alpha_1) - \dot{s}_2\cos(\varphi) = \dot{x}_B - \dot{x}_A + \dot{\varphi}(y_B - y_A) \\ \dot{s}_1\sin(\varphi + \alpha_2 - \alpha_1) - \dot{s}_2\sin(\varphi) = \dot{y}_B - \dot{y}_A - \dot{\varphi}(x_B - x_A) \end{cases} \tag{6.170}$$

The x and y components of the velocities of points D, Q, P, and C, also of interest, are given by the following equations. Again, equalities (6.164) through (6.168) have been applied in each case:

$$\begin{cases} \dot{x}_D = \dot{x}_A + \dot{\varphi}\cdot AD\cos(\varphi + \alpha_2 - \alpha_1) = \dot{x}_A + \dot{\varphi}(y_A - y_D) \\ \dot{y}_D = \dot{y}_A + \dot{\varphi}\cdot AD\sin(\varphi + \alpha_2 - \alpha_1) = \dot{y}_A - \dot{\varphi}(x_A - x_D) \end{cases} \tag{6.171}$$

$$\begin{cases} \dot{x}_K = \dot{x}_D - \dot{\varphi}\cdot DK\sin(\varphi + \alpha_2 - \alpha_1) = \dot{x}_D + \dot{\varphi}(y_D - y_K) \\ \dot{y}_K = \dot{y}_D + \dot{\varphi}\cdot DK\cos(\varphi + \alpha_2 - \alpha_1) = \dot{y}_D - \dot{\varphi}(x_D - x_K) \end{cases} \tag{6.172}$$

$$
\begin{cases}
\dot{x}_Q = \dot{x}_B + \dot{s}_2\cos(\varphi) - \dot{\varphi}\cdot s_2\cdot\sin(\varphi) = \dot{x}_B + \dot{s}_2\cdot\cos(\varphi) + \dot{\varphi}(y_B - y_Q) \\
\dot{y}_Q = \dot{y}_B + \dot{s}_2\sin(\varphi) + \dot{\varphi}\cdot s_2\cdot\cos(\varphi) = \dot{y}_B + \dot{s}_2\cdot\sin(\varphi) - \dot{\varphi}(x_B - x_Q)
\end{cases}
\tag{6.173}
$$

$$
\begin{cases}
\dot{x}_P = \dot{x}_Q + \dot{\varphi}\cdot PQ\cdot\sin(\varphi) = \dot{x}_Q - \dot{\varphi}\cdot(y_P - y_Q) \\
\dot{y}_P = \dot{y}_Q - \dot{\varphi}\cdot PQ\cdot\cos(\varphi) = \dot{y}_Q + \dot{\varphi}\cdot(x_P - x_Q)
\end{cases}
\tag{6.174}
$$

$$
\begin{cases}
\dot{x}_C = \dot{x}_D + \dot{s}_1\cdot\cos(\varphi + \alpha_2 - \alpha_1) - \dot{\varphi}\cdot s_1\cdot\sin(\varphi + \alpha_2 - \alpha_1) \\
\quad = \dot{x}_D - \dot{s}_1\cdot\dfrac{(x_D - x_K)}{DK} - \dot{\varphi}(y_C - y_D) \\
\dot{y}_C = \dot{y}_D + \dot{s}_1\cdot\sin(\dot{\varphi} + \alpha_2 - \alpha_1) + \dot{\varphi}\cdot s_1\cdot\cos(\varphi + \alpha_2 - \alpha_1) \\
\quad = \dot{y}_D - \dot{s}_1\cdot\dfrac{(y_D - y_K)}{DK} + \dot{\varphi}(x_C - x_D)
\end{cases}
\tag{6.175a}
$$

or

$$
\begin{cases}
\dot{x}_C = \dot{x}_Q - \dot{\varphi}\cdot QC\cdot\sin(\varphi + \alpha_2) = \dot{x}_Q - \dot{\varphi}(y_C - y_Q) \\
\dot{y}_C = \dot{y}_Q + \dot{\varphi}\cdot QC\cdot\cos(\varphi + \alpha_2) = \dot{y}_Q + \dot{\varphi}(x_C - x_Q)
\end{cases}
\tag{6.175b}
$$

To solve the acceleration problem, we begin by differentiating Equations 6.170, which yields a set of two linear equations in the unknowns $\ddot{s}_1$ and $\ddot{s}_2$:

$$
\begin{cases}
\ddot{s}_1\cdot\cos(\varphi + \alpha_2 - \alpha_1) - \ddot{s}_2\cdot\cos(\varphi) = \ddot{x}_B - \ddot{x}_A \\
\quad + \ddot{\varphi}\cdot(y_B - y_A) + \dot{\varphi}\cdot(\dot{y}_B - \dot{y}_A) + \dot{s}_1\cdot\dot{\varphi}\cdot\sin(\varphi + \alpha_2 - \alpha_1) - \dot{s}_2\cdot\dot{\varphi}\cdot\sin(\varphi) \\
\ddot{s}_1\cdot\sin(\varphi + \alpha_2 - \alpha_1) - \ddot{s}_2\cdot\sin(\varphi) = \ddot{y}_B - \ddot{y}_A \\
\quad - \ddot{\varphi}\cdot(x_B - x_A) - \dot{\varphi}\cdot(\dot{x}_B - \dot{x}_A) - \dot{s}_1\cdot\dot{\varphi}\cdot\cos(\varphi + \alpha_2 - \alpha_1) + \dot{s}_2\cdot\dot{\varphi}\cdot\cos(\varphi)
\end{cases}
\tag{6.176}
$$

The x and y components of the accelerations of points D, Q, P, and C are obtained by differentiating Equations 6.171 through 6.175b:

$$
\begin{cases}
\ddot{x}_D = \ddot{x}_A + \ddot{\varphi}(y_A - y_D) + \dot{\varphi}(\dot{y}_A - \dot{y}_D) \\
\ddot{y}_D = \ddot{y}_A - \ddot{\varphi}(x_A - x_D) - \dot{\varphi}(\dot{x}_A - \dot{x}_D)
\end{cases}
\tag{6.177}
$$

$$
\begin{cases}
\ddot{x}_K = \ddot{x}_D + \ddot{\varphi}(y_D - y_K) + \dot{\varphi}(\dot{y}_D - \dot{y}_K) \\
\ddot{y}_K = \ddot{y}_D - \ddot{\varphi}(x_D - x_K) - \dot{\varphi}(\dot{x}_D - \dot{x}_K)
\end{cases}
\tag{6.178}
$$

$$\begin{cases} \ddot{x}_Q = \ddot{x}_B + \ddot{s}_2 \cdot \cos(\varphi) - \dot{s}_2 \cdot \dot{\varphi} \cdot \sin(\varphi) + \ddot{\varphi}(y_B - y_Q) + \dot{\varphi}(\dot{y}_B - \dot{y}_Q) \\ \ddot{y}_Q = \ddot{y}_B + \ddot{s}_2 \cdot \sin(\varphi) + \dot{s}_2 \cdot \dot{\varphi} \cdot \cos(\varphi) - \ddot{\varphi}(x_B - x_Q) - \dot{\varphi}(\dot{x}_B - \dot{x}_Q) \end{cases} \tag{6.179}$$

$$\begin{cases} \ddot{x}_P = \ddot{x}_Q - \ddot{\varphi} \cdot (y_P - y_Q) - \dot{\varphi} \cdot (\dot{y}_P - \dot{y}_Q) \\ \ddot{y}_P = \ddot{y}_Q + \ddot{\varphi} \cdot (x_P - x_Q) + \dot{\varphi} \cdot (\dot{x}_P - \dot{x}_Q) \end{cases} \tag{6.180}$$

$$\begin{cases} \ddot{x}_C = \ddot{x}_D - \ddot{s}_1 \cdot \dfrac{(x_D - x_K)}{DK} - \ddot{\varphi}(y_C - y_D) - \dot{s}_1 \cdot \dfrac{(\dot{x}_D - \dot{x}_K)}{DK} - \dot{\varphi}(\dot{y}_C - \dot{y}_D) \\ \ddot{y}_C = \ddot{y}_D - \ddot{s}_1 \cdot \dfrac{(y_D - y_K)}{DK} + \ddot{\varphi}(x_C - x_D) - \dot{s}_1 \cdot \dfrac{(\dot{y}_D - \dot{y}_K)}{DK} + \dot{\varphi}(\dot{x}_C - \dot{x}_D) \end{cases} \tag{6.181a}$$

or

$$\begin{cases} \ddot{x}_C = \ddot{x}_Q - \ddot{\varphi}\left(y_C - y_Q\right) - \dot{\varphi}\left(\dot{y}_C - \dot{y}_Q\right) \\ \ddot{y}_C = \ddot{y}_Q + \ddot{\varphi}\left(x_C - x_Q\right) + \dot{\varphi}\left(\dot{x}_C - \dot{x}_Q\right) \end{cases} \tag{6.181b}$$

These equations have been implemented in procedure **R_T_T** part of unit **LibAssur** with the following heading:

```
R_T_T(Color:Word; xA,yA, vxA,vyA, axA,ayA, xB,yB, vxB,vyB,
axB,ayB, Phi,dPhi,ddPhi, AD,DK,PQ,QC, Alpha1,Alpha2:double;
var xC,yC, vxC,vyC, axC,ayC, xD,yD, vxD,vyD, axD,ayD, xK,yK,
vxK,vyK, axK,ayK, xP,yP, vxP,vyP, axP,ayP, xQ,yQ, vxQ,vyQ,
axQ,ayQ:double; var OK:Boolean);
```

The correspondence between the formal parameters of procedure **R_T_T** and the notations used in Figure 6.29 and in Equations 6.161 through 6.181b is summarized next:

Input parameters of procedure **R_T_T**:

0...16	xA	yA	$\dot{x}A$	$\dot{y}A$	$\ddot{x}A$	$\ddot{y}A$	xB	yB	$\dot{x}B$	$\dot{y}B$	$\ddot{x}B$	$\ddot{y}B$
Color	xA	yA	vxA	vyA	axA	ayA	xB	yB	vxB	vyB	axB	ayB

φ	$\dot{\varphi}$	$\ddot{\varphi}$	AD	DK	PQ	QC	α_1	α_2
Phi	dPhi	ddPhi	AD	DK	PQ	QC	Alph1	Alph2

Output parameters of procedure **R_T_T**:

xC	yC	$\dot{x}C$	$\dot{y}C$	$\ddot{x}C$	$\ddot{y}C$	xD	yD	$\dot{x}D$	$\dot{y}D$	$\ddot{x}D$	$\ddot{y}D$	xK	yK	$\dot{x}K$
xC	yC	vxC	vyC	axC	ayC	xD	yD	vxD	vyD	axD	ayD	xK	yK	vxK

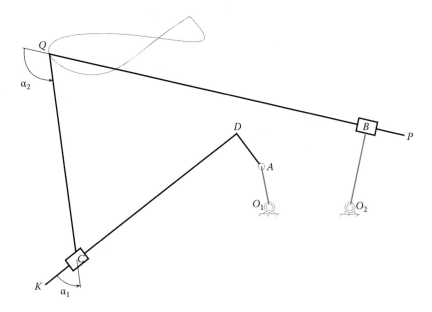

FIGURE 6.30   A two-DOF mechanism consisting of an R_T_T isomer of the RTT dyad, driven by a crank and a rocker. See also animation file **F6_30.GIF**.

$\dot{y}K$	$\ddot{x}K$	$\ddot{y}K$	xP	yP	$\dot{x}P$	$\dot{y}P$	$\ddot{x}P$	$\ddot{y}P$	xQ	yQ	$\dot{x}Q$	$\dot{y}Q$	$\ddot{x}Q$	$\ddot{y}Q$
vyK	axK	ayK	xP	yP	vxP	vyP	axP	ayP	xQ	yQ	vxQ	vyQ	axQ	ayQ

Note that the displacements s_1 and s_2 of the two slider rods relative to the sleeve and their first and second time derivatives are not returned by the procedure. If of interest, they can be easily calculated by calling procedure **VarDist** in unit **LibMec2D**.

Figure 6.30 is one of the simulation frames of a mechanism form with an R_T_T dyadic isomer done using program **P6_30.PAS** listed in Appendix B. All these frames have been assembled in the animation file **F6_30.GIF** available with the book. In addition to recording the locus of point Q and of labeling all joint centers, the program also labels the two fixed angles α_1 and α_2.

6.13.2 RT_T_ Dyadic Isomer: Procedure **RT_T_**

This section discusses the RTT dyad configured as shown in Figure 6.31, symbolized RT_T_ for short. In a kinematic analysis, the parameters listed next are assumed known at any moment of the simulation:

- Coordinates xA, yA, of potential joint A relative to the fixed reference frame OXY.

- Coordinates xP, yP and xQ, yQ of the ends of the rod that guides to the inside sliding block B.

- Offset AC of the pin joint relative to slider C.

- Length BD of the rod supporting slider block C.

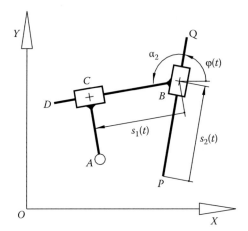

FIGURE 6.31   Notations used in solving the kinematics of the RT_T_ dyadic isomer.

- The value of the constant angle α_2 measured as shown in the figure.

- Projections $\dot{x}A$ and $\dot{y}A$ of the velocity of joint center A onto the fixed reference frame.

- Projections $\ddot{x}A$ and $\ddot{y}A$ of the accelerations of A onto the fixed reference frame.

- Projections $\dot{x}P$ and $\dot{y}P$ of the velocity of joint center P onto the fixed reference frame.

- Projections $\ddot{x}P$ and $\ddot{y}P$ of the accelerations of P onto the fixed reference frame.

- Projections $\dot{x}Q$ and $\dot{y}Q$ of the velocity of joint center Q onto the fixed reference frame.

- Projections $\ddot{x}Q$ and $\ddot{y}Q$ of the accelerations of Q onto the fixed reference frame.

The purpose of the analysis is to determine

- Displacements s_1 and s_2 of sliders C and B measured as shown and their time derivatives $\dot{s}_1$, $\dot{s}_2$, $\ddot{s}_1$, and $\ddot{s}_2$.

- Coordinates xB and yB of point B relative to the OXY reference frame.

- Projections $\dot{x}B$ and $\dot{y}B$ of the velocity of B onto the axes of the fixed reference frame.

- Projections $\ddot{x}B$ and $\ddot{y}B$ of the acceleration of B onto the axes of the fixed reference frame.

- Coordinates xC and yC of sliding block C relative to the OXY reference frame.

- Projections $\dot{x}C$ and $\dot{y}C$ of point C onto the axes of the fixed reference frame.

- Projections $\ddot{x}C$ and $\ddot{y}C$ of the acceleration of C onto the axes of the fixed reference frame.

- Coordinates xD and yD of point D relative to the OXY reference frame.

- Projections $\dot{x}D$ and $\dot{y}D$ of the velocity of D onto the axes of the fixed reference frame.

- Projections $\ddot{x}D$ and $\ddot{y}D$ of the acceleration of D onto the axes of the fixed reference frame.

The vector-loop equation of the dyad is

$$\mathbf{OA} + \mathbf{AC} = \mathbf{OP} + \mathbf{PB} + \mathbf{BC} \tag{6.182}$$

which projects on the x- and y-axes of the fixed reference frame as

$$\begin{cases} x_A + AC \cdot \cos\left(\varphi + \alpha_2 - \dfrac{\pi}{2}\right) = x_P + s_2 \cdot \cos(\varphi) + s_1 \cdot \cos(\varphi + \alpha_2) \\[3mm] y_A + AC \cdot \sin\left(\varphi + \alpha_2 - \dfrac{\pi}{2}\right) = y_P + s_2 \cdot \sin(\varphi) + s_1 \cdot \sin(\varphi + \alpha_2) \end{cases} \tag{6.183}$$

We separate unknowns s_1 and s_2 to the left and obtain the following set of two linear equations:

$$\begin{cases} s_1 \cdot \cos(\varphi + \alpha_2) + s_2 \cdot \cos(\varphi) = x_A - x_P + AC \cdot \sin(\varphi + \alpha_2) \\[2mm] s_1 \cdot \sin(\varphi + \alpha_2) + s_2 \cdot \sin(\varphi) = y_A - y_P - AC \cdot \cos(\varphi + \alpha_2) \end{cases} \tag{6.184}$$

Equations 6.184 have real solutions for any coordinates x_A, y_A, x_P, and y_P, as long as the following inequality holds:

$$\cos(\varphi + \alpha_2) \cdot \sin(\varphi) - \sin(\varphi + \alpha_2) \cdot \cos(\varphi) \neq 0 \quad \text{equivalent to} \quad \alpha_2 \neq 0 \tag{6.185}$$

Once slider displacements s_1 and s_2 become available, the coordinates of points B, D, and C needed when plotting the dyad can be calculated as follows:

$$\begin{cases} x_B = x_P + s_2 \cdot \cos(\varphi) \\[2mm] y_B = y_P + s_2 \cdot \sin(\varphi) \end{cases} \tag{6.186}$$

$$\begin{cases} x_D = x_B + BD \cdot \cos(\varphi + \alpha_2) \\[2mm] y_D = y_B + BD \cdot \sin(\varphi + \alpha_2) \end{cases} \tag{6.187}$$

$$\begin{cases} x_C = x_A + AC \cdot \cos\left(\varphi + \alpha_2 - \dfrac{\pi}{2}\right) = x_A + AC \cdot \sin(\varphi + \alpha_2) \\[3mm] y_C = y_A + AC \cdot \sin\left(\varphi + \alpha_2 - \dfrac{\pi}{2}\right) = y_A - AC \cdot \cos(\varphi + \alpha_2) \end{cases} \tag{6.188a}$$

or

$$\begin{cases} x_C = x_B + s_1 \cdot \cos(\varphi + \alpha_2) \\[2mm] y_C = y_B + s_1 \cdot \sin(\varphi + \alpha_2) \end{cases} \tag{6.188b}$$

The velocity problem requires finding the unknown relative velocities $\dot{s}_1$ and $\dot{s}_2$ occurring when differentiating with respect to time Equations 6.184:

$$\begin{cases} \dot{s}_1 \cos(\varphi + \alpha_2) + \dot{s}_2 \cos(\varphi) = \dot{x}_A - \dot{x}_P + \dot{\varphi}[AC\cos(\varphi + \alpha_2) + s_1 \sin(\varphi + \alpha_2) + s_2 \sin(\varphi)] \\ \dot{s}_1 \sin(\varphi + \alpha_2) + \dot{s}_2 \sin(\varphi) = \dot{y}_A - \dot{y}_P + \dot{\varphi}[AC\sin(\varphi + \alpha_2) - s_1 \cos(\varphi + \alpha_2) - s_2 \cos(\varphi)] \end{cases} \quad (6.189)$$

By applying Equations 6.184 again, these equations simplify to

$$\begin{cases} \dot{s}_1 \cos(\varphi + \alpha_2) + \dot{s}_2 \cos(\varphi) = \dot{x}_A - \dot{x}_P + \dot{\varphi}(y_A - y_P) \\ \dot{s}_1 \sin(\varphi + \alpha_2) + \dot{s}_2 \sin(\varphi) = \dot{y}_A - \dot{y}_P - \dot{\varphi}(x_A - x_P) \end{cases} \quad (6.190)$$

The x and y components of the velocities of points B, D, and C are given by

$$\begin{cases} \dot{x}_B = \dot{x}_P + \dot{s}_2 \cos(\varphi) - \dot{\varphi}s_2 \sin(\varphi) = \dot{x}_P + \dot{s}_2 \cos(\varphi) - \dot{\varphi}(y_B - y_P) \\ \dot{y}_B = \dot{y}_P + \dot{s}_2 \sin(\varphi) + \dot{\varphi}s_2 \cos(\varphi) = \dot{y}_P + \dot{s}_2 \sin(\varphi) + \dot{\varphi}(x_B - x_P) \end{cases} \quad (6.191)$$

$$\begin{cases} \dot{x}_D = \dot{x}_B - \dot{\varphi}BD\sin(\varphi + \alpha_2) = \dot{x}_B + \dot{\varphi}(y_B - y_D) \\ \dot{y}_D = \dot{y}_B + \dot{\varphi}BD\cos(\varphi + \alpha_2) = \dot{y}_B - \dot{\varphi}(x_B - x_D) \end{cases} \quad (6.192)$$

$$\begin{cases} \dot{x}_C = \dot{x}_A + \dot{\varphi}AC\cos(\varphi + \alpha_2) = \dot{x}_A + \dot{\varphi}(y_A - y_C) \\ \dot{y}_C = \dot{y}_A + \dot{\varphi}AC\sin(\varphi + \alpha_2) = \dot{y}_A - \dot{\varphi}(x_A - x_C) \end{cases} \quad (6.193a)$$

or

$$\begin{cases} \dot{x}_C = \dot{x}_B + \dot{s}_1 \cos(\varphi + \alpha_2) - \dot{\varphi}s_1 \sin(\varphi + \alpha_2) = \dot{x}_B + \dot{s}_1 \cos(\varphi + \alpha_2) + \dot{\varphi}(y_B - y_C) \\ \dot{y}_C = \dot{y}_B + \dot{s}_1 \sin(\varphi + \alpha_2) + \dot{\varphi}s_1 \cos(\varphi + \alpha_2) = \dot{y}_B + \dot{s}_1 \sin(\varphi + \alpha_2) - \dot{\varphi}(x_B - x_C) \end{cases} \quad (6.193b)$$

To solve for the relative accelerations $\ddot{s}_1$ and $\ddot{s}_2$, we differentiate Equations 6.190 and obtain:

$$\begin{cases} \ddot{s}_1 \cos(\varphi + \alpha_2) + \ddot{s}_2 \cos(\varphi) = \ddot{x}_A - \ddot{x}_P + \ddot{\varphi}(y_A - y_P) + \dot{\varphi}(\dot{y}_A - \dot{y}_P) \\ \qquad + \dot{\varphi}\dot{s}_1 \sin(\varphi + \alpha_2) + \dot{\varphi}\dot{s}_2 \sin(\varphi) \\ \ddot{s}_1 \sin(\varphi + \alpha_2) + \ddot{s}_2 \sin(\varphi) = \ddot{y}_A - \ddot{y}_P - \ddot{\varphi}(x_A - x_P) - \dot{\varphi}(\dot{x}_A - \dot{x}_P) \\ \qquad s - \dot{\varphi}\dot{s}_1 \cos(\varphi + \alpha_2) - \dot{\varphi}\dot{s}_2 \cos(\varphi) \end{cases} \quad (6.194)$$

The x and y components of the accelerations of points B, D, and C, also of interest, can be calculated with

$$\begin{cases} \ddot{x}_B = \ddot{x}_P + \ddot{s}_2 \cos(\varphi) - 2\dot{\varphi}\dot{s}_2 \sin(\varphi) - \ddot{\varphi}s_2 \sin(\varphi) - \dot{\varphi}^2 s_2 \cos(\varphi) \\ \ddot{y}_B = \ddot{y}_P + \ddot{s}_2 \sin(\varphi) + 2\dot{\varphi}\dot{s}_2 \cos(\varphi) + \ddot{\varphi}s_2 \cos(\varphi) - \dot{\varphi}^2 s_2 \sin(\varphi) \end{cases} \tag{6.195}$$

$$\begin{cases} \ddot{x}_D = \ddot{x}_B - \ddot{\varphi}BD\sin(\varphi+\alpha_2) - \dot{\varphi}^2 BD\cos(\varphi+\alpha_2) \\ \ddot{y}_D = \ddot{y}_B + \ddot{\varphi}BD\cos(\varphi+\alpha_2) - \dot{\varphi}^2 BD\sin(\varphi+\alpha_2) \end{cases} \tag{6.196}$$

$$\begin{cases} \ddot{x}_C = \ddot{x}_A + \ddot{\varphi}AC\cos(\varphi+\alpha_2) - \dot{\varphi}^2 AC\cdot\sin(\varphi+\alpha_2) \\ \ddot{y}_C = \ddot{y}_A + \ddot{\varphi}AC\sin(\varphi+\alpha_2) + \dot{\varphi}^2 AC\cdot\cos(\varphi+\alpha_2) \end{cases} \tag{6.197a}$$

or

$$\begin{cases} \ddot{x}_C = \ddot{x}_B + \ddot{s}_1 \cos(\varphi+\alpha_2) - 2\dot{\varphi}\dot{s}_1 \sin(\varphi+\alpha_2) - \ddot{\varphi}s_1 \sin(\varphi+\alpha_2) - \dot{\varphi}^2 s_1 \cos(\varphi+\alpha_2) \\ \ddot{y}_C = \ddot{y}_B + \ddot{s}_1 \sin(\varphi+\alpha_2) + 2\dot{\varphi}\dot{s}_1 \cos(\varphi+\alpha_2) + \ddot{\varphi}s_1 \cos(\varphi+\alpha_2) - \dot{\varphi}^2 s_1 \sin(\varphi+\alpha_2) \end{cases} \tag{6.197b}$$

obtained by differentiating with respect to time the velocity Equations 6.191, 6.192, and 6.193a and b, respectively.

The variable angle φ and its derivatives $\dot{\varphi}$ and $\ddot{\varphi}$ occurring earlier can be calculated by calling procedure **AngPVA** with the arguments x and y coordinates of points P and Q and their first and second time derivatives. With this observation, the equations derived above have been implemented in procedure **RT_T_** part of unit **LibAssur**:

```
RT_T_(Color:Word; xA,yA, vxA,vyA, axA,ayA, xP,yP, vxP,vyP,
axP,ayP, xQ,yQ, vxQ,vyQ, axQ,ayQ, AC,BD, Alpha2:double; var xB,yB,
vxB,vyB, axB,ayB, xC,yC, vxC,vyC, axC,ayC, xD,yD, vxD,vyD,
axD,ayD:double; var OK:Boolean);
```

The correspondence between the formal parameters and the notations used in Equations 6.183 through 6.197 and in Figure 6.31 is summarized in the following tables:

Input parameters of procedure **RT_T_** :

0...16	x_A	y_A	$\dot{x}_A$	$\dot{y}_A$	$\ddot{x}_A$	$\ddot{y}_A$	x_P	y_P	$\dot{x}_P$	$\dot{y}_P$	$\ddot{x}_P$	$\ddot{y}_P$
Color	xA	yA	vxA	vyA	axA	ayA	xP	yP	vxP	vyP	axP	ayP

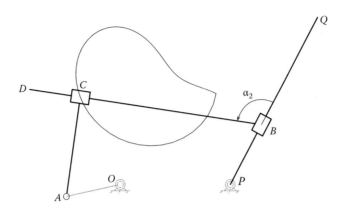

FIGURE 6.32 A two-DOF mechanism consisting of an RT_T_ dyadic isomer driven by two rockers. See also animation file **F6_32.GIF**.

xQ	yQ	$\dot{x}Q$	$\dot{y}Q$	$\ddot{x}Q$	$\ddot{y}Q$	AC	BD	α_2
xQ	yQ	vxQ	vyQ	axQ	ayQ	AC	BD	Alph2

Output parameters of procedure **RT_T_**:

xB	yB	$\dot{x}B$	$\dot{y}B$	$\ddot{x}B$	$\ddot{y}B$	xC	yC	$\dot{x}C$	$\dot{y}C$	$\ddot{x}C$	$\ddot{y}C$
xB	yB	vxB	vyB	axB	ayB	xC	yC	vxC	vyC	axC	ayC

xD	yD	$\dot{x}D$	$\dot{y}D$	$\ddot{x}D$	$\ddot{y}D$	$\theta_1 \neq \theta_2$
xD	yD	vxD	vyD	axD	ayD	OK

Note that the displacements s_1 and s_1 of the two slider blocks and their first and second time derivatives are not returned by procedure **RT_T_**. If of interest, s_1 and s_1 can be determined by calling procedure **VarDist** with arguments set equal to the x and y coordinates of points B, C and P, B and to their first and second time derivatives.

The simulation of the sample mechanism in Figure 6.32 has been done using program **P6_32.PAS** listed in Appendix B. The frames in the animation file **F6_32.GIF** were produced using the **DXF** file output by the same program. As visible in Figure 6.32, the mechanism consists of two rockers, one being the actual link PQ of the dyad and the other one driving the potential pin joint A (see also Figure 6.31).

6.13.3 R_TT_ Dyadic Isomer: Procedure **R_TT_**

The subject of this section is the RTT dyad configured as shown in Figure 6.33 and noted R_TT_. It can be seen as a crossbreed between the R_T_T and RT_T_ dyadic isomers. Consequently, some of the kinematic equations derived earlier will apply for this current embodiment of the RTT dyad.

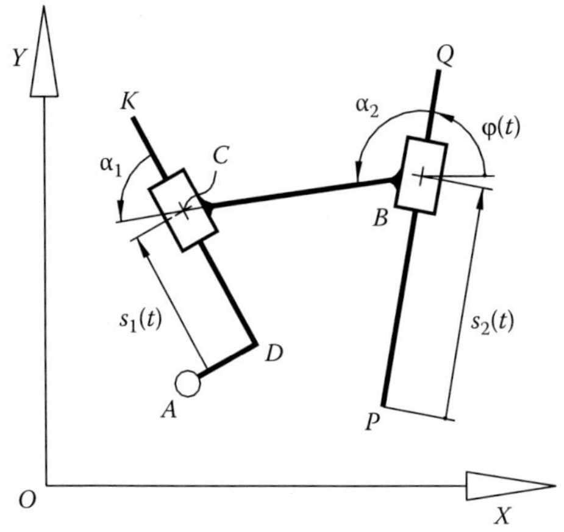

FIGURE 6.33   Notations used in solving the kinematics of the R_TT_ dyadic isomer.

In a kinematic analysis, the following parameters are assumed known at any moment of time:

- Coordinates xA, yA, of potential joint A relative to the fixed reference frame OXY.

- Coordinates xP, yP and xQ, yQ of the ends of the rod that guides to the inside sliding block B.

- Length BC of the spacer rod joining slider blocks B and C.

- Lengths AD and DK of the two sides of the L-shaped link supporting slider C.

- The value of the constant angles α_1 and α_2 measured as shown in Figure 6.33.

- Projections $\dot{x}A$ and $\dot{y}A$ of the velocity of joint center A onto the fixed reference frame.

- Projections $\ddot{x}A$ and $\ddot{y}A$ of the accelerations of point A onto the fixed reference frame.

- Projections $\dot{x}P$ and $\dot{y}P$ of the velocity of joint center P onto the fixed reference frame.

- Projections $\ddot{x}P$ and $\ddot{y}P$ of the accelerations of point P onto the fixed reference frame.

- Projections $\dot{x}Q$ and $\dot{y}Q$ of the velocity of joint center Q onto the fixed reference frame.

- Projections $\ddot{x}Q$ and $\ddot{y}Q$ of the accelerations of Q onto the fixed reference frame.

The purpose of a kinematic analysis is to determine:

- Displacements s_1 and s_2 of sliders C and B measured as shown and their time derivatives $\dot{s}_1$, $\dot{s}_2$, $\ddot{s}_1$, and $\ddot{s}_2$.

- Coordinates xB and yB of point B relative to the OXY reference frame.

- Projections $\dot{x}B$ and $\dot{y}B$ of the velocity of B onto the axes of the fixed reference frame.

- Projections $\ddot{x}B$ and $\ddot{y}B$ of the acceleration of B onto the axes of the fixed reference frame.

- Coordinates xC and yC of sliding block C relative to the OXY reference frame.

- Projections $\dot{x}C$ and $\dot{y}C$ of point C onto the axes of the fixed reference frame.

- Projections $\ddot{x}C$ and $\ddot{y}C$ of the acceleration of C onto the axes of the fixed reference frame.

- Coordinates xD and yD of point D relative to the OXY reference frame.

- Projections $\dot{x}D$ and $\dot{y}D$ of the velocity of point D onto the axes of the fixed reference frame.

- Projections $\ddot{x}D$ and $\ddot{y}D$ of the acceleration of point D onto the axes of the fixed reference frame.

- Coordinates xK and yK of point K relative to the OXY reference frame.

- Projections $\dot{x}K$ and $\dot{y}K$ of the velocity of point K onto the axes of the fixed reference frame.

- Projections $\ddot{x}K$ and $\ddot{y}K$ of the acceleration of point K onto the axes of the fixed reference frame.

The vector-loop equation of the R_TT_ isomer is

$$\mathbf{OD} + \mathbf{DC} = \mathbf{OP} + \mathbf{PB} + \mathbf{BC} \tag{6.198}$$

which projects on the x- and y-axes of the fixed reference frame as

$$\begin{cases} x_A + AD\cos\left(\varphi + \alpha_2 - \alpha_1 - \dfrac{\pi}{2}\right) + s_1\cos(\varphi + \alpha_2 - \alpha_1) = \\ \quad x_P + s_2\cos(\varphi) + BC\cos(\varphi + \alpha_2) \\ y_A + AD\sin\left(\varphi + \alpha_2 - \alpha_1 - \dfrac{\pi}{2}\right) + s_1\sin\left(\varphi + \alpha_2 - \alpha_1\right) = \\ \quad y_P + s_2\cdot\sin(\varphi) + BC\cdot\sin(\varphi + \alpha_2) \end{cases} \tag{6.199}$$

After separating the unknowns s_1 and s_2, the following set of linear equations is obtained:

$$\begin{cases} s_1\cos(\varphi + \alpha_2 - \alpha_1) - s_2\cos(\varphi) = x_P - x_A - AD\sin(\varphi + \alpha_2 - \alpha_1) + BC\cos(\varphi + \alpha_2) \\ s_1\sin(\varphi + \alpha_2 - \alpha_1) - s_2\sin(\varphi) = y_P - y_A + AD\cos(\varphi + \alpha_2 - \alpha_1) + BC\sin(\varphi + \alpha_2) \end{cases} \tag{6.200}$$

Equations 6.200 have solutions for any coordinates x_A, y_A, x_P, and y_P, provided that the following inequality holds:

$$\sin(\varphi + \alpha_2 - \alpha_1) \cdot \cos(\varphi) - \cos(\varphi + \alpha_2 - \alpha_1) \cdot \sin(\varphi) \neq 0 \quad \text{equivalent to} \quad \alpha_2 \neq \alpha_1 \qquad (6.201)$$

Once slider displacements s_1 and s_2 are calculated, the coordinates of points B, D, C, and K, required to represent graphically the dyad, can be calculated using Equations 6.186, 6.164, 6.168a, and 6.165, respectively. Alternatively, the coordinates of point C can be calculated with

$$\begin{cases} x_C = x_B + BC\cos(\varphi + \alpha_2) \\ y_C = y_B + BC\sin(\varphi + \alpha_2) \end{cases} \qquad (6.202)$$

The velocity problem requires solving for the unknown relative velocities $\dot{s}_1$ and $\dot{s}_2$ among equations

$$\begin{cases} \dot{s}_1 \cos(\varphi + \alpha_2 - \alpha_1) - \dot{s}_2 \cos(\varphi) = \dot{x}_P - \dot{x}_A \\ \quad -\dot{\varphi}[AD\cos(\varphi + \alpha_2 - \alpha_1) + BC\sin(\varphi + \alpha_2) - s_1 \sin(\varphi + \alpha_2 - \alpha_1) + s_2 \sin(\varphi)] \\ \dot{s}_1 \sin(\varphi + \alpha_2 - \alpha_1) - \dot{s}_2 \sin(\varphi) = \dot{y}_P - \dot{y}_A \\ \quad -\dot{\varphi}[AD\sin(\varphi + \alpha_2 \quad \alpha_1) - BC\cos(\varphi + \alpha_2) + s_1 \cos(\varphi + \alpha_2 - \alpha_1) - s_2 \cos(\varphi) \end{cases} \qquad (6.203)$$

obtained by differentiating with respect to time Equations 6.200. By applying the same equation one more time, we obtain

$$\begin{cases} \dot{s}_1 \cos(\varphi + \alpha_2 - \alpha_1) + \dot{s}_2 \cos(\varphi) = \dot{x}_P - \dot{x}_A - \dot{\varphi}(y_A - y_P) \\ \dot{s}_1 \sin(\varphi + \alpha_2 - \alpha_1) + \dot{s}_2 \sin(\varphi) = \dot{y}_P - \dot{y}_A + \dot{\varphi}(x_A - x_P) \end{cases} \qquad (6.204)$$

The x and y components of the velocities of points B, D, C, and K can be calculated using Equations 6.191, 6.171, 6.175a, and 6.172, respectively. The scalar components of the velocity of point C can be also obtained by differentiating Equation 6.202, that is,

$$\begin{cases} \dot{x}_C = \dot{x}_B - \dot{\varphi}BC\sin(\varphi + \alpha_2) \\ \dot{y}_C = \dot{y}_B + \dot{\varphi}BC\cos(\varphi + \alpha_2) \end{cases} \qquad (6.205)$$

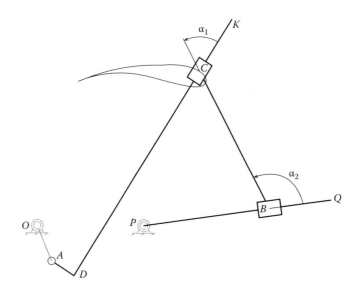

FIGURE 6.34   A two-DOF mechanism consisting of an R_TT_ dyadic isomer driven by two rockers. See also animation file **F6_34.GIF**.

The relative accelerations $\ddot{s}_1$ and $\ddot{s}_2$ are solutions of the equations

$$\begin{cases} \ddot{s}_1 \cos(\varphi + \alpha_2 - \alpha_1) + \ddot{s}_2 \cos(\varphi) = \ddot{x}_P - \ddot{x}_A + \ddot{\varphi}(y_A - y_P) + \dot{\varphi}(\dot{y}_A - \dot{y}_P) \\ \qquad + \dot{\varphi}\dot{s}_1 \sin(\varphi + \alpha_2 - \alpha_1) + \dot{\varphi}\dot{s}_2 \sin(\varphi) \\ \ddot{s}_1 \sin(\varphi + \alpha_2 - \alpha_1) + \ddot{s}_2 \sin(\varphi) = \ddot{y}_P - \ddot{y}_A - \ddot{\varphi}(x_A - x_P) - \dot{\varphi}(\dot{x}_A - \dot{x}_P) \\ \qquad - \dot{\varphi}\dot{s}_1 \cos(\varphi + \alpha_2 - \alpha_1) - \dot{\varphi}\dot{s}_2 \cos(\varphi) \end{cases} \qquad (6.206)$$

obtained by differentiating Equations 6.204 with respect to time.

The x and y components of the accelerations of points B, D, C, and K can be determined using Equations 6.195, 6.177, 6.181a, and 6.178, respectively. Another form of the x and y components of the acceleration of point C can be obtained by differentiating Equations 6.205:

$$\begin{cases} \ddot{x}_C = \ddot{x}_B - \ddot{\varphi}BC \sin(\varphi + \alpha_2) - \dot{\varphi}^2 BC \cos(\varphi + \alpha_2) \\ \ddot{y}_C = \ddot{y}_B + \ddot{\varphi}BC \cos(\varphi + \alpha_2) - \dot{\varphi}^2 BC \sin(\varphi + \alpha_2) \end{cases} \qquad (6.207)$$

The variable angle φ and its derivatives $\dot{\varphi}$ and $\ddot{\varphi}$ occurring the above equations can be determined by calling procedure **AngPVA** with its arguments set equal to x and y coordinates of points P and Q and to their first and second time derivatives.

These kinematic equations have been implemented in procedure **R_TT_** part of unit **LibAssur** with the following heading:

```
R_TT_(Color:Word; xA,yA, vxA,vyA, axA,ayA, xP,yP, vxP,vyP,
axP,ayP, xQ,yQ, vxQ,vyQ, axQ,ayQ, AD,DK,BC, Alpha2:double;
var xB,yB, vxB,vyB, axB,ayB, xC,yC, vxC,vyC, axC,ayC, xD,yD,
vxD,vyD, axD,ayD, xK,yK, vxK,vyK, axK,ayK:double; var OK:Boolean);
```

The correspondence between the formal parameters of the procedures and the notations used in Figure 6.31 and in these kinematic equations is summarized in the following tables:

Input parameters of procedure **R_TT_** :

0...16	xA	yA	$\dot{x}A$	$\dot{y}A$	$\ddot{x}A$	$\ddot{y}A$	xP	yP	$\dot{x}P$	$\dot{y}P$	$\ddot{x}P$	$\ddot{y}P$
Color	xA	yA	vxA	vyA	axA	ayA	xP	yP	vxP	vyP	axP	ayP

xQ	yQ	$\dot{x}Q$	$\dot{y}Q$	$\ddot{x}Q$	$\ddot{y}Q$	AC	BD	α_2
xQ	yQ	vxQ	vyQ	axQ	ayQ	AC	BD	Alph2

Output parameters of procedure **RT_T_** :

xB	yB	$\dot{x}B$	$\dot{y}B$	$\ddot{x}B$	$\ddot{y}B$	xC	yC	$\dot{x}C$	$\dot{y}C$	$\ddot{x}C$	$\ddot{y}C$
xB	yB	vxB	vyB	axB	ayB	xC	yC	vxC	vyC	axC	ayC

xD	yD	$\dot{x}D$	$\dot{y}D$	$\ddot{x}D$	$\ddot{y}D$	xC	yC	$\dot{x}C$	$\dot{y}C$	$\ddot{x}C$	$\ddot{y}C$	$\theta_1 \neq \theta_2$
xD	yD	vxD	vyD	axD	ayD	xC	yC	vxC	vyC	axC	ayC	OK

Note that the displacements s_1 and s_1 of the two slider blocks and their first and second time derivatives are not returned by procedure **R_TT_** . They can be however evaluated by calling procedure **VarDist** with arguments set equal to x and y coordinates of points D, C and P, B and to their first and second time derivatives.

The simulation of the sample mechanism shown in Figure 6.34 has been done using program **P6_34.PAS** listed in Appendix B. The frames used to generate the animation file **F6_34.GIF** were produced using the **DXF** file output by **P6_34.PAS**. As visible in the figure, the mechanism consists of two rockers, one being the actual link PQ of the dyad and the other one driving the potential pin joint A of the dyad (see Figure 6.33).

6.13.4 RT__ Dyadic Isomer: Procedure **RT__T**

The last possible embodiment of the RTT dyad has its configuration as shown in Figure 6.35 and it is symbolized RT__T. This fourth isomer can be interpreted as a crossbreed between the R_T_T and RT_T_ dyadic isomers, and therefore the three will share part of their kinematic equations.

In an analysis, the following parameters will be assumed known at any given time:

- Coordinates xA, yA, of potential joint A relative to the fixed reference frame OXY.

- Coordinates xB, yB of the center of slider block B measured relative to the fixed reference frame.

- Projections $\dot{x}A$ and $\dot{y}A$ of the velocity of joint center A onto the fixed reference frame.

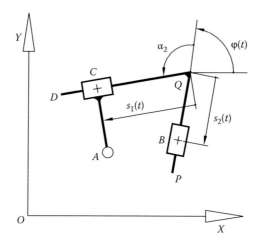

FIGURE 6.35   Notations used in solving the kinematics of the RT__T dyad.

- Projections $\ddot{x}A$ and $\ddot{y}A$ of the accelerations of point A onto the fixed reference frame.

- Projections $\dot{x}B$ and $\dot{y}B$ of the velocity of joint center B onto the fixed reference frame.

- Projections $\ddot{x}B$ and $\ddot{y}B$ of the accelerations of point B onto the fixed reference frame.

- Orientation angle φ of slider B and its first and second time derivatives $\dot{\varphi}$ and $\ddot{\varphi}$.

- The value of the constant angles α_2 measured as shown in Figure 6.35.

- Lengths PQ and QD of the two sections of the V-shaped link supporting slider blocks B and C.

- Offset AC of the pin joint relative to slider block C.

The purpose of the kinematic analysis is to determine:

- Displacements s_1 and s_2 of sliders C and B measured as shown and their time derivatives $\dot{s}_1$, $\dot{s}_2$, $\ddot{s}_1$, and $\ddot{s}_2$.

- Coordinates xP and yP of point P relative to the OXY reference frame.

- Projections $\dot{x}P$ and $\dot{y}P$ of the velocity of point P onto the axes of the fixed reference frame.

- Projections $\ddot{x}P$ and $\ddot{y}P$ of the acceleration of point P onto the axes of the fixed reference frame.

- Coordinates xC and yC of sliding block C relative to the OXY reference frame.

- Projections $\dot{x}C$ and $\dot{y}C$ of point C onto the axes of the fixed reference frame.

- Projections $\ddot{x}C$ and $\ddot{y}C$ of the acceleration of C onto the axes of the fixed reference frame.

- Coordinates xD and yD of point D relative to the OXY reference frame.

- Projections $\dot{x}D$ and $\dot{y}D$ of the velocity of point D onto the axes of the fixed reference frame.

- Projections $\ddot{x}D$ and $\ddot{y}D$ of the acceleration of point D onto the axes of the fixed reference frame.

- Coordinates xQ and yQ of point Q relative to the OXY reference frame.

- Projections $\dot{x}Q$ and $\dot{y}Q$ of the velocity of point Q onto the axes of the fixed reference frame.

- Projections $\ddot{x}Q$ and $\ddot{y}Q$ of the acceleration of point Q onto the axes of the fixed reference frame.

The vector-loop equation of the RT__T isomer is

$$\mathbf{OA} + \mathbf{AC} = \mathbf{OB} + \mathbf{BQ} + \mathbf{QC} \tag{6.208}$$

This is equivalent to the following scalar equations:

$$
\begin{cases}
x_A + AC\cos\left(\varphi + \alpha_2 - \dfrac{\pi}{2}\right) = x_B - s_2 \cos(\varphi) + s_1 \cos(\varphi + \alpha_2) \\[2mm]
y_A + AC\sin\left(\varphi + \alpha_2 - \dfrac{\pi}{2}\right) = y_B - s_2 \sin(\varphi) + s_1 \sin(\varphi + \alpha_2)
\end{cases}
\tag{6.209}
$$

After separating the unknowns s_1 and s_2, the following set of linear equations is finally obtained:

$$
\begin{cases}
s_1 \cos(\varphi + \alpha_2) - s_2 \cos(\varphi) = x_A - x_B + AC\sin(\varphi + \alpha_2) \\
s_1 \sin(\varphi + \alpha_2) - s_2 \sin(\varphi) = y_A - y_B - AC\cos(\varphi + \alpha_2)
\end{cases}
\tag{6.210}
$$

Equations 6.210 have solutions for any coordinates x_A, y_A, x_B, y_B and angle φ, as long as the following inequality holds:

$$\sin(\varphi + \alpha_2)\cos(\varphi) - \cos(\varphi + \alpha_2)\sin(\varphi) \neq 0 \quad \text{equivalent to} \quad \alpha_2 \neq 0 \tag{6.211}$$

Once slider displacements s_1 and s_2 become known, the coordinates of points P, Q, and C can be calculated using Equations 6.167, 6.166, and 6.188a, respectively. The x and y coordinates of point C can be also calculated using equations

$$
\begin{cases}
x_C = x_Q + s_1 \cos(\varphi + \alpha_2) \\
y_C = y_Q + s_1 \sin(\varphi + \alpha_2)
\end{cases}
\tag{6.212}
$$

while the coordinates of point D can be calculated with

$$
\begin{cases}
x_D = x_Q + DQ\cos(\varphi + \alpha_2) \\
y_D = y_Q + DQ\sin(\varphi + \alpha_2)
\end{cases}
\tag{6.213}
$$

The velocity problem requires determining the unknown relative velocities $\dot{s}_1$ and $\dot{s}_2$ by solving simultaneously the following equations:

$$\begin{cases} \dot{s}_1 \cos(\varphi+\alpha_2) - \dot{s}_2 \cos(\varphi) = \dot{x}_A - \dot{x}_B + \dot{\varphi}[AC\cos(\varphi+\alpha_2) + s_1\sin(\varphi+\alpha_2) - s_2\sin(\varphi)] \\ \dot{s}_1 \sin(\varphi+\alpha_2) - \dot{s}_2 \sin(\varphi) = \dot{y}_A - \dot{y}_B + \dot{\varphi}[AC\sin(\varphi+\alpha_2) - s_1\cos(\varphi+\alpha_2) + s_2\cos(\varphi)] \end{cases} \tag{6.214}$$

obtained by differentiating with respect to time the position Equations 6.210. By reapplying the position Equations 6.214, we further get

$$\begin{cases} \dot{s}_1 \cdot \cos(\varphi+\alpha_2) - \dot{s}_2 \cdot \cos(\varphi) = \dot{x}_A - \dot{x}_B + \dot{\varphi}(y_A - y_B) \\ \dot{s}_1 \cdot \sin(\varphi+\alpha_2) - \dot{s}_2 \cdot \sin(\varphi) = \dot{y}_A - \dot{y}_B - \dot{\varphi}(x_A - x_B) \end{cases} \tag{6.215}$$

The scalar components of the velocities of points P, Q, and C can be calculated using Equations 6.174, 6.173, and 6.175a, respectively. The components of the velocity of point C can be also obtained by differentiating with respect to time Equations 6.212:

$$\begin{cases} \dot{x}_C = \dot{x}_Q + \dot{s}_1 \cos(\varphi+\alpha_2) - \dot{\varphi}s_1\sin(\varphi+\alpha_2) = \dot{x}_Q + \left(\dfrac{\dot{s}_1(x_D - x_Q)}{DQ} \right) - \dot{\varphi}(y_C - y_Q) \\ \dot{y}_C = \dot{y}_Q + \dot{s}_1 \sin(\varphi+\alpha_2) + \dot{\varphi}s_1\cos(\varphi+\alpha_2) = \dot{y}_Q + \left(\dfrac{\dot{s}_1(y_D - y_Q)}{DQ} \right) + \dot{\varphi}(x_C - x_Q) \end{cases} \tag{6.216}$$

Likewise, the x and y components of the velocity of point D result from the differentiation of Equation 6.213:

$$\begin{cases} \dot{x}_D = \dot{x}_Q - \dot{\varphi}DQ\sin(\varphi+\alpha_2) \\ \dot{y}_D = \dot{y}_Q + \dot{\varphi}DQ\cos(\varphi+\alpha_2) \end{cases} \tag{6.217}$$

Regarding acceleration problem, we must first determine linear accelerations $\ddot{s}_1$ and $\ddot{s}_2$. These are the solutions of the following set of equations:

$$\begin{cases} \ddot{s}_1 \cos(\varphi+\alpha_2) + \ddot{s}_2 \cos(\varphi) = \ddot{x}_A - \ddot{x}_P + \ddot{\varphi}(y_A - y_P) + \dot{\varphi}(\dot{y}_A - \dot{y}_P) \\ \quad + \dot{\varphi}\dot{s}_1 \sin(\varphi+\alpha_2) + \dot{\varphi}\dot{s}_2 \sin(\varphi) \\ \ddot{s}_1 \sin(\varphi+\alpha_2) + \ddot{s}_2 \sin(\varphi) = \ddot{y}_A - \ddot{y}_P - \ddot{\varphi}(x_A - x_P) - \dot{\varphi}(\dot{x}_A - \dot{x}_P) \\ \quad - \dot{\varphi}\cdot\dot{s}_1 \cdot\cos(\varphi+\alpha_2) - \dot{\varphi}\dot{s}_2 \cos(\varphi) \end{cases} \tag{6.218}$$

obtained by differentiating Equations 6.204 with respect to time.

The x and y components of the accelerations of points P, Q, and C can be calculated using Equations 6.180, 6.179, and 6.181a, respectively. The x and y components of the acceleration of point C can be also obtained by differentiating Equations 6.216:

$$\begin{cases} \ddot{x}_C = \ddot{x}_Q + \left(\dfrac{\ddot{s}(x_D - x_Q)}{DQ} \right) + \left(\dfrac{\dot{s}(x_D - x_Q)}{DQ} \right) - \ddot{\varphi}(y_C - y_Q) - \dot{\varphi}(\dot{y}_C - \dot{y}_Q) \\[4mm] \ddot{y}_C = \ddot{y}_Q + \left(\dfrac{\ddot{s}(y_D - y_Q)}{DQ} \right) + \left(\dfrac{\dot{s}(y_D - y_Q)}{DQ} \right) - \ddot{\varphi}(x_C - x_Q) - \dot{\varphi}(\dot{x}_C - \dot{x}_Q) \end{cases} \tag{6.219}$$

Likewise, by differentiating Equations 6.217, we obtain the components of the acceleration of point D:

$$\begin{cases} \ddot{x}_D = \ddot{x}_Q - \ddot{\varphi}DQ\sin(\varphi + \alpha_2) - \dot{\varphi}^2 DQ\cos(\varphi + \alpha_2) \\[2mm] \ddot{y}_D = \ddot{y}_Q + \ddot{\varphi}DQ\cos(\varphi + \alpha_2) - \dot{\varphi}^2 DQ\sin(\varphi + \alpha_2) \end{cases} \tag{6.220}$$

These kinematic equations have been implemented in procedure **RT__T** in unit **LibAssur** with the heading

```
RT__T(Color:Word; xA,yA,vxA,vyA,axA,ayA, xB,yB,vxB,vyB,axB,ayB,
Phi,dPhi,ddPhi, AC,PQ,QD, Alpha2:double; var xP,yP,vxP,vyP,axP,ayP,
xQ,yQ,vxQ,vyQ,axQ,ayQ, xD,yD,vxD,vyD,axD,ayD:double; var OK:Boolean);
```

The correspondence between the formal parameters and the notations used in these equations and in Figure 6.35 is summarized in the following tables:

Input parameters of procedure **RT__T**:

0...16	x_A	y_A	$\dot{x}_A$	$\dot{y}_A$	$\ddot{x}_A$	$\ddot{y}_A$	x_B	y_B	$\dot{x}_B$	$\dot{y}_B$	$\ddot{x}_B$	$\ddot{y}_B$
Color	xA	yA	vxA	vyA	axA	ayA	xB	yB	vxB	vyB	axB	ayB

φ		$\dot{\varphi}$		$\ddot{\varphi}$		AC		PQ		QD		α_2
Phi		dPhi		ddPhi		AC		PQ		QD		Alph2

Output parameters of procedure **RT__T**:

x_P	y_P	$\dot{x}_P$	$\dot{y}_P$	$\ddot{x}_P$	$\ddot{y}_P$	x_Q	y_Q	$\dot{x}_Q$	$\dot{y}_Q$	$\ddot{x}_Q$	$\ddot{y}_Q$
xP	yP	vxP	vyP	axP	ayP	xQ	yQ	vxQ	vyQ	axQ	ayQ

x_D		y_D		$\dot{x}_D$		$\dot{y}_D$		$\ddot{x}_D$		$\ddot{y}_D$		$\alpha_2 \neq 0$
xD		yD		vxD		vyD		axD		ayD		OK

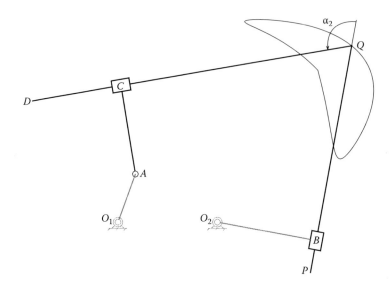

FIGURE 6.36   A two-DOF mechanism consisting of an RT__T dyadic isomer driven by two rockers. See also animation file **F6_36.GIF**.

Note that the displacements s_1 and s_2 of the two slider blocks and their first and second time derivatives are not returned by procedure **RT__T**. They can however be calculated by calling procedure **VarDist** with its arguments set equal to x and y coordinates of points D, C and P, B and to their first and second time derivatives.

The simulation of the sample mechanism in Figure 6.36 has been done using program **P6_36.PAS** listed in Appendix B, which outputs a **DXF** file that was used to generate the animation file **F6_36.GIF**. As visible from the figure, the mechanism consists of two rockers, one being the actual link PQ of the dyad and the other one driving the potential pin joint A. The program includes, with the simulation, the locus of point Q and labels of each joint and of the constant angle at α_2.

***

All five Assur groups with two links and three joints have been analyzed in their most general configurations, adding up to 11 *dyadic isomers*. The kinematic equations obtained have been implemented in a number of procedures gathered in unit **LibAssur** available with the book. Starting from one or more actuators, these procedures can be called in the same order in which the mechanism has been formed, and the position, velocity, and acceleration of its links or points of interest can be determined. In addition, the mechanism can be represented graphically on the computer screen and also exported to **DXF**. More applications of the procedures in unit **LibAssur** are discussed in Chapter 9. As Figure 6.3 shows, many practical mechanisms employ oftentimes simplified dyads or dyadic isomers, which have some link lengths and eccentricities equal to zero.

REFERENCES AND FURTHER READINGS

For general mechanism theory including mobility analysis, vector-loop equations, and kinematic analysis, see

Cleghorn, W. L. (2005). *Mechanics of Machines*. New York: Oxford University Press.

Norton, R. L. (2011). *Design of Machinery*. New York: McGraw Hill.

Uicker, J. J. Jr., Pennock, G. R., and Shigley, J. E. (2003). *Theory of Machines and Mechanisms*. New York: Oxford University Press.

Waldron, K. J. and Kinzel, G. L. (2003). *Kinematics, Dynamics, and Design of Machinery*. Hoboken, NJ: John Wiley & Sons.

Wilson, C. E. and Sadler, J. P. (2003). *Kinematics and Dynamics of Machinery*. Upper Saddle River, NJ: Prentice Hall.

For additional works on Assur group kinematics, see

Galetti, C. U. (1986). A note on modular approaches to planar linkage kinematic analysis. *Mechanism and Machine Theory, 21(5)*, 385–391.

Hansen, M. R. (1996). A general method for analysis of planar mechanisms using a modular approach. *Mechanism and Machine Theory, 31(8)*, 1155–1166.

Verho, A. (1971). An extension of the concept of group. *Mechanism and Machine Theory, 8(2)*, 249–256.

CHAPTER 7

Design and Analysis of Disk Cam Mechanisms

Sᴀᴍᴇ ᴀs ʟɪɴᴋᴀɢᴇs, cam mechanisms are capable of converting continuous rotational motion into rectilinear or rotary motion that alternate between a lower and an upper limit. As opposed to linkages, however, the correlation between the input and output motions can be precisely programmed and can include one or more dwells, while the mechanism itself may result of smaller size than the equivalent linkage. On the downside, cam mechanisms have lower reliability and are noisier and the follower can bounce. Figure 7.1 shows the schematic of the cam mechanisms considered in this chapter. Cam profile generation as follower envelope in an inverted motion (i.e., the follower rotates around the cam that is held stationary) and the problem of kinematic analysis of a given cam-follower pair will be studied in this chapter. The preliminary problem of synthesizing the follower motion is also tackled using **AutoCAD** interpolating functions and the **Util~DXF** and **Util~TXT** programs.

7.1 SYNTHESIS OF FOLLOWER MOTION

The choice of follower motion in a cam-follower mechanism influences the magnitude of the contact forces and therefore the wear rate of the mechanism. If not properly selected, it can cause follower bounce that is associated with increased noise and vibrations level during operation. The reader may have experience with elevators, some being less comfortable to ride than others, depending on the type of motion programmed to their cars. Likewise, if the follower motion is jerked at start-up, then large contact forces will develop, while if the motion ends abruptly and the retaining force is small (i.e., provided by a soft spring or gravity only), then the contact between the follower and the cam will be lost. These effects are also influenced by the type of cam and follower materials, the presence of lubricant, contact geometry, joint clearances, etc.

Let us assume a follower displacement $\delta_F(\delta_C)$ that consists of a lower dwell for the first 10% and last 10% of total cam cycle, a follower rise from 10% to 45%, upper dwell of magnitude 1 from 45% to 57.5%, and follower return from 57.5% to 90% of the cam cycle

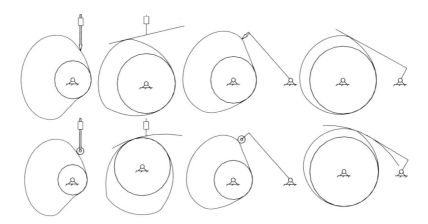

FIGURE 7.1   Disk cam-follower mechanisms subject of this chapter.

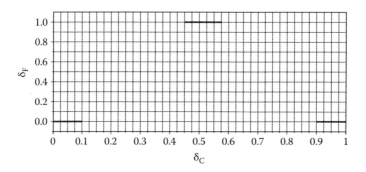

FIGURE 7.2   Prescribed follower motion with 20% lower dwell, 12.5% upper dwell, 35% rise, and 32.5% fall (percentages of total cam displacement). Configuration file to redo this plot **F7_02.CF2**.

(see Figure 7.2). Note that the diagram has been normalized with respect to cam and follower motion ranges, which allows both the rotational and translational motions of the cam and of the follower to be obtained through scaling.

Synthesizing the motion program $\delta_F(\delta_C)$ of a cam mechanism requires connecting the prescribed dwells such that the follower does not exhibit theoretically infinite accelerations and decelerations (which translate in large contact forces and are associated with follower bounce, respectively). These may occur at the beginning and at the end of the follower motion and should be avoided, particularly for high-speed cam mechanisms. The third derivative of the follower displacement called *jerk* is also monitored during the design process of high-speed cams and should also remain finite. For slow-speed applications, however, infinite theoretical accelerations are considered acceptable.

Of the numerous types of follower motions described in literature, spline rise and fall will be considered. Specifically, the upper and lower dwells in Figure 7.2 were connected with nonuniform rational B-splines (NURBS) with four control points each, and horizontal end tangents, produced using the **AutoCAD** *spline* command (see Figure 7.3). The complete follower displacement curve has been then exported to a **R12 DXF** file, which caused the splines to be approximated with successions of short line segments. The resulting **DXF** file was then

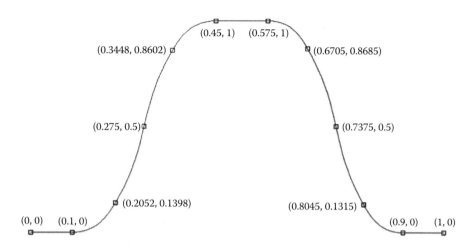

FIGURE 7.3   NURBS normalized motion program produced with **AutoCAD**. The end tangents of the splines, that is, segments (0, 0)–(0.1, 0), (0.45, 1)–(0.575, 1), and (0.9, 0)–(1, 0), are aligned with the respective dwells.

reimported into **AutoCAD,** where the original dwell lines and the newly occurring approximating segments were manually reassembled into a single polyline. After that, a second **R12 DXF** file has been generated (see file **F7_03.DXF**) and was opened using the **Util~DXF** application available with the book. The vertices of this polyline were then exported to ASCII file **F7_03.XY**. Additional points, linearly interpolated between the existing data points, were added to **F7_03.XY** using the **Util~TXT** application (see data file **F7_04-0.XY** and configuration file **F7_04-0.CON**). Through numerical differentiation of the **F7_04-0.XY** data done using the same **Util~TXT** program, two more ASCII files have been generated, that is, **F7_04-1.XY** and **F7_04-2.XY** (see configuration files **F7_04-12.CON** and **F7_04. CF2**). These three files were then used to plot the normalized displacement $\delta_F(\theta)$ and its first and second derivatives $\delta'_F(\theta) = d\delta/d\theta$ and $\delta''_F(\theta) = d^2\delta/d\theta^2$ graphed in Figure 7.4.

The more irregular appearance of the $\delta''_F(\theta)$ curve is due to the approximations performed when the **AutoCAD** NURBS have been converted to **R12 DXF** line segments, amplified by the numerical calculation of the derivatives. The fact that there are no acceleration spikes at the beginning and at the end of the follower displacement is however a good indication that the $\delta_F(\delta_C)$ motion program in Figure 7.4 is suitable for high-speed cam applications.

The aforementioned follower displacement data file **F7_04-0.XY** renamed **dFvdC. XY** is read by a number of Pascal programs that call procedures from units **LibAssur,** **LibCams,** and **LibMec2D,** and served to synthesize the profiles of cam mechanisms of the type shown in Figure 7.1. A second follower motion file named **dFvdC_L.XY,** which contains every 10th data point extracted from file **dFvdC.XY,** was used to produce the frames of the animated **GIF**s that accompany this chapter.

Important: The last entry from both these normalized follower motion files has been edited, so that the initial and final cam profile points do not coincide. Without this, the cam-follower kinematic analysis procedures in unit **LibCams** may produce spikes at the beginning and at the end of the follower motion.

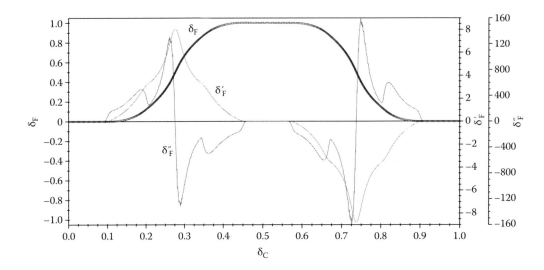

FIGURE 7.4   Follower displacement δ and its first derivative $\delta' = d\delta/d\theta$ and second derivative $\delta'' = d^2\delta/d\theta^2$ calculated using finite differences. Configuration file **F7_04.CF2**.

7.2 SYNTHESIS AND ANALYSIS OF DISC CAMS WITH TRANSLATING FOLLOWER, POINTED OR WITH ROLLER

The first mechanism considered is the disk cam with translating follower ending with a knife edge or with a roller. The two are directly related in that for the same input–output displacement function $s(\theta)$, the profile of the cam designed to operate with a roller follower is the offset of the cam designed for the knife-edge follower. Therefore, the synthesis of cam mechanisms with pointed translating follower will be primarily discussed in this chapter. Any offset cam profile can be obtained inside **AutoCAD** using the *offset* command. Later in the chapter, the procedure **EnvelopeOfCircles** will be introduced, which allows the profile of the cam working with a roller follower to be directly generated.

In the disk cam profile synthesis problem, in addition to a normalized motion $\delta_F(\delta_C)$, the duration of the cam cycle $\Delta\theta$ and follower amplitude Δs must be also specified. This allows the follower displacement to be prescribed as function of the cam angle as

$$s(\theta) = s_0 + \Delta s \cdot \delta_F\left(\frac{\theta}{\Delta\theta}\right) \quad \text{with } 0 \leq \theta \leq \Delta\theta \qquad (7.1)$$

Most common is for cam cycle to be $\Delta\theta = 2\pi$, although cams that perform less than a full rotation exist. Cams that generate two or three follower cycles per turn can be designed by letting $\Delta\theta = \pi$ or $\Delta\theta = 2\pi/3$, respectively, and then repeating the partial profile obtained as a polar array about the cam center. Additional parameters that have to be specified in a design problem are (see Figure 7.5) roller radius r, follower eccentricity e_F (can be either positive or negative), and follower bias s_0 (which is recommended to be 2–3 times bigger than the follower amplitude). Note that this last recommendation has not been strictly observed throughout this chapter.

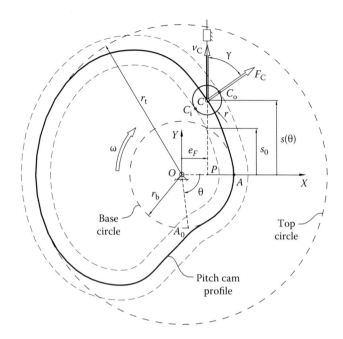

FIGURE 7.5 A disk cam with translating follower. C_i and C_o are the contact points between the roller and the inner and outer offset cam profiles represented in thick, dashed lines.

The pitch cam profile will be generated as a collection of discrete points (x_j, y_j) relative to a reference frame with the origin coincident with the center of rotation of the cam. Of additional interest are the cam curvature ρ and the pressure angle γ between the pitch cam and the follower—both evaluated at the same points (x_j, y_j). Pressure angle γ is defined as the angle between the cam-follower contact force vector F_C and the velocity vector v_C of the tip of the follower. For this particular cam-follower the velocity vector v_C will always be aligned with the direction of the follower axis. For most cam-follower mechanisms, it is recommended that the pressure angle does not to exceed 30° (Norton 2002). Because the direction of the contact force vector F_C is always through the center of curvature of the cam at the point of contact C, the problems of cam curvature and pressure angle analysis are related. Once the direction of the normal to the pitch cam profile is determined (this is also the direction of the force vector F_C—see Figure 7.5), the inner and outer offset points C_i and C_o can be computed as intersections between the normal to the cam profile at point C and the circle centered at C and of radius r. In the programs that accompany this chapter, this has been done using procedure **DoubleOffset**:

```
procedure DoubleOffset(x,y,DnX,DnY,Rho,r:double;
var xi,yi,xo,yo:double);
```

available from unit **LibCams**. The procedure returns the coordinates of inward point (**xi**, **yi**) and outward point (**xo**, **yo**) collinear with point (**x**, **y**) situated along the direction (**DnX**, **DnY**) at distance ±**r** from (**x**, **y**). It can be shown that the pressure angle γ calculated for the knife-edge follower is identical to the pressure angle of an offset cam profile operating in conjunction with the same follower, equipped with a roller of radius r equal to the offset distance (see the closed curves shown in dashed line in Figure 7.5). The base circle radius r_b and top circle radius r_t are important design parameters. For the cam mechanisms like the one in Figure 7.5, these can be calculated with the following formulae:

$$r_b = \sqrt{e_F^2 + s_0^2} \tag{7.2}$$

$$r_t = \sqrt{e_F^2 + \left(s_0 + \Delta s\right)^2} \tag{7.3}$$

The pitch cam profile can be traced accurately as the locus of point C in a motion inversion (Waldron and Kinzel 2003). This method of motion inversion will be employed exclusively throughout this chapter to generate the cam profiles, where the cam is maintained immobile and the follower is rotated in the opposite direction to the normal operation of the cam. The motion-inversion setup (see Figure 7.6a) consists of crank OA connected with a linear actuator positioned perpendicularly to OA and offset by the amount $e_F = OP$. The actuator has been programmed to extend according to the prescribed follower motion $s(\theta)$. If the cam is intended to operate with a roller-follower and generate the same function $s(\theta)$, its profile will be either the inner or the outer envelope of the roller in a motion inversion. These envelopes can be produced as loci of points C_i and C_o (Figure 7.5) or by offsetting the pitch cam profile an amount r using the **AutoCAD** *offset* command.

Program **P7_06.PAS** (see Appendix B) implements a motion-inversion cam profile synthesis strategy as explained earlier. If constant **Color** on line #**12** equals 1, the program plots the crank and the linear actuator of the generating mechanism together with the locus of roller center C with displacement data read from the lower-resolution file **dFvdC_L.XY** (Figure 7.6a). If constant **Color** equals 0, the program plots the roller only and the locus of point C (see Figure 7.6b), with motion data read from the higher-resolution file **dFvdC.XY**.

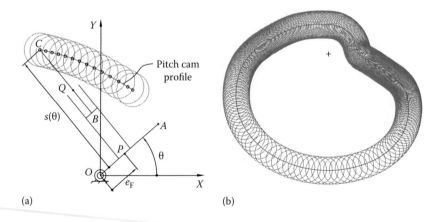

(a)　　　　　　　　　　　　　　(b)

FIGURE 7.6　Cam profile generation in a motion inversion of a knife-edge translating follower (a) and complete profile and roller envelope generated for $\delta_F(\delta_C)$ in Figure 7.4 with $\Delta\theta = 2\pi$, $\Delta s = 1$, $e_F = 0.2$, $s_0 = 0.25$, and $r = 0.15$ (b). See also animation file **F7_06a.GIF**.

Before the actual simulation begins, the program reports the base circle and top circle radii calculated using Equations 7.2 and 7.3—see lines #**31** and #**32**.

By running program **P7_06.PAS** with the follower bias s_0 set to smaller values, the pitch profile may exhibit regions with concavities where the radius of curvature ρ turns negative. When offset to accommodate a roller, the radii of curvature (in absolute value) around these concavities could become smaller than the radius of the roller. Consequently, the prescribed motion will not be reproduced as intended when the follower travels over these areas. One remedy is to redesign the follower motion program $\delta_F(\delta_C)$; the other is to increase the follower bias s_0 that will cause an increase of the base circle radius r_b. Enlarging the base circle radius has the additional favorable effect of reducing the maximum deviation of the pressure angle γ from its ideal value of zero degrees.

Program **P7_07.PAS** (see Appendix B) performs a follower displacement analysis, pressure angle and curvature analysis, and offset cam profile extraction of disk cams with translating knife-edge follower. The cam profile is read from data file **Cam06.D2D** output by program **P7_06.PAS**. If constant **Anim** on line #**15** is set equal to 1, the analysis is accompanied by an animation of the mechanism. If **Anim** equals 0, then only a progress report will be displayed on the computer screen every tenth cam position point (see also line #**36** and the use of variable **Skip**).

Given the cam profile as discrete points (x_j, y_j) and assuming the axis of rotation of the cam is at (0, 0), its base circle radius and top circle radius can be evaluated numerically using procedure **GetProfileRadii** called from unit **LibCams** (see line #**25** of program **P7_07.PAS**). This procedure with the heading

```
procedure GetProfileRadii(FxyName:PathStr; var Rmin,Rmax:double);
```

reads the ASCII or **D2D** file **FxyName** containing the cam profile points and calculates the distance from the center of the cam (0, 0) to each of these points. At the end, it returns the minimum and maximum of these distances assigned to variables **Rmin** and **Rmax**, reported by program **P7_07.PAS** as the base and top circle radii.

Procedure **RotCamTransPointed** called on line #**47** from unit **LibCams** evaluates the intersection point between the pitch cam profile rotated by angle **Theta** and the vertical line $x = e_F$ along which the follower translates. The procedure's heading is

```
procedure RotCamTransPointed(FxyName:PathStr; OP,Theta:double;
var s, xC,yC, DnX,DnY, Rho:double);
```

where **FxyName** is the file name from where the cam profile points centered at (0, 0) are read, **OP** is the eccentricity e_F, and **Theta** is the current cam angle measured clockwise (Figure 7.5), that is, opposite to the direction shown in Figure 7.6a. The procedure returns the coordinates **xC** and **yC** of contact point C, the components **DnX** and **DnY** of the normal to the cam profile at point C, and the radius of curvature **Rho** (i.e., ρ) around that same point C. Procedure **RotCamTransPointed** first identifies the (x_j, y_j) point of the cam rotated clockwise by angle **Theta** having its coordinate y_j positive and its coordinate x_j

the closest to the follower eccentricity e_F. A parabola is then fit through this point and its two neighbors (x_{j-1}, y_{j-1}) and (x_{j+1}, y_{j+1})—see Equation B.15. The intersection between this parabola and the vertical line $x = e_F$ is an improved approximation of the contact point between the cam profile and the tip of the follower. The radius of the circle circumscribed to the same points (x_{j-1}, y_{j-1}), (x_C, y_C), and (x_{j+1}, y_{j+1}) is reported as the radius of curvature ρ of the pitch cam profile around contact point C (see Appendix A). The line connecting the center of this circle with contact point (x_C, y_C) is the normal to the cam profile at point C. Also returned by procedure **RotCamTransPointed** to the calling program are the projections **DnX** and **DnY** of the line connecting contact point C and the center of curvature of the pitch cam profile at (x_C, y_C).

Using projections **DnX** and **DnY**, procedure **DoubleOffset** called from unit **LibCams** on line **#48** then calculates the coordinates of points C_i and C_o belonging to the two offset cam profiles visible in Figure 7.5. Program **P7_07.PAS** also animates the pitch cam as it rotates, showing in addition the contact point C, velocity vector v_C of the follower, and the normal vector F_C to the cam profile at point C (see Figure 7.7). Note that the vectors v_C and F_C have arbitrarily assigned magnitudes. The angle between these two vectors is evaluated using procedure **U2dirs2D90** called from unit **LibGe2D** (see line **#49**) and is written to ASCII file **F7_07.TXT** as the pressure angle γ of the mechanism. Also written to **F7_07. TXT** are the cam angle θ, follower displacement s, contact point coordinates x_C and y_C, cam profile radius of curvature ρ at point C, and the x and y coordinates of the inner and outer offset points C_i and C_o. The program also displays on the top-left corner of the screen a short report consisting of the values of the current cam angle θ, pressure angle γ, and radius of curvature ρ (see Figure 7.7). The data from ASCII file **F7_07.TXT** was then use to generate the graphs in Figure 7.8. The coordinates of points C, C_i, and C_o in the same file served to generate the three cam profiles in Figure 7.5 using the **D_2D** program.

A companion program named **P7_07BIS.PAS** available with the book performs the same type of analysis (less the ASCII file output) and in addition plots on the screen the inner and outer cam profiles as comet loci of points C_i and C_o. In the animation file **F7_07bis.GIF**

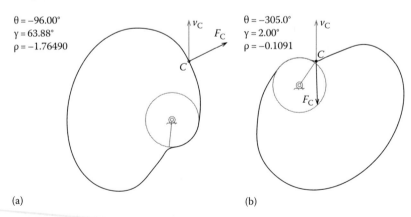

(a)   (b)

FIGURE 7.7 Analysis done using program **P7_07.PAS** of a disk cam equipped with pointed follower, shown in the positions where maximum pressure angle (a) and minimum radius of curvature (b) occur. See also animation files **F7_07.GIF**.

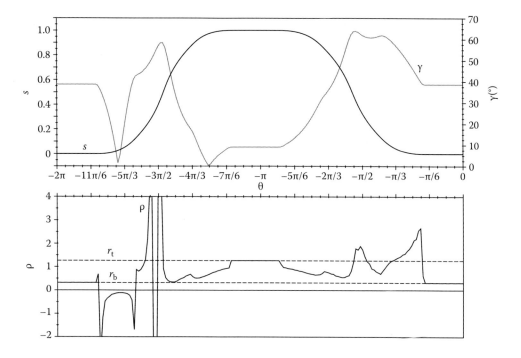

FIGURE 7.8   Plots of the follower displacement s, pressure angle γ, radius of curvature ρ, and base circle r_b and top circle radii r_t of the cam in Figure 7.6. Configuration files **F7_08UP.CF2** and **F7_08DN.CF2**.

generated with this program, an *undercut* of the outer cam becomes visible, that is, the cam profile folds over itself around the point of minimum curvature. This undercut is less apparent when the **AutoCAD** *offset* command is employed because the software will remove the mentioned fold, leaving only a cusp in that region.

Noticeably, the performance of the mechanism in Figure 7.7 can be improved. For example, increasing the base circle radius r_b will cause a reduction in the maximum pressure angle γ. This will also increase the minimum radius of curvature ρ of the cam, with the possibility of maintaining this curvature positive over the entire cam profile. Similar effects upon pressure angle γ and radius of curvature ρ will have an increase of the follower bias s_0 or a reduction of the follower offset e_F.

7.3 SYNTHESIS AND ANALYSIS OF DISC CAMS WITH OSCILLATING FOLLOWER, POINTED OR WITH ROLLER

This is the second type of disk cams that can be equipped with either knife-edge follower or roller follower. Same as earlier, the internal or external cam profiles intended to operate with a roller can be obtained by offsetting the pitch cam profile an amount equal to the roller radius (see Figure 7.9). For this reason, the synthesis of disk cam mechanisms with knife-edge follower will be primarily discussed in this section.

In a design problem where a normalized motion program $\delta_F(\delta_C)$ is prescribed, the amplitude of the cam displacement $\Delta\theta$ and amplitude of the follower displacement $\Delta\varphi$

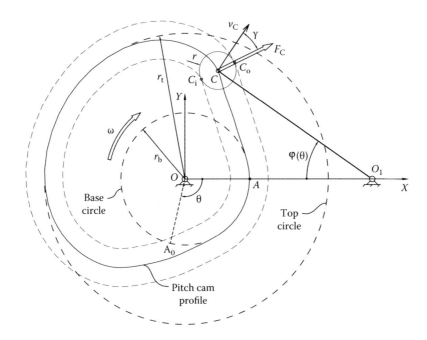

FIGURE 7.9 Main parameters of a disk cam with oscillating follower. C_i and C_o are the contact points between the roller and the inner and outer offset cam profiles, represented in thick, dashed lines.

must be specified, together with ground joint distance OO_1, follower bias φ_0 (corresponding to point C being located on the base circle), and roller radius r. Giving these parameters, the follower displacement vs. cam angle $\varphi(\theta)$ is

$$\varphi(\theta) = \varphi_0 + \Delta\varphi \cdot \delta_F(\theta/\Delta\theta) \quad \text{with } 0 \le \theta \le \Delta\theta \tag{7.4}$$

The pitch cam profile will be the locus of point C in an inverted motion, where the cam is maintained fixed, and pin joint O_1 of the follower is rotated opposite to the normal direction of rotation of the cam. Pitch cam profile radius of curvature ρ and pressure angle γ, that is, the angle between the velocity and the contact force vectors v_C and F_C, are also of interest and should be evaluated in a number of discrete positions as the cam rotates. The direction of the contact force vector F_C coincides with line C_iC_o, irrespective of the cam profile considered, that is, inner, outer, or pitch cam profile. The velocity vector of the contact point will be different however: it will be perpendicular to line O_1C_i for external cams, perpendicular to line O_1C_o for internal cams, and perpendicular to line O_1C for knife-edge follower cam mechanisms (this third one is vector v_C shown in Figure 7.9). Consequently, for a given cam angle θ, the pressure angles on the external, internal, and pitch cam profiles will be different.

The base circle radius r_b and top circle radius r_t of the cam with oscillating knife-edge follower can be calculated using the following equations:

$$r_b = \sqrt{OO_1^2 + O_1C^2 - 2 \cdot OO_1 \cdot O_1C \cdot \cos(\varphi_0)} \tag{7.5}$$

$$r_t = \sqrt{OO_1^2 + O_1C^2 - 2 \cdot OO_1 \cdot O_1C \cdot \cos(\varphi_0 + \Delta\varphi)} \tag{7.6}$$

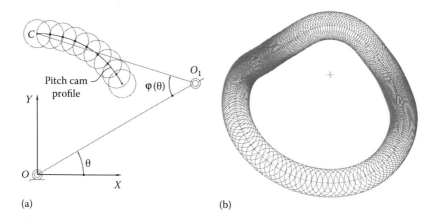

FIGURE 7.10   Cam profile generation in a motion inversion (a) and complete pitch profile and roller locus generated for $\delta_F(\theta)$ in Figure 7.4 with $OO_1 = 2$, $O_1C = 2$, $\varphi_0 = 20°$, $\Delta\varphi = 25°$, and $r = 0.2$ (b). See also animation file **F7_10a.GIF**.

Program **P7_10.PAS** (see Appendix B) implements a motion-inversion approach to the generation of cam profiles operating with oscillating knife-edge followers. If on line #**12** constant **Color** is set equal to one, the program simulates the inverted mechanism and plots the locus of point C using follower motion data read from the lower-resolution file **dFvdC_L.XY** (see Figure 7.10a). If constant **Color** is zero, the program will plot only the roller and the locus of its center C (i.e., the tip of the knife-edge follower) using motion data read from the higher-resolution file **dFvdC.XY** (see Figure 7.10b) and also saves the pitch cam profile to data file **Cam10.D2D** (see line #**58**).

If the follower bias φ_0 is too small, the cam may develop regions of negative curvature (concavities). When such a cam profile is offset, the radii of curvature along these concavities can become smaller than the radius of the roller, and the motion of the follower equipped with a roller will be different than the prescribed function $\varphi(\theta)$. Other than redesigning the follower motion $\delta_F(\delta_C)$, increasing the base circle radius r_b can eliminate such occurrences. This will also reduce the maximum pressure angle γ, which is also desirable. Increasing ground joint distance O_1O and follower length O_1C may have comparable effects.

To analyze the cam mechanism obtained through synthesis (i.e., the pitch cam profile recorded to file **Cam10.D2D**), program **P7_11.PAS** has been written and its listing is included in Appendix B. For a given cam angle θ between 0 and 2π, the program determines the profile point (x_j, y_j), which is located relative to joint center O_1 at a distance closest to the follower arm length O_1C. A better approximation of the contact point of coordinates (x_C, y_C) is then determined as the intersection between the circle centered at O_1 and radius O_1C and a parabola through points (x_{j-1}, y_{j-1}), (x_j, y_j), and (x_{j+1}, y_{j+1})—see also Appendix A. This algorithm is implemented in procedure **RotCamOscilPointed** called from unit **LibCams** on line #**52** of program **P7_11.PAS**. The procedure has the heading:

```
procedure RotCamOscilPointed(FxyName:PathStr; OO1,O1C,Theta:double;
var Phi, xC,yC, DnX,DnY, Rho:double);
```

In addition to the coordinates **xC** and **yC** of the contact point, the procedure also returns the follower angle **Phi** and the radius of curvature **Rho** of the cam around the contact point approximated with the radius of the circle circumscribed to points (x_{j-1}, y_{j-1}), (x_C, y_C), and (x_{j+1}, y_{j+1}). It also returns the components of the normal to the pitch cam profile **Dnx** and **Dny**, calculated as ox and oy the projections of the line that connects contact point (x_C, y_C) with the center of the circle through the same three points (x_{j-1}, y_{j-1}), (x_C, y_C), and (x_{j+1}, y_{j+1}).

Program **P7_11.PAS** then calls procedure **DoubleOffset** (line #53) to determine the coordinates of offset points C_i and C_o (Figure 7.9), which later serve to evaluate the pressure angles γ_i and γ_o between the roller and the inner and outer contact cam profiles. The pressure angle γ is calculated as the angle between the normal to the pitch cam profile which has the directions **Dnx** and **Dny**, and line O_1C rotated by 90° (see line #54). Similar approaches are implemented to calculate γ_i and γ_o (lines #56 and #58). Lines #70 to #73 write to ASCII file **F7_11.TXT** the current cam angle **Theta**; the corresponding follower angle **Phi**; the values of the three pressure angles γ, γ_i, and γ_o; the coordinates of points C, C_i, and C_o rotated back to the reference position of the cam; and the base circle and top circle radii of the pitch cam profile. This ASCII file served to generate the plots in Figure 7.11, as well as the pitch cam profile and offset cam profiles in Figure 7.9.

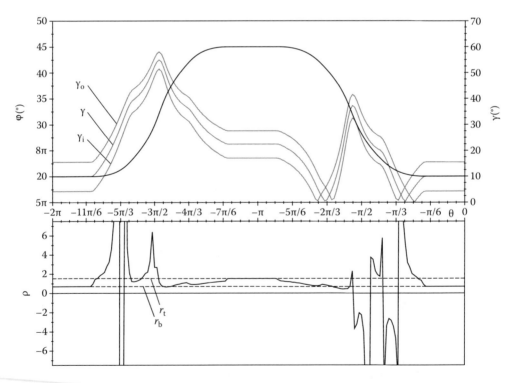

FIGURE 7.11 Plot of the follower displacement φ, radius of curvature ρ, and pressure angles γ, γ_i, and γ_o on the pitch cam, inner cam, and outer cam, respectively, generated with program **F7_11.PAS**. Also plotted in dashed lines are the base circle and top circle radii r_b and r_t. Configuration files **F7_11UP.CF2** and **F7_11DN.CF2**.

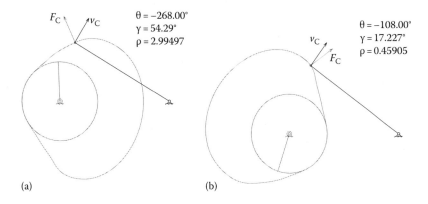

FIGURE 7.12 The cam in Figure 7.10 in the position where maximum pressure angle (a) and minimum radius of curvature (b) occur. See also animation file **F7_12.GIF**.

When operating in conjunction with its roller follower, the inner offset cam exhibits an overall reduced pressure angle γ_i compared to angle γ of the pitch cam with knife-edge follower, while the outer offset cam will exhibit an increased pressure angle γ_o.

Program **F7_11.PAS** also simulates the cam and follower motions (see Figure 7.12), showing the contact point C between the knife-edge and the cam. It also shows the displacement vector (which is parallel to the velocity vector v_C), and the normal direction to the cam profile (which coincides with the contact force vector F_C). If parameter **Anim** on line #**17** of the program equals 0, then only a progress report will be displayed every 10th position of the cam.

A related program named **P7_07BIS.PAS** available with the book allows the same type of cam-follower analysis, with the additional plotting of the inner and outer cam profiles as comet loci of points C_i and C_o. Animation file **F7_12bis.GIF** has been generated using the **F7_11BIS.DXF** file output by this program.

In all the aforementioned figures and simulation, the follower was represented as a straight-line O_1C. Evidently, in practice, the follower has to be shaped such that its body does not interfere with the cam.

7.4 SYNTHESIS AND ANALYSIS OF DISC CAMS WITH TRANSLATING FLAT-FACED FOLLOWER

Disk cams with translating flat-faced follower are commonly used in applications where a simple, compact arrangement is required, like in sidevalve engines and fuel-injection pumps. As opposed to translating knife-edge or roller-follower cam mechanisms, in this case, follower eccentricity does not influence the input–output kinematics. This eccentricity however has an effect upon the magnitude of the reaction moment at the sliding joint and consequently upon the overall mechanical efficiency of the mechanism.

When performing the synthesis of a disk cam with translating flat-faced follower for which a normalized input–output motion $\delta_F(\delta_C)$ is prescribed, several additional parameters must be specified, that is, cam rotational cycle $\Delta\theta$, follower displacement amplitude Δs, follower bias s_0, and the angle of the face of the follower γ measured from a parallel

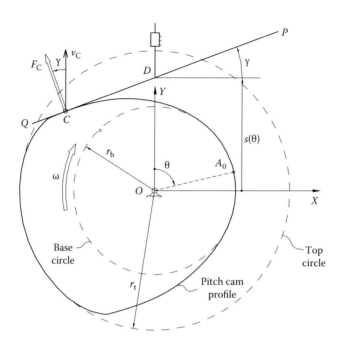

FIGURE 7.13 Main parameters of a disk cam with translating flat-faced follower.

to OX (it is assumed that the follower slides along a vertical line as shown in Figure 7.13). Because the contact force vector $\boldsymbol{F}_C$ has the direction of the normal to the face of the follower, and because the velocity vector $\boldsymbol{v}_C$ is aligned with its direction of sliding, the angle between these two vectors (i.e., the pressure angle) is constant and equal to the follower face angle.

To extract the follower envelope (i.e., the cam profile) in an inverted motion approach, a number of radial lines originating from a polar point situated inside the cam contour will be employed. The intersection between the follower face PQ and these radial lines will be done for every inverted motion position of the follower. As follower face PQ intersects each of these polar lines, they will be progressively shortened towards the polar point. In the end, these outer points will be connected together to form the sought-after cam profile. Such a strategy has been implemented in procedure **EnvelOfLines** available from unit **LibCams**:

```
procedure EnvelOfLines(Color,n:Integer; x0,y0, R00, xA,yA,
xB,yB:double);
```

When the procedure is first called, it generates **n** equally spaced radial lines originating from polar point (**x0,y0**), each of length **R00**. The procedure then reduces the length of these lines as they are intersected by the face of the follower, that is, the segment that connects points (**xA,yA**) and (**xB,yB**). Depending on the value of parameter **Color** in procedure **EnvelOfLines** (either positive, negative, or zero) all **n** radial lines, every 10th radial line, or neither line will be plotted on the screen as their length is adjusted. After the desired number of intersections with segments AB representing the face of the follower has been performed, the ends of these **n** radial lines will be connected to form a polyline. This is done by calling procedure **EndEnvelopes** from the same unit **LibCams**:

```
procedure EndEnvelopes(Name8:NameStr; Color:Word);
```

In this procedure, **Name8** is the name of the **D2D** file where the coordinates of the enveloping polyline will be written. Color information **Color** will also be written to this **D2D** file, coded using multipliers of the **InfD** constant as explained in Chapter 1. The actual extension of the cam profile data file is **$2D** (same extension as for locus files as discussed in Chapter 6). To retain this envelope file at the end of the simulation, procedure **CloseMecGraph** must be called with its argument set to TRUE.

The cam profile thus obtained must be analyzed for curvature at each vertex (x_j, y_j). Being a flat-faced follower mechanism, it is essential for the cam profile not to have rectilinear portions (i.e., ρ should not be infinity). If this happens, the follower will step over the respective areas, and the specified input–output motion $\delta_F(\delta_C)$ will not be satisfied. Also detrimental to a reliable operation of the cam is the occurrence of cusps (i.e., $\rho = 0$), where large contact stresses will develop during operation. Note that the occurrences of both flat portions and the cusps are indicative of the cam profile being undercut.

The base and top circle radii of this cam are important parameters and can be calculated analytically using the following equations:

$$r_b = s_0 \cdot \cos(\gamma) \tag{7.7}$$

$$r_t = (s_0 + \Delta s) \cdot \cos(\gamma) \tag{7.8}$$

Figure 7.14a shows a motion-inversion setup of a disk cam with translating flat-faced follower, consisting of crank OA aligned with a linear actuator that expands by $s(\theta)$ according to Equation 7.1. Program **P7_14.PAS** in Appendix B implements this approach to

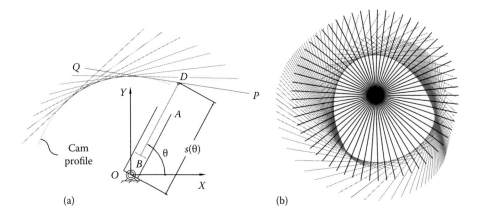

(a) (b)

FIGURE 7.14 Cam profile generation in a motion inversion (a) and complete profile and follower envelope produced for $\delta_F(\theta)$ in Figure 7.4 with $s_0 = 1.5$, $\Delta s = 1.0$, and $\gamma = 20°$ (b). See also animation files **F7_14a.GIF** and **F7_14b.GIF**.

generate the envelope of follower face *PQ*. It employs procedure **EnvelOfLines** with a family of radial lines originating from point (**xPC,yPC**) to extract the cam profile to a polyline (see Figure 7.14b and lines #**54** and #**56** of the program). As a new follower face *PQ* is drawn during the inverted motion, its intersection with all radial lines originating from polar point (**xPC,yPC**) is recalculated and the free ends of these radial lines updated—see animation file **F7_14b.GIF**. Lastly, the outer ends of these radial lines are connected in a polyline by calling procedure **EndEnvelopes** (lines #**69** and #**71**).

If constant **Color** on line #**13** in program **P7_14.PAS** equals one, the follower displacement data are read from the lower-resolution file **dFvdC_L.XY**. If **Color=0**, then the full resolution file **dFvdC.XY** is utilized instead. In all these cases, the cam profile is extracted to a polyline and its vertices are written to data file **Cam14.D2D**. If constant **Anim** on line #**14** equals one, the follower and its driving mechanism are animated in their inverted motion (see Figure 7.14a). If **Anim=0** and **Color=1**, the program animates the follower face only, together with the radial lines as they get shortened (see Figure 7.14b). If **Color=0**, then the cam profile is output directly, without any animation.

A second program named **P7_15.PAS** (see Appendix B) allows for a kinematic analysis of flat-faced follower cam mechanisms with the cam profile read from data file **Cam14.D2D** output by program **P7_14.PAS**. The same follower angle γ and cam rotational amplitude Δθ as in the synthesis program **P7_14.PAS** are specified (see lines #**18** and #**19**). The program calls procedure **RotCamTransFlat** on line #**45** from unit **LibCams** with the heading

```
procedure RotCamTransFlat(FxyName:PathStr; Theta,Gamma:double;
var s,xC,yC,Rho:double);
```

The procedure reads the cam profile from file **FxyName**, and for a given cam rotation angle **Theta** and follower assumed risen above the cam, it identifies the cam profile point (x_j, y_j) that is closest to the follower face. Parameter **Gamma** is the follower face angle in radians, measured as shown in Figure 7.13. It is assumed that the follower is infinitely long in both directions and that it translates along the *OY* axis (see Figure 7.13). The procedure then fits a parabola through this point (x_j, y_j) and neighboring points (x_{j-2}, y_{j-2}) and (x_{j+2}, y_{j+2}). By calling procedure **TangOfSlopem2Parab** from unit **LibGe2D**, procedure **RotCamTransFlat** then calculates the point where the tangent to this parabola has an angle equal to γ (see Appendix A). This tangent point will be returned to the calling program as the contact point (x_C, y_C) between the cam and the follower. Lastly, by calling procedure **Circ4Pts** from unit **LibGe2D**, the radius of curvature of the cam around contact point (x_C, y_C) is evaluated as the radius of the circle circumscribed to profile points (x_{j-2}, y_{j-2}) and (x_{j+2}, y_{j+2}) and a fictitious point of coordinates $x = 0.5(x_{j-1} + x_{j+1})$ and $y = 0.5(y_{j-1} + y_{j+1})$. This is equivalent to solving the following simultaneous equations in the unknowns x_K, y_K, and ρ:

$$(x_{j-2} - x_K)^2 + (y_{j-2} - y_K)^2 = \rho^2$$

$$(x_{j+2} - x_K)^2 + (y_{j+2} - y_K)^2 = \rho^2 \tag{7.9}$$

$$[0.5(x_{j-1} + x_{j+1}) - x_K]^2 + [0.5(y_{j-1} + y_{j+1}) - y_K]^2 = \rho^2$$

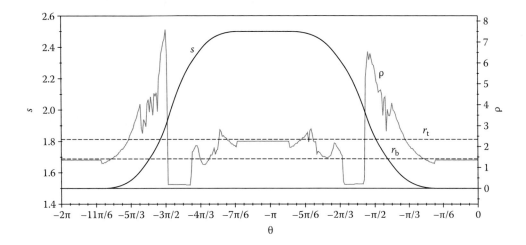

FIGURE 7.15 Plot of the follower displacement s and radius of curvature ρ of the cam in Figure 7.14b. Also shown in dashed lines are the base circle and top circle radii r_b and r_t. Configuration file **F7_15.CF2**.

Using the data in ASCII file **F7_15.TXT** output by program **P7_15.PAS**, the plot in Figure 7.15 has been generated. The horizontal lines on this graph are the base circle and top circle radii r_b and r_t, evaluated by calling procedure **GetProfileRadii** on line **#25** of the program.

Simulation frames of the positions where the follower contacts the cam at its minimum and maximum radius of curvature are available in Figure 7.16. In the previous two cam mechanisms examined, exactly calculated profile points were available to evaluate the radii of curvature of the cam. This time, however, the cam profile points were obtained as intersections of the follower face with a finite number of radial lines, in an inverted motion. This explains the more noisier appearance of the radius of

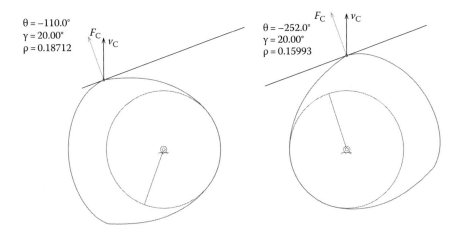

FIGURE 7.16 The cam in Figure 7.14b in the positions where the minimum radii of curvature occur. See also animation file **F7_16.GIF**.

curvature graph $\rho(\theta)$ in Figure 7.15, compared to similar graphs in Figures 7.7 and 7.11, as well as the slight departure of ρ from the exactly calculated radii r_b and r_t over the circular portions of the cam profile.

7.5 SYNTHESIS AND ANALYSIS OF DISC CAMS WITH OSCILLATING FLAT-FACED FOLLOWER

Probably the most widely used cam mechanisms are the disk cams with oscillating flat-faced follower. These are commonly employed in the design of valve trains of overhead internal combustion engines, including the variable timing models.

Same as before, the cam profile will be determined as the envelope of the follower face PQ in an inverted motion, using procedure `EnvelOfLines`. In a design problem, along-side follower motion $\delta_F(\delta_C)$, the cam and follower amplitudes $\Delta\theta$ and $\Delta\varphi$, follower bias φ_0 (i.e., follower angle when in contact with the base circle), and follower face eccentricity e_F (either positive or negative) should be specified—see Figure 7.17. Part of the design process, the radius of curvature ρ of the cam and the pressure angle γ at the contact point with the follower must be evaluated for a number of discrete positions of the cam as it rotates. For proper operation, the synthesized cam profile should not exhibit rectilinear portions ($\rho = \infty$) or cusps ($\rho = 0$). In case they occur, other than modifying the follower motion $\delta_F(\delta_C)$, increasing the base circle radius r_b of the cam or reducing the ratio $\Delta\varphi/\varphi_0$ will both work towards eliminating such defects.

The base circle radius r_b and top circle radius r_t of the cam profile can be exactly calculated using the following two equations:

$$r_b = OO_1 \cdot \sin(\varphi_0) + e_F \tag{7.10}$$

$$r_t = OO_1 \cdot \sin(\varphi_0 + \Delta\varphi) + e_F \tag{7.11}$$

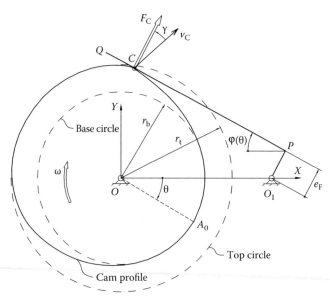

FIGURE 7.17 Parameters of a disk cam with oscillating flat-faced follower. As drawn, follower eccentricity e_F is considered positive.

In disk cams with oscillating flat-faced follower, the contact force vector F_C remains perpendicular to the follower face PQ (i.e., has the direction of the normal to the cam profile), while the velocity vector v_C will always remain perpendicular to line O_1C (Figure 7.17). Because during operation the contact point C changes location along the face of the follower, the angle between vectors F_C and v_C (i.e., the pressure angle γ) will also change, less for eccentricity $e_F = 0$ when the pressure angle γ will remain zero, irrespective of the cam angle.

Figure 7.18a shows a motion-inversion setup of a disk cam with translating flat-faced follower, consisting of a crank OO_1 driven as shown, in series with a second crank that rotates relative to the first one according to Equation 7.4. Program **P7_18.PAS** listed in Appendix B generates the cam profile as follower envelope in an inverted motion. Procedure **EnvelOfLines** called on lines #54 and #56 updates the lengths of an array of radial lines originating from polar point (**xPC,yPC**) as they are intersected by follower face line PQ, until these ends approximate the cam profile. Procedure **EndEnvelopes** called on line #68 and #70 of the program then connects the ends of these radial lines into a closed polyline and writes its vertices to data file **Cam18.D2D**.

If on lines #14 and #15 constants **Color=1** and **Anim=1**, then the program animates the cam-follower mechanism in a motion inversion similar to Figure 7.18a. If **Color=1** and **Anim=0**, then the program animates the follower face only, together with the radial lines as they are progressively shortened to extract the follower envelope (see Figure 7.18b). If **Color=0**, then the cam profile is output without animation. Note that either the reduced follower motion file **dFvdC_L.XY** or the full size file **dFvdC_L.XY** is used as input based on the values of the same constants **Anim** and **Color**.

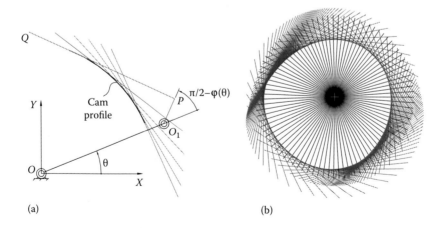

(a)   (b)

FIGURE 7.18 Cam profile generation in a motion inversion (a) and complete profile and follower envelope generated for $\delta_F(\theta)$ in Figure 7.4 with $\varphi_0 = 35°$ and $\Delta\varphi = 15°$ (b). See also animation files **F7_19a.GIF** and **F7_19b.GIF**.

The companion program **P7_19.PAS** (see Appendix B) reads the cam profile file **Cam18. D2D**, and for a given follower face eccentricity e_F and joint distance OO_1, it performs a kinematic analysis of the mechanism. Procedure **RotCamOscilFlat** with the heading

```
procedure RotCamOscilFlat(FxyName:PathStr; OO1,O1P,Theta:double;
var Phi,xC,yC,Rho:double);
```

is called from unit **LibCams** on line #**49** of the program. The procedure identifies, for a given cam angle **Theta**, the profile point (x_j, y_j) from where a tangent to a circle centered at O_1 and of radius e_F has the highest slope (i.e., angle φ in Figure 7.18a has a maximum value). Once this point is identified, a parabola is fit through points (x_{j-1}, y_{j-1}), (x_j, y_j), and (x_{j+1}, y_{j+1}), and by calling procedure **TangComParabCirc** from unit **LibGe2D**, the common tangent to this parabola and to the circle centered at O_1 and of radius e_F is determined as explained in Appendix A. The tangent point on this parabola will then be returned to the calling program as the contact point (x_C, y_C) between the cam and its follower. Procedure **RotCamOscilFlat** also calculates the radius of curvature of the cam profile around the same point (x_C, y_C) by employing procedure **Circ4Pts** available from unit **LibGe2D**.

Using the data in ASCII file **F7_19.TXT** output by program **P7_19.PAS**, the graphs in Figure 7.19 have been generated. Notice the noisy appearance of the pressure angle

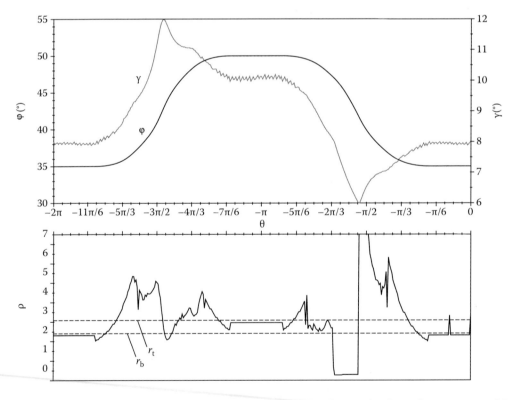

FIGURE 7.19   Plot of the follower displacement φ, pressure angle γ, and radius of curvature ρ of the cam in Figure 7.18b. Also shown by the dashed lines are the base circle and top circle radii r_b and r_t. Configuration files **F7_19A.CF2** and **F7_19B.CF2**.

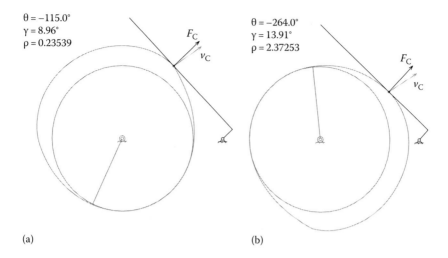

$\theta = -115.0°$
$\gamma = 8.96°$
$\rho = 0.23539$

$\theta = -264.0°$
$\gamma = 13.91°$
$\rho = 2.37253$

(a)   (b)

FIGURE 7.20   The cam in Figure 7.18b in the positions where minimum radius of curvature (a) and maximum pressure angle (b) occur. See also animation file **F7_20.GIF**.

and radius of curvature on these two graphs. This is because the cam profile points were extracted with some approximation using the follower envelope, compared to the exactly generated cam profiles operating with pointed follower discussed earlier in the chapter. Simulation frames of the positions where the follower contacts the cam at its minimum radius of curvature and maximum pressure angle are available in Figure 7.20.

After numerous trials, it was found that for cam mechanisms with flat-faced follower, both translating and oscillating, the radius of curvature along the base circle and top circle of the cam profile is more accurately evaluated using Equation 7.9 and procedure **Circ4Pts**. In the same respects, it was also found that it is better when the radial lines originate from the center of the cam (0, 0), explicable because the base circle and top circle of the cam are centered at this point. Since the radii of curvature along the rise and fall sections of the cam profile are not precisely known, the conclusion cannot be immediately extended to these other sections of the radius of curvature graphs $\rho(\theta)$ in Figures 7.16 and 7.20.

7.6 SYNTHESIS OF DISC CAMS WITH CURVILINEAR-FACED FOLLOWER

In spite of their apparent practical advantage, there has been little work done on the design of disk cams with curvilinear-faced followers. Concave follower cams experience reduced wear rate due to better contact stresses, while if equipped with convex follower, they can be made smaller than the equivalent flat-faced follower cams, without the danger of their profile becoming undercut. In this last section, the synthesis of two types of cams will be discussed, that is, one where the face of the follower is a circular arc and the other where the follower is a smooth curve approximated by short line segments.

7.6.1 Synthesis of Disk Cams with Arc-Shaped Follower

As mentioned earlier, the profile of a cam intended to operate with a roller follower can be obtained from its pitch cam profile using the **AutoCAD** *offset* command. The same offset profile can be generated as the envelope of the roller follower in a motion-inversion process. To illustrate this other method, programs **F7_21.PAS** and **F7_22.PAS** have been written and are available with the book. The first program generates the inner cam of a translating roller-follower mechanism in a motion inversion, and the second program generates the inner cam intended to operate with an oscillating roller follower. Both programs call procedure **EnvelOfCircles** for every position of the follower in its inverted motion. The heading of the procedure is

```
procedure EnvelOfCircles(Color,n:Integer; x0,y0, R00, x1,y1,
x2,y2, x3,y3:double);
```

When called for the first time, the procedure draws a family of **n** radial lines (**n** cannot exceed 1000) in color **Color** and of length **R00** that originate from point (**x0,y0**). To expedite the simulation, if **Color** is zero, no line will be drawn, while if **Color** is negative, then only part of these lines will be drawn. After these **n** radial lines are generated, the procedure intersects them with the circle circumscribed to points (**x1, y1**), (**x2, y2**), and (**x3, y3**). Of each pair of intersection points between this circle representing the roller, and the **n** radial lines, the point that is closest to (**x0,y0**) is retained and then used to adjust the length of the respective polar line—see animation files **F7_21a.GIF** and **F7_22a.GIF**. After all intersections between the roller circles and the polar lines have been evaluated, procedure **EndEnvelopes** is called to connect the outer ends of these lines into a closed polyline (see Figures 7.21 and 7.22).

Evidently, the accuracy with which the roller envelopes are extracted depends on the number **n** of polar lines in procedure **EnvelOfCircles**. Other factors are the number of cam positions in the motion inversion and the number of significant digits used to record the follower motion. Also influencing is the location of point (**x0,y0**) from where the **n** radial lines originate. It appears that if polar point (**x0,y0**) coincides with the center of rotation of the cam, then the dwell portions of the cam profiles are more accurately generated as envelope, while if (**x0,y0**) is selected close to the centroid of the cam, then the same

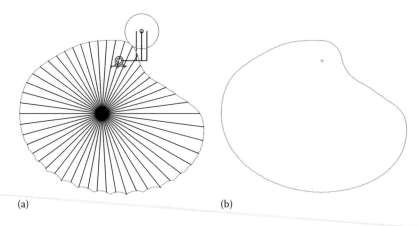

(a)   (b)

FIGURE 7.21 Inner cam profile generation as envelope of a translating roller follower in a motion inversion. Lower-resolution cam (see animation file **F7_21a.GIF**) (a) and higher-resolution cam (b).

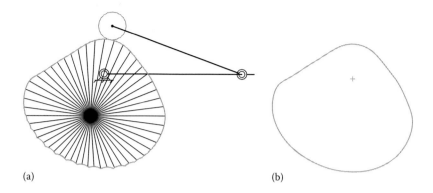

(a)   (b)

FIGURE 7.22   Inner cam profile generation as envelope in a motion inversion of an oscillating roller follower. Lower-resolution cam (see animation file **F7_22a.GIF**) (a) and higher-resolution cam (b).

holds true for the rise and fall portions of the cam. With such limitations, offsetting the pitch cam profile obtained as described in Sections 7.2 and 7.3 is the preferred method of roller-follower cam profile synthesis.

The true benefit of procedure **EnvelOfCircles** is that it can be used to synthesize the profile of disk cams that operate with arc-shaped followers (either concave or convex) as it will be explained in the remainder of this section.

P7_23.PAS listed in Appendix B is a modification of program **P7_14.PAS**, where the face of the follower has a circular arc attached to it. This arc is specified by three points noted 1, 2, and 3, the coordinates of which are given relative to a reference frame that moves together with the follower. This mobile reference frame has its origin at D and its x-axis oriented towards point P of the follower (see Figures 7.13 and 7.23 and lines #**25**, #**26**, and #**27** of program **P7_23.PAS**). Note that the program must be first run with constant **Color** and **Anim** (lines #**14** and #**15**) set both equal to zero so that a higher-resolution file with the cam profile points **Cam23.D2D** is generated. When the program is run with the same two constants equal to one, an animation of the inverted follower motion is produced (see Figure 7.23a and b and the corresponding animated **GIF**s). When **Color=0** and **Anim=1**, program **P7_23.PAS** (see Appendix B) animates the follower envelope extraction process with the radial lines visible. In all these cases, the screen will be copied to file **F7_23.DXF** (either in separate layers or not), but only for **Anim=0**, the cam profile will be written to the **Cam23.D2D** data file (see line #**84**).

Figure 7.23a shows a representative frame of the motion-inversion simulation recorded file **F7_23.DXF**, generated for points 1, 2, and 3 having their coordinates equal to (2, –0.4), (0, 0), and (–2, –4), respectively. Figure 7.23b shows a similar motion-inversion frame produced for points 1, 2, and 3 of coordinates (2, 0.4), (0, 0), and (–2, 4). This is the same arc-faced follower, but in a convex orientation. The companion Figure 7.23c and d illustrate the follower envelope extraction process, the result of calling procedure **EnvelOfCircles**. Note that the radial lines used to extract the cam profile are intersected with the entire circle through points 1, 2, and 3, not only by the portion of this circle representing the face of the follower.

To synthesize disk cam profiles intended to operate with oscillating arc-faced follower (Figure 7.24), program **P7_18.PAS** has been modified into program **P7_24.PAS**

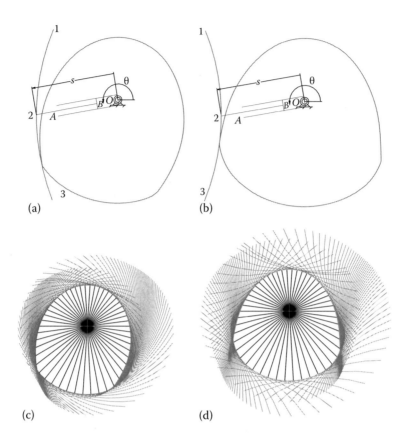

FIGURE 7.23 Motion inversion of disk cams with arc-shaped translating followers in concave (a) and (c) and convex (b) and (d) arrangements. See also animation files **F7_23a.GIF** to **F7_23d.GIF**.

available with the book. The flat-faced follower in program **P7_18.PAS** has now an arc of circle attached to it, again specified by three points 1, 2, and 3. The coordinates of these three distinct points are given relative to a local reference frame with the origin coincident with point P and its x-axis oriented toward point Q of the follower (see Figures 7.17 and 7.24). The program calls procedure **EnvelOfCircles** to extract the envelope of this arc of circle as the follower is driven according to Equation 7.4 in an inverted motion. Sample output by program **P7_24.PAS** generated for both concave and convex followers are given in Figures 7.24. Same as in the case of **P7_23.PAS**, the program must be first run with constants **Color** and **Anim** set equal to zero. This will generate a higher-resolution cam profile and will write its points to file **Cam24.D2D**. The data from this file will then be used to represent the cam in any subsequent simulations done with **P7_24.PAS**.

7.6.2 Synthesis of Disk Cams with Polygonal-Faced Follower

Two more computer programs available in Appendix B will be briefly discussed, that is, **P7_25.PAS** (derived from program **P7_14.PAS**) and **P7_26.PAS** (derived from program **P7_18.PAS**). These programs allow disk cam profiles with curvilinear

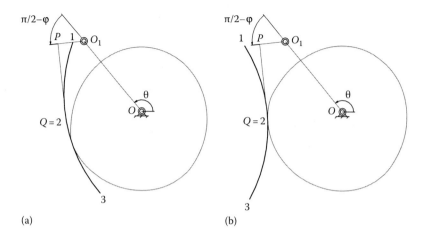

FIGURE 7.24   Motion inversion of disk cams with arc-shaped oscillating followers in concave (a) and convex (b) arrangements. See also animation files **F7_24a.GIF** to **F7_24d.GIF**.

translating and oscillating followers to be synthesized. The face of the follower must be supplied as an ASCII file of *x* and *y* coordinates of the vertices of a polyline, similar to the *shape* files discussed in Chapter 5. File **FFace.XY** read by these two programs (see line #16 of **P7_25.PAS**) consists of 127 vertices that approximate an arc of an ellipse that was first drawn in **AutoCAD**. The center of the ellipse was located at (0, −0.625), its major and minor radii were 3.0 and 0.625, and the start and end angles were equal to 184° and 356°, respectively. An arc of this ellipse was then saved to **R12 DXF**, and in the process, it was converted to a polyline. Finally, using **Util~DXF**, the vertices of the polyline were extracted to file **FFace.XY** (note that the header generated automatically by **Util~DXF** had to be deleted).

The polyline red from ASCII file **FFace.XY** is attached to the flat-faced follower of an inverted cam mechanism like the one in Figure 7.14 (see line #54 of program **P7_25. PAS**) or the mechanism in Figure 7.18 in case of program **P7_26.PAS**. The actual follower envelope extraction was done by calling procedure **EnvelOfPlynes** from unit **LibCams** (see lines #56 and #60 of program **P7_25.PAS**). This procedure has the following heading:

```
procedure EnvelOfPlynes(Color, n:Integer; xO,yO, R00, xA,yA,
xB,yB:double;  FxyName:PathStr);
```

It reads vertex file **FxyName** and aligns the respective polyline with a reference frame centered at (**xA,yA**), having its positive *x*-axis oriented in the direction of point (**xB,yB**). Internally, **EnvelOfPlynes** calls procedure **EnvelOfLines** for each line segment that forms this polyline and trims the outer ends of the same family of **n** polar lines originating from point (**xO,yO**), having their initial lengths equal to **R00**.

The polar line trimming by procedure **EnvelOfLines** is repeated for every position of the follower in a motion inversion. At the end, the cam profile is extracted to a temporary locus file of extension **$2D** by calling procedure **EndEnvelopes**. The cam profile is drawn to a separate layer named "Cam25" (see line #**70**) or to the last numerical layer of file **F7_25.DXF** (see line #**70**). At the end of the program, procedure **CloseMecGraph** is called with either a TRUE or FALSE argument, depending on the value of variable **Anim** (see line #**75** of program **P7_25.PAS**). In the former case, the temporary file with the cam profile points will be preserved, by changing its extension to **D2D**. This will be the output cam profile **Cam25.D2D**.

Results obtained using simulation programs **P7_25.PAS** and **P7_26.PAS** are available in Figures 7.25 and 7.26 and the animated **GIF** files that accompany these figures. Note that animations of the follower in an inverted motion showing the radial lines (similar to Figure 7.23) are also available for Figures 7.24 through 7.26. In the absence of specific kinematic analysis programs, you can verify the curvature of the cam profiles intended to operate with curvilinear follower using program **P7_15.PAS** or **P7_19.PAS** described earlier. The pressure angle and follower motion information output by these programs should not be substituted to the case where the respective cams operate the intended arc-faced or curvilinear-faced followers. The follower motion could be however relatively easily verified

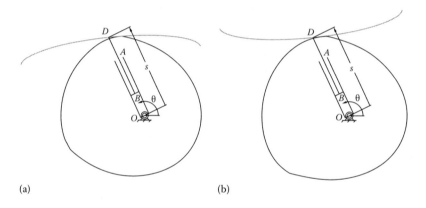

(a)   (b)

FIGURE 7.25   Motion inversion of disk cams with curvilinear translating followers in concave (a) and convex (b) arrangements. See also animation files **F7_25a.GIF** to **F7_25d.GIF**.

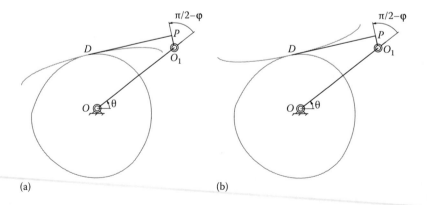

(a)   (b)

FIGURE 7.26   Motion inversion of disk cams with curvilinear oscillating followers in concave (a) and convex (b) arrangements. See also animation files **F7_26a.GIF** to **F7_26d.GIF**.

by simulating the respective mechanisms using **Working Model 2D** software, which is capable of evaluating the contact between planar bodies.

***

The problem of designing the profile of the most common disk cam-follower mechanisms through motion inversion has been discussed in this chapter. Iterative kinematic analysis of the same mechanisms and of pressure angle determination was also discussed. The follower motion considered throughout the chapter was synthesized using **AutoCAD** software, a promising alternative to the generation of the desired follower motion analytically using standardized functions.

REFERENCES AND FURTHER READINGS

For cam design theory, see

Angeles, J. and López-Cajún, C. S. (1991). *Optimization of Cam Mechanisms*. Berlin, Germany: Springer.

Chen, F. Y. (1982). *Mechanics and Design of Cam Mechanisms*. Amsterdam, the Netherlands: Pergamon Press/Elsevier.

Norton, R. L. (2002). *Cam Design and Manufacturing Handbook*. New York: Industrial Press.

Rothbart, H. (2003). *Cam Design Handbook: Dynamics and Accuracy*. New York: McGraw-Hill.

Waldron, K. J. and Kinzel, G. L. (2003). *Kinematics, Dynamics, and Design of Machinery*. Hoboken, NJ: John Wiley & Sons.

For curvature of planar curves, see

Weisstein, E. W. (2013). Curvature. From *MathWorld—A Wolfram Web Resource*. http://mathworld. wolfram.com/Curvature.html.

CHAPTER 8

Spur Gear Simulation Using **Working Model 2D** and **AutoLISP**

T HERE IS A GOOD AMOUNT OF SIMILARITY between disk cam mechanisms with oscillating followers and gear pairs, where the tooth of the pinion acts as a cam, while the active tooth of the driven gear is the follower. The first obvious difference between cam mechanisms and gears is that during the meshing process, constantly new teeth (the equivalent of cam–follower pairs) make contact, while others separate. The other difference is that the angular velocity of the gear over that of the pinion must remain constant, although noncircular gears can be designed, where this velocity ratio is some given function of the pinion angle. In this chapter, several **Working Model 2D** and **AutoLISP** applications will be described, which can be used to demonstrate how involute gears operate and how their profiles are generated. **Working Model 2D**, or **WM 2D** in short, available from Design Simulation Technologies (www.design-simulation.com), is a planar multibody software capable of performing kinematic and dynamic simulation of interconnected bodies subject to constraints. **WM 2D** allows for **DXF** import/export and has scripting capabilities through formula and **WM Basic** language systems.

8.1 INVOLUTE-GEAR THEORY

Involute gears are the most widely used in practice, being preferred to cycloidal and circular profile gears owing to the following desirable properties:

- The transmission ratio between two involute gears is not sensitive to center distance modification.

- The same cutting tool (rack or hob cutter) can be used to manufacture gears of any number of teeth (the proportions of their teeth, described through the module or diametral pitch, will be the same however).

- The rack or hob cutting tools used to fabricate involute gears can be conveniently mass produced because their cutting edges are straight and therefore easy to sharpen.

As the name suggests, an involute gear has the active flanks of its teeth shaped as involute curves of a common circle called base circle. Geometrically, the involute curve can be generated by attaching a taut, inextensible string to the base circle, and recording the locus of its free end as it is unwound off this circle. The concept is illustrated in Figure 8.1, where r_b is the radius of the base circle, BC is the string, and the involute curve is represented in thick line. Note that the involute curve can only exist outside the base circle.

Because the string is inextensible, the length of the circular arch AB subtended by angle t is equal to the length BC of the sting according to equation

$$BC = \rho = r_b \cdot \tan(\varphi) = \text{arc}(AB) = r_b \cdot t = r_b \cdot (\beta + \varphi) \tag{8.1}$$

which yields

$$\beta = \tan(\varphi) - \varphi = \text{inv}(\varphi) \tag{8.2}$$

Length BC is also the radius of curvature of the involute around point C, while the corresponding center of curvature is located at point B.

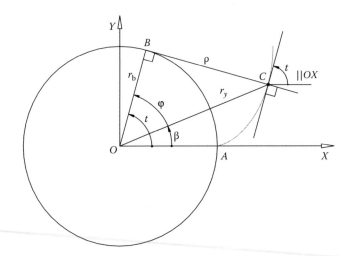

FIGURE 8.1 The involute of a circle of radius r_b generated using program **F8_01.PAS**. Additional editing has been done inside **AutoCAD**.

In order to derive the scalar equation of the involute, we project vector equation **OC** = **OB**−**CB** on the axes of the *OXY* reference frame to obtain

$$\begin{cases} x_C = r_b \cdot \cos(t) - r_b \cdot t \cdot \cos\left(t + \dfrac{\pi}{2}\right) \\ y_C = r_b \cdot \sin(t) - r_b \cdot t \cdot \sin\left(t + \dfrac{\pi}{2}\right) \end{cases}$$

(8.3)

which after rearranging terms become

$$\begin{cases} x_C = r_b \left[\cos(t) + t \cdot \sin(t)\right] \\ y_C = r_b \left[\sin(t) - t \cdot \cos(t)\right] \end{cases}$$

(8.4)

Equations 8.4 have been implemented in the program **F8_01.PAS** listed in Appendix B. The program calls procedures **InitDXFfile**, **ExpectDXFplines**, **AddVertexPline**, **DXFplineEnd**, **CloseDXFfile**, and **Fcircle** from unit **LibDXF** and was used to generate Figure 8.1. Note that unlike earlier programs discussed in this book, **F8_01.PAS** writes directly to the **R12 DXF** file, without plotting the image on the computer screen.

The equations of the involute can be also expressed using the polar angle β (see Figure 8.1):

$$\beta = t - \varphi = t - \arctan\left(\frac{BC}{OB}\right) = t - \arctan\left(\frac{r_b \cdot t}{r_b}\right) = t - \arctan(t)$$

(8.5)

and polar radius r_y:

$$r_y = \sqrt{OB^2 + BC^2} = r_b \sqrt{1 + t^2}$$

(8.6)

which yield the following alternative set of scalar equations

$$\begin{cases} x_C = r_b \sqrt{1 + t^2} \cdot \cos(t - \arctan(t)) \\ y_C = r_b \sqrt{1 + t^2} \cdot \sin(t - \arctan(t)) \end{cases}$$

(8.7)

Figure 8.2 shows the main parameters of external and internal involute gears. The size of their teeth is standardized through the diametral pitch *P*, which is the number of teeth of the gear per inch of its pitch diameter, that is,

$$P = \frac{0.5N}{r}$$

(8.8)

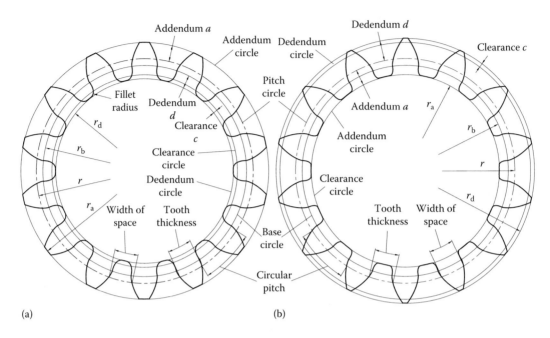

(a)   (b)

FIGURE 8.2   External (a) and internal (b) involute-gear terminology and notations.

For metric gears, the equivalent standardized parameter is the module m, defined as

$$m = \frac{2r}{N} \tag{8.9}$$

The circular pitch p is defined as the distance between teeth measured along the pitch circle and can be calculated with any of the following equations:

$$p = \left(\frac{2\pi r}{N}\right); \quad p = \left(\frac{\pi}{P}\right) \quad \text{or} \quad p = \pi m \tag{8.10}$$

Note that on a pitch circle the tooth thickness and width of space are both equal to $p/2$.

Additional important parameters used to specify teeth proportions are the addendum a and dedendum d. These are measured radially from the pitch circle to the addendum and to the dedendum circles, respectively. Both a and d as well as the full-depth $a + d$ and clearance c are defined in terms of diametral pitch P or module m as listed in Table 8.1. The clearance c is the amount by which the dedendum of the gear exceeds the addendum of the pinion and vice versa, when no backlash between their teeth is allowed.

TABLE 8.1   Standard Proportions of Involute-Gear Teeth

Teeth Proportions	Addendum, a	Dedendum, d	Whole depth, $a + b$	Clearance, $c = b - a$
Full depth ($\phi = 14.5°$)	$1/P$	$1.157/P$	$2.157/P$	$0.157/P$
Stub ($\phi = 20°$)	$0.8/P$	$1/P$	$1.8/P$	$0.2/P$
Full depth ($\phi = 20°$ or $\alpha = 20°$)	$1/P$ or $1m$	$1.25/P$ or $1.25m$	$2.25/P$ or $2.25m$	$0.25/P$ or $0.25m$
Full depth ($\phi = 25°$)	$1/P$	$1.25/P$	$2.25/P$	$0.25/P$

Although in theory $P = 1/m$, SI system (i.e., metric gears) and US customary system gears are not interchangeable. Also note that neither m nor P can be measured directly on the gear. There are indirect ways to estimate what module m or diametral pitch P a gear is however. One method is to try to mesh the unknown gear with gears of the known module or diametral pitch. The other is to measure the whole depth of the unknown gear, and assuming, for example, that it is a full-depth tooth, divide this amount by 2.25 to obtain m or $1/P$.

8.2 INVOLUTE PROFILE MESH

Figure 8.3 shows that a pair of external involute gears of teeth numbers N_1 and N_2 is equivalent to a crossbelt transmission with pulleys of radii r_{b1} and r_{b2} (the base radii of the two gears). Similarly, Figure 8.4 shows that one external and one internal gear pair is equivalent to a regular belt transmission. Both are also equivalent to two friction wheels of radii r_1 and r_2 (pitch radii) or r_{w1} and r_{w2} (rolling radii in case the center distance is modified)

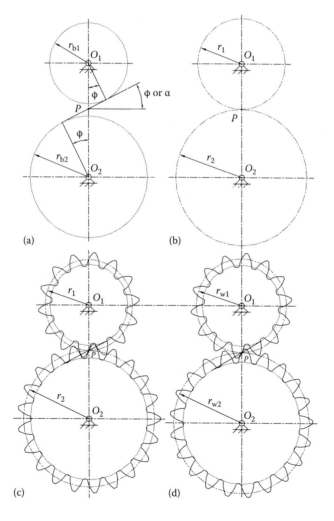

FIGURE 8.3 Equivalence between a crossbelt transmission (a), a pair of friction wheels (b), and a pair of external gears without (c) and with (d) backlash.

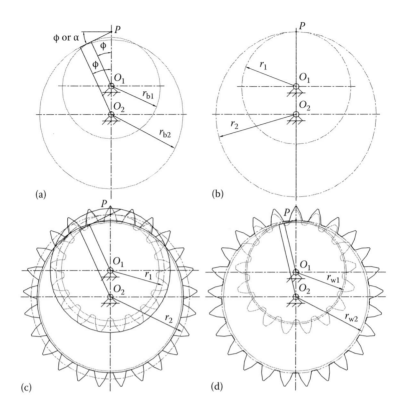

FIGURE 8.4 Equivalence between a belt transmission (a), a pair of friction wheels (b), and a pair of external–internal gears, without (c) and with (d) backlash.

of the two gears. By simultaneously recording the locus of a point on the belt relative to a plane that rotates together with gear one, and also relative to a second plane that rotates together with gear two, the involute curves forming the flanks of the meshing teeth of the two gears are obtained.

The belt transmission equivalence explains why the transmission ratio remaining constant as the two involutes profiles mesh, thus satisfying the *fundamental law of tooth gearing*, which states that "the common normal to the two involutes at the point of contact—which is the common tangent to the two base circles—will always intersects the line of centers O_1O_2 at the pitch point P" (Uicker et al. 2003).

The angle formed by the belt perpendicular to the line of centers O_1O_2 is the pressure angle ϕ between the teeth of the two gears when their point of contact coincides with the pitch point P. Note that in metric gear terminology, the pressure angle is noted α. If the center distances of the (cross)belt transmission and of the gear pair with zero backlash are modified from a standard center distance D to an operating center distance D_w, the pitch radii r_1 and r_2 will remain the same, but the pressure angle will change its value according to equation:

$$\phi_w = \cos^{-1}\left(\frac{r_{b2} \pm r_{b1}}{D_w}\right) = \cos^{-1}\left(\frac{D}{D_w}\cos\phi\right) \qquad (8.11)$$

Important: In Equation 8.11 and throughout this chapter, the upper sign will correspond to external gears and the lower sign to internal gears.

To maintain the same input–output speed ratio ω_1/ω_2, the equivalent friction wheel transmission with modified center distance D_w will have to be equipped with new wheels of radii r_{w1} and r_{w2} (these are the *rolling radii* of the gear pair) calculated using the following equations:

$$r_{w1} = \frac{r_{b1}}{\cos\phi_w} = D_w \frac{N_1}{N_2 \pm N_1} \tag{8.12a}$$

$$r_{w2} = \frac{r_{b2}}{\cos\phi_w} = D_w \frac{N_2}{N_2 \pm N_1} \tag{8.12b}$$

Overall, the following equalities should hold between the angular velocity of the input and output gears, their number of teeth N_1 and N_2, and the radii of their base circles, pitch circles, and rolling circles:

$$\frac{\omega_1}{\omega_2} = \frac{N_2}{N_1} = \frac{r_{b2}}{r_{b1}} = \frac{r_2}{r_1} = \frac{r_{2w}}{r_{1w}} \tag{8.13}$$

The meshing of involute profiles and the insensitivity of the transmission ratio of involute gears to center distance modification is illustrated by **WM 2D** simulations named **InvPairExt.WM2** and **InvPairInt.WM2** provided with the book. Using the program **P8_01.PAS** mentioned earlier, two polygonal bodies representing the two involute curves connected to their base circles of radii r_{b1} = 3 m and r_{b2} = 4 m have been created inside **AutoCAD** (see Figure 8.5) and then exported to **WM 2D** via the **DXF** format. A slider control allows the user to adjust the distance between the centers of these two base circles (i.e., distance O_1O_2) within the limits 9–10.75 for the external involutes and 1.1 and 1.5 for the internal–external involutes. A pair of rotary motors drives separately the two involutes and are imposed correlated oscillatory motions of 0.4 and 0.3 radians amplitude according to equations

$$\theta_1 = -0.4\sin(t)$$

$$\theta_2 = \mp\theta_1 \frac{r_{b1}}{r_{b2}} \tag{8.14}$$

The initial positions of the two involutes, that is, at time $t = 0$, are such that the contact point C is collinear with gear centers O_1 and O_2. Irrespective of the value of center distance D_w, the following equalities hold:

$$PO_1 = \frac{(D_w \cdot r_{b1})}{(r_{b1} \mp r_{b1})}$$

$$PO_2 = \frac{(D_w \cdot r_{b2})}{(r_{b2} \mp r_{b2})} \tag{8.15}$$

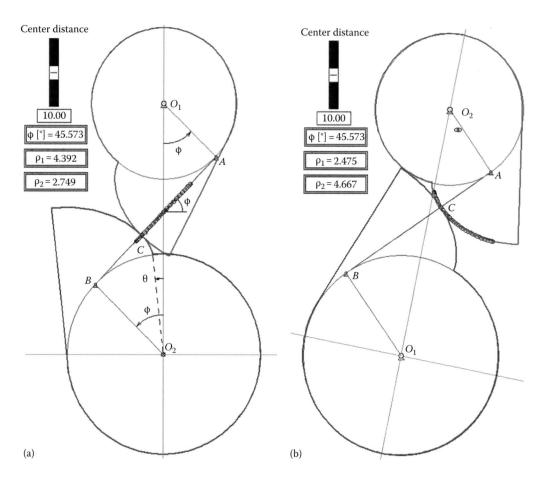

FIGURE 8.5 Screenshots of **WM 2D** simulations of two mutually enveloping external involute curves recorded relative to the ground (a) and relative to a reference frame attached to involute number two (b). See also movie files **F8_5A.MP4** and **F8_5B.MP4**.

As the two involute profiles mesh, their contact point C moves along the line of action AB, which is the common normal to the two profiles and the common tangent to the base circles of the two gears. In any position, the coordinates of point C can be determined by projecting vector equation **BC** + **OB** = **OC** on the axes of the reference frame (see Figure 8.5a). For the involute curve rotated by angle θ, we get

$$x_C = \rho_1 \cos(\phi) - r_{b1} \sin(\phi) = -\rho_2 \cos(\phi) + r_{b2} \sin(\phi)$$

$$y_C = \rho_1 \sin(\phi) + r_{b1} \cos(\phi) = D_w - \rho_2 \sin(\phi) - r_{b2} \cos(\phi)$$

(8.16)

where the radii of curvature ρ_1 and ρ_1 of the two involutes around point C are given by equations

$$\rho_1 = r_b \left(\phi - \theta_1 \right) \quad \text{and} \quad \rho_2 = r_b \left(\phi - \theta_2 \right)$$

(8.17)

If in any of these two **WM 2D** simulations the observer's reference frame is moved to one of the involutes, the locus of the contact point *C* traces the respective involute (Figures 8.5b and 8.6b).

According to the *Aronhold–Kennedy theorem of the three instant centers* (Uiker et al. 2003), a pure rolling between the two involute profiles occurs only when contact point *C* coincides with the pitch point *P*. Moreover, the farthest away from point *P* the contact between the two teeth takes place, the higher the amount of relative sliding between the two involutes is. In case of an actual gear transmission, the sliding between teeth causes power losses, which will be higher for gears with smaller diametral pitch *P* or bigger module *m*.

In simulations **InvPairExt.WM2** and **InvPairInt.WM2**, in order to maintain contact between the two involutes as the operating center distance O_1O_2 is modified, the initial angles of the two oscillating polygons has been programmed using **WM 2D** formula language such that each changes the amount

$$\Delta\theta = \sqrt{\frac{D_w^2}{\left(r_{b2} \mp r_{b1}\right)^2} - 1} - \arctan\left(\sqrt{\frac{D_w^2}{\left(r_{b2} \mp r_{b1}\right)^2} - 1}\right) \tag{8.18}$$

This equation has been obtained by eliminating parameter *t* between Equations 8.5 and 8.6 with r_y set equal to either OP_1 or OP_2 and then applying Equations 8.15.

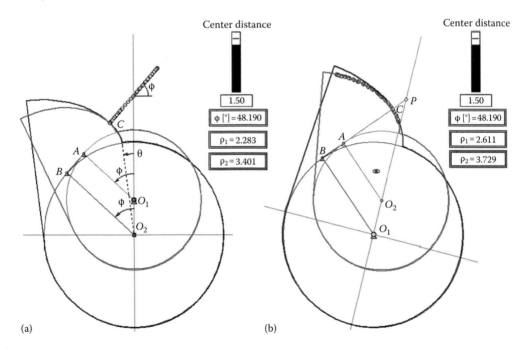

FIGURE 8.6 Screenshots of **WM 2D** simulations of two mutually enveloping involute curves (one external and one internal) recorded relative to the ground (a) and a reference frame attached to involute number two (b). See also movie files **F8_6A.MP4** and **F8_6B.MP4**.

The pressure angle displayed with these simulations (see Figures 8.5 and 8.6) is the angle between the velocity vector of the contact point and the normal force to the two profiles, when the contact point coincides with pitch point P, that is,

$$\phi = \arccos\left(\frac{(r_{b2} \mp r_{b1})}{D_w}\right) \tag{8.19}$$

Also displayed with these simulations are the radii of curvatures ρ_1 and ρ_2 of the two involutes around contact point C, calculated with Equations 8.17.

8.3 INVOLUTE-GEAR MESH

In order to demonstrate additional properties that involute gears have, simulations **GearPairExt.WM2** and **GearPairInt.WM2** have been produced and are available with the book—see Figures 8.7 and 8.8 and movie file **F8_7.MP4** and **F8_8.MP4**. These simulations consist of two *standard gears* with adjustable center distance. Standard gears are zero profile shift gears, that is, their addendum modification coefficient x equals zero; see Section 8.4 for details. The first of these simulations depicts two external gears with $N_1 = 15$ and $N_2 = 17$ teeth; the other simulation consists of one external and one internal gear with $N_1 = 17$ and $N_2 = 25$ teeth.

Since each gear has a number of identical involute curves equally spaced around the base circle, the concepts introduced earlier with reference to Figures 8.5 remain valid for any two gears in mesh. Therefore, as the center distance is modified, the pressure angle changes as well, while the transmission ratio remains the same. To maintain contact between the teeth of the two gears in these two simulations, as their center distance is modified, the initial angle of the two gears is adjusted an amount calculated using Equation 8.18.

The pressure angle between the two involute profiles varies as the contact point between the two gears moves along the line of action. The magnitude of the pressure angle also changes as the center distance of the two gears is increased or decreased and is also function of the direction in which the torque is transmitted—either from gear 1 to gear 2 or vice versa. The only position in which the pressure angle is not dependent of which gear is the driving gear is the one where the contact point and the pitch point P coincide. This is the same *pressure angle* of the gear pair discussed earlier with reference to Figures 8.3 and 8.4 and is noted α in the SI system and ϕ in the US customary system.

The minimum center distance of an external gear pair is limited by their meshing teeth making double contact; this is known as the *zero-backlash* gear pair arrangement (Figure 8.7). Conversely, in case of internal–external gear pairs, there is a maximum center distance limited by their teeth making double contact (Figure 8.8). The center distance corresponding to a zero-backlash arrangement of standard gears is called *standard center distance* and is calculated with equation

$$D = 0.5\left(N_2 - N_1\right) \cdot m \tag{8.20a}$$

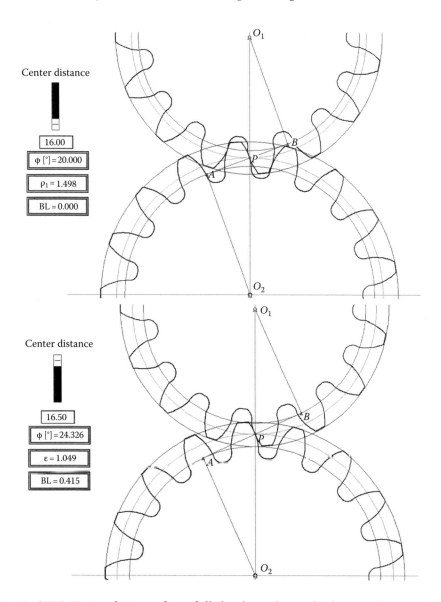

FIGURE 8.7 **WM 2D** simulations of two full-depth tooth standard external gear pairs shown in their reference center distance configuration (top) and in a configuration where the center distance is increased (bottom). See also simulation files **GearPairExt.WM2** and movie file **F8_7.MP4**.

or for US customary gears

$$D = \frac{0.5(N_2 - N_1)}{P} \tag{8.20b}$$

One property of *standard gear* pairs is that when the operating center distance D_w equals the standard center distance D, the pressure angle between the two gears equals half the

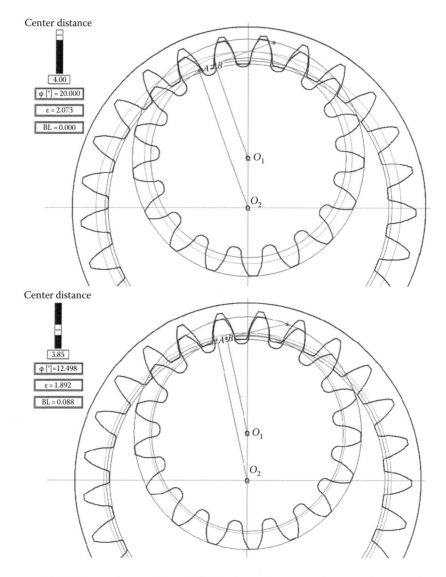

FIGURE 8.8 **WM 2D** simulation of two full-depth tooth standard external–internal gear pairs, shown in reference center distance configuration (top) and in a configuration where center distance $D_w < D$, backlash is nonzero, and contact ratio is diminished (bottom). See also files **GearPairInt. WM2** and **F8_8.MP4**.

angle between the flanks of the teeth of the *basic rack*. The basic rack of a gear is obtained by hypothetically making the number of teeth of the respective gear equal infinitely.

In addition to the pressure angle at the pitch point calculated with Equation 8.19, two more parameters are displayed with the **WM 2D** simulations in Figures 8.7 and 8.8. One is the contact ratio ε of the two gears, defined as the average number of teeth in contact and calculated with

$$\varepsilon = \frac{\sqrt{r_{a1}^2 - r_{b1}^2} \pm \sqrt{r_{a2}^2 - r_{b2}^2} \mp D_w \sin \phi_w}{p \cos \phi} \tag{8.21}$$

where

$$r_{a1} = \frac{\left(0.5N_1 + x_1 + a\right)}{P}$$

$$r_{a2} = \frac{\left(0.5N_2 + x_2 + a\right)}{P}$$

$$r_{b1} = \frac{\left(0.5N_1 \cos\alpha\right)}{P}$$

$$r_{b2} = \frac{\left(0.5N_2 \cos\alpha\right)}{P}$$

(8.22a)

or for metric gears

$$r_{a1} = m\left(0.5N_1 + x_1 + a\right)$$

$$r_{a2} = m\left(0.5N_2 + x_2 + a\right)$$

$$r_{b1} = 0.5mN_1 \cos\alpha$$

$$r_{b2} = 0.5mN_2 \cos\alpha$$

(8.22b)

The third parameter displayed with these simulations is the backlash *BL*, defined as the width of the gap between two meshing teeth measured along their rolling circles. The backlash will be equal to the circular pitch measured on the rolling circles p_w of any of the two gears, minus the teeth thickness of gear one s_{1w} and of gear two s_{2w} measured along their respective rolling circles. The mentioned parameters p_w, s_{1w}, and s_{2w} can be calculated using the following equations:

$$p_w = \frac{\pi}{P} \cdot \frac{\cos\phi}{\cos\phi_w}$$

$$s_{1w} = \frac{1}{P} \cdot \left(\frac{\pi}{2} + 2x_1 \tan\phi\right) \cdot \frac{\cos\phi}{\cos\phi_w} - 2r_{1w}\left(\text{inv}\phi_w - \text{inv}\phi\right)$$

$$s_{2w} = \frac{1}{P} \cdot \left(\frac{\pi}{2} \pm 2x_2 \tan\phi\right) \cdot \frac{\cos\phi}{\cos\phi_w} \mp 2r_{2w}\left(\text{inv}\phi_w - \text{inv}\phi\right)$$

(8.23a)

or

$$p_w = \pi \cdot m \frac{\cos\alpha}{\cos\alpha_w}$$

$$s_{1w} = m \cdot \left(\frac{\pi}{2} + 2x_1 \tan\alpha\right) \cdot \frac{\cos\alpha}{\cos\alpha_w} - 2r_{1w}\left(\text{inv}\alpha_w - \text{inv}\alpha\right)$$

$$s_{2w} = m \cdot \left(\frac{\pi}{2} \pm 2x_2 \tan\alpha\right) \cdot \frac{\cos\alpha}{\cos\alpha_w} \mp 2r_{2w}\left(\text{inv}\alpha_w - \text{inv}\alpha\right)$$

(8.23b)

With these parameters known, the backlash between the teeth in mesh of the two gears when their center distance is modified from D to D_w becomes

$$BL = -2m \cdot \left(x_1 \pm x_2\right) \frac{\sin\phi}{\cos\phi_w} - 2\left(r_{1w} \pm r_{2w}\right)\left(\mathrm{inv}\phi_w - \mathrm{inv}\phi\right) \tag{8.24a}$$

or for US customary gears

$$BL = -2\frac{1}{P} \cdot \left(x_1 \pm x_2\right) \frac{\sin\alpha}{\cos\alpha_w} - 2\left(r_{1w} \pm r_{2w}\right)\left(\mathrm{inv}\alpha_w - \mathrm{inv}\alpha\right) \tag{8.24b}$$

Simulation **GearPairExt.WM2** and **GearPairInt.WM2** reveal that as the center distance is modified from its standard value, the contact ratio is reduced because of the shortening of the length of action, while the backlash between the two gears will increase. In practice, a small amount of backlash between gears is essential in order to allow for thermal expansion and for the slight deflection of the teeth as they mesh under load. Too much backlash is undesirable however because it reduces the contact ratio of the two gears, and as a consequence teeth are loaded more when they first make contact. Also, if the direction of rotation of the gears is reversed, impact loads or unacceptable position inaccuracies can occur. If the center distance is imposed a value other than the standard center distance D, the ensuing backlash or interference can be reduced or eliminated by employing profile shifted gears discussed in the next section.

8.4 **WORKING MODEL 2D** SIMULATIONS OF INVOLUTE PROFILE GENERATION

There are several ways of manufacturing involute gears. Of these, the shaping process using a pinion cutter and a rack cutter will be illustrated in the remainder of this chapter. **WM 2D** simulations of these two type of gear generation processes have been produced and are available with the book, that is, **GearGen0.WM2** to **GearGen4.WM2**.

In both processes, the cutter is first fed into the gear blank until the reference line of the rack or the pitch circle of the pinion cutter becomes tangent to the pitch circle of the future gear. After that, with each cutting stroke, the reference line of the rack cutter or the pitch circle of the pinion cutter will slowly roll without slip on the pitch circle of the gear blank. The process ends when the last tooth of the gear is fully formed. It is called *reference line* of the generating rack, the line along which the tooth thickness and width of space of the rack are equal. A similar line can be defined for the *basic rack*, which is the rack obtained by making the number of teeth of the gear equal to infinity.

If the aforementioned rolling without slip takes place between the reference line of the rack cutter (or the pitch circle of the pinion cutter) and the pitch circle of the blank, a *zero profile shift* gear is generated. The teeth of such a gear are said to have *no correction* (see Figure 8.9a). If this rolling without slip occurs between the pitch circle of the blank and a different line of the rack cutter, or between different circles of the blank and pinion cutter, a *modified* or *profile shifted* gear is obtained instead. Specifically, when the cutter is displaced radially outwards from the zero shift position, a gear with *positive profile shift* ($X > 0$) as shown in Figure 8.9b is generated. Conversely, a radially inward displacement of

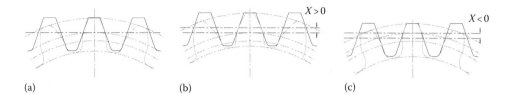

FIGURE 8.9   Standard gear with $N = 18$ teeth and zero profile shift (a), with positive addendum modification, that is, rack is retracted during the generation process (b), and with negative addendum modification, that is, rack is approached during the generation process (c).

the rack or pinion cutter results in a gear that has a *negative profile shift* ($X < 0$) as seen in Figure 8.9c. The ratio between tool displacement X and its module m or the inverse of its diametral pitch P is called *profile shift coefficient* and is symbolized x. Note that the base circle of the future gear as well as its pitch circle remains the same, irrespective of the magnitude of the profile shift coefficient x.

The involute-gear generation methods using a pinion cutter and a rack cutter were implemented in **WM 2D** simulations **GearGen0.WM2** to **GearGen4.WM2** available with the book. **GearGen0.WM2** can be used to simulate the generation of an entire involute gear, either with external or internal teeth (see Figures 8.10 and 8.11). Because the pinion cutter in

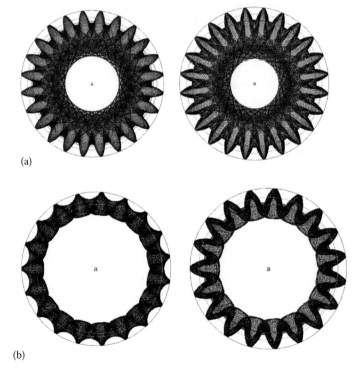

FIGURE 8.10   Full-depth $\phi = 20°$ internal gears generated using **GearGen0.WM2** with $N = 24$ teeth (top) and $N = 19$ teeth (bottom) and with zero profile shift (a) and $x = 0.5$ profile shift (b). See also movie file **F8_10.MP4**.

(a)   (b)

FIGURE 8.11   Full-depth $\phi = 20°$ internal gears generated using **GearGen0.WM2** with $N = 18$ and $N = 11$ teeth having zero profile shift (a) and $x = 0.5$ profile shift (b). See also movie file **F8_10.MP4**.

this simulation has 18 teeth, internal standard gears with at least 24 teeth can be generated without undercut. If a profile shift is applied, then gears with down to 19 teeth can be generated. Figure 8.10 shows four internal gears generated with **GearGen0.WM2**, two having $N = 24$ teeth and the other two having $N = 19$ teeth. Both the zero profile shift gears and addendum modified gears by $x = 0.5$ are shown in this figure. As anticipated, the zero profile shift gear with $N = 19$ teeth appear severally undercut. The companion internal gear with $N = 19$ and positive profile shift has its teeth better formed, but they result shortened because of the interference with the tip of the generating pinion. The figure also shows that a positive profile shift results in an internal gear that has an increase width of space.

Figure 8.11 are four full-depth $\phi = 20°$ external gears with $N = 18$ and $N = 11$ teeth. The zero profile shift gear with $N = 18$ exhibits no undercut, but once the number of teeth is reduced below 18, undercut starts to occur. This is clearly visible on the $N = 11$ teeth gear that appears severely undercut. The companion $x = 0.5$ profile shifted gears illustrate the effect of addendum modification upon tooth shape and undercut occurrence. Notice how for positive profile shifted gears, the tooth becomes pointed while its root thickens while for $x < 0$, the effect is opposite and can result in undercut teeth. In practice, it is recommended that the tooth thickness on the addendum circle be no less than 0.3 times m or $1/P$. Stub teeth can be employed when there is no other way of avoiding the teeth from becoming pointed.

WM 2D applications **GearGen1.WM2–GearGen4.WM2** illustrate how one complete tooth of an external involute gear of module $m = 1$ mm with number of teeth N and addendum modification coefficient x can be generated using a rack cutter. **GearGen1.WM2** simulates $\alpha = 20°$ full-depth tooth involute profiles, **GearGen2** simulates $\alpha = 20°$ stub-tooth involute profiles, **GearGen3.WM2** simulates $\alpha = 25°$ full-depth tooth involute profiles, and **GearGen4.WM2** simulates $\alpha = 14.5°$ full-depth tooth involute profiles.

Sample tooth profiles generated with these four **WM 2D** simulations are available for comparison in Figure 8.12. Unfortunately, once these simulations have been performed, there is no convenient way to export the cutter envelopes to **AutoCAD** (same applies for the simulations done using **GearGen0.WM2**). This is because **WM 2D** can export to **DXF** only one animation frame at a time. In addition, the entities whose visibility has been intentionally turned *off* will also be exported to **DXF**, making the task of extracting cutter envelopes even more tedious. To overcome these drawbacks and allow the user to generate accurate involute-gear profiles, the **AutoLISP** application **Gears.LSP** has been written and is available with the book.

8.5 INVOLUTE PROFILE GENERATION USING **Gears.LSP**

The **AutoLISP** application **Gears.LSP** allows one to generate as polylines, internal or external gears with any number of teeth and any addendum modification coefficient x. For the convenience of input data management, the module m (or diametral pitch P) of the gear will be equal to one. Any desired module or diametral pitch can be easily obtained at the end through scaling. Note that the gear will result centered at origin and must be produced one at a time always starting in a new drawing. To launch the program, issue the *appload* command, load **Gears.LSP** from its directory, and then type at the command line either "external" or "internal," depending on the gear profile you want to generate. You will then be asked to input the number of teeth N and profile shift coefficient x and will be prompted by the program to confirm the advancement through the involute profile generation steps shown in Figure 8.13.

These steps are as follows (see Figure 8.13): (i) Draw the addendum, dedendum, base, and pitch circles of the future gear and half of the generating rack and its reference line. (ii) Copy the generating rack in a number of positions to form the envelopes of the left flank of the top gear, similar to **WM 2D** simulation **GearGen1.WM2**. (iii and iv) Extract the tooth flank using an array of parallel lines that are trimmed from the right. (v) Mirror the tooth flank to the right and draw the top land of the tooth. (vi) Generate the entire gear profile as a polar array of the single tooth produced earlier, and connect these teeth into a polyline.

If of interest, you can use the left side of the rack cutter and the reference line from **Gears.LSP** to manually produce an entire rack (see Figures 8.13 and 8.14). First, mirror

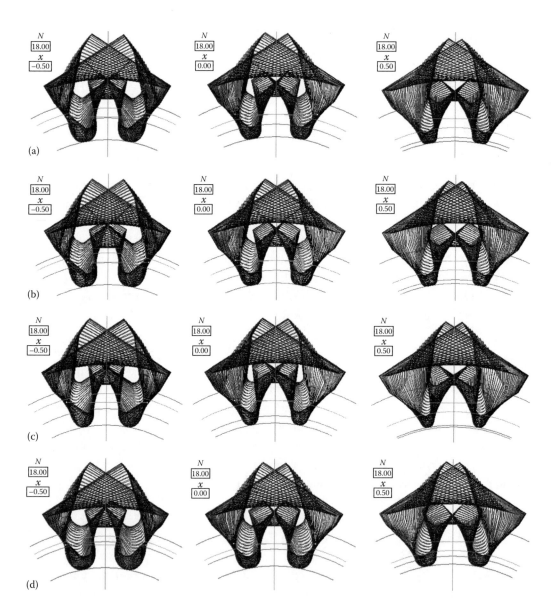

FIGURE 8.12 Profile shift effects upon an external gear with 18 teeth and $\alpha = 20°$ full-depth tooth (a), $\alpha = 20°$ stub tooth (b), $\alpha = 25°$ full-depth tooth (c), and $\alpha = 14.5°$ full-depth tooth (d). From left to right, the addendum modification coefficient equals to $x = -0.5$, $x = 0$ and $x = +0.5$. See also movie files **F8_12A.MP4** to **F8_12D.MP4**.

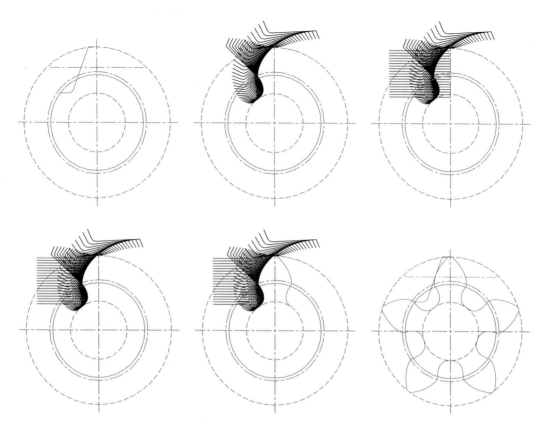

FIGURE 8.13 Steps in generating a standard gear with $N = 5$ teeth and $x = 0.2$ addendum modification using the application **Gears.LSP**.

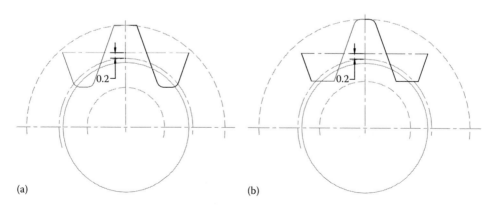

(a)   (b)

FIGURE 8.14 How to obtain a complete section of the generating rack after running the program **Gears.LSP** for external (a) and internal (b) gears. Also shown is the profile shift coefficient $x = 0.2$ that can be measured directly on the drawing.

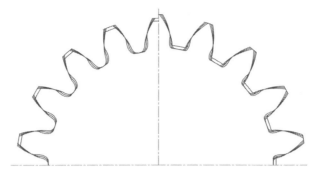

FIGURE 8.15   External gears (left) and internal gears (right) with 18 teeth and $x = 0$, full-depth tooth with $\alpha = 20°$, stub tooth with $\alpha = 20°$, full-depth tooth with $\alpha = 25°$, and full-depth tooth $\alpha = 14.5°$, separately generated using **Gears.LSP**. Note the different undercut and tooth land width.

the left side of the rack about the *OX* axis. Then connect the two outer ends with the ends of the reference line of the rack, and complete the necessary filleting. You can then multiply any number of times the tooth and space thus obtained.

It is also possible to generate using **Gears.LSP** stub gears, or gears having the angle of the generating rack other than 20° (see Figure 8.15). To do this, edit the addendum coefficient **aa** or angle **Phi** on the last lines of the file **Gears.LSP** prior to loading it into **AutoCAD**. Note in the following excerpt that it is also possible to modify the fillet radius **FilletR** of the generating rack or the number of enveloping positions **Ncuts** required to extract the first tooth.

```
;;;;;;;;;;;;;;;;;;;;;;;;;;;;;;;;;;;;;;;;;;;;;;;;;;;;;;;;;;;;;;;;;;;;;;;;;;;
   (setq Ncuts 100)      ;Number of rack cuts around one tooth
   (setq NRscans 30)     ;Number of scan lines to extract the involute

   (setq FilletR 0.25)   ;generating rack filled radius
   (setq Phi 20.0)       ;rack angle in degrees
   (setq aa  1.00)       ;addendum coefficient
   (setq dd  1.25)       ;dedendum coefficient
;;;;;;;;;;;;;;;;;;;;;;;;;;;;;;;;;;;;;;;;;;;;;;;;;;;;;;;;;;;;;;;;;;;;;;;;;;;
```

★★★

A review of the involute-gear theory has been presented, illustrated with **WM 2D** simulations. Additional simulations demonstrate how gear profiles can be generated using gear and rack cutters. **AutoLISP** application **Gears.LSP** available with the book allows one to generate as **AutoCAD** polylines accurate involute profiles—both external and internal. For convenience, a summary of involute-gear nomenclature and geometric equations used in this chapter is made available in Table 8.2. Most of the equations in this table have been entered in the spreadsheet **InvGearCalc.XLS** also available with the book.

TABLE 8.2   Summary of Involute-Gear Formulae

	Nomenclature	Notations	Comments (The Lower Sign Is for Internal Gears [N_2 Only] or External–Internal Pairs)
1	Pinion tooth number	N_1	$N_1 \geq 17$ (less than 17 possible with nonstandard gears)
2	Gear tooth number	N_2	$N_2 \geq N_1$
3	Generating rack angle	ϕ	20° standard (other values in use are 14.5°, 22.5°, and 25°)
4	Module for metric gears	m [mm]	1, 1.25, 1.5, 2, 2.5, 3, 4, 5, 6, 8, 10, 12, 16, 20, 25, 32, 40, 50 (first choice) or 1.125, 1.375, 1.75, 2.25, 2.75, 3.5, 4.5, 5.5, 7, 9, 11, 14, 18, 22, 28, 36, 45 (second choice)
5	Diametral pitch for US customary gears	P [in^{-1}]	1, 1¼, 1½, 1¾, 2, 2¼, 2½, 3, 4, 6, 8, 10, 12, 16 (coarse) or 20, 24, 32, 40, 48, 64, 80, 96, 120, 150, 200 (fine)
6	Transmission ratio	i_{12}	$i_{12} = N_2/N_1$
7	Pitch circle radii	r_1, r_2	$r_1 = 0.5 \cdot N_1 \cdot m$; $r_2 = 0.5 \cdot N_2 \cdot m$ or $r_1 = 0.5 \cdot N_1/P$; $r_2 = 0.5 \cdot N_2/P$
8	Base circle radii	r_{b1}, r_{b2}	$r_{b1} = r_1 \cdot \cos\phi$; $r_{b2} = r_2 \cdot \cos\phi$
9	Standard center distance	D	$D = r_2 \pm r_1$
10	Operating distance	D_w	$D_w \in [D...D \pm m]$ OR $D_w \in [D...D \pm 1/P]$
11	Pressure angle at pitch point	ϕ_w	$\phi_w = \cos^{-1}(D\cos\phi/D_w)$
12	Pressure angle on a circle of radius r_y	ϕ_y	$\phi_y = \cos^{-1}(r_b/r_y)$
13	Circular pitch	p	$p = \pi m$ or $p = \pi/P$
14	Circular pitch on a circle of radius r_y	p_y	$p_y = p(\cos\phi/\cos\phi_y)$
15	Profile shift coefficients	x_1, x_2	To avoid undercut: $x_{1,2} \geq (17 - N_1)/17$
16	Total profile shift coefficient	$x = x_1 \pm x_2$	$x = \dfrac{N_2 \pm N_1}{2\tan\phi}\cos\left(\mathrm{inv}\phi_w - \mathrm{inv}\phi\right)$
17	Profile shift distance	X_1, X_2	$X_1 = x_1 \cdot m$; $X_2 = x_2 \cdot m$ or $X_1 = x_1/P$; $X_2 = x_2/P$
18	Addendum	a	Full-depth tooth: $a = m$ or $1/P$; stub tooth, $0.8/P$;
19	Dedendum	d	Full-depth tooth: $d = 1.25m$ or $1.25/P$; stub tooth, $d = 1/P$;
20	Clearance	c	$c = d - a$
21	Dedendum radius	r_{d1}, r_{d2}	$r_{d1} = r_1 + X_1 - d$; $r_{d2} = r_2 + X_27d$
22	Addendum radius	r_{a1}, r_{a2}	$r_{a1} = r_1 + X_1 + a$; $r_{a2} = r_2 + X_2 \pm a$ Rack and pinion: $r_{a1} = r_1 + X_1 + a$; $r_{a2} = \infty$
23	Pitch circle radii of meshing gears	r_{w1}, r_{w2}	$r_{w1,2} = \dfrac{r_{b1,2}}{\cos\phi_w} = \dfrac{N_{1,2}}{N_2 \pm N_1}D_w$
24	Tooth thickness on the pitch circle	s_1, s_2	$s_{1,2} = 0.5 \cdot p72 \cdot X_{1,2} \cdot \tan\phi$
25	Tooth thickness on a circle of radius r_y	s_{y1}, s_{y2}	$s_{y1,2} = s_{1,2}\dfrac{\cos\phi}{\cos\phi_w} \mp r_{y1,2}\left(\mathrm{inv}\phi_y - \mathrm{inv}\phi\right)$
26	Contact ratio	$\varepsilon > 1$	$\varepsilon = \left(\sqrt{r_{a1}^2 - r_{b1}^2} \pm \sqrt{r_{a2}^2 - r_{b2}^2} \mp D_w\sin\phi_w\right)/\left(p\cos\phi\right)$ Rack and pinion: $\varepsilon = \left(\sqrt{r_{a1}^2 - r_{b1}^2} - r\sin\phi\right)/\left(p\cos\phi\right) + (1-x)/(0.5p\sin 2\phi)$
27	Backlash		$BL = -2m \cdot \left(x_1 \pm x_2\right)\dfrac{\sin\phi}{\cos\phi_w} - 2\left(r_{1w} \pm r_{2w}\right)\left(\mathrm{inv}\phi_w - \mathrm{inv}\phi\right)$

REFERENCES AND FURTHER READINGS

AGMA 913-A98. (1992). *Method for Specifying the Geometry of Spur and Helical Gears.* Alexandria, VA: The American Gear Manufacturers Association.

Dudley, D. (1994). *Handbook of Practical Gear Design.* Boca Raton, FL: CRC Press.

Litvin, F. L. and Fuentes, A. (2004). *Gear Geometry and Applied Theory.* New York: Cambridge University Press.

Uicker, J. J. Jr., Pennock, G. R., and Shigley, J. E. (2003). *Theory of Machines and Mechanisms.* New York: Oxford University Press.

Weisstein, E.W. (2013). Involute. From *MathWorld—A Wolfram Web Resource.* http://mathworld.wolfram.com/Involute.html (last accesses June 2014).

CHAPTER 9

More Practical Problems and Applications

I N THIS FINAL CHAPTER, a number of applications of the programs and procedures introduced earlier in this book are presented. The first three of these applications are from Dynamics and Vibrations solved using the **D_2D** program. The next examples are of curve fitting through minimization, graphical representation of single-valued functions of three or more variables and random number generation (including plotting them as histograms). Additional applications of the procedures in unit **LibAssur** are then presented (i.e., kinematic simulation of dwell and quick-return mechanisms and of mechanisms with repetitive topology and animation of the fixed and moving centrodes of a four-bar linkage), followed by two **Working Model 2D** simulations of planetary gears. Also presented is a program to purge unwanted files from current directory. After submitting the first draft of manuscript to the publisher, several more applications have been added to this chapter as follows: plotting implicit functions, direct and inverse kinematics of SCARA robots, rope shovel and excavator motion simulation, multilink suspension analysis and flywheel design of a punch press.

9.1 DUFFING OSCILLATOR

To illustrate the usefulness of arrow markers and the ability of **D_2D** program to generate comet plots and to handle large input data files, the case of the Duffing nonlinear oscillator is considered next:

$$\ddot{x} + \delta\dot{x} + \alpha x + \beta x^3 = \gamma\cos(\omega t) \tag{9.1}$$

Equation 9.1 describes a class of damped oscillators with a harmonic forcing term and viscous friction coupling. This equation has been extensively studied in the past (Kovacic and Brennan 2011) because of the interesting dynamic behavior it exhibits for certain combinations of parameters α, β, γ, δ, and ω. The plot in Figure 9.1, known as *phase path*, is a solution of Duffing equation with $\omega = 1.0$, $\alpha = -1.0$, $\beta = 1.25$, $\gamma = 0.3$, and $\delta = 0.15$ and initial

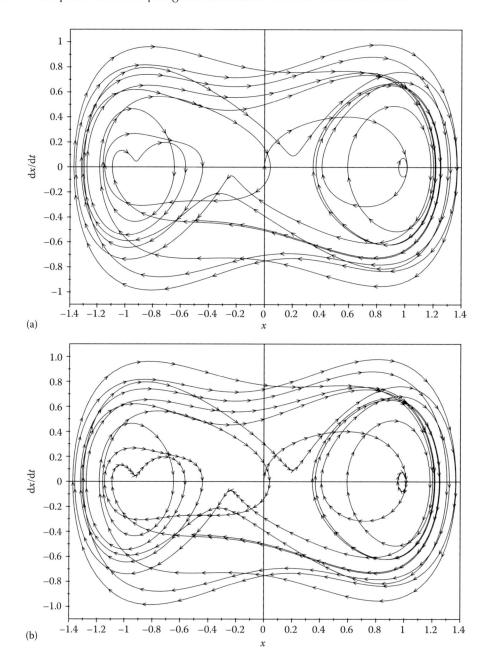

FIGURE 9.1   Phase path of Duffing oscillator plotted with equally spaced arrow markers (a) and arrow markers placed at every five data points, the static equivalent an animated comet-like **F9_01C. GIF** (b). See also configuration files **F9_01A.CF2**, **F9_01B.CF2**, and **F9_01C.CF2**.

conditions $x(0) = 0$ and $\dot{x}(0) = 0.000001$. It has what is known as *chaotic behavior*, where the time response of the system is bounded but not periodic. Figure 9.1 illustrates how the arrow markers available in the **D_2D** program were used to indicate the time evolution of the system. If these arrow markers are equally spaced along the plot curve like in Figure 9.1a, the direction of the process is revealed, but not its velocity. If the data points used to plot the phase path are

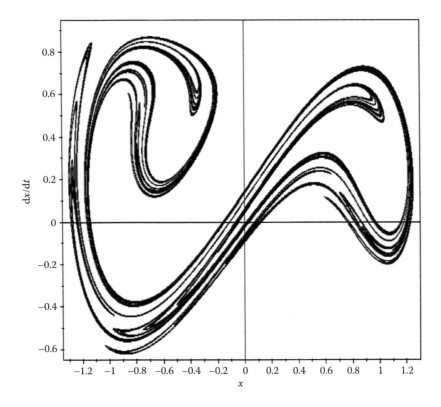

FIGURE 9.2   Poincaré map of the Duffing oscillator in Figure 9.1. Configuration file **F9_02.CF2**.

generated at a constant time step, then by placing an arrow marker at every certain number of data points, a sense of the speed at which the process occurs is also conveyed with the graph (Figure 9.1b). This latter graph is comparable to representing the respective line as a comet plot (see animation file **F9_01c.GIF** available with the book). The use of comet plots is restricted however by the availability for their display of an interactive environment.

Figure 9.2, known as a *Poincaré map* or *first recurrence map*, reveals some regularity in the chaotic behavior of the Duffing oscillator. These maps are phase configurations of the oscillations recorded for discrete time values $t = 2\pi n$ with $n = 0, 2, 3, 4$, and so on. The number of instances n recorded and plotted as dots in Figure 9.2 equals one million and is read by **D_2D** from file **F9_02.D2D**.

Data file **F9_02.D2D** together with ASCII file **F9_01.TXT** used to plot Figure 9.1 has been generated using program **F9_01.PAS** listed in Appendix B. Note that in order to output either the **F9_01.TXT** file or the **F9_02.D2D** file, constant **Poincare** must be set equal to FALSE or to TRUE, respectively (see line #**6** of the program).

9.2 FREE OSCILLATION OF A SPRING–MASS–DASHPOT SYSTEM

This section discusses the simulation of a spring–mass system with viscous damping. In case of this single degree-of-freedom (DOF) system, Newton's second law

$$\Sigma F_y = ma_y \tag{9.2}$$

writes (see Figure 9.3a)

$$mg - c\dot{y} - F_s = m\ddot{y} \tag{9.3}$$

After substituting the spring force, the differential equation of motion is obtained as

$$\ddot{y} = g - \frac{c}{m}\dot{y} - (y - l_0)k \tag{9.4}$$

Two Pascal programs have been written to integrate the equation of motion (9.4) of the system for $0 \le t \le$ **tend**, using Euler–Taylor algorithm (see Appendix A). Of these, program **P9_03.PAS** listed in Appendix B writes the displacement $y(t)$ and the velocity $dy(t)/dt$ of the mass to two separate data files named **F9_03LONG.DTA** and **F9_03SHRT. DTA**. The first of these files receives **nPoz** data points result of the numerical integration, while the second one receives every **Skip** data points (see lines #75 to #78 and lines #79 to #81). Depending on the step size **h** defined on line #15, the integration can be done over more points, but only **nPoz** points are recorded to file **F9_03LONG.DTA**. For the same reduced number of points that are recorded to file **F9_03SHRT.DTA**, the program also draws on the screen and to the multilayer **DXF** file **F9_03.DXF** a circle representing

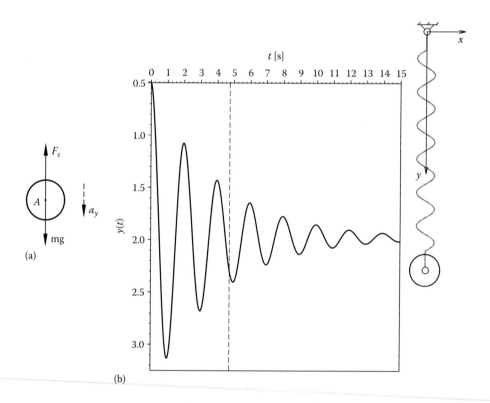

FIGURE 9.3 Free-body diagram of the mass (a), and time response of an underdamped spring–mass system (b). See also animation file **F9_03b.GIF** and configuration file **F9_03.CF2**. Note that plot has been mirrored and scaled inside **AutoCAD** to match the motion of the spring.

the mass, the helical spring, and its two attachment—see the use of procedures **Shape**, **Spring**, **PutGPoint**, and **PutPoint**.

A spring of free length l_0 = 1.0 m and constant k = 10 N/m and a suspended mass m = 1 kg were assumed. For the damping coefficient c = 0.5 Ns/m corresponding to an underdamped system and for initial conditions $y(0)$ = 0.5 m and $dy(0)/dt$ = 0, the system's time–response graph looks as shown in Figure 9.3b. The companion animation file **F9_03b.GIF** has been generated by combining inside **AutoCAD** the multilayer **DXF** file **F9_03.DXF** and a **D_2D** scan line graph generated using files **F9_03LONG.DTA** (for the background curve) and **F9_03SHRT.DTA** (for scan line points).

The second program named **P9_04.PAS** (listing not included in appendix) is structured similarly to the two-DOF spring-pendulum simulation program **P3_04.PAS** discussed in Chapter 3. The program generates vector **_t** of the independent variable (i.e., time) and the displacement and velocity vectors **_y** and **_dy**. These vectors then serve to plot the displacement and velocity graphs of the mass (see Figure 9.4). The sample animation frame output by this program in Figure 9.4 corresponds to the same initial conditions and parameters l_0, k, m, and c. Additional damping coefficient values have been considered (i.e., c = 0, c = 6.32456 Ns/m = $2(mk)^{1/2}$, and c = 10 Ns/m, corresponding to an undamped, critically damped, and overdamped system, respectively), and animation files **F9_03a.GIF** through **F9_03d.GIF** available with the book have been generated.

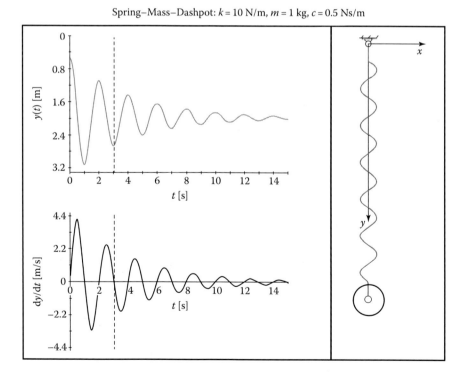

Spring–Mass–Dashpot: k = 10 N/m, m = 1 kg, c = 0.5 Ns/m

FIGURE 9.4 Underdamped spring–mass system simulation generated with **P9_04.PAS**. See also animation files **F9_04a.GIF** to **F9_04d.GIF**.

9.3 FREQUENCY AND DAMPING RATIO ESTIMATION OF OSCILLATORY SYSTEMS

Often times, dynamics and vibrations problems require for a solution, determining the *damped period of motion* τ_d, the corresponding *frequency* ($\omega_d = 2\pi/\tau_d$ and $f_d = 1/\tau_d$), and the amount of *damping* present in a system. For an underdamped, single DOF system for which a time–response curve $y(t)$ is available, the period of motion τ_d can be determined by measuring the time interval between two successive maximum or minimum displacement values or the time it takes for the system to pass twice through its equilibrium position. In turn, the amount of damping in the system, quantified by the *damping ratio* ζ, is traditionally determined by employing the *logarithmic decrement method*, which however requires accurate knowledge of the equilibrium position of the system.

In this section, an exponential-curve fit approach to damping ratio determination will be described, facilitated by the ability of the **D_2D** program to export to file the coordinates of the extrema of plots. Figure 9.5 is the time–response curve $y(t)$ of a spring–mass–dashpot system with mass $m = 1$ kg, spring rate $k = 10$ N/m, and viscous damping coefficient $c = 0.5$ N·s/m. This curve has been produced with data from file **F9_05.D2D** generated by program **P9_05.PAS** (see Appendix B). This program implements Equations 9.5 through 9.10 (Rao 2013), with the initial conditions $\dot{y}_0 = 0$ and $y_0 = 1.5 - y_\infty = 0.5$ m, where $y_\infty = y(\infty) = 1.0$ m is the displacement of the mass at equilibrium (i.e., the static displacement):

$$y(t) = y_\infty + A_0 e^{-\zeta\omega_n t} \cdot \sin(\omega_d t + \psi) \tag{9.5}$$

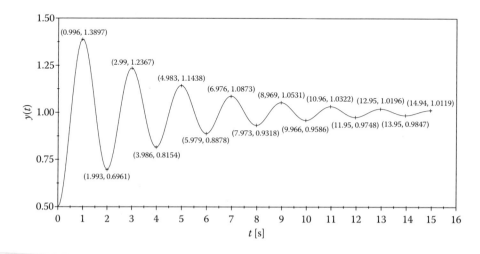

FIGURE 9.5 Time response of an underdamped spring–mass–dashpot system, showing the minimum and maximum displacement values. The plot consists of 200 data sets, and the peak values were interpolated parabolically over three points that bracket a local extrema. Configuration file **F9_05.CF2**.

where the maximum amplitude is

$$A_0 = \sqrt{y_0^2 + \left(\frac{\dot{y}_0 + \zeta\omega_n y_0}{\omega_d}\right)^2} = 0.50156986 \text{ m} \qquad (9.6)$$

phase angle is

$$\psi = \tan^{-1}\left(\frac{y_0\omega_d}{\dot{y}_0 + \zeta\omega_n x_0}\right) = 4.63324946 \text{ rad} \qquad (9.7)$$

damped circular frequency is

$$\omega_d = \omega_n\sqrt{1-\zeta^2} = 3.15238005 \text{ rad/s} \qquad (9.8)$$

natural frequency is

$$\omega_n = \sqrt{k/m} = 3.16227766 \text{ rad/s} \qquad (9.9)$$

damping ratio is

$$\zeta = \frac{c}{2m\omega_n} = 0.07905694 \qquad (9.10)$$

and damped period of motion is

$$\tau_d = \frac{2\pi}{\omega_n\sqrt{1-\zeta^2}} = 1.99315603 \text{ s} \qquad (9.11)$$

Using the coordinates of the minimum and maximum points extracted to file from Figure 9.5, the damped period of motion τ_d can be calculated as the time interval between two successive maximum or minimum displacements. Better precision has been obtained when the time interval between the first and the last maximum recorded values was divided by the number of in-between complete oscillations (see spreadsheet file **F9_05. XLS** available with the book). In this case, the damped period of motion was found to be 1.99314475 s, corresponding to a relative error of −0.0006%. When the first and the last minimum values were used instead, the damped period of motion was found to be 1.99320023 s, translating into a relative error of 0.0022%. Averaging these two values yields $\tau_d = 1.99317249$ s with a relative error of 0.0008%.

Regarding the amount of damping in the system, the common way to estimate it is to evaluate the *logarithmic decrement* using the time–response curve $y(t)$. The *logarithmic decrement* is defined as the natural logarithm of the ratio of two distinct peak displacements (either minimum or maximum) noted p and q, measured from the equilibrium position:

$$\delta = \frac{1}{q-p} \ln\left[\frac{(y_{\max})_p - y_{\infty}}{(y_{\max})_q - y_{\infty}}\right] = \frac{1}{q-p} \ln\left[\frac{(y_{\min})_{q-1} - y_{\infty}}{(y_{\min})_q - y_{\infty}}\right] \quad (9.12)$$

Alternatively, both the minimum and maximum values can be combined in calculating the logarithmic decrement, according to the following formula:

$$\delta = \frac{2}{q-p} \ln\left|\frac{(y_{\text{peak}})_p - y_{\infty}}{(y_{\text{peak}})_q - y_{\infty}}\right| = \frac{2\pi\zeta}{\sqrt{1-\zeta^2}} \quad (9.13)$$

where these minimum and maximum values are numbered successively using the same index.

The application of Equations 9.12 and 9.13 is limited by the knowledge of the static displacement y_{∞} of the mass. If the equilibrium position of the mass is not exactly known, y_{∞} can be determined together with the product $\zeta\omega_n$ and amplitude A_0 in a curve fitting process, as the minimum of an objective function of the type

$$\text{Fobj}_1(\zeta\omega_n, A_0, y_{\infty}) =$$

$$\underset{k}{\text{Max}}\left(\left|y_{\infty} + A_0 \cdot \exp(-\zeta\omega_n \cdot t_{\max k}) - y_{\max k}\right|, \left|y_{\infty} - A_0 \cdot \exp(-\zeta\omega_n \cdot t_{\min k}) - y_{\min k}\right|\right) \quad (9.14)$$

where $t_{\min k}$ and $t_{\max k}$ are the moments of time where the peak values $y_{\max k}$ and $y_{\min k}$ of the time–response curve occur. Program **P9_06.PAS** listed in Appendix B implements this approach to determine product $\zeta\omega_n$, displacement at equilibrium y_{∞}, and maximum amplitude A_0.

For $y_{\max}$ and $y_{\min}$ extracted through parabolic interpolation from a time–response curve with 200 data points (Figure 9.5), the following numerical results were obtained:

$$A_0 = 0.50005500 \text{ m}$$

$$\zeta \cdot \omega_n = 0.25002500 \text{ rad/s} \quad (9.15)$$

$$y_{\infty} = 1.00000099 \text{ m}$$

These results have been output by program **P9_06.PAS** to data file **F9_06.REZ**, together with the value of the objective function at minimum, that is, 0.00002106. This value is the maximum deviation in absolute value between the time–response curve and its exponentially decaying envelope shown overlapped in Figure 9.6—the envelope curve in this figure has been produced using data file **F9_06.REZ**.

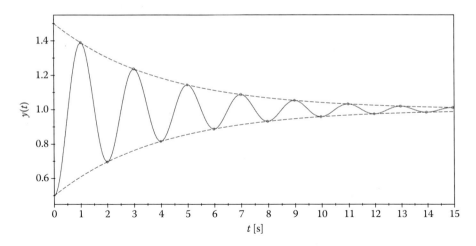

FIGURE 9.6　Combined plot of the spring–mass–dashpot system time response and peak envelopes obtained through minimization of the objective function (9.14). Configuration file **F9_06.CF2**.

Knowing that the undamped natural circular frequency of the system is

$$\omega_d = \frac{2\pi}{\tau_d} = \omega_n \sqrt{1-\zeta^2} \tag{9.16}$$

and for τ_d determined as explained earlier, the damping ratio of the system was found to be

$$\zeta = \sqrt{\frac{\left(\zeta \cdot \omega_n\right)^2 \cdot \left(\tau_d\right)^2}{4\pi^2 + \left(\zeta \cdot \omega_n\right)^2 \cdot \left(\tau_d\right)^2}} = 0.0790654468. \tag{9.17}$$

This represents an error of only 0.0108% compared to the exactly calculated value 0.07905694 in Equation 9.10.

9.4　NONLINEAR CURVE FIT TO DATA

One problem frequently encountered in numerical data analysis that can be solved using optimization techniques is adjusting the coefficients of a function (in particular, a polynomial), so that this chosen function best approximates a set of n data pairs (x_i, y_i). Such a problem was discussed in Section 9.3 where an exponential curve was fit to some experimentally determined points. A similar example will be discussed next, where supplementary the effect of rounding off the computed coefficients is addressed right from within the optimization problem. The example that will be considered in this section refers to adjusting the coefficients C_1 through C_5 of the function

$$\sigma(\varepsilon) = \frac{2\varepsilon(3+3\varepsilon+\varepsilon^2)}{(1+\varepsilon)^4}\Big[2C_5\varepsilon^5 + (10C_5 + 3C_3)\varepsilon^4 + (14C_5 + 4C_4 + 9C_3 + C_2)\varepsilon^3$$

$$+ (6C_5 + 6C_4 + 6C_3 + 3C_2 + C_1)\varepsilon^2 + (3C_2 + 2C_1)\varepsilon + C_2 + C_1\Big] \tag{9.18}$$

for which the sigmoidal stress–strain curve of the elastomeric material graphed in Figures 3.16 and 3.17 is best approximated.

The corresponding minimax approximation problem requires solving the following objective function in five variables (Weisstein 2013):

$$\text{Fobj}_2(C_{1...5}) = \underset{i=1}{\overset{n}{\text{Max}}} |\sigma(\varepsilon_i) - \sigma_i| \qquad (9.19\text{a})$$

where ε_i and σ_i are data pairs extracted from the plot in Figure 3.17. Other forms of the objective function (9.19) are possible, like the sum of squared deviations:

$$\text{Fobj}_2(C_{1...5}) = \sum_{i=1}^{n} (\sigma(\varepsilon_i) - \sigma_i)^2 \qquad (9.19\text{b})$$

or the sum of absolute values of the deviations:

$$\text{Fobj}_2(C_{1...5}) = \sum_{i=1}^{n} |\sigma(\varepsilon_i) - \sigma_i| \qquad (9.19\text{c})$$

Of these three, an objective function like the one in Equation 9.19a can ensure that the departure between the given data points and the approximating curve will be evenly spread along $\sigma(\varepsilon)$ (see Figure 9.7).

It is not unusual in curve fitting problems for the found coefficients (like $C_{1...5}$ in Equation 9.18) to be rounded off their computed values without verifying the effect upon the accuracy of the approximation. This situation can be addressed from within the search algorithm, as it has been done in program **P9_07.PAS** listed in Appendix B. As shown, procedure **NelderMead** that implements the Nelder–Mead searching algorithm is called 100 times, each time using a different initial guess (see the *for* loop between lines **#59** and **#79**). After each iteration, the value of the objective function is evaluated, and if a

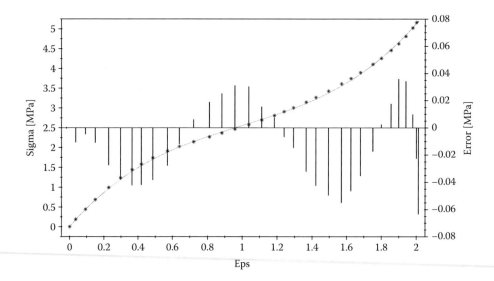

FIGURE 9.7 Plot of the original data points (∗), analytical curve (solid line), and error bars. Configuration file: **F9_07.CF2**.

smaller value has been found, then coefficients **C1–C5** are rounded to their third decimal (see line #**71** of the program). Objective function **Fobj2** is evaluated one more time, and only if the improvement over the best optimum found that far is preserved after roundup, then the variables **Xbest** and **vFbest** are updated. Data pairs ε_i and σ_i of the elastomeric material are read from ASCII file **F9_07.DTA**, which is a copy of file **F3_20.XY** from Chapter 3, with the first two lines removed. Note that the number of input data points **nPts** in program **P9_07.PAS** is equivalent to parameter n in Equation 9.19.

One of the best results returned by program **P4_18.PAS** is

```
Max Deviation = 0.049003131;
C1=0.8510;   C2=0.0200;   C3=-0.1620;  C4=0.0470;   C5=0.0570;
```

These coefficients were used to plot the best-fit curve and the corresponding error bars in Figure 9.7. The almost equal in magnitude negative and positive deviations are a first indication that a good solution has been found.

9.5 PLOTTING FUNCTIONS OF MORE THAN TWO VARIABLES

So far we dealt with graphical representation of function of one and two variables using line, surface, or level-curve diagrams. Occasionally, there is an interest in visualizing single-valued functions of more than two variables. Analytical functions of the form $F(x_1, x_2, x_3)$ can be explored graphically by maintaining constant one variable, for example, x_3, while scanning the remaining two variables within some prescribed limits, for the purpose of generating the data file needed for plotting projected level-curve or 3D surface diagrams. If several such plots are generated for ordered values of x_3, then these can be displayed successively as computer animations, where time plays the role of variable x_3.

Let us consider the following function called the generalized Rosenbrock's function

$$R_n(x_1 \ldots x_n) = \sum_{i=1}^{n-1} \left[100 \cdot (x_{i+1} - x_i^2)^2 + (1 - x_i)^2 \right] \tag{9.20}$$

used in evaluating the performance of optimization algorithms. Its global minimum equals 0 and occurs for $x_i = 1$. For $n = 2$, this function is known as Rosenbrock's Banana function and its plot looks as shown in Figure 9.8.

To visualize the $n = 3$ version of Rosenbrock's function as animation, program **P9_09.PAS** has been written and is listed in Appendix B. The program outputs ASCII file **F9_09.T3D** consisting of multiple columns of $R_3(x_1, x_2, x_3)$ values produced for various x_3's, preceded by the grid sizes n_{x_1}, n_{x_2} and limits $x_{1\min}$, $x_{1\max}$, $x_{2\min}$, $x_{2\max}$. The animation frames generated for x_3 equal to −2.0, −1.0, 0.0, 1.0, and 2.0, and $x_{1\min} = x_{2\min} = -2.5$ and $x_{1\max} = x_{2\max} = 2.5$ are available in Figure 9.9.

When the single-valued function of interest has more than three variables, the animation method described earlier can no longer be applied. Of the various dimension reduction techniques applicable to functions of the form $F(x_1, x_2, \ldots, x_n)$, the *partial minimax method* (Simionescu and Beale 2004) will be illustrated and applied to visualizing R_n in Equation 9.20. This is a method of projecting hyperfunctions from n dimensions down to 3D or 2D,

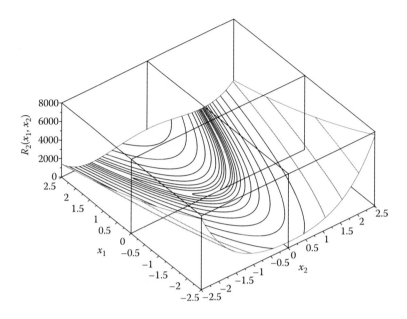

FIGURE 9.8 Graph of Rosenbrock's function with $n = 2$. See also program **P9_08.PAS** available with the book used to generate the **D3D** file for this plot. Configuration file **F9_08.CF3**.

where two of the function variables, for example, x_1 and x_2, are scanned at constant step as in a regular 3D or level-curve plot (these are the *scan variables*), while the remaining ones are the *search variables* in the following global-minimization and global-maximization problems:

$$F_{\downarrow 3\ldots n}(x_1, x_2) = \text{Global} \min_{x_3 \ldots x_n} F(x_1 \ldots x_n)$$

$$\text{subjected to } x_{j\min} \leq x_j \leq x_{j\max} \quad \text{with } j = 3 \ldots n \tag{9.21}$$

and

$$F_{\uparrow 3\ldots n}(x_1, x_2) = \text{Global} \max_{x_3 \ldots x_n} F(x_1 \ldots x_n)$$

$$\text{subjected to } x_{j\min} \leq x_j \leq x_{j\max} \quad \text{with } j = 3 \ldots n \tag{9.22}$$

$F\downarrow(x_1, x_2)$ and $F\uparrow(x_1, x_2)$ are called *partial minima* and *partial maxima functions* and are the lower and the upper envelopes of the hypersurface of the original function $F(x_1, x_2,\ldots, x_n)$ when projecting it from $n + 1$ dimension space $(x_1, x_2,\ldots, x_n, F)$ down to three dimensions, for example (x_1, x_2, F). Also of interest are the plots of the x_3–x_n values at these partial minima and partial maxima. These are called *lower bound* and *upper bound* paths and are noted $x_{3\downarrow}, x_{4\downarrow}, \ldots, x_{n\downarrow}$ and $x_{3\uparrow}, x_{4\uparrow}, \ldots, x_{n\uparrow}$, respectively.

Program **P9_10.PAS** listed in Appendix B implements the *partial minimax method* to plot the generalized Rosenbrock's function with $n = 5$. The program outputs ASCII file **F9_10.T3D**, with separate columns for $F\downarrow(x_1, x_2)$ and $F\uparrow(x_1, x_2)$ and for the corresponding $x_{3\downarrow}, x_{4\downarrow}, x_{5\downarrow}$ and $x_{3\uparrow}, x_{4\uparrow}, x_{5\uparrow}$ values. When represented graphically, these partial minima and partial maxima functions appear as shown in Figure 9.10. Note that the narrow valley exhibited by Rosenbrock's function of two variables is also present in the 3D projection of its $n = 5$ generalization.

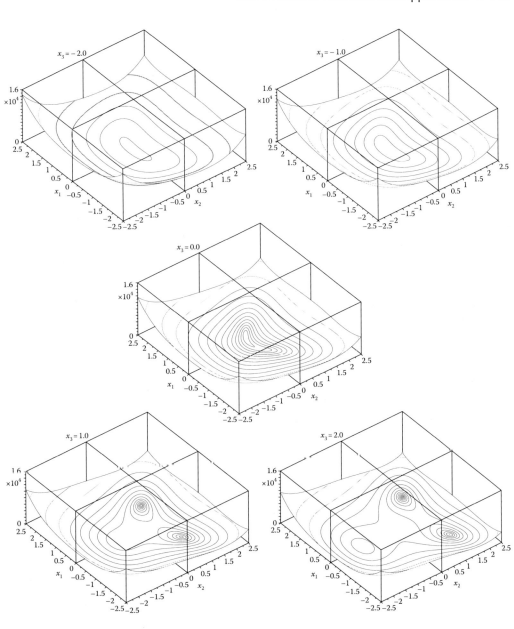

FIGURE 9.9 Some of the frames in the animation file **F9_09.GIF** of the generalized Rosenbrock's function with $n = 3$, where time is associated to variable x_3 and it is listed on the top of each frame.

The first-order discontinuity in the graphs of the lower bound paths $x_{3\downarrow}$, $x_{4\downarrow}$, and $x_{5\downarrow}$ visible in Figure 9.10 is indicative that for $n = 5$, the generalized Rosenbrock's function has more than one minima. This can be verified by visualizing the Euclidean norm of the gradient of R_5

$$|\nabla R_n| = \sqrt{\left(\frac{\partial R_n}{\partial x_1}\right)^2 + \left(\frac{\partial R_n}{\partial x_2}\right)^2 + \cdots + \left(\frac{\partial R_n}{\partial x_n}\right)^2} \tag{9.23}$$

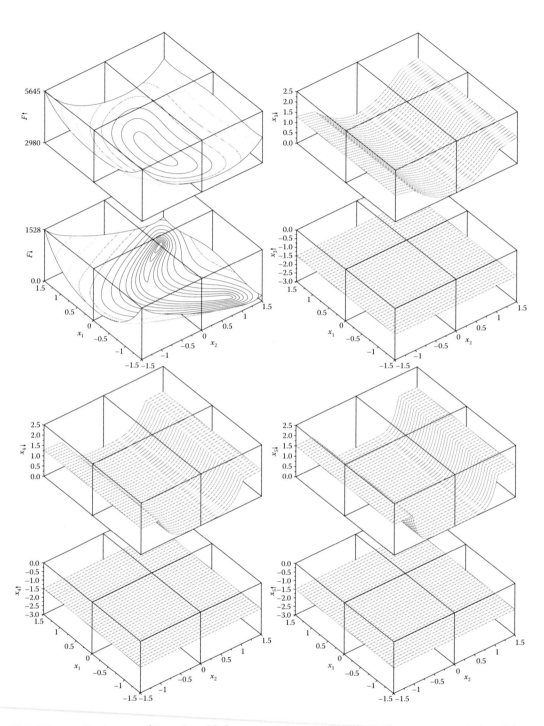

FIGURE 9.10 Projection of Rosenbrock's function with $n = 5$ variables to the 3D space and plot of the lower bound and upper bound paths. Configuration files **F9_10_1.CF3** to **F9_10_8.CF3**.

using the same *partial minimax method*, where

$$\frac{\partial R_n}{\partial x_i} = 200x_1(x_i - x_{i-1}^2) - 400x_i(x_{i+1} - x_i^2) - 2(1 - x_i) \quad \text{for } 2 < i < n-1$$

$$\frac{\partial R_n}{\partial x_1} = -400x_1(x_2 - x_1^2) - 2(1 - x_1) \quad \text{and} \quad \frac{\partial R_n}{\partial x_n} = 200(x_n - x_{n-1}^2) \quad (9.24)$$

Using a new program named **P9_11.PAS** available with the book, data files **F9_11_12.D3D** to **F9_11_45.D3D** have been generated and served to produce the graphs in Figure 9.11a and b. Based on these graphs, it can be inferred that for $n = 5$, the generalized Rosenbrock's function has one global minimum at $(1, 1, 1, 1, 1)$ and one local

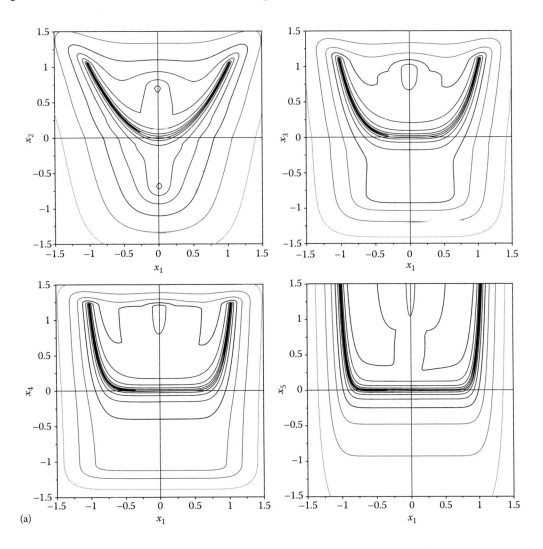

(a)

FIGURE 9.11   (a) Magnitude of the gradient of the generalized Rosenbrock's function with $n = 5$ variables projected down to 3D for scan variables (x_1, x_2), (x_1, x_3), (x_1, x_4), and (x_1, x_5). Configuration files **F9_11A12.CF3** to **F9_11A15.CF3**. *(Continued)*

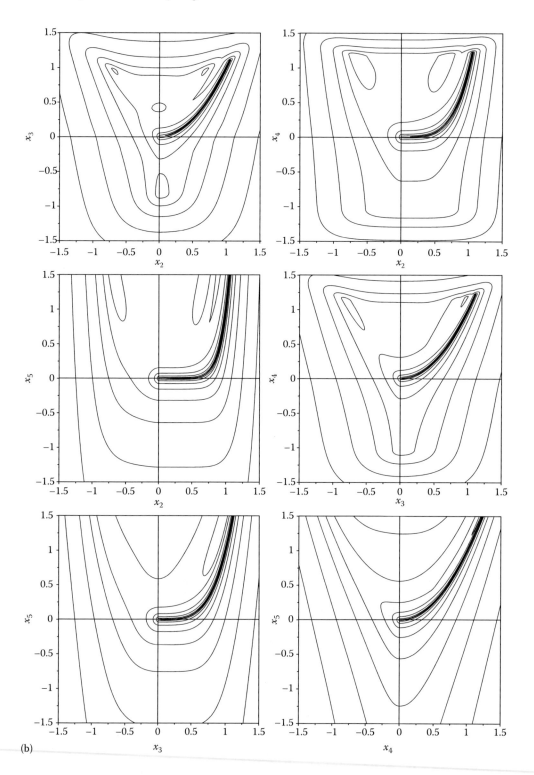

FIGURE 9.11 (*Continued*)   (b) Magnitude of the gradient of the generalized Rosenbrock's function with $n = 5$ variables projected down to 3D for scan variables (x_2, x_3), (x_2, x_4), (x_2, x_5), (x_3, x_4), (x_3, x_5), and (x_4, x_5). Configuration files **F9_11B23.CF3** to **F9_11B35.CF3**.

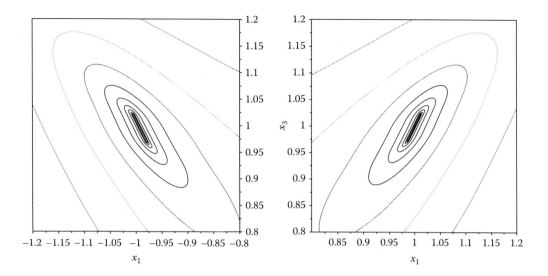

FIGURE 9.12   Details of the magnitude of the gradient of the generalized Rosenbrock's function with $n = 5$ variables around minima, projected down to 3D. Configuration files **F9_12A.CF3** and **F9_12B.CF3**.

minimum at $(-1, 1, 1, 1, 1)$. Narrowing the range of the scan variables to $[-1...1]$ indeed confirms that for $n = 5$, the function in Equation 9.20 has one global minimum and one local minimum (see Figure 9.12). The data files **F9_12A.D3D** and **F9_12B.D3D** used to produce these (Figure 9.12) were generated by program **P9_12.PAS** also available with the book.

Note that if different combinations of the scan variables and search variables are chosen, then the appearance of the *partial minimax* graphs and of the respective *lower bound* and *upper bound paths* will change. Also note that modifying the limits of both the scan variables and search variables will affect the appearance of these graphs.

Undoubtedly it is very time-consuming to perform the repeated minimizations and maximizations required to project hypersurfaces down to the 3D space as explained earlier. Fortuitously, the location of the partial minima and partial maxima values do not change significantly when moving to the next pair of scan variables, particularly when the scan-variable grid is tight. With this in mind, the previously found solution can be used as initial guess for the next search, as it was actually done in programs **P9_09.PAS**, **P9_10.PAS**, **P9_11.PAS**, and **P9_12.PAS**.

Depending on the optimization algorithm employed, finding the actual partial global minima is not guaranteed. The *partial minimax* projection method is however inherently suited to parallel processing. This, together with the fact that increasingly powerful heuristic searching algorithms are constantly being developed, will facilitate the practical implementation of the dimension reduction method described.

Finally, if the graphs produced exhibit a noisy appearance or have unexpected discontinuities, then these are signs that the search algorithm employed converged prematurely and must be readjusted or a different algorithm should be employed. This suggests that the ranking of different optimization algorithms for speed and robustness can be done by employing them in generating *partial minimax* projections of carefully selected hypersurfaces, and then compare the appearance of the graphs obtained and the time required to generate these graphs for the given scan-variable grid sizes.

9.6 RANDOM NUMBER GENERATION AND HISTOGRAM PLOTS

Random number generators have numerous applications in computer games, cryptography, search algorithms, various numerical simulations, etc. Most programming languages include functions capable of providing a random number that is uniformly distributed between certain limits. In the case of **Turbo Pascal**, the system function **Random(range)** returns with each call a uniform random value between 0 and **range**. When called without the argument, **range** is assumed to be equal to 1. The sequence produced by calling the **Random** function will always be the same, however, unless the internal number generator is initialized

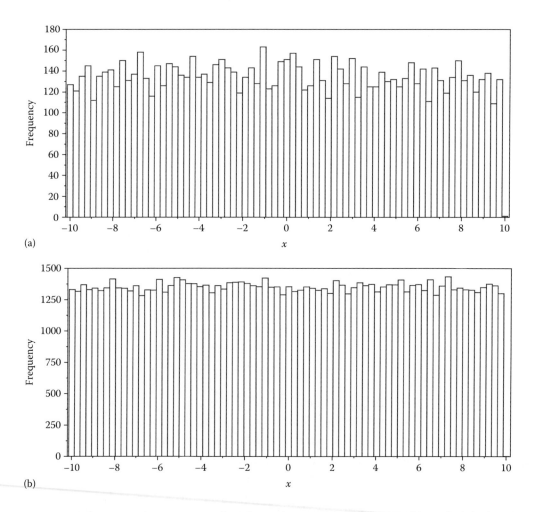

FIGURE 9.13   Frequency histograms with 75 bins of 10,000 (a) and 100,000 (b) uniformly distributed data points read by **D_2D** from files produced with program **P9_13.PAS**. Configuration files **F9_13A.CF3** and **F9_13B.CF3**.

by calling the **Randomize** function first. This uses a seed value obtained from the system clock to assist the function **Random** to produce numbers that is close to being true random.

Program **P9_13.PAS** (see Appendix B) uses the **Random** function to generate data files **F9_13A.DAT** with 10,000 values and **F9_13B.DAT** with 100,000 values that are uniformly distributed within the interval [−10…10]. These files in turn were used to produce the frequency histograms in Figure 9.13 using the **D_2D** program. For the same number of bins (i.e., 75), the top land of the graph generated using a larger number of samples has a visibly smoother appearance.

A second computer program named **P9_14.PAS** (see listing in Appendix B) served to generate the data files used to produce the frequency histograms in Figure 9.14. This program implements the method of Box and Muller to generate pairs of Gaussian (normally) distributed pseudorandom numbers, starting from a source of uniformly distributed values produced by calling the **Turbo Pascal Random** function.

The closeness of the randomly generated values by **P9_14.PAS** to a true normal distribution is illustrated by the plots in Figure 9.15. It shows overlapped a *relative frequency*

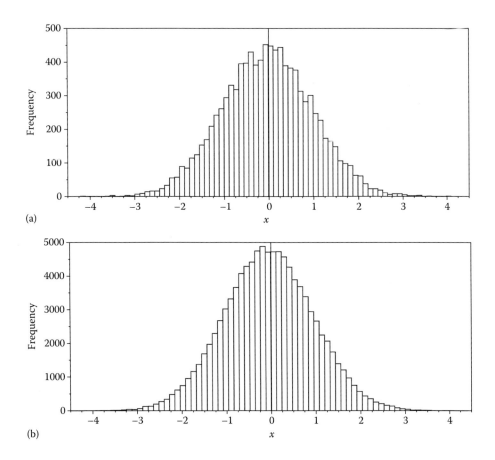

FIGURE 9.14 Frequency histograms with 75 bins of 10,000 (a) and 100,000 (b) Gaussian distributed data points, generated using program **P9_14.PAS**. Configuration files **F9_14A.CF3** and **F9_14B.CF3**.

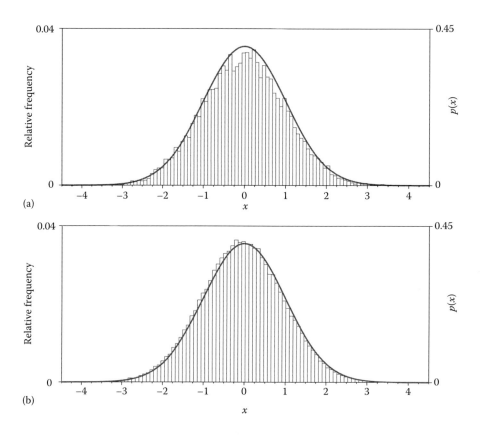

FIGURE 9.15 Relative frequency histograms with 100 bins of 10,000 (a) and 100,000 (b) data points (same data as in Figure 9.14), overlapped with a Gaussian probability density function $p(x)$. Data file to plot $p(x)$ has been generated by program **P9_15.PAS**. Configuration files **F9_15A.CF3** and **F9_15B.CF3**.

histogram with 100 bins of the same two data files used to generate Figure 9.14 and a plot of the normal probability density function:

$$p(x) = \frac{1}{\sigma\sqrt{2\pi}} e^{-0.5(x-\mu)^2/\sigma^2} \tag{9.25}$$

with $\mu = 0$ and $\sigma = 1$. A relative frequency histogram is one where the number of occurrences in each bin is divided by the total number of data points, which makes its appearance less sensitive to the number of the input values. You can experiment with different number of bins by editing the configuration file **F9_15B.CF3** and with different input data file sizes by rerunning program **P9_14.PAS** with constant **N** set to different values.

9.7 DWELL MECHANISM ANALYSIS

Dwell mechanisms have the property that for a constant rotary input, their output link remains (quasi) stationary for a portion of the motion cycle. Such a property is required by some manufacturing, textile, and packaging equipment applications. Cam and follower mechanisms, Geneva wheels, gear linkages, and linkage mechanisms (like the Stephenson III linkage that will be discussed in this paragraph) are typical dwell motion generators used in practice. Of these, linkage mechanisms have better dynamic properties, but are more difficult to synthesize and usually result larger in size. If they employ pin joints only (which can be sealed and greased for life), dwell linkage mechanisms may benefit from increased reliability.

Program **P9_16.PAS** listed in Appendix B simulates the motion of a Stephenson III linkage, which comprises a four-bar path generator of input link *OA*, amplified with an RRR dyad (see Figure 9.16). The link lengths of the four-bar *OABC* are selected such that a portion of the coupler curve of point *D* is close in shape to an arch of a circle. In turn, the *DEF* dyad is sized such that link length *DE* equals the radius of the almost circular portion of the coupler curve, while joint *E* is located at its center of curvature. In addition to the **DXF** frame file **P9_16.DXF**, program **P9_16.PAS** writes to ASCII file **F9_16.DTA** angle θ_1 of input link *OA*, and angle θ_6 of the output link *EF* (see Figure 9.16), together with the angular velocity $d\theta_1/d\theta_6$ of the same link *EF*.

During the first cycling of the *repeat–until* loop (lines #**35** to #**68**), the program generates the output file **F9_16.DTA** and updates the workspace limits, but without animating

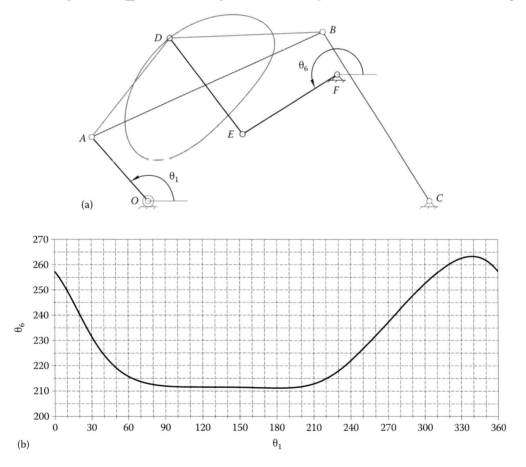

FIGURE 9.16 Simulation of Stephenson III dwell mechanism done with program **P9_16.PAS** (a) and the input–output displacement diagram of the mechanism (b). See also animation file **F9_16. GIF** and configuration file **P9_16.CF2**.

the mechanism. These limits are then used when calling procedure **OpenMecGraph** on lines #**40**. During the second cycling of the *repeat–until* loop, when the number of positions is reduced from **nPozDTA** to **nPozDXF**, the animation frames are written to file **F9_16.DXF**. After that, the simulation repeats itself with no file output.

Using the same data file **F9_16.DTA** as input, the plot in Figure 9.17 has been generated. Next, the minimum and maximum θ_6 values on this graph are exported by **D_2D** to file. Using these values, the range of the output link displacement was calculated as

$$\Delta\theta_6 = 263.30° - 211.19° = 52.115° \tag{9.26}$$

Since link *EF* does not remain exactly immobile, the duration of the dwell is evaluated based on some accepted deviation from the limit position of the output link. One approach is to assume that link *EF* is still dwelling when departed from its limit positions only a small fraction *r* of its entire motion range $\Delta\theta_6$. Assuming this amount to be $r = 0.02$, the dwell range of the mechanism in Figure 9.16 has been determined to be $\Delta\theta_1=120.6°$. This has been done by editing the displacement curve in Figure 9.16 using **AutoCAD** software as shown in Figure 9.17.

The second method of measuring the duration of the dwell is to assume that it lasts as long as the velocity of the output link remains less than a chosen amount. When both the input and output links perform a rotary motion, this deviation can be defined as a percentage of the input link angular velocity. For $d\theta_6/d\theta_1 = 0.05$ corresponding to the output link velocity $d\theta_6/dt$ being 5% of the input link velocity $d\theta_1/dt$, the duration of the dwell was found to be $\Delta\theta_1=108.04°$. Again, **AutoCAD** software has been used to graphically solve the intersection between the deviation boundaries shown in dashed lines and the output velocity curve as shown in Figure 9.17. Note that in order to simplify the analysis, in program **F9_16.DTA**, the input link velocity $d\theta_1/dt$ has been set equal to unity (see line #**45**).

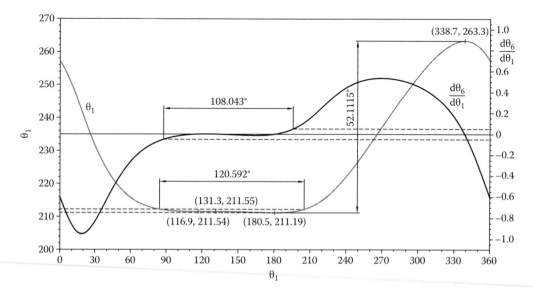

FIGURE 9.17 Output displacement and velocity diagrams of the dwell mechanism in Figure 9.16, edited using **AutoCAD** to extract the dwell range. Configuration file **F9_17.CF2**.

9.8 TIME RATIO EVALUATION OF A QUICK-RETURN MECHANISM

The quick-return mechanism in Figure 9.18 is a classical example of RTR dyad use. It is named *quick return* because slider D moves slower in one direction than it does in reverse, as shown on the input–output diagram in Figure 9.18b. Such a mechanism is used in shaper machine tools, where the faster inactive stroke allows for an increased productivity, as compared, for example, to the slider–crank mechanism.

For a constant rotational input, the time ration TR of a quick-return mechanism like the one in Figure 9.18a with input crank length OA and ground-joint center distance OB is (Cleghorn 2005)

$$TR = \frac{\pi}{\arccos(OA/OB)} - 1 = \frac{\pi}{\arccos(0.1/0.25)} - 1 = 1.7099 \tag{9.27}$$

Therefore, the time ratio can be interpreted as the duration of the fast stroke divided by the duration of the slow stroke.

Alternatively, the time ratio can be calculated using the coordinates of the minimum and maximum points available on the kinematic diagram in Figure 9.18b:

$$TR = \frac{360° - (336.4° - 203.5°)}{336.4° - 203.5°} = 1.7088 \tag{9.28}$$

Note that this latter method can be applied to any mechanism with constant rotational input for which kinematic diagrams like the one in Figure 9.18b are available.

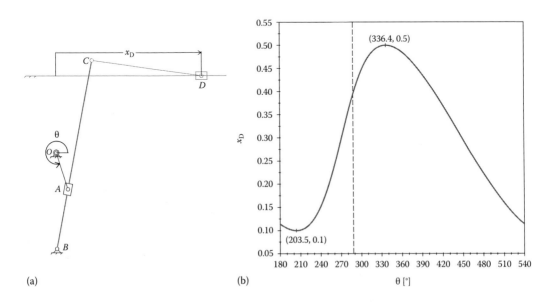

(a)   (b)

FIGURE 9.18   Quick-return mechanism (a) and its output slider displacement diagram (b). See also animation file **F9_18.GIF** and the **D_2D** configuration file **F9_18B.CF2**.

9.9 EXAMPLES OF ITERATIVE USE OF THE PROCEDURES IN UNIT **LibAssur**

Programs **P9_19.PAS** to **P9_22.PAS** listed in Appendix B illustrate how the procedures in unit **LibAssur** can be called repetitively. Of these, programs **P9_19.PAS** and **P9_20.PAS** simulate the motion of radial piston engines, and programs **P9_21.PAS** and **P9_22.PAS** simulate the motion of a mechanical iris. All four programs were written such that any number of equally spaced cylinders or iris vanes can be specified, including one cylinder or one vane only.

Figure 9.19 illustrate the cases of one, three, seven, and nine cylinder engines with stationary cylinder blocks, while Figure 9.20 show the corresponding engines with rotational cylinder blocks of the Gnome type (also known as rotary engines—see also animation files **F9_19.GIF** and **F9_20.GIF**). In both programs, when the number of cylinders is set equal to three or less, the piston axis and the pin joints (other than the piston pin) are labeled as shown in the Figures 9.19 and 9.20.

Further examples of iterative use of procedure **RRT_** are the iris mechanisms in Figures 9.21 and 9.22. Of the different designs used in practice, the mechanism considered here consists of an array of half-ring-shaped vanes that are fitted with pin joints at one end, noted P, while their other end, noted Q, can slide along equally spaced radial directions OA. The ends Q of these vanes are designed as pin-in-slot joints, with OA being the slots. The iris can operate either with its pin joints P stationary and slots OA rotating together

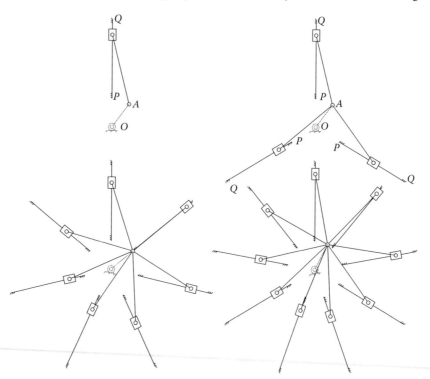

FIGURE 9.19 Single-piston and radial engines with three, seven, and nine cylinders simulated with program **P9_19.PAS**. See also animation file **F9_19.GIF**.

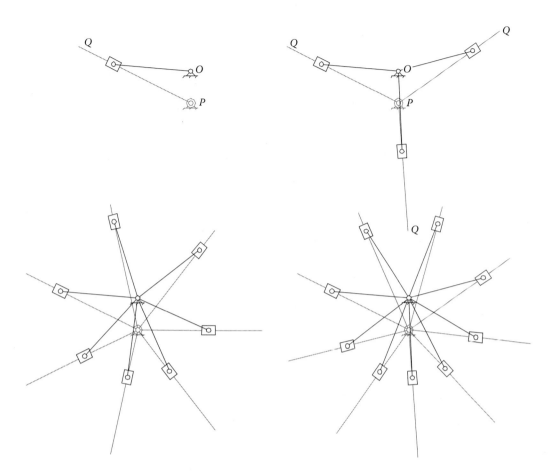

FIGURE 9.20 Rotary engines of the Gnome type with one, three, seven, and nine cylinders simulated with programs **P9_20.PAS**. See also animation file **F9_20.GIF**.

(Figure 9.21) or as inversions, that is, the slots maintain their direction stationary and pins *P* rotate about the center of the iris *O* (Figure 9.22).

Same as before, if the number of vanes in these two simulation programs is set equal to three or less, the underlying mechanisms are displayed, and their sliding axes *OA* and joints *P* and *Q* are labeled (see Figures 9.21 and 9.22). Otherwise, procedures **gCrank** and **RRT_** are called with their color parameter set equal to 0 or the **BGI** constant **Black**, so that vanes only are displayed.

Note in these four programs the extensive use of procedures **gCrank**, **RRT_** and of the generic variable _ preassigned to 10^{100} and defined in the interface section of unit **LibMath**.

Regarding the actual vanes in these last two simulation programs, they are polylines, the vertices of which are read by procedure **Shape** from the same file named **VANE.XY**. This ASCII file has been generated as follows: One vane only was drawn in **AutoCAD** with point *P* at origin and point *Q* of coordinates (9, 0) (see file **Vane.DWG** available with the book). This drawing was then plotted to file **Vane.PLT** and opened with program **Util~PLT**. The *x* and *y* limits inside **Util~PLT** were then edited such that the vane has

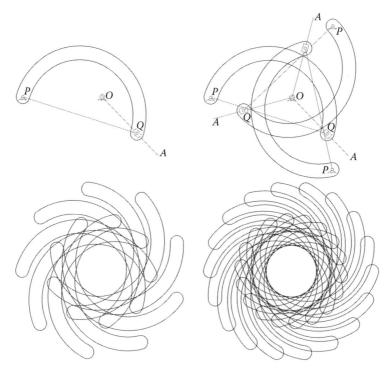

FIGURE 9.21 Iris mechanisms simulated with program **P9_21.PAS**. See also animation file **F9_21.GIF**.

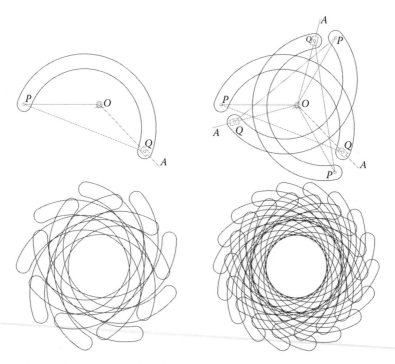

FIGURE 9.22 Iris mechanisms simulated with program **P9_22.PAS**. See also animation file **F9_22.GIF**.

the same size and origin as in the original **DWG** file (i.e., $x_{min} = -0.5$, $x_{max} = 9.5$, $y_{min} = -0.5$, and $y_{max} = 5$). Only then the file was exported to files **PLT-0001.XY** and **PLT-0001. DXF**. Because the **PLT-0001.XY** file has too many vertices, and because these vertices may result out of sequences, in the iris mechanism simulation programs a lower-resolution file has been utilized instead. To generate such a low resolution ASCII file named **VANE. XY**, file **PLT-0001.DXF** was opened using **Util~DXF.EXE** and its polylines extracted to file **POLY0001.XY** file. Prior to extracting the vertices of the polyline in **PLT-0001.DXF** to file, the **DXF** colinearity parameter was increased from its default value, and thus the number of vertices from **PLT-0001.DXF** was further reduced. In the end, the ASCII file **POLY0001.XY** thus obtained was renamed **VANE.XY**.

9.10 SIMULATION OF A FOUR-BAR LINKAGE AND OF ITS FIXED AND MOVING CENTRODES

In this section, it is shown how procedure **Shape** from unit **LibMec2D** can be used to animate shapes that change their configuration during animation. The case of the fixed centrode and moving centrode of a drag-link four-bar linkage will be considered as example. The fixed centrode will be animated using the **CometLocus** procedure, while the moving centrode (which moves together with the coupler) will be modeled as a shape that gains (x, y) points as the animation progresses. Therefore, the animations done inside **AutoCAD** using the **M_3D.LSP** application will be very similar to the one displayed on the computer screen and recorded as **PCX** frames (Figure 9.23).

Program **P9_23.PAS** listed in Appendix B animates a drag-link four-bar linkage, having crank *OA* as input, and with *AB* the coupler, and *BC* the second link jointed to the ground. The coordinates **xIC** and **yIC** of the *instant center of rotation* (*IC* in short) of the coupler relative to the ground are calculated as the intersection of lines *OA* and *BC*. This is done on line #**47** of the program by calling procedure **Int2Lns** from

4-bar left—fixed and moving centrodes   4-bar right—fixed and moving centrodes

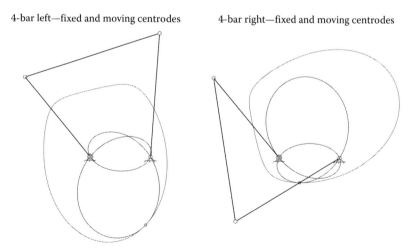

FIGURE 9.23 Drag-link four-bar linkage in its left-hand and right-hand configurations with the *fixed centrodes* and *moving centrodes* of the coupler shown, produced using **P9_23.PAS**. See also animation files **F9_23.GIF**, **F9_23L_PCX.GIF**, and **F9_23R_PCX.GIF**.

unit **LibGe2D**. Together, these points of coordinates (**xIC,yIC**) form the *fixed centrode* of the coupler, plotted by calling procedure **CometLocus** on line #**56**. The coordinates of the instant center recorded relative to coupler *AB* will form the *moving centrode*. These coordinates noted **xICm** and **yICm** are calculated on line #**49** by calling procedure **RT2D**. Point (**xIC,yIC**) is first translated to *A* and then rotated by the angle formed by coupler *AB* with the ground. As they are calculated during the first simulation cycle, coordinates **xICm** and **yICm** are written to files **ICF.XY**. This file is then used as input to procedure **Shape** called on line #**55**.

Also note in program **P9_23.PAS** the use of procedure **SetTitle** (lines #**32** and #**39**) to display the title of the simulation and of procedure **InitGr** with zero argument (line #**28**) to display the animation on white background. As you noticed from previous chapters, by default, the animation is done on black background.

Two types of animations are possible using the files output by program **P9_23.PAS**. One is using the **M_3D.LSP** application with **F9_23L.DXF** or **F9_23R.DXF** as input, which resulted in animation file **F9_23.GIF**. The other possibility is to use the screenshots exported to **PCX** by the program (see line #**58**) that were used to produce animation files **F9_23L_PCX.PCX** and **F9_23R_PCX.PCX**. Note that not all simulation frames are exported to **PCX** or **DXF** layers, but rather every fifth screen. The vertices of the *fixed* and *moving centrodes* however are updated every frame so that they will have a smooth appearance in the respective animations.

9.11 PLANETARY GEAR KINEMATIC SIMULATION USING WORKING MODEL 2D

As compared with fixed-axis transmissions, planetary gear trains have gears (called *planets*), the axes of which move on a circular path while meshing with at least two central gears called *sun gears* or *central gears*. The simplest of these transmissions have two DOFs and are known as basic planetary gear trains. There are 12 known such two DOF basic planetary gear trains (Lévai 1968), with those shown in Figures 9.24 and 9.25 being the most commonly used.

Figures 9.24 and 9.25 are screenshots of two **Working Model 2D (WM 2D)** simulations created to illustrate the correlation that exists between the rotational velocities of the central gears, planet gears, and planet carrier of the respective basic planetary gear trains. Because these gear trains have two DOFs, the rotational speed of any of their two bodies must be specified—usually the motion of the central gears or of one central gear and of the planet carrier. If the speeds of the other two bodies are not correctly calculated, then the gears will interfere with each other during simulation.

According to the motion-inversion method due to Willis (Wilson and Sadler 2003), for the planetary unit in Figure 9.24, the following relations hold between the angular velocities ω of the carrier, sun, planet, and ring gears and the number of teeth *N*:

$$\frac{\omega_{Sun} - \omega_{Carrier}}{\omega_{Planet} - \omega_{Carrier}} = -\frac{N_{Planet}}{N_{Sun}} = -\frac{13}{15} \tag{9.29a}$$

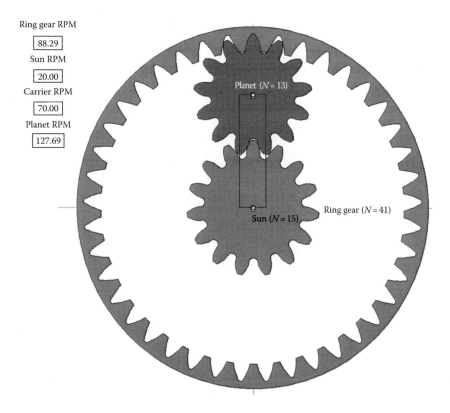

Ring gear RPM

88.29

Sun RPM

20.00

Carrier RPM

70.00

Planet RPM

127.69

Planet ($N = 13$)

Ring gear ($N = 41$)

Sun ($N = 15$)

FIGURE 9.24 **WM 2D** simulation of a basic planetary gear train consisting of one sun gear with 15 teeth, one ring gear with 41 teeth, planet carrier, and a simple planet gear with 13 teeth. See simulation file **PlanetGear1.WM2** and movie file **F9_24.MP4**.

$$\frac{\omega_{\text{Ring}} - \omega_{\text{Carrier}}}{\omega_{\text{Planet}} - \omega_{\text{Carrier}}} = \frac{N_{\text{Planet}}}{N_{\text{Ring}}} = \frac{13}{41} \tag{9.29b}$$

$$\frac{\omega_{\text{Sun}} - \omega_{\text{Carrier}}}{\omega_{\text{Ring}} - \omega_{\text{Carrier}}} = -\frac{N_{\text{Ring}}}{N_{\text{Sun}}} = -\frac{41}{15} \tag{9.29c}$$

Likewise, for the planetary unit in Figure 9.25, the following relations hold:

$$\frac{\omega_{\text{Sun}} - \omega_{\text{Carrier}}}{\omega_{\text{Planet}} - \omega_{\text{Carrier}}} = -\frac{N_{\text{Planet1}}}{N_{\text{Sun}}} = -\frac{15}{13} \tag{9.30a}$$

$$\frac{\omega_{\text{Ring}} - \omega_{\text{Carrier}}}{\omega_{\text{Planet}} - \omega_{\text{Carrier}}} = \frac{N_{\text{Planet2}}}{N_{\text{Ring}}} = \frac{11}{39} \tag{9.30b}$$

$$\frac{\omega_{\text{Sun}} - \omega_{\text{Carrier}}}{\omega_{\text{Ring}} - \omega_{\text{Carrier}}} = -\frac{N_{\text{1Planet}}}{N_{\text{Sun}}} \cdot \frac{N_{\text{Ring}}}{N_{\text{2Planet}}} = -\frac{15}{13} \cdot \frac{39}{11} \tag{9.30c}$$

Note that Equations 9.29 and 9.30 are only two-by-two independent.

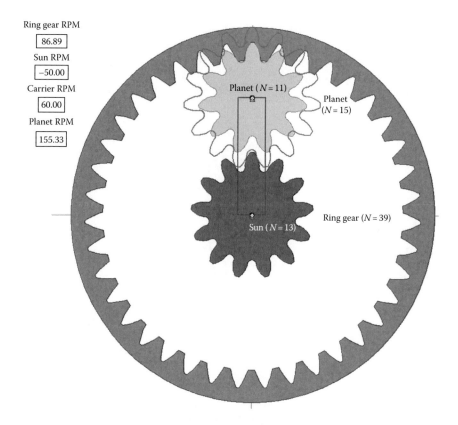

Ring gear RPM

| 86.89 |

Sun RPM

| −50.00 |

Carrier RPM

| 60.00 |

Planet RPM

| 155.33 |

Planet ($N = 11$)

Planet ($N = 15$)

Ring gear ($N = 39$)

Sun ($N = 13$)

FIGURE 9.25 **WM 2D** simulation of a basic planetary gear train consisting of one sun gear with 13 teeth, one ring gear with 39 teeth, and planet carrier and a compound planet with 15 and 11 teeth, respectively. Simulation file **PlanetGear2.WM2** and movie file **F9_25.MP4**.

For the number of teeth of the gear wheels in these simulations as labeled over the respective **WM 2D** bodies visible in Figures 9.24 and 9.25, there is an infinite number of rotational speeds of the respective gears and planet carrier that satisfies these equations. These combinations can include assigning zero rotations per minute (RPM) to one of the central gear or to the planet carrier. Note that if Equations 9.28 and 9.29 are not satisfied, then the respective gears will interfere and overlap as they rotate.

The involute-gear generation application **GearGen0.WM2** introduced in Chapter 8 is actually an extension of the **PlanetGear1.WM2** simulation considered here. To illustrate the concept, the visibility of the carrier, sun gear, and ring gear have been turned off, while the *track outline* of the planet has been turned on and the modified simulation file saved under the name **PlanetGear1x.WM2**.

To generate an external gear with 15 teeth (i.e., the sun gear—see Figure 9.26a), the rotational velocities of the carrier and of the planet must be selected in **PlanetGear1x.WM2** such that the following equality holds:

$$\frac{-\omega_{\text{Carrier}}}{\omega_{\text{Planet}} - \omega_{\text{Carrier}}} = -\frac{N_{\text{Planet}}}{N_{\text{Sun}}} = -\frac{13}{15} \quad \text{or} \quad \frac{\omega_{\text{Planet}}}{\omega_{\text{Carrier}}} = \frac{28}{13} \qquad (9.31)$$

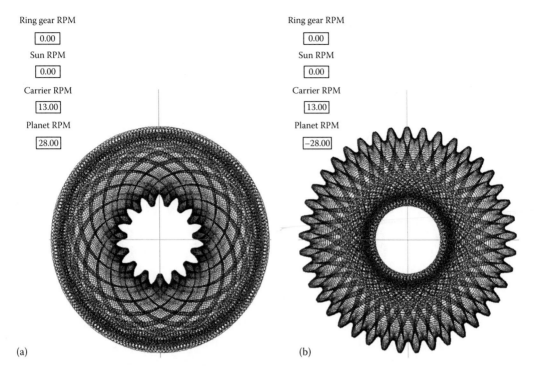

Ring gear RPM
[0.00]
Sun RPM
[0.00]
Carrier RPM
[13.00]
Planet RPM
[28.00]

Ring gear RPM
[0.00]
Sun RPM
[0.00]
Carrier RPM
[13.00]
Planet RPM
[−28.00]

(a) (b)

FIGURE 9.26 Modification of **PlanetGear1.WM2** simulation demonstrating how external gears (a) and internal gears (b) can be generated using a wheel tool. See also simulation **PlanetGear1x. WM2** and animation files **F9_26A.MP4** and **F9_26B.MP4** available with the book.

To generate an internal gear with 41 teeth, that is, the ring gear (see Figure 9.26b), the rotational velocities of the carrier and of the planet must satisfy equalities (9.32) instead:

$$\frac{-\omega_{\text{Carrier}}}{\omega_{\text{Planet}} - \omega_{\text{Carrier}}} = \frac{N_{\text{Planet}}}{N_{\text{Ring}}} = \frac{13}{41} \quad \text{or} \quad \frac{\omega_{\text{Planet}}}{\omega_{\text{Carrier}}} = -\frac{28}{13} \tag{9.32}$$

What is missing from **PlanetGear1x.WM2**, but are implemented in the **WM 2D** simulation **GearGen0.WM2** discussed earlier, is the possibility of adjusting the length of the carrier such that the generation of involute gears (both internal and external) with any number of teeth can be simulated.

9.12 IMPLICIT FUNCTION PLOT

A function defined by an equation that cannot be solved for its variables analytically is called implicit. The following are two such examples:

$$y^3 - x^3 - 10xy + 1 = 0 \tag{9.33}$$

$$xy \cdot \cos(x^2 + y^2) - 1 = 0 \tag{9.34}$$

further referred to generically as $f(x, y) = 0$. One way of representing implicit functions graphically is to first find all roots y of the equation $f(x, y) = 0$ using procedure **ZeroGrid** discussed in Chapter 4, for a number n_x of discrete values x_i within the interval $[x_{min}...x_{max}]$. Comparable results are obtained if, the equation $f(x, y) = 0$ is solved for variable x, assuming a number of n_y discrete values y_j within the interval $[y_{min}...y_{max}]$. The sets (x_i, y) or (x, y_j) thus obtained can then be plotted as point clouds using the **D_2D** program. For increased accuracy, both sets (x_i, y) and (x, y_j) can be generated and plotted together on the same graph.

Program **P9_27.PAS** listed in Appendix B implements this latter strategy and was used to produce data files **F9_27A.D2D** and **F9_27B.D2D** that served to plot the graphs in Figure 9.27. On lines #**23** and #**24** of the program, grid sizes **nX** and **nY** are defined, together with the plot intervals over the respective axes. Depending on the name assigned to the output program (line #**8**), either the function in Equation 9.33 or 9.34 is transmitted to the **ZeroGrid** procedure (see lines #**11** to #**17**). The drawbacks of this implicit-function graphing method is that the plot points are not assembled into polylines and that some points can occur twice, that is, both as a (x_i, y) pair and as a (x, y_j) pair. If only the x variable is scanned at a constant step, the graph may exhibit discontinuities in areas where the tangent to the curve is aligned with $x = x_i$ line. Same may occur if only the y variable is scanned at constant step, not both variables like in program **P9_27.PAS**. Also note that the number of multiple roots of equation $f(x_i, y) = 0$ or $f(x, y_i) = 0$ may exceed the value of the constant **Nmax** defined in the interface section of unit **LibMath**, the case in which the plot will appear truncated.

A different, more efficient method to implicit function plotting is to graph the function $z = f(x, y)$ as top-view level curves with only one level-curve place at $z = 0$ (see Figure 9.28). Data files **F9_28A.D3D** and **F9_28B.D3D** used to generate these two graphs have been output by program **P9_28.PAS** listed in Appendix B. Same as before

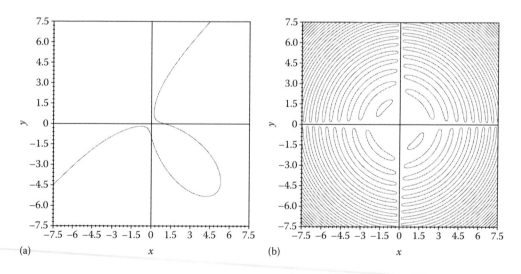

(a)   (b)

FIGURE 9.27   Graphs of the implicit functions in Equation 9.33 (a) and Equation 9.34 (b) produced with **D_2D** (configuration files **F9_27A.CF2** and **F9_27B.CF2**). Data files to produce these plots were generated using program **P9_27.PAS**.

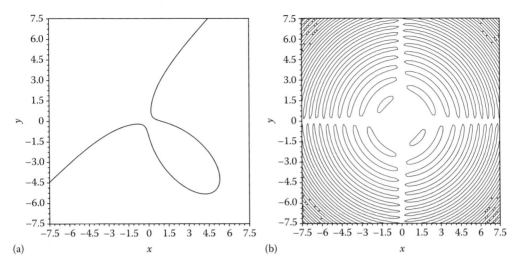

FIGURE 9.28 Plot of the implicit functions in Equation 9.33 (a) and Equation 9.34 (b) produced with **D_3D** (configuration files **F9_28A.CF3** and **F9_28B.CF3**). Data files were generated using program **P9_28.PAS**.

the name assigned to the output program (line #7) controls which function is used to generate the **D3D** data file (see lines #**12** and #**15**).

In terms of the actual **D_3D** program use, there are two ways of producing single-level-curve plots of zero elevation: One possibility is to edit the **CF3** file of the respective plot such that only the value **0.00** is appended to it, and then choose to read the level-curve heights from file. The other possibility is to select evenly spaced level curves, then set their number to one. When only one level curve is specified, **D_3D** will calculate its elevation as the average of the limits over the z-axis. Therefore, with this second approach, the limits over the z-axis must be edited such that they are equal in magnitude but of opposite sign. These two methods have been implemented in configuration files **F9_28A.CF3** and **F9_28B.CF3** used to generate the plots in Figure 9.28. Note that in case of the plot in Figure 9.28a, because of the relatively low resolution at which the function has been sampled, the graph exhibits bridge-like defects and also lacks smoothness at several different places. Increasing the **nX** and **nY** values will reduce or eliminate such artifacts. The fact that the plot consists of continuous lines rather than individual points is a net advantage of this second implicit function plotting method.

9.13 INVERSE AND DIRECT KINEMATICS OF 5R AND 2R ASSEMBLY ROBOTS

This section deals with the inverse and direct kinematics of 5R parallel and 2R serial SCARA robots, like the RP and RH families of micro-assembly robots from Mitsubishi Electric (Figure 9.29).

First, a method of designing the robot endeffector path will be presented. The actual inverse and direct kinematics problems will then be solved using the **RRR**, **gCrank**, and **Crank** procedures from units **LibAssur** and **LibMec2D**. Note that only the **J1** and **J2**

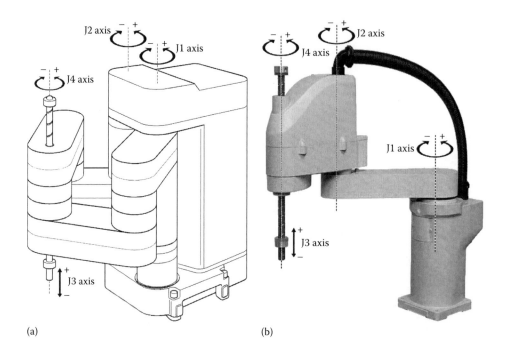

(a)   (b)

FIGURE 9.29   SCARA robots of the 5R (a) and 2R (b) type. Courtesy of **Mitsubishi Electric**.

axes motions shown in Figure 9.29 will be considered, which allow the kinematics problems to be solved in two dimensions. With additional programming effort, however, the remainder degrees of freedom can be accounted for, and accurate 3D models of these robots can be simulated using **AutoCAD** and the **M_3D.LSP** application.

The path to be traced by the endeffector was assumed identical to the shape of the vanes of the iris mechanisms in Section 9.9. Similarly, we begin with a plot file, that is, **VANE0. PLT** of the original drawing **VANE0.DWG**, and open this file using **UTIL~PLT.EXE**. **VANE0.PLT**, with its four semicircles converted to polylines, was exported to **DXF** (file name **VANE1.DXF**). In order to bring the file **VANE1.DXF** back to the origin and proportions of **VANE0.DWG**, prior to **DXF** export, the limits inside **UTIL~PLT** were edited such that $x_{min} = -0.5, x_{max} = 9.5, y_{min} = -0.5$, and $y_{max} = 5.0$. The file **VANE1.DXF** was then opened inside **AutoCAD** (see file **Vane2.DWG**), and its constituent polylines were connected into a single polyline. After that, the drawing was exported back to **R12 DXF** under the name **VANE2.DXF**. This new **DXF** file was then opened using **UTIL~DXF.EXE**, and the vertices of its only polyline exported to ASCII—the ASCII file name has been changed from its default value to **VANE2.XY**. Using **UTIL~TXT.EXE**, file **VANE2.XY** was further edited as follows (see configuration file **VANE.CON**): First, linearly interpolated points were added such that the distance between vertices is decreased to about 0.03 units. This resulted in the intermediate file **DELETE.ME**. Every fourth data point of this intermediate file was extracted to file **RoboPath.XY**, a plot of which is available in Figure 9.30. In addition, every 14th data point from **DELETE.ME** was written to a second file named **VERTEX.XY**, which can be used interchangeably by the **Shape** procedure inside programs **P9_21.PAS** and **P9_22.PAS**, discussed earlier.

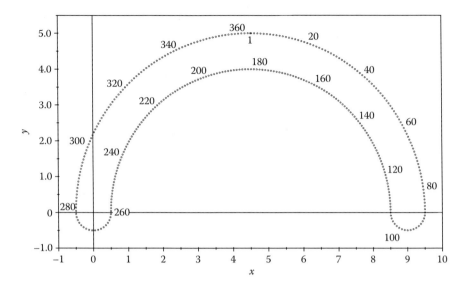

FIGURE 9.30   Plot of the prescribed endeffector path read from file **RoboPath.XY**. Configuration file **F9_30.CF2**.

The ASCII file **RoboPath.XY** thus obtained served as input to the inverse kinematic analysis program **P9_31.PAS** and to the program **P9_34.PAS** discussed in Appendix B. The program drives pin joint C shared by two RRR dyads (i.e., A_1B_1C and A_2B_2C in Figure 9.30) through the points read from file **RoboPath.XY**. Using the assumed velocity **vC** of the endeffector defined on line #**28**, the program calculates the time required for joint C to travel between every two successive path points (line #**52**), and then writes these accumulated time values, starting with **t** = 0, to file **F9_31.DaTA** (line #**72**). Also, the outputs to the file **F9_31.DTA** are the joint angles θ_{A_1}, θ_{A_2}, θ_{B_1}, and θ_{B_2}, defined as shown in Figure 9.31b, which were then used to plot the graphs in Figure 9.31b. In addition, the program writes to **DXF** every fourth frame of the simulation (see line #**48**). This output **R12 DXF** file, named **F9_31.DXF**, was then used to generate the animated **GIF** file **F9_31.GIF**, also shown Figure 9.31a.

The joint angle values recorded to file **F9_31.DTA** served as input to programs **P9_32.PAS** and **P9_33.PAS** listed in Appendix B. The first of these programs performs a direct kinematic analysis of the 5R parallel robot (Figure 9.32), while the second program performs the same type of simulation of the 2R serial robot (Figure 9.33).

Program **P9_32.PAS** reads from file **F9_31.DTA** link lengths $A_1B_1 = A_2B_2 = AB$ and $B_1C = B_2C = BC$, and ground joint coordinates (x_{A_1}, y_{A_1}) and (x_{A_2}, y_{A_2})—see lines #**27** and #**28** of the program and Figure 9.32. During the simulation cycle, it then reads angle values θ_{A_1} and θ_{A_2} (line #**35**) and uses them as input to cranks A_1B_1 and A_2B_2. The 5R robot mechanism is completed using the RRR dyad B_1CB_2 on lines #**44**, #**45**, and the locus of its middle joint is recorded on the screed and to the **DXF** output file (line #**56**).

Program **P9_33.PAS** performs a direct kinematic analysis of a 2R robot. Depending on the value of parameter **LftRgt** set on line #**15**, during the simulation cycle the program uses either angles θ_{A_1} and θ_{B_1} with the ground-joint centered at (x_{A_1}, y_{A_1}), or angles θ_{A_2} and θ_{B_2} and ground joint at (x_{A_2}, y_{A_2}). These values, read from file **F9_31.DTA** (see lines #**39** to #**46**),

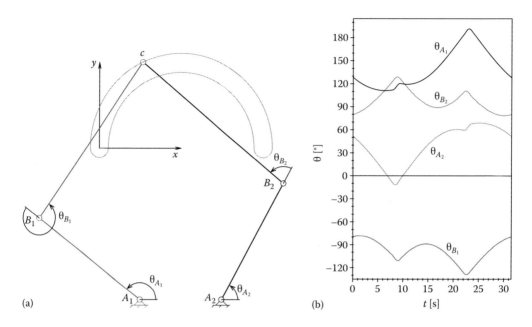

(a)   (b)

FIGURE 9.31 (a) Inverse kinematic analysis of a 5R robot done using two RRR dyads running in parallel, and (b) plot of the joint angle values recorded by the **P9_31.PAS** program. See also animation file **F9_31.GIF** and **D_2D** configuration file **F9_31.CF2**.

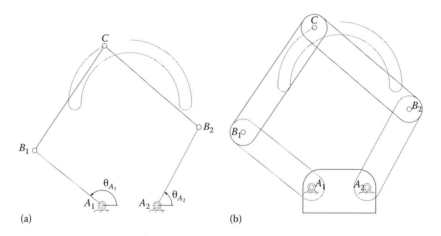

(a)   (b)

FIGURE 9.32 Direct kinematic analysis of a 5R robot modeled as two cranks, that is, A_1B_1 and A_2B_2, amplified with an RRR dyad, that is, B_1-C-B_2. Both a simplified representation of the robot (a) and a more realistic representation using the **Link** and **Base** procedures (b) are shown. See also animated **GIF** files **F9_32a.GIF** and **F9_32b.GIF**.

correspond to the left-hand and right-hand orientation of the robot, respectively (Figure 9.33). The two angles chosen are then used to drive cranks AB and BC, as shown in Figure 9.33. Same as in program **P9_32.PAS**, also read from file **F9_31.DTA** are the crank lengths AB and BC of the robot (see lines #**30** and #**31**).

Note that in both these direct kinematic analysis programs, by setting the **Sticks** constant to zero on line #**15**, the **Link** procedures are called with the width parameter set to

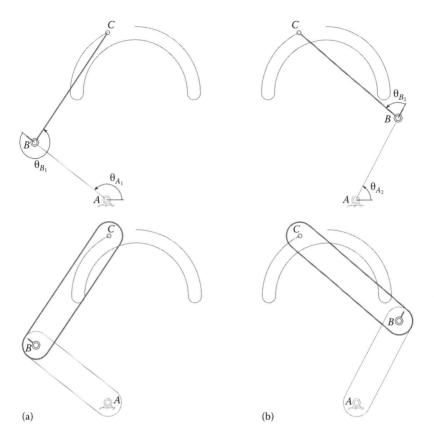

FIGURE 9.33 Direct kinematics of 2R robots modeled as two cranks (*AB* and *BC*) in series, in the left hand (a) and right hand (b) configurations. Both simplified and more realistic representations (i.e., using the **Link** procedure) are shown. See also animated **GIF** files **F9_33a.GIF** and **F9_33b.GIF**.

zero, which will cause the links to be represented as lines, rather than filleted rectangles. Also note that the two input angles are visualized only in the **Sticks = 0** mode.

If only the 2R serial robot kinematics is of concern, program **P9_31.PAS** can be modified such that only one RRR dyad is driven through the endeffector path points. The direct kinematic analysis program **P9_33.PAS** must also be modified, such that at each iteration step, only a pair of angle is read from the input data file.

9.14 INVERSE AND DIRECT KINEMATICS OF THE RTRTR GEARED PARALLEL MANIPULATOR

The discussion on the kinematics of SCARA robots is continued in this section, where the case of the RTRTR kinematic chain configured as shown in Figure 9.34a will be considered. This rack-and-pinion actuated planar parallel manipulator appears to be of a new configuration, not yet described in literature.

Two computer programs will be introduced in Appendix B, that is, **P9_34.PAS** and **P9_35.PAS**. The first program is an inverse kinematic analysis program similar to **P9_31. PAS**. It reads the *x* and *y* coordinates of the path in Figure 9.30 (see lines **#13** and **#54**), which

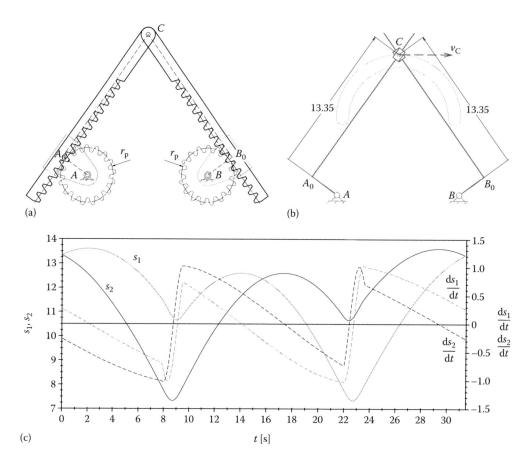

FIGURE 9.34 Geared RTRTR parallel manipulator (a), animation frame output by program **P9_34.PAS** (b), and plot of the linear motor displacements s_1, s_2 and velocities ds_1/dt, ds_2/dt required for constant endeffector speed $v_C = 1$ (c). See also animation file **F9_34b.GIF** and **D_2D** configuration file **F9_34C.CF2**.

are then used to calculate, using the Pythagoras theorem, linear motor displacements s_1 and s_2 (lines #**55** and #**56**). Using the imposed **vC** endeffector velocity defined on line #**16**, the program calculates, beginning with the second position point, the time increase **dt** and the corresponding linear actuator velocities ds_1/dt and ds_2/dt (see lines #**58** to #**61**). Lines #**63** to #**75** of the program serve to animate an RTRTR kinematic chain using the calculated linear motor displacements s_1, s_2, and provide some visual feedback to the user, including the display of the locus of point C and its constant velocity vector (see Figure 9.34b). These lines can be eliminated, however, the output to data file **F9_34.DTA** of parameters s_1, s_2, ds_1/dt and ds_2/dt being essential, together with the time value done (line #**78**). A plot of these linear motor input parameters is available in Figure 9.34c.

The companion program **P9_35.PAS** reads from the file **F9_34.DTA** (produced with **P9_34.PAS**) the RTRTR linear actuator displacements s_1, s_2 and velocities ds_1/dt and ds_2/dt, as well as the corresponding time t (see lines #**15** and #**55**). Also read from this data file are the ground joint coordinates (x_A, y_A), (x_B, y_B) and linear actuator eccentricities A_0A and B_0B (see line #**40**). Required in the analysis is the pitch radius r_p of the two input pinions (see Figure 9.34a and line #**28** of the program). The program calls procedure **RTRTR** for every position read from the input **DTA** file, and using the linear motor displacements and position angles returned by the procedure (see variable **Phi1** and **Phi2**), it calculates the required input pinion angles using the following equations:

$$\theta_1 = \frac{\varphi_1 - s_1}{r_p + \varphi_{10}}$$

$$\theta_2 = \frac{\varphi_2 + s_2}{r_p + \varphi_{20}}$$

(9.35)

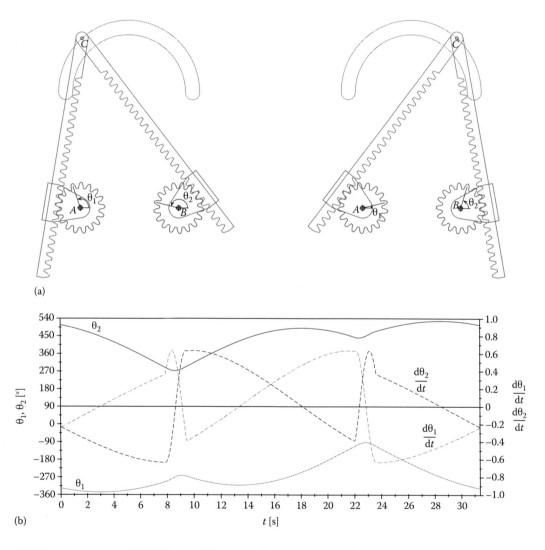

(a)

(b)

FIGURE 9.35 Geared RTRTR parallel manipulator animation frames generated by the program **P9_34.PAS** (a), and plot of the required pinion angular displacements θ_1, θ_2 and angular velocities θ_1/dt and θ_2/dt required for a constant endeffector speed $v_C = 1$ (b). See also animation file **F9_35a. GIF** and **D_2D** configuration file **F9_35B.CF2**.

where the corresponding variables to angles θ_1 and θ_2 are **Thta1** and **Thta2**. Although in program **P9_35.PAS** they are both equal to zero (see lines #**56** and #**61**), depending on the orientation of the pinion when the shape file **RTRTR0.XY** has been generated, nonzero constant angles φ_{10} and φ_{20} might be required in Equation 9.34, in order to properly align the pinions with their rakes. Using finite differences, the time derivatives of the pinion angles (variable names **dThta1** and **dThta2**) are also calculated and are written to the output file **F9_35.DTA**. This data file is then used to generate the plot of the input pinions angular displacement and angular velocity in Figure 9.35b.

In addition to numerical calculations, the program animates the mechanism using polygonal shapes read from the following ASCII files: **RTRTR0.XY** (pinion), **RTRTR1.XY** (left rack), **RTRTR2.XY** (right rack), and **RTRTR3.XY** (pinion bracket). Similar to the way the endeffector path file **RoboPath.XY** was generated, these shapes were first drawn inside **AutoCAD**, then they were printed to **PLT** to convert arches of circles to polylines. Next, using the **Util~PLT** program, these **PLT** files were converted to **DXF** and were opened with **AutoCAD**. From there they were exported to **R12 DXF**, and, finally, using the **Util~DXF** program, the **XY** shape files were generated (see the files of the form **RTRTR*.*** available with the book).

9.15 KINEMATIC ANALYSIS OF A HYDRAULIC EXCAVATOR AND OF A ROPE SHOVEL

The subject of this section is the kinematic simulation of the digging mechanisms of a hydraulic excavator and of a rope shovel. The yaw motion associated with dumping the load will not be considered, which allows these simulations to be performed in two dimensions using the procedures available from units **LibMecIn**, **LibAssur**, and **LibMec2D**.

A compact hydraulic excavator similar to model **27D** from **John Deere**, or model **301** from **Caterpillar** (Figures 9.36 and 9.37) will be analyzed first. The excavator

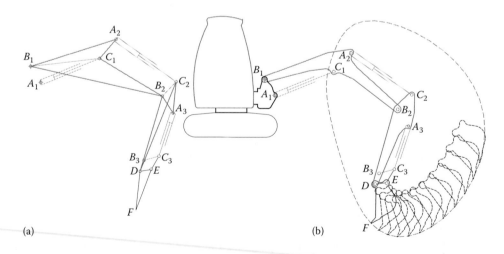

(a) (b)

FIGURE 9.36   Excavator arm modeled using three RTRR actuators, six offset points, and one RRR dyad (a) and motion simulation of the same arm done using shapes attached to the moving links (b). See also animated **GIF** files **F9_36a.GIF** and **F9_36b.GIF**.

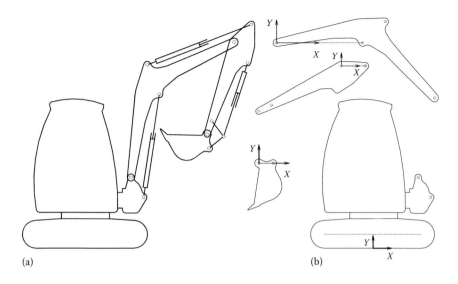

FIGURE 9.37 Schematic of a compact hydraulic excavator (a) and its main components represented as polylines defined relative to the local reference frames shown (b). See also files of the form **EX*.*** available with the book.

arm was modeled using three RTRR actuators (A_1–C_1–B_1, A_2–C_2–B_2, and A_3–C_3–B_3) arranged in series, interconnected via offset points A_2, B_2 of link B_1C_1 and offset points A_3, B_3 of link B_2C_2. One RRR, dyad that is, the bucket-swing amplifier D–E–C_3, was also included. Note that pin joint D is an offset point of link B_2C_2, while the tip of the bucket (the locus of which has been recorded during simulation) is an offset point of link DE (see Figure 9.36a).

Program **P9_36A.PAS**, available with the book, performs the kinematic analysis of the excavator arm, as shown in Figure 9.36a. The companion program **P9_36A.PAS** listed Appendix B is an extension of the former, where shapes read from files **EXbody. XY**, **EXboom.XY**, **EXstick.XY**, and **EXbucket.XY** (lines #52, #55, #62, #89 and #94) are added to the model. Of these, the excavator body shape file **EXbody.XY** consists of four polylines, three of them having their color set from within the actual file. These polylines associated with the excavator body and with the moving parts of the digging arm, together with the joint location, were extracted from the raster images of a compact excavator as follows: The raster image file was opened inside **AutoCAD** and was scaled to match the overall dimensions of the real excavator. Then the coordinates of the joint center were marked with small circles. Polylines representing the excavator parts were then overlaid to the raster image, and then each was extracted to a separate **DWG** file. After orienting them as shown in Figure 9.35b, they were exported to **PLT**, so that arches of circles are converted to vertex polylines. Using the **Util~PLT** application, these shapes were converted back to **DXF**, and then were opened inside **AutoCAD**. The constituent polylines were joined together using the *pedit* command (if it was the case), were scaled back to their original size, and were positioned relative to the world coordinate system of the drawing, as shown in Figures 9.37b. From inside **AutoCAD**, one more export to **R12 DXF** has been performed, and then

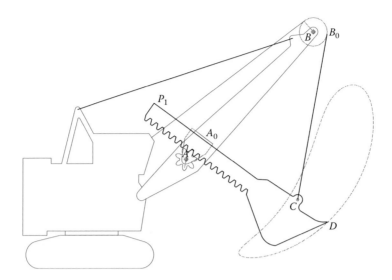

FIGURE 9.38 Simulation of a rope shovel performed using the program **P9_38.PAS**. See also animation file **P9_38.GIF** and drawing file **RopeShovel.DWG**.

using the **Util~DXF** program, the **XY** shape files used by program **P9_36B.PAS** (see Appendix B) have been finally produced.

The motions s_1, s_2, and s_3 of the three actuators of the excavator are harmonic functions of time. Different paths can be obtained by changing the phase angle and amplitudes on lines #**49**, #**50**, and #**51** of program **P9_36B.PAS**, and their effect upon the workspace of the excavator links and bucket and locus of point D observed.

The second part of this section explores a similar problem of the kinematic analysis of a rope shovel used in surface mines and quarries (Figure 9.38). Same as for the hydraulic excavator discussed earlier, the shapes of the stationary body and of the moving

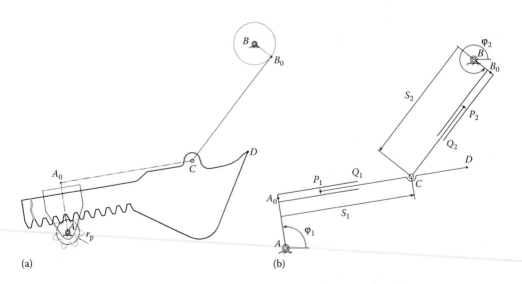

FIGURE 9.39 Rope shovel mechanism (a) and the equivalent RTRTR kinematic chain (b).

links have been extracted to ASCII files of extension **XY** (see the corresponding files **RopeShovel.DWG** and **RS*.***).

As can be seen in Figure 9.39, the digging mechanism of the rope shovel can be easily modeled as a RTRTR kinematic chain using either the **RTRT** or **RTRTc** procedures in unit **LibMecIn** (see also line #36 of program **P9_38.PAS** listed in Appendix B). The linear motor inputs s_1 and s_2 are harmonic functions defined on lines #34 and #35 of this program. Then, using angle values φ_1 and φ_2 (see Figure 9.39) evaluated on lines #43 and #47, crank angles **Theta1** and **Theta2** are calculated. These serve to insert the pinion shape and also to show the position angle of the pinion and of the rope sheave. This way more realistic simulations can be generated, as shown in the animated **GIF** file **P9_38.GIF** available with the book.

9.16 KINEMATIC ANALYSIS OF INDEPENDENT WHEEL SUSPENSION MECHANISMS OF THE MULTILINK AND DOUBLE-WISHBONE TYPE

Suspension systems of automobiles are complex 3D mechanisms. They are tuned to satisfy the multiple requirements associated with the motion of car wheels relative to the chassis, and of the chassis relative to the ground, during acceleration, braking, and turning maneuvers. In this section, the displacement of the wheel relative to the car body of *five-link*, *four-link*, and *double-wishbone* suspension mechanisms with rectilinear steering input will be analyzed. The wheel track, camber, and toe angle variations of such mechanisms will be determined in an iterative approach following a method described in Simionescu and Beale (2002). The problem will be formulated for the general *five-link* suspension mechanism as schematized in Figure 9.40, of which the *four-link* and the more commonly used *double-wishbone* suspensions are particular embodiments, obtained by making ball joints B_4 and B_5 and/or B_2 and B_3 coincident, respectively.

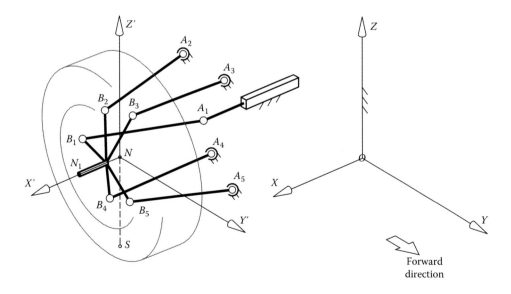

FIGURE 9.40   Five-link suspension mechanisms with translational steering input.

Without the trivial rotations of the connecting links around their own axes, a *five-link* rear wheel suspension has only one degree of freedom. In front-wheel suspension mechanisms, a second degree of freedom is available, corresponding to the steering input. It means that the position of the wheel can be uniquely specified by the coordinate z_N relative to the fixed reference frame $Oxyz$ attached to the car body and by the rack-end displacement x_{A_1}. The remaining position parameters (i.e., wheel-center coordinates x_N, y_N and angles α, β, and γ of the moving frame $Nx'y'z'$ attached to the wheel carrier relative to the fixed reference frame) can be determined by solving the following system of five simultaneous equations:

$$\left(x_{A_i} - x_{B_i}\right)^2 + \left(y_{A_i} - y_{B_i}\right)^2 + \left(z_{A_i} - z_{B_i}\right)^2 = l_i^2 \quad (i = 1,\ldots,5), \tag{9.36}$$

where l_i is the length of link A_iB_i. The coordinates x_{B_i}, y_{B_i}, and z_{B_i} in Equation 9.36 result from applying a rotation, followed by a translation to the coordinates x'_{B_i}, y'_{B_i}, and z'_{B_i} of ball joint B_i originally specified in the $Nx'y'z'$ moving frame according to the equation

$$\begin{bmatrix} x_{B_i} \\ y_{B_i} \\ z_{B_i} \end{bmatrix}_{Oxyz} = \left[R_{\beta\alpha\gamma}\right] \cdot \begin{bmatrix} x'_{B_i} \\ y'_{B_i} \\ z'_{B_i} \end{bmatrix}_{Nx'y'z'} + \begin{bmatrix} x_N \\ y_N \\ z_N \end{bmatrix}_{Oxyz}. \tag{9.37}$$

In this equation, matrix $[R_{\beta\alpha\gamma}]$ transforms the $Nx'y'z'$ reference frame into a frame parallel to $Oxyz$, by rotating it by angles β, α, and γ (in this order). With the notations $c\alpha = \cos \alpha$, $s\alpha = \sin \alpha$, and so forth, this transformation matrix can be written as

$$\left[R_{\beta\alpha\gamma}\right] = \begin{bmatrix} c\alpha \cdot c\beta & -s\alpha & c\alpha \cdot s\beta \\ s\alpha \cdot c\beta \cdot c\gamma + s\beta \cdot s\gamma & c\alpha \cdot c\gamma & s\alpha \cdot s\beta \cdot c\gamma + c\beta \cdot s\gamma \\ s\alpha \cdot c\beta \cdot s\gamma + s\beta \cdot c\gamma & c\alpha \cdot s\gamma & s\alpha \cdot s\beta \cdot s\gamma + c\beta \cdot c\gamma \end{bmatrix}. \tag{9.38}$$

For a given value of the wheel vertical displacement z_N and steering rack input x_{A_1}, the system of five equations (9.36) in the unknowns α, β, γ, x_N, and y_N can be conveniently solved by minimizing the following objective function:

$$F\left(\alpha,\beta,\gamma,x_N,y_N\right) = \sum_{i=1}^{5} \left[\left(x_{A_i} - x_{B_i}\right)^2 + \left(y_{A_i} - y_{B_i}\right)^2 + \left(z_{A_i} - z_{B_i}\right)^2 - l_i^2\right]^2. \tag{9.39}$$

Once parameters α, β, γ, x_N, and y_N are determined for successive z_N and/or x_{A_1} values, the change of the wheel track Δy_S, wheel base Δx_S, camber angle $\Delta\delta$, and toe angle $\Delta\varphi$ can then be calculated. The first two parameters require evaluating the coordinates of the contact patch center S using an equation similar to (9.37). The camber angle is calculated as the angle between the Oz-axis and the projection on the vertical plane Oxy of line NN_1 (i.e., the wheel axis). In turn, the toe angle is determined as the angle between the Ox-axis of the fixed frame and the projection on the horizontal plane Oxy of the same line NN_1.

This strategy has been implemented in the kinematic analysis program **An_5link.PAS** available with the book. The program reads from input data files **5link.AN**, **4link.AN**, and **3link.AN** the values of the following parameters (see Figure 9.40):

- Wheel base length over wheel track length of the vehicle used to calculate the angle of steer of the left wheel versus that of the right wheel according to the Ackermann principle.

- The rebound and jounce limits $\Delta z_{N\min}$ and $\Delta z_{N\max}$ of the wheel, measured from the static position corresponding to the center of the wheel N being located at point $\left(z_{N_0}, x_{N_0}, y_{N_0}\right)$.

- The horizontal, lock-to-lock travel of the steering rack $\Delta x_{A_1\max}$.

- Link lengths l_1 to l_5.

- Outer radius and length of the wheel hub, assumed to be a cylinder starting at the middle point N of the wheel and extending inwards.

- Radius of ball joints A_1 to A_5 and B_1 to B_5, considered all identical.

- Wheel radius, equal to the distance NS.

- Coordinates z_{N_0}, x_{N_0}, and y_{N_0} of the center of the wheel in the straight ahead, static position of the vehicle.

- Coordinates x_{A_i}, y_{A_i}, and z_{A_i} $(i = 1,\ldots,5)$ of the ball joints attached to the chassis relative to the $Oxyz$ reference frame.

- Coordinates x'_{B_i}, y'_{B_i}, and z'_{B_i} $(i = 1,\ldots,5)$ of the ball joints attached to the wheel carrier relative to the moving reference frame $Nx'y'z'$.

Note that it is not essential to provide the link lengths l_{1-5} since they can be calculated as the distance between the joint centers A_i and B_i for the wheel in its reference position. For such an option, you must replace the corresponding numbers in the input **AN** file with nonnumeric characters. Conversely, you can use the **An_5link.PAS** program to verify the effect of altering the length of any of these five links upon the kinematics of the mechanism.

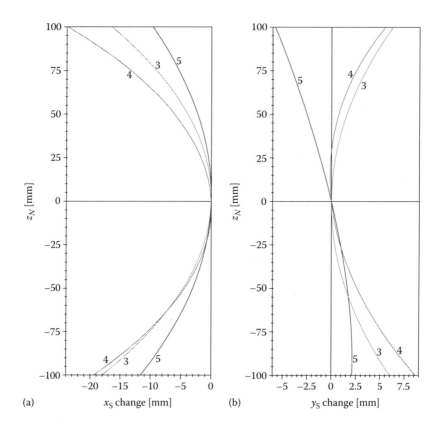

FIGURE 9.41 Wheel track alteration Δx_S (a) and recessional wheel motion Δy_S (b) during jounce and rebound of the *five-link* (curve 5), *four-link* (curve 4), and *double-wishbone* (curve 3) suspension mechanisms with geometry read from files **5link.AN**, **4link.AN**, and **3link.AN**, respectively. The configuration files to redo these plots are **F9_41A.CF2** and **F9_41B.CF2**.

Program **An_5link.PAS** outputs two ASCII files with the same name (specified by the user) and extensions **DTA** and **M3D**. The first file contains the kinematic analysis data, while the second file, readable by the **M_3D.LSP** application, allows the motion of the mechanism to be simulated inside **AutoCAD**. The drawing **Wheel.DWG** available with the book should be used with these **M3D** files because it contains a block named "wheel" required by these simulations.

Using the aforementioned files of extension **AN** as inputs, files **5link_H.DTA**, **5link_V.DTA**; **4link_V.M3D**, **4link_H.M3D**; and **3link_V.M3D**, **3link_H.M3D** have been generated. Of these, the he ***_V.DAT** files (corresponding to the wheel performing vertical motion for z_N changing value between −100 and 100 mm around z_{N_0}) served to plot the graphs of the wheel track change Δx_S, recessional wheel motion Δx_S, camber angle change $\Delta \delta$, and toe angle change $\Delta \varphi$ available in Figures 9.41 and 9.42. The ***_H.DAT** files (wheel performing steering motion caused by changing x_{A_1} between −70 and 70 mm around its reference position) were used to generate the wheel steer graphs in Figure 9.43. The corresponding **M3D** files served to produce Figures 9.43 through 9.46 and the companion animated **GIF** files available with the book.

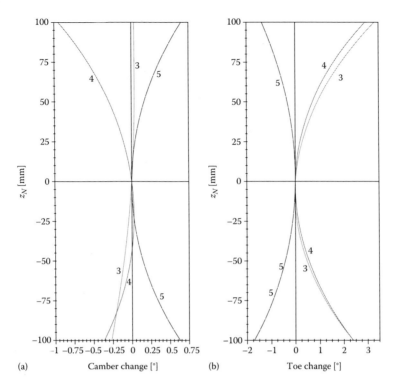

FIGURE 9.42   Camber alteration $\Delta\delta$ (a) and toe angle alteration $\Delta\varphi$ (b) during jounce and rebound companion to the graphs in Figure 9.42. Configuration files **F9_42A.CF2** and **F9_42B.CF2**.

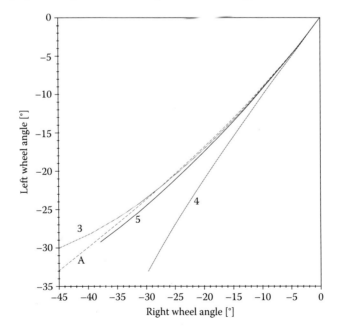

FIGURE 9.43   Wheel steer angle correlation of the *five-link* (curve 5), *four-link* (curve 4), and *double-wishbone* (curve 3) suspension mechanisms in Figures 9.41 and 9.42, overlapped with the Ackermann law (curve *A*). Configuration file **F9_43.CF2**.

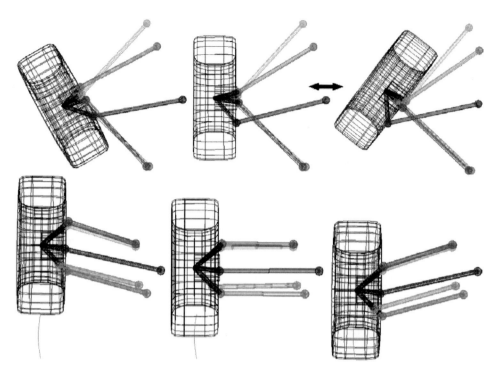

FIGURE 9.44   Limit positions of the five-link suspension mechanism with the geometry read from file **5link.AN**. See also animated **GIF** files **F9_44a.GIF** and **F9_44b.GIF**.

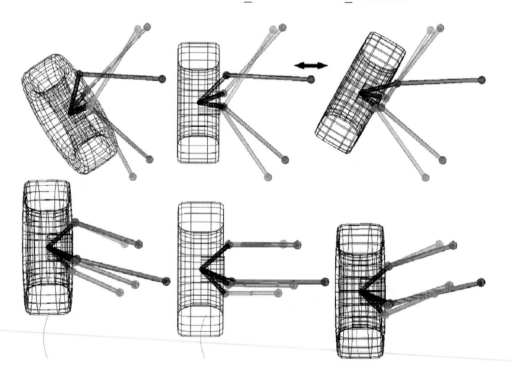

FIGURE 9.45   Limit positions of the four-link suspension mechanism with the geometry read from file **4link.AN**. See also animated **GIF** files **F9_45a.GIF** and **F9_45b.GIF**.

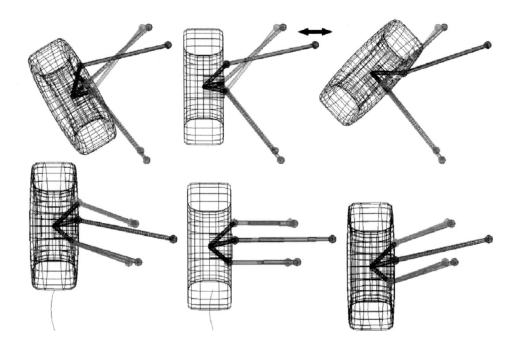

FIGURE 9.46 Limit positions of the double-wishbone suspension mechanism with the geometry read from file **3link.AN**. See also animated **GIF** files **F9_46a.GIF** and **F9_46b.GIF**.

9.17 FLYWHEEL SIZING OF A PUNCH PRESS

Flywheels are used to reduce the speed fluctuations during the working cycle of a machine. They increase their rotational speed when there is an excess of energy and decrease their rotational speed to release energy when there is not enough available. Flywheels serve a function similar to accumulators used in pneumatic or hydraulic circuits, which maintain nearly constant fluid pressure while the demand varies. In piston engines, the flywheel compensates for the strokes when energy is consumed rather than created during the engine cycle, thus allowing the crankshaft torque to be delivered at close to constant speed. In case of punch presses, like the one in Figure 9.47, the actual punching occurs for only a small fraction of the machine cycle, causing a strongly fluctuating load torque. To limit the size of the motor, and also to alleviate its speed fluctuation (electric motors are known to work best at certain rpm), the energy delivered during the actual punch is supplemented by the energy released by the flywheel as it slows down from a maximum angular velocity ω_{max} right before the punch, to a minimum angular velocity ω_{min} right after the punch ends.

In this section, the problem of selecting the electric motor and that of sizing the flywheel required by a punch press will be discussed. The flywheel is assumed mounted on the crankshaft, which is driven by the motor via a speed reducer as shown in Figure 9.47. The mechanism of the press is a centric crank–slider with the crank length $OA = 0.15$ m, coupler length $AB = 0.5$ m, and punch-head length $BP = 0.15$ m. The press punches $d = 65$ mm diameter holes into $h = 20$ mm thick aluminum stock of shear strength $S_{Sy} = 140$ MPa, at a rate of $np = 80$ holes per minute. The punch begins when the displacement s of the punch

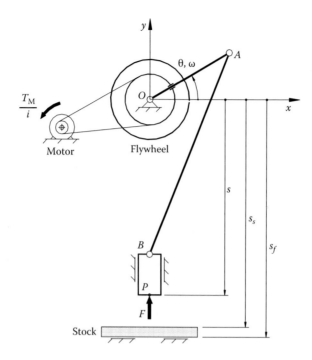

FIGURE 9.47   Schematic of a crank–slider punch press.

head equals s_s and ends when s equals s_f (see Figure 9.47). We assume that $s_f = 0.75$ m and correspondingly $s_s = s_f - h = 0.73$ m. Using the *principle of virtual work*, we will find the resisting torque at the crankshaft as a function of the crank angle θ, and then we will evaluate the required average motor torque and its corresponding power. Then we will calculate the moment of inertia I of a flywheel for which the motor speed fluctuation ranges between 1740 and 1580 rpm. A **Working Model 2D** simulation that validates the calculations is also provided and is available with the book.

The problem will be solved under the following simplifying assumptions:

- The stock material exhibits an ideal plastic behavior.

- The friction in the joints of the crank–slider mechanism and between the punch and the aluminum stock is considered negligible.

- The output torque of the driving motor is assumed constant and independent of speed.

- The transmission between the motor and the crankshaft is 100% efficient, and its speed ratio i remains unchanged.

- The inertias of the moving links of the press and of the motor armature are neglected.

We define the average rotational speed n_0 of the crankshaft of the press as

$$n_0 = \frac{1}{2}\left(n_{max} + n_{min}\right) = \frac{1}{2}\left(\frac{1740}{i} + \frac{1580}{i}\right) \text{rpm.} \qquad (9.40)$$

This is also equal to the imposed number of punches per minute n_p, which yields the ratio of the transmission between the motor and the crankshaft $i = 20.75$. Correspondingly, the average speed of the electric motor is 1660 rpm, while the minimum and maximum speeds of the crankshaft will be $n_{max} = 83.855$ rpm and $n_{min} = 76.145$ rpm.

The coefficient of speed fluctuation C_S of the press will be

$$C_S = \frac{n_{max} - n_{min}}{n_0} = \frac{83.855 - 76.145}{80} = 0.0964. \tag{9.41}$$

C_S is recommended to be around 0.1 for punch presses, 0.005 for electric generators, and 0.2 for rock crushers. It means that the coefficient of speed fluctuation in Equation 9.41 is satisfactory.

The maximum resisting force F_{max} opposing the punch occurs when the plate material begins to yield. Assuming a shear stress τ versus punch penetration like the one in Figure 9.48a with a zero elastic range, the punch force F will peak right after the punch head makes contact with the stock and decreases to zero as the penetration progresses. The maximum punch force depends on the shear strength of the material and on the side area of the hole A_S (i.e., the shear area) according to the equation

$$F_{max} = S_{Sy} \cdot A_S = S_{Sy}(\pi d \cdot h) = 140 \cdot 10^6 \cdot \pi \cdot 0.065 \cdot 0.02 = 5.72 \cdot 10^5 \text{ N}. \tag{9.42}$$

Since A_S varies from a maximum value to zero, the punch force will decrease linearly with the punch-head penetration (see also Figure 9.48b), that is,

$$F(s) = \begin{cases} \dfrac{s_f - s}{s_f - s_s} F_{max}, & \text{for } s_s \leq s \leq s_f, \\ 0, & \text{otherwise.} \end{cases} \tag{9.43}$$

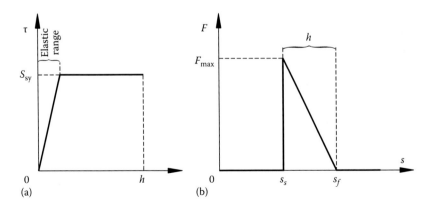

FIGURE 9.48 Diagrams of the shear stress τ function of punch penetration (a) and punch force F function of punch displacement for a material with zero elastic range (b).

In order to apply the *principle of virtual work* and evaluate how the punch force in Equation 9.43 translates into the resisting torque at the crankshaft, the kinematic analysis program **P9_49.PAS** has been written and is listed in Appendix B.

The program reads from lines #14 to #17 and #23 to #27 a number of parameters of the punch press and workpiece material. It uses these values to perform a position and velocity analysis, accompanied by an animation of the mechanism. The program calculates, for **nPoz** discrete crank positions, the punch force according to Equation 9.43—see lines #51 to #55 of the program. Then, using the punch velocity **vyP** returned by procedure **RRT_** (labeled **ds/dt** in the output data file **F9_49.DAT**), it calculates the load torque transmitted to the crank as

$$T(\theta) = \frac{ds}{d\theta} F = \frac{ds/dt}{\omega} F. \tag{9.44}$$

In addition to data file **F9_49.DAT**, program **P9_49.PAS** outputs the multilayer **DXF** file **F9_49.DXF** that was used to generate Figure 9.49a and the animated **GIF** file **F9_49a.GIF**. Data file **F9_49.DAT** served to produce the plots in Figure 9.49b, showing the areas under the $T(\theta)$ and $F(s)$ curves labeled "integrals." Using these integral values, the work W_P required to punch one hole and the average crankshaft torque T_M were determined:

$$W_P = 5782.8 \text{ J}, \tag{9.45}$$

$$T_M = \frac{5782.8}{2\pi} = 920.36 \text{ N m}. \tag{9.46}$$

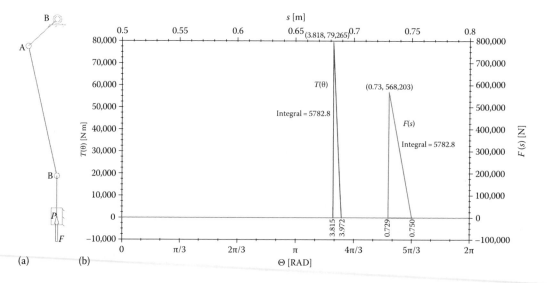

(a) (b)

FIGURE 9.49 Simulation frame generated with program **P9_49.PAS** (a) and diagrams of the load torque T versus crank angle θ and punch force F versus punch displacement s (b), obtained by overlapping inside **AutoCAD** the plots done using configuration files **F9_49B1.CF2** and **F9_49B2.CF2**.

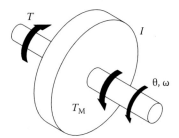

FIGURE 9.50   Free-body diagram of the flywheel of the punch press.

The corresponding electric motor power to be used in conjunction with a flywheel will therefore be

$$P_{M} = \frac{n_{P}}{60} \cdot W_{P} = \frac{80}{60} \cdot 5782.8 = 7.71 \text{ kW}$$

$$= T_{M} \cdot \frac{n_{0} \cdot \pi}{30} = 920.36 \cdot \frac{80 \cdot \pi}{30}, \tag{9.47}$$

where $\omega_{0} = 80\pi/30 = 8.3776$ rad/s is the average angular velocity of the crankshaft.

Figure 9.50 is a free-body diagram of the flywheel removed from the crankshaft. Its equation of motion is

$$T_{M} - T(\theta) = I \frac{d^{2}\theta}{dt^{2}}, \tag{9.48}$$

equivalent to

$$T_{M} - T(\theta) = I \frac{d\omega}{dt} = I \frac{d\theta}{dt} \cdot \frac{d\omega}{d\theta} = I\omega \frac{d\omega}{d\theta}. \tag{9.49}$$

We integrate this second equation between crank angles θ_{1} and θ_{2} when the maximum and minimum angular velocities of the crankshaft occur,

$$\int_{\theta_{1}}^{\theta_{2}} (T_{M} - T)d\theta = \int_{\omega_{max}}^{\omega_{min}} I\omega d\omega, \tag{9.50}$$

and obtain

$$\Delta E = 0.5I \left(\omega_{min}^{2} - \omega_{max}^{2} \right). \tag{9.51}$$

This second equation relates the change in kinetic energy of the system ΔE to the required mass moment of inertia I of the flywheel for which the angular velocity of the crankshaft fluctuation is limited between ω_{min} and ω_{max}. Using the average angular velocity of the crankshaft ω_0 and the coefficient of speed fluctuation C_S, Equation 9.51 can be rewritten as

$$I = \frac{\Delta E}{C_S \cdot \omega_0^2}. \tag{9.52}$$

The change in kinetic energy ΔE of the punch press during one cycle equals the area (in absolute value) situated above or below the horizontal axis of the curve $T - T_M$ versus ω. This was conveniently obtained by editing inside **D_2D** the lower limit of the vertical axis of the and $T(\omega)$ plot in Figure 9.49b and redoing the graph. According to Figure 9.51, this change in kinetic energy is 5781.8 J. Correspondingly, the required mass moment of inertia of the flywheel is

$$I = \frac{\Delta E}{C_S \cdot \omega_0^2} = \frac{5781.8}{0.0964 \cdot 8.3776^2} = 854.6 \text{ kg m}^2. \tag{9.53}$$

A **Working Model 2D** simulation of the punch press has been prepared and is available with the book (see file **Punch_Press.WM2** and Figure 9.52). Using the formula language of the software, a conditional force is applied to the punch head when it engages the stock according to Equation 9.43. The maximum punch force F_{max} calculated with Equation 9.42, stock location s_f, and stock thickness h must be specified using the text boxes provided. Also input via text box controls are the constant crankshaft torque T_M and the moment of inertia of the flywheel I. The "crank initial rpm" value must be selected by the user in order for the press to operate around the required $n_P = 80$ punches per minute. The simulation confirms that the motor torque T_M and flywheel moment of inertia I were properly calculated, in that the crank holds its rotational speed between 76 and 83 rpm as intended. A slight tendency of the punch rate to increase is visible on Figure 9.52, which can be eliminated by fine-tuning the crankshaft torque T_M. The user may want to repeat the simulation for different flywheel moment of inertia and input torque and observe their effect upon the crank speed change.

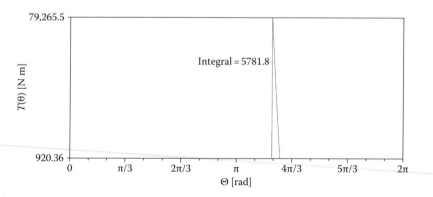

FIGURE 9.51 Plot done with **D_2D** that was used to calculate ΔE. Configuration file **F9_51.CF2**.

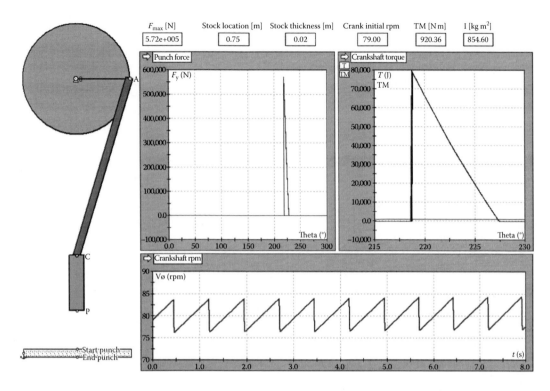

F_{max} [N]	Stock location [m]	Stock thickness [m]	Crank initial rpm	TM [N m]	I [kg m^2]
5.72e+005	0.75	0.02	79.00	920.36	854.60

FIGURE 9.52 **Working Model 2D** simulation of the punch press.

9.18 A PROGRAM FOR PURGING FILES FROM THE CURRENT DIRECTORY

This final section refers to the program **Purge.PAS** (see Appendix B), which can be used to automatically delete certain files in the current directory. When launching the executable **~Purge.EXE**, it will delete without confirmation all the files with extensions **$XY**, **$2D**, **$3D**, **OLD**, and **BAK**; all files of the type **~POLY*.TMP**; and the **AutoCAD** error files **acad. err** and **acadstk.dmp**. With confirmation it will delete all **BMP**, **PCX**, and **SCR** files from the directory where **~Purge.EXE** is located. This program is useful to keep directories clean, following stream **PCX** or **BMP** export and can be easily modified to fit other needs.

***

A number of applications and practical problems that complement the material in earlier chapters have been presented. Same as for the rest of the chapters, the source codes of these programs and the respective **Working Model 2D** simulations are available upon request from the author.

REFERENCES AND FURTHER READINGS

Box, G. E. P. and Muller, M. E. (1958). A note on the generation of random normal deviates. *The Annals of Mathematical Statistics, 29(2)*, 610–611.

Weisstein, E. W. (2013). Minimax approximation. *MathWorld—A Wolfram Web Resource*. http:// mathworld.wolfram.com/MinimaxApproximation.html (last accessed June 2014).

For vibration of single and multiple degree of freedom systems see:
Rao, S. S. (2010). *Mechanical Vibrations.* Upper Saddle River, NJ: Prentice Hall.

For the Duffing equation see:
Kovacic, I. and Brennan, M. J. (2011). *The Duffing Equation: Nonlinear Oscillators and Their Behaviour.* Hoboken, NJ: John Wiley & Sons.

For plotting singe-valued functions of more than two variables using partial mini-max method see:
Simionescu, P. A. and Beale, D. G. (2004). Visualization of hypersurfaces and multivariable (objective) functions by partial global optimization. *The Visual Computer, 20(10)*, 665–681.

For random number generation algorithms see:
Knuth, D. E. (1997). *Art of Computer Programming,* Volume 2: *Seminumerical Algorithms.* Boston, MA: Addison-Wesley.
Press, W. H., Flannery, B. P., Teukolsky, S. A., and Vetterling, W. T. (1989). *Numerical Recipes in Pascal: The Art of Scientific Computing.* New York: Cambridge University Press.

For time ratio in mechanisms see:
Cleghorn, W. L. (2005). *Mechanics of Machines.* New York: Oxford University Press.

For dwell mechanisms and fixed and moving centrodes in planar motion see:
Artobolevsky, I. I. (1975). *Mechanisms in Modern Engineering Design.* Moscow, Russia: Mir Publishers.
Cleghorn, W. L. (2005). *Mechanics of Machines.* New York: Oxford University Press.
Uicker, J. J. Jr., Pennock, G. R., and Shigley, J. E. (2003). *Theory of Machines and Mechanisms.* New York: Oxford University Press.

For more information on planetary gears see:
Lévai, Z. (1968). Structure and analysis of planetary gear trains. *Journal of Mechanisms, 3(3),* 131–148.
Wilson, C. E. and Sadler, J. P. (2003). *Kinematics and Dynamics of Machinery.* Upper Saddle River, NJ: Prentice Hall.

For more information on serial and parallel robots and manipulators see:
Merlet, J. P. (2005). *Parallel Robots.* 2nd edn. Dordrecht, the Netherlands, Springer.
Taghirad, H. D. (2013). *Parallel Robots: Mechanics and Control.* Boca Raton, FL: CRC Press.
Tsai, L. W. (1999). *Robot Analysis: The Mechanics of Serial and Parallel Manipulators.* New York: Wiley.

For more information on steering and suspension of automobiles see:
Bastow, D., Howard, G., and Whitehead, J. P. (2014). *Car Suspension and Handling.* Warrendale, PA: SAE International.
Dixon, J. C. (1996). *Tires, Suspension, and Handling.* Warrendale, PA: SAE International.
Harrer, H. and Pfeffer, P. (2014). *Steering Handbook.* Berlin, Germany: Springer.
Reimpell, J. and Stoll, H. (1996). *The Automotive Chassis: Engineering Principles.* Warrendale, PA: SAE International.
Simionescu, P. A. and Beale, D. G. (2002). Synthesis and analysis of the five-link rear suspension system used in automobiles. *Mechanism and Machine Theory, 37(9),* 815–832.
Wolfe, W. A. (1959). Analytical design of an Ackermann steering linkage. *Transactions of the ASME Journal of Engineering for Industry, 11(1),* 11–14.

For more information on flywheel design see:

Boswirth, L. and Bof, E. (1976). Calculation and design of the flywheel for small and medium compressors. In *International Compressor Engineering Conference*, Paper 193. http://docs.lib.purdue.edu/icec/193.

Genta, G. (1985). *Kinetic Energy Storage: Theory and Practice of Advanced Flywheel Systems*. Oxford, U.K.: Butterworth-Heinemann.

Norton, R. L. (2001). *Design of Machinery*. New York: McGraw-Hill.

Appendix A: Useful Formulae

A.1 EQUATIONS OF A LINE

The equation of the line through points A (x_A, y_A) and B (x_B, y_B) (see Figure A.1) in determinant form is:

$$\begin{vmatrix} x & y & 1 \\ x_A & y_A & 1 \\ x_B & y_B & 1 \end{vmatrix} = 0 \qquad (A.1)$$

equivalent to

$$y = \frac{y_A - y_B}{x_A - x_B} x + \frac{x_A y_B - x_B y_A}{x_A - x_B} \qquad (A.2)$$

and also to

$$\frac{x - x_A}{x_B - x_A} = \frac{y - y_A}{y_B - y_A} \qquad (A.3)$$

The equation of a line through point A (x_A, y_A) and of slope m is:

$$y = m(x - x_A) + y_A \quad \text{or} \quad y = mx + (mx_A + y_A) \qquad (A.4)$$

The equation of a line of slope m and OY-intercept n is:

$$y = mx + n \qquad (A.5)$$

The equation of a line of OX-intercept p and OY-intercept n is:

$$\frac{x}{p} + \frac{y}{n} = 1 \qquad (A.6)$$

The parametric equation of a line through points A (x_A, y_A) and B (x_B, y_B) is:

$$x = x_A(t-1) + x_B t$$

$$y = y_A(t-1) + y_B t \qquad (A.7)$$

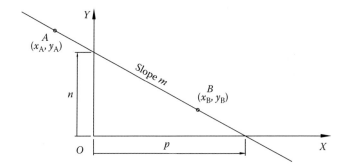

FIGURE A.1   One line in 2D.

equivalent to

$$\begin{pmatrix} x \\ y \end{pmatrix} = t \begin{bmatrix} 1 & 0 \\ 0 & 1 \end{bmatrix} \begin{pmatrix} x_B - x_A \\ y_B - y_A \end{pmatrix} + \begin{pmatrix} x_A \\ y_A \end{pmatrix} \tag{A.8}$$

Note that for $t \in [0,1]$, (x, y) spans the portion of the line from A to B only.

A.2 CONDITION FOR TWO LINES TO BE PERPENDICULAR

Two lines of equations $y = m_1 \cdot x + n_1$ and $y = m_2 \cdot x + n_2$ are perpendicular if

$$m_1 = \frac{-1}{m_2} \tag{A.9}$$

A.3 CONDITION FOR TWO LINES TO BE PARALLEL

Two lines of equations $y = m_1 \cdot x + n_1$ and $y = m_2 \cdot x + n_2$ are parallel if

$$m_1 = m_2 \tag{A.10}$$

A.4 ANGLE BETWEEN TWO LINES

The angle between two lines of equations $y = m_1 \cdot x + n_1$ and $y = m_2 \cdot x + n_2$ is

$$\tan \theta = \frac{m_1 - m_2}{1 + m_1 m_2} \tag{A.11}$$

A.5 POINT COLINEAR WITH OTHER TWO AT A PRESCRIBED LOCATION

Given two points A (x_A, y_A) and B (x_B, y_B), find the coordinates of a third point P (x, y) collinear with them, located at a specified distance AP (Figure A.2).

The following double equality should hold:

$$\frac{x - x_A}{x_B - x_A} = \frac{y - y_A}{y_B - y_A} = \frac{AP}{AB} \tag{A.12}$$

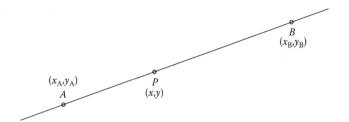

FIGURE A.2   Three collinear points A, B, and P.

where AB is the distance between the two given points, that is,

$$AB = \sqrt{(x_A - x_B)^2 + (y_A - y_B)^2}$$ (A.13)

From Equation A.12, we get

$$x = x_A + \frac{AP}{AB}(x_B - x_A)$$

$$y = y_A + \frac{AP}{AB}(y_B - y_A)$$ (A.14)

Not that for AP negative or $AP > AB$, point P will located outside $A - B$.

A.6 CONDITION FOR A POINT COLLINEAR WITH OTHER TWO TO BE LOCATED BETWEEN THEM

Given three collinear points $A\,(x_A, y_A)$, $B\,(x_B, y_B)$, and $P\,(x, y)$, the condition for point P to be located between A and B is (see Figure A.2)

$$(x_A - x)(x_B - x) + (y_A - y)(y_B - y) < 0$$ (A.15)

A.7 DISTANCE FROM A POINT TO A LINE

Given a line through points $A\,(x_A, y_A)$ and $B\,(x_B, y_B)$ and a third point $C\,(x_C, y_C)$ not collinear with them, find distance d between point C and line $A - B$ and the coordinates (x, y) of the projection P of point C onto line $A - B$ (Figure A.3).

The following two relations should hold simultaneously:

$$(x_B - x_A)(x_C - x) + (y_B - y_A)(y_C - y) = 0$$ (A.16)

$$\frac{x - x_A}{x_B - x_A} = \frac{y - y_A}{y_B - y_A}$$ (A.17)

The first equation is the dot product between vectors **AB** and **CP**, which must be equal to zero, and the second equation is the condition for points A, B, and P to be collinear.

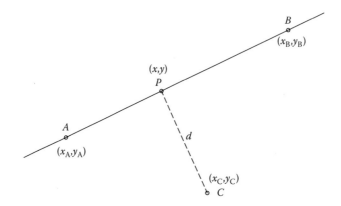

FIGURE A.3   Distance from a point to a line in 2D.

Expanding the two equations and rearranging terms yields a set of two linear equations in the unknowns x and y that is easy to solve:

$$\begin{cases} (x_B - x_A)x + (y_B - y_A)y = (x_B - x_A)x_C + (y_B - y_A)y_C \\ (y_B - y_A)x - (x_B - x_A)y = (y_B - y_A)x_A - (x_B - x_A)y_A \end{cases} \tag{A.18}$$

Once the coordinates of point P are found, the sought for distance d can be calculated with the equation:

$$d = \sqrt{(x_C - x)^2 + (y_C - y)^2} \tag{A.19}$$

A.8 ORIENTATION OF A TRIANGULAR LOOP

Given three noncolinear points $A(x_A, y_A)$, $B(x_B, y_B)$, and $C(x_C, y_C)$, if the cross product $AB \times AC > 0$ or

$$(x_2 - x_1)(y_3 - y_1) - (y_2 - y_1)(x_3 - x_1) > 0 \tag{A.20}$$

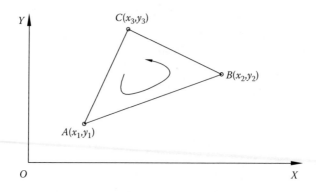

FIGURE A.4   A triangle in 2D.

then the triangular loop ABC is oriented counterclockwise as shown in Figure A.4. Otherwise, the triangular loop is oriented clockwise.

A.9 AREA OF A TRIANGLE

Given points $A(x_1, y_1)$, $B(x_2, y_2)$, and $C(x_3, y_3)$ that are not collinear, the area of triangle ABC is given by the following equation:

$$A = \frac{1}{2} \begin{vmatrix} 1 & 1 & 1 \\ x_1 & x_2 & x_3 \\ y_1 & y_2 & y_3 \end{vmatrix} = \frac{1}{2}(x_2 y_3 - x_3 y_2 - x_1 y_3 + x_3 y_1 + x_1 y_2 - x_2 y_1) \quad (A.21)$$

A.10 CONDITION OF THREE POINTS TO BE COLLINEAR

Given points $A(x_1, y_1)$, $B(x_2, y_2)$, and $C(x_3, y_3)$, they are collinear if

$$\begin{vmatrix} 1 & 1 & 1 \\ x_1 & x_2 & x_3 \\ y_1 & y_2 & y_3 \end{vmatrix} = x_2 y_3 - x_3 y_2 - x_1 y_3 + x_3 y_1 + x_1 y_2 - x_2 y_1 = 0 \quad (A.22)$$

A.11 CIRCLE CIRCUMSCRIBED TO THREE POINTS

Given points $A(x_1, y_1)$, $B(x_2, y_2)$, and $C(x_3, y_3)$, find the center (x, y) and radius r of the circle through these three points. The following relations should hold simultaneously:

$$\begin{cases} (x - x_1)^2 + (y - y_1)^2 = r^2 \\ (x - x_2)^2 + (y - y_2)^2 = r^2 \\ (x - x_3)^2 + (y - y_3)^2 = r^2 \end{cases} \quad (A.23)$$

which after expanding, the squares become

$$\begin{cases} x^2 - 2x_1 x + x_1^2 + y^2 - 2y_1 y + y_1^2 = r^2 \\ x^2 - 2x_2 x + x_2^2 + y^2 - 2y_2 y + y_2^2 = r^2 \\ x^2 - 2x_3 x + x_3^2 + y^2 - 2y_3 y + y_2^2 = r^2 \end{cases} \quad (A.24)$$

Subtracting the first equation from the other two, we obtain a set of two linear equations in the unknowns x and y that is easy to solve:

$$\begin{cases} 2(x_1 - x_2)x + 2(y_1 - y_2)y = x_1^2 + y_1^2 - x_2^2 - y_2^2 \\ 2(x_1 - x_3)x + 2(y_1 - y_3)y = x_1^2 + y_1^2 - x_3^2 - y_3^2 \end{cases} \quad (A.25)$$

Radius of the circle r can now be obtained by substituting x and y back into any of the original equations A.23, for example,

$$r = \sqrt{(x-x_1)^2 + (y-y_1)^2}$$ (A.26)

A.12 INTERSECTION BETWEEN A CIRCLE AND A LINE

Given a circle centered at $O(x_O, y_O)$ and of radius r and a line through arbitrary points $A(x_A, y_A)$ and $B(x_B, y_B)$, find the coordinates of the intersection point(s) between the given line and circle (see Figure A.5). The problem can have two distinct solutions, a unique solution when line $A - B$ is tangent to the circle, or no real solution when the line does not intersect the circle.

The following relations should be satisfied simultaneously:

$$(x_O - x)^2 + (y_O - y)^2 = r^2$$ (A.27)

$$\frac{x_A - x}{x_A - x_B} = \frac{y_A - y}{y_A - y_B}$$ (A.28)

To simplify the analysis, we translate the figure such that $x_O = 0$ and $y_O = 0$, and the coordinates of points A and B become $x_A = x_A - x_O$, $y_A = y_A - y_O$, $x_B = x_B - x_O$, and $y_B = y_B - y_O$. After this transformation, Equation A.27 becomes

$$x^2 + y^2 = r^2$$ (A.29)

To avoid a possible division by zero, we compare differences $x_A - x_B$ and $y_A - y_B$, and if $|x_A - x_B| > |y_A - y_B|$, we write Equation A.28 as

$$y = \frac{y_B - y_A}{x_B - x_A}x + \frac{x_B y_A - x_A y_B}{x_B - x_A}$$ (A.30)

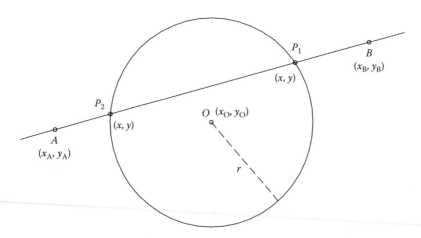

FIGURE A.5   Intersection between a circle centered at O and of radius r and a line through points A and B.

Substituting y in Equation A.29, a quadratic equation in the unknown x is obtained:

$$[(x_B - x_A)^2 + (y_B - y_A)^2]x^2 + 2(y_B - y_A)(x_B y_A - x_A y_B)x$$
$$+ (x_B y_A - x_A y_B)^2 - (x_B - x_A)^2 r^2 = 0 \qquad (A.31)$$

with solutions

$$x = \frac{-(y_B - y_A)(x_B y_A - x_A y_B) \pm \sqrt{\Delta}}{(x_B - x_A)^2 + (y_B - y_A)^2} \qquad (A.32)$$

where the discriminant is

$$\Delta = (x_B - x_A)^2[(x_B - x_A)^2 r^2 + (y_B - y_A)^2 r^2 - (x_A y_B - x_B y_A)^2] \qquad (A.33)$$

If instead we have $|y_A - y_B| > |x_A - x_B|$, then we rewrite Equation A.30 as

$$x = \frac{x_B - x_A}{y_B - y_A} y + \frac{x_A y_B - x_B y_A}{y_B - y_A} \qquad (A.34)$$

and the corresponding quadratic equation becomes

$$[(x_B - x_A)^2 + (y_B - y_A)^2]y^2 + 2(x_B - x_A)(x_A y_B - x_B y_A)y$$
$$+ (x_A y_B - x_B y_A)^2 - (y_B - y_A)^2 r^2 = 0 \qquad (A.35)$$

with solutions

$$y = \frac{-(x_B - x_A)(x_A y_B - x_B y_A) \pm \sqrt{\Delta}}{(x_B - x_A)^2 + (y_B - y_A)^2} \qquad (A.36)$$

and discriminant

$$\Delta = (y_B - y_A)^2[(x_B - x_A)^2 r^2 - (y_B - y_A)^2 r^2 - (x_A y_B - x_B y_A)^2] \qquad (A.37)$$

The actual intersection points are $x = x + x_O$ and $y = y + y_O$, obtained by translating of the figure back to its original location.

A.13 TANGENT FROM A POINT TO A CIRCLE

Given a circle centered at $O(x_O, y_O)$ and of radius r and external point $A(x_A, y_A)$, find the coordinates of point $P(x, y)$ on the line passing through (x_A, y_A) that is tangent to the circle. The problem has two solutions, represented in Figure A.6 in solid and dashed lines,

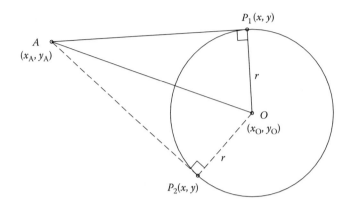

FIGURE A.6   Dual solution of the tangent line from point A to a circle centered at O and of radius r.

respectively. The proper point P_1 or P_2 has to be selected based on other considerations, like the orientation of the triangles AOP_1 and AOP_2 or magnitude of its x or y coordinates.

As stated, the problem is equivalent to the following two equations:

$$(x_O - x)^2 + (y_O - y)^2 = r^2 \tag{A.38}$$

$$(x_A - x)^2 + (y_A - y)^2 = (x_A - x_O)^2 + (y_A - y_O)^2 - r^2 \tag{A.39}$$

To simplify the analysis, we translate the entire figure such that $x_O = 0$ and $y_O = 0$. External point A will have its new coordinates $x_A = x_A - x_O$ and $y_A = y_A - y_O$. With these transformations, the previous equations become

$$x^2 + y^2 = r^2 \tag{A.40}$$

$$(x_A - x)^2 + (y_A - y)^2 = x_A^2 + y_A^2 - r^2 \tag{A.41}$$

After squaring terms, Equation A.41 becomes

$$x^2 - 2x_A x + y^2 - 2y_A y = -r^2 \tag{A.42}$$

If we subtract this new equation from Equation A.40, we get

$$x_A x + y_A y = r^2 \tag{A.43}$$

To avoid a possible division by zero, we must compare coordinates x_A and y_A. If $|x_A| > |y_A|$, then we extract x from Equation A.43 as

$$x = \frac{(r^2 - y_A y)}{x_A} \tag{A.44}$$

which when substituted in A.40 yields a quadratic equation in the unknown y:

$$(x_A^2 + y_A^2)y^2 - 2r^2 y_A y + r^4 - x_A^2 r^2 = 0 \tag{A.45}$$

with roots

$$y_{1,2} = \frac{r^2 y_A \pm r x_A \sqrt{x_A^2 + y_A^2 - r^2}}{x_A^2 + y_A^2} \tag{A.46}$$

If $|y_A| > |x_A|$, we conversely have

$$y = \frac{(r^2 - x_A x)}{y_A} \tag{A.47}$$

which when substituted in Equation A.40 yields a new quadratic equation

$$(x_A^2 + y_A^2)x^2 - 2r^2 x_A x + r^4 - y_A^2 r^2 = 0 \tag{A.48}$$

with roots

$$y_{1,2} = \frac{r^2 x_A \pm r y_A \sqrt{x_A^2 + y_A^2 - r^2}}{x_A^2 + y_A^2} \tag{A.49}$$

The actual solution to the problem is obtained by translating the figure back to its original location, that is, letting $x = x + x_O$ and $y = y + y_O$.

A.14 TANGENT OF A GIVEN SLOPE TO A CIRCLE

Find the equation of the tangent of slope m to the circle centered at (x_O, y_O) and of radius r (see Figure A.7). This is equivalent to the following relations holding simultaneously:

$$\frac{y - y_O}{x - x_O} = -\frac{1}{m} \tag{A.50}$$

$$(x_O - x)^2 + (y_O - y)^2 = r^2 \tag{A.51}$$

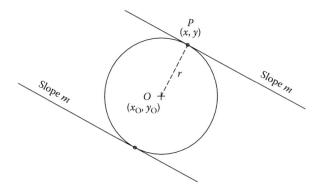

FIGURE A.7   Dual solution of the tangent of given slope m to a circle.

From Equation A.50, we get

$$x = x_O - my + my_O \tag{A.52}$$

which substituted in the second equation yields

$$(m^2 + 1)y - 2y_O(m^2 + 1)y + (m^2 + 1)y_O^2 - r^2 = 0 \tag{A.53}$$

This last equation has solutions

$$y = y_O \pm \frac{r}{\sqrt{m^2 + 1}} \tag{A.54}$$

In turn, the x coordinate of the tangent point(s) P become

$$x = x_O \mp \frac{mr}{\sqrt{m^2 + 1}} \tag{A.55}$$

Applying now Equation A.4, the equation of the tangent line will finally be

$$y = mx \pm r\sqrt{m^2 + 1}\left(\frac{m^2 - 1}{m^2 + 1}\right) + mx_O + y_O \tag{A.56}$$

A.15 PARABOLA THROUGH THREE POINTS

The equation of the parabola through points (x_1, y_1), (x_2, y_2), and (x_3, y_3) (see Figure A.8) is

$$y = ax^2 + bx + c \tag{A.57}$$

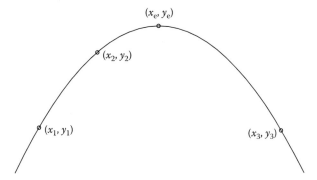

FIGURE A.8 A parabola through points (x_1, y_1), (x_2, y_2), and (x_3, y_3). Also shown is the extremum point (x_e, y_e).

where coefficients a, b, and c are solutions of the following set of three linear equations:

$$\begin{cases} x_1^2 a + x_1 b + c = y_1 \\ x_2^2 a + x_2 b + c = y_2 \\ x_3^2 a + x_3 b + c = y_3 \end{cases} \qquad (A.58)$$

The extremum point (minimum or maximum) of this parabola has the coordinates

$$x_e = -\frac{b}{2a}$$
$$y_e = \frac{b^2}{4a} - \frac{h^2}{2a} + c \qquad (A.59)$$

A.16 CUBIC PARABOLA THROUGH FOUR POINTS

Given points (x_1, y_1), (x_2, y_2), (x_3, y_3), and (x_4, y_4), the equation of the cubic parabola passing through these four points is (see Figure A.9)

$$y = ax^3 + bx^2 + cx + d$$

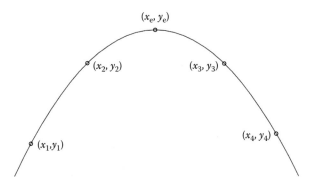

FIGURE A.9 A cubic parabola through points (x_1, y_1), (x_2, y_2), (x_3, y_3), and (x_4, y_4). Also shown is the extremum point (x_e, y_e).

where coefficients a, b, and c are solutions of the following set of four linear equations:

$$\begin{cases} x_1^3 a + x_1^2 b + x_1 c + d = y_1 \\ x_2^3 a + x_2^2 b + x_2 c + d = y_2 \\ x_3^3 a + x_3^2 b + x_3 c + d = y_3 \\ x_4^3 a + x_4^2 b + x_4 c + d = y_4 \end{cases} \tag{A.60}$$

The extremum point(s) of this cubic parabola has the coordinates

$$x_e = \frac{-b \pm \sqrt{b^2 - 3ac}}{3a}$$

$$y_e = ax_e^3 + bx_e^3 + cx_e + d \tag{A.61}$$

Note that there can be two points of extrema, and the one outside the interval $[x_1, x_4]$ should be excluded.

A.17 TANGENT OF A GIVEN SLOPE TO A PARABOLA

Given a parabola of equation $y = ax^2 + bx + c$, find coordinates (x, y) of point P where the tangent to the parabola has a slope m (see Figure A.10).

The following relations should hold simultaneously:

$$y = ax^2 + bx + c \tag{A.62}$$

$$2a \cdot x + b = m \tag{A.63}$$

which yield the tangent point as

$$x = \frac{m - b}{2a}$$

$$y = \frac{(m - b)^2}{4a} x^2 + \frac{mb - b^2}{2a} + c \tag{A.64}$$

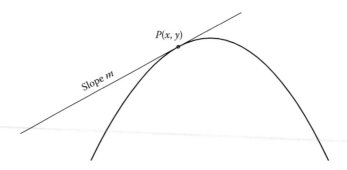

FIGURE A.10 Tangent of slope m to a parabola.

A.18 TANGENT FROM A POINT TO A PARABOLA

Given a parabola of equation $y = ax^2 + bx + c$ and an external point (x_A, y_A), find the line through point A that is tangent to the parabola. Specifically, find the coordinates (x, y) of the tangency points P_1 and P_2 (see Figure A.11).

The problem is equivalent to the following three equations in the unknowns x, y, and m:

$$y = m(x - x_A) + y_A \tag{A.65}$$

$$y = ax^2 + bx + c \tag{A.66}$$

$$2ax + b = m \tag{A.67}$$

Equating y from the first two equations yields

$$m(x - x_A) + y_A - ax^2 - bx - c = 0 \tag{A.68}$$

and after substituting m from Equation A.67, we obtain a quadratic equation in the unknown x:

$$ax^2 - 2ax_A x - bx_A + y_A - c = 0 \tag{A.69}$$

with solutions

$$x = x_A \pm \sqrt{x_A^2 + \frac{bx_A - y_A + c}{a}} \tag{A.70}$$

Slope m and the y coordinate of the tangent point can then be calculated using Equations A.67 and A.66, respectively.

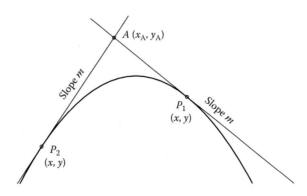

FIGURE A.11   Dual solution to the tangent to a parabola from an external point A.

A.19 INTERSECTION BETWEEN A CIRCLE AND A PARABOLA

Given a parabola of equation $y = ax^2 + bx + c$ and a circle centered at (x_O, y_O) and of radius r, find the coordinates (x, y) of their intersection point(s) (see Figure A.12). This is equivalent to the following two equations being satisfied simultaneously:

$$y = ax^2 + bx + c \qquad (A.71)$$

$$(x_O - x)^2 + (y_O - y)^2 = r^2 \qquad (A.72)$$

We substitute y from the first equation into the second equation:

$$x^2 - 2x_O x + x_O^2 + (ax^2 + bx + c - y_O)^2 - r^2 = 0 \qquad (A.73)$$

and after expanding terms, we obtain a fourth-degree equation

$$f(x) = a^2 x^4 + 2ab\, x^3 + [2a(c - y_O) + b^2 + 1]x^2 + [2b(c - y_O) - x_O]x$$
$$+ (c - y_O)^2 + x_O^2 - r^2 = 0 \qquad (A.74)$$

which can be solved iteratively. If the *Newton–Raphson method* is used with the iteration

$$x_j = x_{j-1} - f(x)\frac{f(x_{j-1})}{f'(x_{j-1})} \qquad (A.75)$$

then the first derivative of $f(x)$ is:

$$f'(x) = 4a^2 x^3 + 6ab\, x^2 + [4a(c - y_O) + 2b^2 + 2]x + 2b(c - y_O) - x_O \qquad (A.76)$$

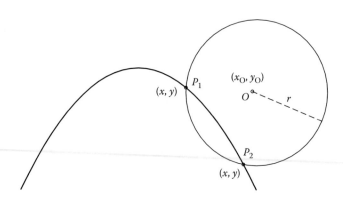

FIGURE A.12   Dual solution of the intersection between a parabola and a circle.

A.20 COMMON TANGENT TO A PARABOLA AND A CIRCLE

Given a circle centered at (x_O, y_O) and of radius r and a parabola of equation $y = ax^2 + bx + c$, find the coordinates of point (x, y) on the parabola and the coordinates of point (x_A, y_A) on the circle belonging to the common tangent to the parabola and to the circle. The problem has two solution represented in solid and dashed lines in Figure A.13. It implies that the proper points A_1 and P_1 or A_2 and P_2 had to be selected based on other considerations, like the orientation of triangles A_1OP_1 and A_2OP_2 or the magnitude of the x or y coordinates of the solution point.

The aforementioned requirements are equivalent to the following analytical relations:

$$(x_O - x_A)^2 + (y_O - y_A)^2 = r^2 \tag{A.77}$$

$$y = ax^2 + bx + c \tag{A.78}$$

$$2ax + b = -\frac{x_A - x_O}{y_A - y_O} \tag{A.79}$$

$$(x_A - x)^2 + (y_A - y)^2 = (x_O - x)^2 + (y_O - y)^2 - r^2 \tag{A.80}$$

Equation A.79 is the condition of the tangent to the parabola at point (x, y) to be perpendicular to the radius OA_1, and Equation A.80 is the condition of OAP to be a right-angle triangle.

To simplify the analysis, we translate the entire figure such that the center of the circle has the coordinates $x_O = 0$ and $y_O = 0$. This will change the coefficients of the parabola as follows:

$$a = a$$

$$b = 2ax_O + b \tag{A.81}$$

$$c = ax_O^2 + bx_O + c + y_O$$

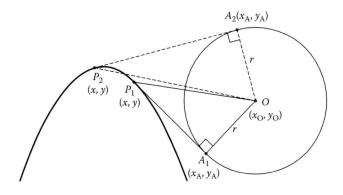

FIGURE A.13   Dual solution to the common tangent to a parabola and a circle.

After translating the entire figure such that the circle becomes centered at (0,0), Equations A.77 through A.80 simplify to

$$x_A^2 + y_A^2 = r^2 \tag{A.82}$$

$$y = ax^2 + bx + c \tag{A.83}$$

$$2ax + b = -\frac{x_A}{y_A} \tag{A.84}$$

$$x_A^2 - 2x_A x + y_A^2 - 2y_A y + r^2 = 0 \tag{A.85}$$

We subtract the first equation from the last one and obtain

$$x_A x + y_A y - r^2 = 0 \tag{A.86}$$

then we substitute y from Equation A.83. The following set of three nonlinear equations is obtained:

$$\begin{cases} f_1(x, x_A, y_A) = x_A^2 + y_A^2 - r^2 = 0 \\[2mm] f_2(x, x_A, y_A) = 2ax + \dfrac{x_A}{y_A} + b = 0 \\[2mm] f_3(x, x_A, y_A) = ax^2 y_A + xx_A + bxy_A + cy_A - r^2 = 0 \end{cases} \tag{A.87}$$

which can be solved iteratively. To apply *Newton's method*, we must evaluate the Jacobian:

$$J = \begin{bmatrix} \dfrac{\partial f_1}{\partial x} & \dfrac{\partial f_1}{\partial x_A} & \dfrac{\partial f_1}{\partial y_A} \\[3mm] \dfrac{\partial f_2}{\partial x} & \dfrac{\partial f_2}{\partial x_A} & \dfrac{\partial f_2}{\partial y_A} \\[3mm] \dfrac{\partial f_3}{\partial x} & \dfrac{\partial f_3}{\partial x_A} & \dfrac{\partial f_3}{\partial y_A} \end{bmatrix} = \begin{bmatrix} 0 & 2x_A & 2y_A \\[2mm] 2a & \dfrac{1}{y_A} & -\dfrac{x_A}{y_A^2} \\[2mm] 2axy_A + x_A + by_A & x & ax^2 + bx + c \end{bmatrix} \tag{A.88}$$

and invert it every iteration step according to the following equation:

$$\begin{bmatrix} x_j \\ x_{Aj} \\ y_{Aj} \end{bmatrix} = \begin{bmatrix} x_{j-1} \\ x_{Aj-1} \\ x_{Aj-1} \end{bmatrix} - J^{-1} \cdot \begin{bmatrix} f_1(x_{j-1}, x_{Aj-1}, y_{Aj}) \\ f_2(x_{j-1}, x_{Aj-1}, y_{Aj}) \\ f_3(x_{j-1}, x_{Aj-1}, y_{Aj}) \end{bmatrix} \tag{A.89}$$

At the end, the coordinates of the solution points will be $x = x + x_O$, $y = y + x_O$, $x_A = x_A + x_O$, $y_A = y_A + x_O$.

A.21 COMMON TANGENT TO TWO CIRCLES

Given one circle centered at (x_{O_1}, y_{O_1}) and of radius r_1 and a second circle centered at (x_{O_2}, y_{O_2}) and of radius r_2, find the coordinates of tangency points (x_1, y_1) and (x_2, y_2) on their common tangent (Figure A.14). In the following analysis, we will assume that $r_1 \geq r_2$. If it is not the case, a relabeling of the points in Figure A.14 is required.

Note that the problem has four solutions, of which only two are shown on Figure A.14, the other two being their mirror image about the line of centers O_1O_2. The solutions where the common tangent intersects the line of centers between points O_1 and O_2 will be called *cross-tangent case*, and the other two where the intersection occurs outside the line segment O_1O_2 will be called *side-tangent case*. We begin with the following notations:

$$O_1O_2 = \sqrt{(x_{O_1} - x_{O_2})^2 + (y_{O_1} - y_{O_2})^2}$$

$$c\theta = \cos(\theta) = \frac{x_{O_2} - x_{O_1}}{O_1O_2} \tag{A.90}$$

$$s\theta = \sin(\theta) = \frac{y_{O_2} - y_{O_1}}{O_1O_2}$$

To simplify the analysis, we translate the whole figure such that O_1 becomes the origin and then rotate it about point O_1 clockwise by angle θ (see Figures A.15 and A.16).

For the *cross-tangent* case in Figure A.15, we can write the following trigonometric identities within the triangle $O_1P_1'O_2$:

$$\cos(\alpha) = \frac{r_1 + r_2}{O_1O_2}$$

$$\sin(\alpha) = \sqrt{1 - \frac{(r_1 + r)^2}{O_1O_2^2}} \tag{A.91}$$

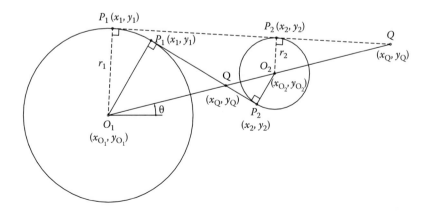

FIGURE A.14   Two of the four solutions to the common tangent to two circles problem.

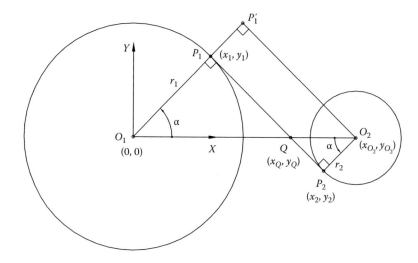

FIGURE A.15   The *cross-tangent case* with O_1 at origin and horizontal center line O_1O_2.

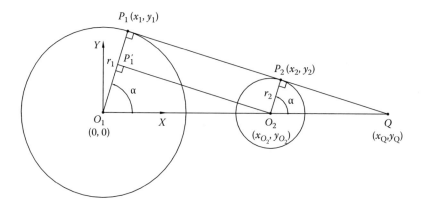

FIGURE A.16   The *side-tangent case* with O_1 at origin and horizontal center line O_1O_2.

With these notations, we have

$$x_1 = r_1 \cdot \cos(\alpha)$$
$$y_1 = \pm r_1 \cdot \sin(\alpha)$$

(A.92)

and

$$x_2 = O_1O_2 - r_2 \cdot \cos(\alpha)$$

$$y_2 = \mp r_2 \cdot \sin(\alpha)$$

(A.93)

where the upper sign corresponds to the solution shown in Figure A.15 and the lower sign to the mirror image (not shown).

For the *side-tangent* case in Figure A.16, we write the following trigonometric identities within triangle $O_1 P_1' O_2$:

$$\cos(\alpha) = \frac{r_1 - r_2}{O_1 O_2}$$

$$\sin(\alpha) = \sqrt{1 - \frac{(r_1 - r)^2}{O_1 O_2^2}}$$

(A.94)

With these notations, we have

$$x_1 = r_1 \cdot c\alpha$$

$$y_1 = \pm r_1 \cdot s\alpha$$

(A.95)

and

$$x_2 = O_1 O_2 + r_2 \cdot c\alpha$$

$$y_2 = \pm r_2 \cdot s\alpha$$

(A.96)

Again, the upper sign corresponds to the solution shown in Figure A.16 and the lower sign to the mirror case, not shown.

The final solutions (x_1, y_1) and (x_2, y_2) to the actual problem are obtained by rotating points P_1 and P_2 counterclockwise by the angle θ and then translating them by the amount x_{O_1} along the OX axis and by amount y_{O_1} along the OY axis.

A.22 TRANSLATIONS AND ROTATIONS IN 2D

The coordinate transformation from reference frame OXY to a translated reference frame $O_1 X_1 Y_1$ (Figure A.17a) is

$$x = x_{O_1} + x_1$$

$$y = y_{O_1} + y_1$$

(A.97)

The inverses transformation is

$$x_1 = x - x_{O_1}$$

$$y_1 = y - y_{O_1}$$

(A.98)

The coordinate transformation from reference frame OXY to a rotated reference frame $OX_1 Y_1$ (Figure A.17b) is

$$x = x_1 \cos\theta - y_1 \sin\theta$$

$$y = x_1 \sin\theta + y_1 \cos\theta$$

(A.99)

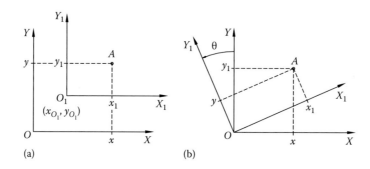

FIGURE A.17   Translation (a) and rotation by angle θ (b) in 2D.

For $\theta = 90°$ and $\theta = -90°$, the transformations are respectively:

$$x = -y_1 \quad \text{and} \quad y = x_1 \tag{A.100}$$

$$x = y_1 \quad \text{and} \quad y = -x_1 \tag{A.101}$$

The inverses general transformation is

$$x_1 = x \cdot \cos\theta + y \cdot \sin\theta$$
$$y_1 = -x \cdot \sin\theta + y \cdot \cos\theta \tag{A.102}$$

For $\theta = 90°$ and $\theta = -90°$, the inverse transformations are respectively:

$$x_1 = y \quad \text{and} \quad y_1 = -x \tag{A.103}$$

$$x_1 = -y \quad \text{and} \quad y_1 = x \tag{A.104}$$

A.23 TRANSLATIONS AND ROTATIONS IN 3D

When changing the coordinates from reference frame $OXYZ$ (Figure A.18a) to a translated reference frame $O_1X_1Y_1Z_1$ (Figure A.18b), the following equations apply:

$$x = x_{O_1} + x_1$$
$$y = y_{O_1} + y_1 \tag{A.105}$$
$$z = z_{O_1} + z_1$$

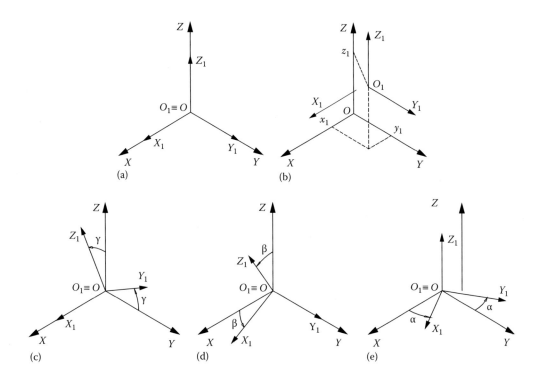

FIGURE A.18 Initially aligned reference frames $OXYZ$ and $O_1X_1Y_1Z_1$ (a) and transformed reference frame $O_1X_1Y_1Z_1$ through translation (b), rotation about OX (c), rotation about OY (d), and rotation about OZ (e).

The inverses coordinate transformation from $O_1X_1Y_1Z_1$ in Figure 8.18b to $OXYZ$ in Figure 8.18a is

$$x_1 = x - x_O$$
$$y_1 = y - y_O \qquad (A.106)$$
$$z_1 = z - z_O$$

The basic 3D transformation matrices that rotate vectors (x_1, y_1, z_1) about the OX, OY, and OZ are as follows. A rotation by angle γ (*roll angle*) about OX axis (see Figure A.18c) is

$$\begin{pmatrix} x \\ y \\ z \end{pmatrix} = \begin{pmatrix} x_1 & y_1 & z_1 \end{pmatrix} \begin{bmatrix} 1 & 0 & 0 \\ 0 & \cos\gamma & -\sin\gamma \\ 0 & \sin\gamma & \cos\gamma \end{bmatrix} = \begin{pmatrix} x_1 & y_1 & z_1 \end{pmatrix} R_x(\gamma) \qquad (A.107)$$

A rotation by angle β (*pitch angle*) about OY axis (see Figure A.18d) is

$$\begin{pmatrix} x \\ y \\ z \end{pmatrix} = \begin{pmatrix} x_1 & y_1 & z_1 \end{pmatrix} \begin{bmatrix} \cos\beta & 0 & \sin\beta \\ 0 & 1 & 0 \\ -\sin\beta & 0 & \cos\beta \end{bmatrix} = \begin{pmatrix} x_1 & y_1 & z_1 \end{pmatrix} R_y(\beta) \qquad (A.108)$$

A rotation by angle α (*yaw angle*) about OZ axis (see Figure A.18e) is

$$\begin{pmatrix} x \\ y \\ z \end{pmatrix} = \begin{pmatrix} x_1 & y_1 & z_1 \end{pmatrix} \begin{bmatrix} \cos\alpha & -\sin\alpha & 0 \\ \sin\alpha & \cos\alpha & 0 \\ 0 & 0 & 1 \end{bmatrix} = \begin{pmatrix} x_1 & y_1 & z_1 \end{pmatrix} R_z(\alpha) \qquad \text{(A.109)}$$

More complex transformations can be obtained through matrix multiplication. For example, the sequence of *roll, pitch,* and *yaw* (in this order) about a fixed reference frame $OXYZ$ is described by

$$R(\gamma, \beta, \alpha) = R_z(\alpha) R_x(\gamma) R_y(\beta) \qquad \text{(A.110)}$$

Because matrix multiplication is not commutative, the end result will depend on the order in which these basic rotation transformations are applied.

A.24 NUMERICAL DIFFERENTIATION

Let $f(x)$ be a continuous function of x that has derivatives up to order n. Below are formulae for the first- and second-order derivatives of $f(x)$, where $O(..)$ is a remainder, which depends on Δx. The smaller Δx, the less error is incurred when $O(..)$ is left apart.

A.24.1 First-Order Differentiation

Forward differentiation

$$\frac{\partial f(x)}{\partial x} = \frac{f(x + \Delta x) - f(x)}{\Delta x} + O(\Delta x) \qquad \text{(A.111)}$$

Backward differentiation

$$\frac{\partial f(x)}{\partial x} = \frac{f(x) - f(x - \Delta x)}{\Delta x} + O(\Delta x) \qquad \text{(A.112)}$$

Centered differentiation

$$\frac{\partial f(x)}{\partial x} = \frac{f(x + \Delta x) - f(x - \Delta x)}{2\Delta x} + O(\Delta x)^2 \qquad \text{(A.113)}$$

Notice that *centered differentiation* has better accuracy.

A.24.2 Second-Order Differentiation

The most common *second-order differentiation* formula is

$$\frac{\partial^2 f(x)}{\partial x^2} = \frac{f(x + \Delta x) - 2f(x) + f(x - \Delta x)}{(2\Delta x)^2} + O(\Delta x)^2 \qquad \text{(A.114)}$$

These equations can be obtained from *Taylor's series* approximation of $f(x)$, that is,

$$f(x) \cong f(x \pm \Delta x) + \frac{\partial f(x \pm \Delta x)}{\partial x}(\pm \Delta x) + \frac{\partial^2 f(x \pm \Delta x)}{\partial x^2} \cdot \frac{(\pm \Delta x)^2}{2!}$$
$$+ \cdots \frac{\partial^n f(x \pm \Delta x)}{\partial x^n} \cdot \frac{(\pm \Delta x)^n}{n!} \tag{A.115}$$

when the function $f(x)$ and its derivatives up to order n are known at point $x \pm \Delta x$. Notice that formulae A.111, A.112, and A.113 are obtained by setting $n = 1$ in Equation A.115, while formula A.114 is obtained for $n = 2$ in the same equation.

A.25 NEWTON–RAPHSON METHOD FOR ROOT FINDING

Newton–Raphson is a fast converging method for finding approximations to the roots (or zeroes) of real-valued functions, which is based on successive *Taylor's approximations* of the function.

Given a real, continuous function $f(x)$ and its derivative $f'(x)$, we begin with an initial guess x_0 for a root r of the function. Assuming that $f'(x_1) \neq 0$, a better approximation x_1 of the root r will be

$$x_1 = x_0 - \frac{f(x_0)}{f'(x_0)} \tag{A.116}$$

The approximating process is repeated as

$$x_{k+1} = x_k - \frac{f(x_k)}{f'(x_k)} \tag{A.117}$$

until $|f(x_{k+1})| < \varepsilon$ or $|x_{k+1} - x_k| < \varepsilon$, where ε is the desired accuracy.

A geometric interpretation of Newton–Raphson method is provided in Figure A.19.

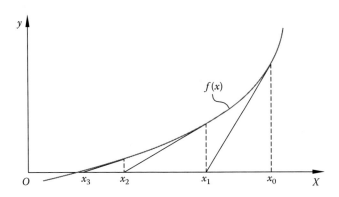

FIGURE A.19   Newton–Raphson iteration.

A.26 AREA UNDER A CURVE USING TRAPEZOIDAL RULE

The area A delimited by a curve of equation $y = f(x)$ and the $[a, b]$ interval of the x-axis is defined by

$$A = \int_a^b f(x)dx = \lim_{n \to \infty} \sum_{i=1}^{n} A_i \qquad (A.118)$$

where A_i is the area of the ith strip (see Figure A.20). We can develop approximation schemes for the value of the integral by assuming a finite number n and adopting simple trapezoidal approximations of these strips.

We assume the top portion of the stripes in Figure A.21 to be straight lines. For area A_i, this approximating straight line is the chord joining points $(x_i, f(x_i))$ and $(x_{i+1}, f(x_{i+1}))$ and has the equation

$$\frac{y - f_i}{x - x_i} = \frac{f_{i+1} - f_i}{x_{i+1} - x_i} \qquad (A.119)$$

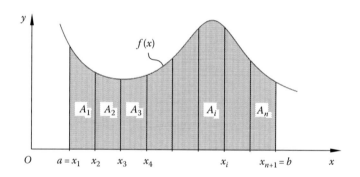

FIGURE A.20   Integration as the area under the curve $f(x)$ over the interval $[a, b]$.

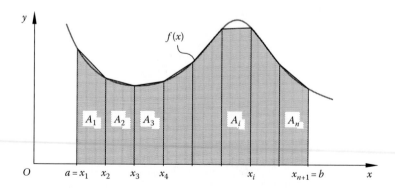

FIGURE A.21   Approximation of the area under a curve using trapezoids.

Using the notation $h = x_{i+1} - x_i$ for the step size in x, we can write the linear approximation to $f(x)$ as

$$y = f_i + \frac{1}{h}(f_{i+1} - f_i)(x - x_i)$$

(A.120)

The area A_i of the ith strip will therefore be the area of a trapezoid, that is,

$$A_i = \frac{1}{h}(f_i + f_{i+1})$$

(A.121)

Figure A.21 shows the OX interval a to b divided into n equal intervals of length h. We can then write the approximate value of the total area A as

$$A = \sum_{i=1}^{n} a_i = \frac{h}{2} \sum_{i=1}^{n} (f_i + f_{i+1})$$

(A.122)

Note that except for the end values, that is, $i = 1$ and $i = n$, each evaluation of $f(x)$ at a node x_i occurs twice. Thus, the approximation to the integral can be written in the simplified form:

$$\int_a^b f(x)dx \approx \frac{h}{2}(f_1 + 2f_2 + \cdots + 2f_n + f_{n+1}) = \frac{h}{2}\big(f(a) + f(b)\big) + h \sum_{i=1}^{n-1} f(a + ih)$$

(A.123)

which is the *trapezoid rule* formula.

In developing the aforementioned approximation, we left some area under the $f(x)$ out of the sum or included some area of trapezoids that lies below $f(x)$—Figure A.21. It can be shown that the error can be expressed in the following form:

$$\varepsilon = \frac{b-a}{12} h^2 f''(\xi)$$

(A.124)

where $f''(\xi)$ is the value of the second derivative of $f(x)$, evaluated at some point ξ within the interval $[a, b]$. It is known that the second derivative relates to the curvature of $f(x)$, confirmed by the largest apparent error in Figure A.21, which coincides with the maximum of $f(x)$ where the curvature appears most extreme. Equation A.124 suggests that if the step size h is reduced to half, the error estimate ε decreases by a factor of four, while the number of calculations required to compute the sum A is only doubled.

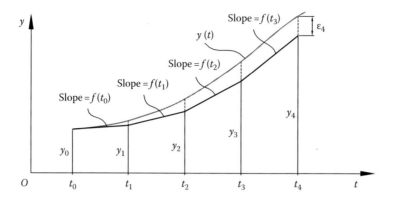

FIGURE A.22   Euler integration steps.

A.27 EULER INTEGRATION OF FIRST-ORDER ORDINARY DIFFERENTIAL EQUATIONS

Consider the first-order differential equation $dy/dt = f(t)$, where the function $f(t)$ may not be readily integrable. Euler method performs numerical integration, that is, find $y(t)$ given initial condition $y_0 = y(t_0)$, by means of a slope-projection technique (see Figure A.22).

Begin at t_0, at which the value y_0 is known. Project the slope over a horizontal subinterval $t_1 - t_0$ and evaluate y_1 as $y_1 = y_0 + f(t_0)(t_1 - t_0)$. Repeat the process at t_2, t_3, t_4 and so forth according to the following equation:

$$y_{k+1} = y_k + f(t_k)(t_{k+1} - t_k) \tag{A.125}$$

until the desired final value of t is reached. For the case shown in Figure A.22, after four steps, the estimate y_4 is less than the true value of the function $y(t)$ at t_4 by the amount ε_4. This error ε is called *algorithm error* and increases as the integration advances. To reduce its effect, it is recommended to begin with a relatively large step $(t_{k+1} - t_k)$ and then steadily decreases its size until the corresponding changes in the integrated result are much smaller than the desired accuracy. However, a step size that is too small can result in an increased *round-off error*, so a trade-off between these two errors must be sought.

A.28 SOLUTIONS OF TWO AND THREE LINEAR EQUATIONS

The set of two linear equations

$$\begin{cases} a_1 x + b_1 y = c_1 \\ a_2 x + b_2 y = c_2 \end{cases} \tag{A.126}$$

with $a_1 b_2 - a_2 b_1 \neq 0$ has the solutions

$$x = \frac{b_2 c_1 - b_1 c_2}{a_1 b_2 - a_2 b_1}$$

$$y = \frac{a_1 c_2 - a_2 c_1}{a_1 b_2 - a_2 b_1} \tag{A.127}$$

The set of three linear equations

$$\begin{cases} a_1 x + b_1 y + c_1 z = d_1 \\ a_2 x + b_2 y + c_2 z = d_2 \\ a_3 x + b_3 y + c_3 z = d_3 \end{cases} \tag{A.128}$$

with $a_1 b_2 c_3 + a_2 b_3 c_1 + a_3 b_1 c_2 - a_3 b_2 c1 - a_1 b_3 c_2 - a_2 b_1 c_3 \neq 0$ has the solutions

$$x = \frac{b_2 c_3 d_1 + b_3 c_1 d_2 + b_1 c_2 d_3 - b_2 c_1 d_3 - b_3 c_2 d_1 - b_1 c_3 d_2}{a_1 b_2 c_3 + a_2 b_3 c_1 + a_3 b_1 c_2 - a_3 b_2 c_1 - a_1 b_3 c_2 - a_2 b_1 c_3}$$

$$y = \frac{a_1 c_3 d_2 + a_2 c_1 d_3 + a_3 c_2 d_1 - a_3 c_1 d_2 - a_1 c_2 d_3 - a_2 c_3 d_1}{a_1 b_2 c_3 + a_2 b_3 c_1 + a_3 b_1 c_2 - a_3 b_2 c_1 - a_1 b_3 c_2 - a_2 b_1 c_3} \tag{A.129}$$

$$z = \frac{a_1 b_2 d_3 + a_2 b_3 d_1 + a_3 b_1 d_2 - a_3 b_2 d_1 - a_1 b_3 d_2 - a_2 b_1 d_3}{a_1 b_2 c_3 + a_2 b_3 c_1 + a_3 b_1 c_2 - a_3 b_2 c_1 - a_1 b_3 c_2 - a_2 b_1 c_3}$$

A.29 TRIGONOMETRIC IDENTITIES

$$\sin(-u) = -\sin u$$

$$\cos(-u) = +\cos u$$

$$\tan(-u) = -\tan u$$

$$\cot(-u) = -\cot u$$

$$\sin(u \pm \pi/2) = \pm\cos u$$

$$\cos(u \pm \pi/2) = \mp\sin u$$

$$\tan(u \pm \pi/2) = -\cot u$$

$$\cot(u + \pi/2) = -\tan u$$

$$\sin(u \pm \pi) = -\sin u$$

$$\cos(u \pm \pi) = -\cos u$$

$$\tan\left(u \pm \pi\right) = -\cot u$$

$$\cot\left(u + \pi\right) = -\tan u$$

$$\sin^2 u + \cos^2 u = 1$$

$$\sin u = \pm\sqrt{1 - \cos^2 u} = \pm\sqrt{\frac{1 - \cos 2u}{2}} = \pm\frac{\tan u}{\sqrt{\left(1 + \tan^2 u\right)}} = \pm\frac{1}{\sqrt{\left(1 + \cot^2 u\right)}}$$

$$\cos u = \pm\sqrt{1 - \sin^2 u} = \pm\sqrt{\frac{1 + \cos 2u}{2}} = \pm\frac{\cot u}{\sqrt{\left(1 + \cot^2 u\right)}} = \pm\frac{1}{\sqrt{\left(1 + \tan^2 u\right)}}$$

$$\sin(u \pm v) = \sin u \cos v \pm \cos u \sin v$$

$$\cos(u \pm v) = \cos u \cos v \mp \sin u \sin v$$

$$\tan(u \pm v) = \frac{\tan u \pm \tan v}{1 \mp \tan u \tan v}$$

$$\cot\left(u \pm v\right) = \frac{\cot u \mp \cot v}{1 \pm \cot u \cot v}$$

$$\sin u \pm \sin v = 2\sin\frac{u \pm v}{2}\cos\frac{u \mp v}{2}$$

$$\cos u + \cos v = 2\cos\frac{u + v}{2}\cos\frac{u - v}{2}$$

$$\cos u - \cos v = -2\sin\frac{u + v}{2}\sin\frac{u - v}{2}$$

$$\tan u \pm \tan v = \frac{\sin(u \pm v)}{\cos u \sin v}$$

$$\cot u \pm \cot v = \frac{\sin(u \pm v)}{\sin u \sin v}$$

$$\sin u \, \sin v = \frac{1}{2}\cos(u - v) - \frac{1}{2}\cos(u + v)$$

$$\cos u \ \cos v = \frac{1}{2}\cos(u-v) + \frac{1}{2}\cos(u+v)$$

$$\sin u \ \cos v = \frac{1}{2}\sin(u-v) + \frac{1}{2}\sin(u+v)$$

$$\tan u \tan v = \frac{\tan u + \sin v}{\cot u + \cot v}$$

$$\cot u \cot v = \frac{\cot u + \cot v}{\tan u + \tan v}$$

$$\sin^2 u = \frac{1 - \cos 2u}{2}$$

$$\cos^2 u = \frac{1 + \cos 2u}{2}$$

$$\sin^2 u - \sin^2 v = \cos^2 v - \cos^2 u = \sin(u+v)\sin(u-v)$$

$$\cos^2 u - \sin^2 v = \cos^2 v - \sin^2 u = \cos(u+v)\cos(u-v)$$

$$\cos^2 u - \cos^2 v = \sin^2 v - \sin^2 u = -\sin(u+v)\sin(u-v)$$

$$\sin\frac{u}{2} = \sqrt{\frac{1-\cos u}{2}} = \frac{\sqrt{1+\sin u}}{2} - \frac{\sqrt{1-\sin u}}{2}$$

$$\cos\frac{u}{2} = \sqrt{\frac{1+\cos u}{2}} = \frac{\sqrt{1+\sin u}}{2} + \frac{\sqrt{1-\sin u}}{2}$$

$$\tan\frac{u}{2} = \sqrt{\frac{1-\cos u}{1+\cos u}} = \frac{1-\cos u}{\sin u} = \frac{\sin u}{1+\cos u}$$

$$\cot\frac{u}{2} = \sqrt{\frac{1+\cos u}{1-\cos u}} = \frac{1+\cos u}{\sin u} = \frac{\sin u}{1-\cos u}$$

$$\sin 2u = 2\sin u \cos u$$

$$\cos 2u = \cos^2 u - \sin^2 u = 1 - 2\sin^2 u = 2\cos^2 u - 1$$

$$\tan 2u = \frac{2\tan u}{1-\tan^2 u} = \frac{2}{\cot u - \tan u}$$

$$\cot 2u = \frac{\cot^2 u - 1}{2\cot u} = \frac{\cot u - \tan u}{2}$$

$$\sin 3u = 3\sin u - 4\sin^3 u$$

$$\cos 3u = 4\cos^3 u - 3\cos u$$

$$\sin 4u = 8\sin u\cos^3 u - 4\sin u\cos u$$

$$\cos 4u = 8\cos^4 u - 8\cos^2 u + 1$$

$$\frac{d}{dt}\sin u(t) = \cos u(t)\,\frac{du(t)}{dt}$$

$$\frac{d}{dt}\cos u(t) = -\sin u(t)\,\frac{du(t)}{dt}$$

$$\frac{d}{dt}\sin^{-1} u(t) = \frac{1}{\sqrt{1-u(t)^2}}\,\frac{du(t)}{dt} \quad \text{for } -\frac{\pi}{2} \le \sin^{-1} u(t) \le \frac{\pi}{2}$$

$$\frac{d}{dt}\cos^{-1} u(t) = -\frac{1}{\sqrt{1-u(t)^2}}\,\frac{du(t)}{dt} \quad \text{for } 0 \le \cos^{-1} u(t) \le \pi$$

$$\frac{d}{dt}\tan^{-1} u(t) = \frac{1}{1+u(t)^2}\,\frac{du(t)}{dt} \quad \text{for } -\frac{\pi}{2} \le \tan^{-1} u(t) \le \frac{\pi}{2}$$

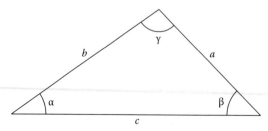

FIGURE A.23 Triangle with sides of lengths a, b, and c and angles of magnitudes α, β, and γ.

With the notations in Figure A.23, the following relations hold:

Law of cosine: $a^2 = b^2 + c^2 - 2bc\cos\alpha$

Law of sines: $\dfrac{a}{\sin\alpha} = \dfrac{b}{\sin\beta} = \dfrac{c}{\sin\gamma}$

The trigonometric equation $a\cos\theta + b\sin\theta = c$ has the solutions

$$\theta = 2\cdot\arctan\left(\frac{b \pm \sqrt{a^2 + b^2 - c^2}}{a+c}\right) \quad\text{or}\quad \theta = \text{Atan2}(b,a) \pm \text{Atan2}\left(\sqrt{a^2 + b^2 - c^2}, c\right)$$

where $\text{Atan2}(y, x) = \arctan(y/x)$ is the arctangent function of two arguments that uses the individual signs of x and y to determine the quadrant of the resultant angle. Atan2 is implemented in many programming languages as well as in **Excel, MATLAB®, Scilab, Mathematica**, etc. Note that some programming languages and computer algebra systems implement the Atan2 function as $\text{Atan2}(y, x)$ while others (like **Excel**) implement it as $\text{Atan2}(x, y)$.

Appendix B: Selected Source Code

```pascal
1    Program P1_01;
2    {============================================================}
3    Generate files F1_01.DTA, F1_01.D2D & F1_01.R2D with nx = 501 data points
4    (for nx=61 see F1_03.PAS) to plot  F(x)=1/(x*x-2x+1.1)+1/(x*x-6x+9.2)-3
5    {============================================================}
6    uses CRT;
7    const  nx=501; xmin=-1.0; xmax=5.0;  {nr. of plot points and limits over x}
8    var FT: Text;              {output ASCII file     }
9        FD: File of double;    {output file of double}
10       FR: File of real;      {output file of real  }
11       x,F: double;   xr,Freal: real;   ix: integer;
12   BEGIN
13     Assign(FT,'F1_01.DTA'); Rewrite(FT);
14     Assign(FD,'F1_01.D2D'); Rewrite(FD);
15     Assign(FR,'F1_01.R2D'); Rewrite(FR);
16     ClrScr;   {Next write ASCII file header ..}
17     WriteLn(FT,'F(x)=1/(x*x-2x+1.1) + 1/(x*x-6x+9.2)  - 3');
18     WriteLn(FT,'  x            F(x)');
19     for ix:=1 to nx do BEGIN
20       x:=xmin+(xmax-xmin)/(nx-1)*(ix-1);  {generate x}
21       WriteLn(ix:3,' ') x= ',x:12:8,'   Fx= ',F:12:8);  {screen echo}
22       F:=1/(x*x-2*x+1.1)+1/(x*x-6*x+9.2)-3;  {evaluate F(x)}
23       WriteLn(FT,x:10:6,' ',F:12:8);  {write x and Fx to ASCII file}
24       Write(FD,x,F);  {write x and Fx to the file of double}
25       xr:=x;
26       Freal:=F;
27       Write(FR,xr,Freal);     {write x and F(x) to the file of reals}
28     END;
29     Close(FT);  Close(FD);  Close(FR);
30     Write('Data files generated successfully. Press <CR>..'); ReadLn;
31   END.
```

```pascal
1   Program P1_02;
2   {=========================================================
3   Generate files  F1_02.DTA  &  F1_02.D2D  to plot the function
4   F'(x) = (2x-2)/Sqr(x*x-2x+1.1) + (2x-6)/Sqr(x*x-6x+9.2)
5   =========================================================}
6   uses CRT;
7   const nx=501; xmin=-1.0; xmax=5.0; {nr. of plot points and limits over x}
8   var FT: Text;                {output ASCII file     }
9       FD: File of double;      {output file of double}
10      x,Fp: double; ix: integer;
11  BEGIN
12    Assign(FT,'F1_02.DTA'); Rewrite(FT);
13    Assign(FD,'F1_02.D2D'); Rewrite(FD);
14    ClrScr; {Next write ASCII file header:}
15    WriteLn(FT,'F''=(2x-2)/Sqr(x*x-2x+1.1)+(2x-6)/Sqr(x*x-6x+9.2)'');
16    WriteLn(FT,'    x     F''(x)');
17    for ix:=1 to nx do BEGIN
18      x:=xmin+(xmax-xmin)/(nx-1)*(ix-1); {generate x}
19      Fp:=(2*x-2)/Sqr(x*x-2*x+1.1)+(2*x-6)/Sqr(x*x-6*x+9.2);
20      WriteLn(FT,x:10:6,' ',Fp:12:8); {write x and Fp to ASCII}
21      Write(FD,x,Fp); {write x and Fp to the file of doubles}
22    END;
23    Close(FT);  Close(FD);
24    Write('Data files generated successfully. Press <CR>..');  ReadLn;
25  END.
```

```
1   Program P1_03;
2   {===========================================================
3   Generate file F1_03.D2D to plot  F(x)=1/(x*x-2x+1.1)+1/(x*x-6x+9.2)-3
4   ===========================================================}
5   uses CRT;
6   const nx=61;   xmin=-1.0; xmax=5.0;  {nr. of plot points and limits over x}
7   var  FD: File of double;   x,F: double;   ix: integer;
8   BEGIN
9     ClrScr;    Assign(FD,'F1_03.D2D');    Rewrite(FD);
10    for ix:=1 to nx do BEGIN
11      x:=xmin+(xmax-xmin)/(nx-1)*(ix-1);
12      F:=1/(x*x-2*x+1.1)+1/(x*x-6*x+9.2)-3;
13      Write(FD,x,F);
14    END;
15    Close(FD);
16    Write('Data file generated successfully.   <CR>..');   ReadLn;
17  END.
```

```
1   Program P1_10;
2   {===========================================================
3   Generate files F1_10.DTA & F1_10.D2D if nx=400, OR F1_11.DTA
4   & F1_11.D2D if nx=401 to plot the function  F(x)=(x*x*x-3x)/(x*x-4)
5   ===========================================================}
6   uses CRT, LibMath;
7   const   nx  = 401;   {number of plot points (either 400 or 401) }
8           xmin =-8.0;  xmax = 8.0;  {limits over x}
9   var FT: Text;         {output ASCII file }
10      FD: File of double;  {output file of double}
11      x,Fx: double;   ix: integer;
12  function F(x: double): double;   {a rational function of degree 3}
13  BEGIN
14    if Abs(x*x-4.0) > EpsD then   {check for division by zero ..}
```

```
15        F:=(x*x*x-3*x)/(x*x-4)
16      else
17        F:=InfD;
18  END;  {.. F(x)}
19  BEGIN
20    if nx = 400 then BEGIN
21      Assign(FT,'F1_10.DTA'); Rewrite(FT);
22      Assign(FD,'F1_10.D2D'); Rewrite(FD);
23    END
24    else BEGIN
25      Assign(FT,'F1_11.DTA'); Rewrite(FT);
26      Assign(FD,'F1_11.D2D'); Rewrite(FD);
27    END;
28    ClrScr;  {Next write ASCII file header ..}
29    WriteLn(FT,'F=(x*x*x-3x)/(x*x-4)');
30    WriteLn(FT,'       x          F(x)');
31    for ix:=1 to nx do BEGIN
32      x:=xmin+(xmax-xmin)/(nx-1)*(ix-1);
33      Fx:=F(x);
34      if (Fx = InfD) then BEGIN  {insert line breakers..}
35        Write(FD,InfD,InfD);            {D2D line breaker}
36        WriteLn(FT,'======');           {ASCII line breaker}
37      END
38      else BEGIN  {write actual data to file..}
39        Write(FD,x,Fx);
40        WriteLn(FT,x:10:6,'  ',Fx:12:8);
41      END;
42    END;
43    Close(FT);  Close(FD);
44    Write('Data files generated successfully. Press <CR>..'); ReadLn;
45  END.
```

```
1   Program P1_12A;
2   {==============================================================
3   Generate files F1_12A.DTA & F1_12A.D2D to plot F(x)=x*(x*x-3)/(x*x-4)
4   ==============================================================}
5   uses CRT, LibMath;
6   const nx=401; xmin=-8.0; xmax=8.0; {nr. of plot points and limits over x}
7   var  FT: Text;                      {output ASCII file  }
8        FD: File of double; {output file of double}
9        Dx,x,Fx,LimLeft,LimRight: double   ix: integer;
10  function F(x: double): double;      {a rational function of degree 3}
11  BEGIN
12    if Abs(x*x-4.0)  <= EpsD then {check for division by zero}
13      F:=InfD
14    else
15      F:=x*(x*x-3)/(x*x-4);
16  END;  {.. F(x)}
17  BEGIN
18    Assign(FT,'F1_12A.DTA');  Rewrite(FT);
19    Assign(FD,'F1_12A.D2D');  Rewrite(FD);
20    ClrScr;   {Next will write ASCII file header:}
21    WriteLn(FT,'F=x*(x*x-3)/(x*x-4)');
22    WriteLn(FT,'    x      F(x)');
23    Dx:=0.1*(xmax-xmin)/(nx-1); {left/right offset to a singular pt.}
24    for ix:=1 to nx do BEGIN
25      x:=xmin+(xmax-xmin)/(nx-1)*(ix-1);
26      Fx:=F(x);
27      if (Fx = InfD) then BEGIN {singular point..}
28        LimLeft:=F(x-Dx);            {limit to the left}
29        Fx:=Round(LimLeft/Abs(LimLeft))*InfD;
30        Write(FD,Fx);
31        WriteLn(FT,x:10:6,' ',Fx:10:6);
```

```
32        Write(FD,InfD,InfD);          {D2D line breaker}
33        WriteLn(FT,'=====');          {ASCII line breaker}
34        LimRight:=F(x+Dx);            {limit to the right}
35        Fx:=Round(LimRight/Abs(LimRight))*InfD;
36        Write(FD,x,Fx);
37        WriteLn(FT,x:10:6,' ',Fx:12:8);
38      END  {..singular point}
39      else BEGIN {regular point..}
40        Write(FD,x,Fx);
41        WriteLn(FT,x:10:6,' ',Fx:12:8);
42      END; {..regular point}
43    END;
44    Close(FT);  Close(FD);
45    Write('Data files generated successfully. Press <CR>..'); ReadLn;
46 END.
```

```
1  Program P1_15;
2  {=====================================================================
3  Generate file F1_15.D2D to plot function
4  H(W_Wn)=1/Sqrt(Sqr(1-Sqr(W_Wn)) + Sqr(2*Zeta*W_Wn) )
5  =====================================================================}
6  uses LibMath;
7  const nx = 251;  xmin = 0.0; xmax = 2.5;  {grid size and limits over Zeta}
8        ny = 10;  ymin = 0.1; ymax = 1.0;  {grid size and limits over W_Wn}
9  var FD: File of double; x,Y, F: double;  ix,iy: integer;
10 function H(W_Wn,Zeta:double): double;
11 BEGIN
12  H:=1.0/Sqrt(Sqr(1-Sqr(W_Wn))+Sqr(2*Zeta*W_Wn));
13 END; {.. H(W_Wn,Zeta) }
14 BEGIN
15  Assign(FD,'F1_15.D2D');  Rewrite(FD);
16  for iy:=1 to ny do BEGIN
```

```
17      y:=ymin+(ymax-ymin)/(ny-1)*(iy-1);   {y=W_Wn}
18  for ix:=1 to nx do BEGIN
19      x:=xmin+(xmax-xmin)/(nx-1)*(ix-1);   {x=Zeta}
20      F:=H(x,y);    Write(FD,x,F);
21      END;
22  Write(FD,InfD,InfD);    {D2D line breaker}
23  END;
24  Close(FD);
25  END.
```

```
1  Program P1_156;
2  {===================================================================
3  Generate files F1_15.D3D & F1_16.D3D to plot using D_3D function:
4  H(Zeta,W_Wn)=1/Sqrt(Sqr(1-Sqr(W_Wn)) + Sqr(2*Zeta*W_Wn))
5  ===================================================================}
6  const Wmin: double = 0.0;  Wmax: double = 2.5;   {W/Wn limits}
7        Zmin: double = 0.1;  Zmax: double = 1.0;   {Zeta limits}
8  { Decomment one of the lines below    (FDnme is the output file name) }
9  { nZ: double = 10;   nW: double = 261;  FDnme = 'F1_15.D3D';}
10     nZ: double = 100;  nW: double = 37;   FDnme = 'F1_16.D3D';
11  var FD: File of double;    W,Z,H: double;    i,j: integer;
12  BEGIN
13     Assign(FD,FDnme);  Rewrite(FD);
14     Write(FD, nZ,nW, Zmin,Zmax, Wmin,Wmax);
15     for i:=1 to Round(nZ) do BEGIN
16        Z:=Zmin+(Zmax-Zmin)/(nZ-1)*(i-1);
17        for j:=1 to Round(nW) do BEGIN
18           W:=Wmin+(Wmax-Wmin)/(nW-1)*(j-1);
19           H:=1/Sqrt(Sqr(1-Sqr(W))+Sqr(2*Z*W));
20           Write(FD,H);
21           END;
```

```
22    END;
23    Close(FD);
24  END.
```

```
1   Program P1_23;
2   {=========================================================================
3   Generate file F1_23.D2D to plot inequality Sqr(sin(x)+sin(y))-(y*x+0.5)>0
4   =========================================================================}
5   uses CRT;
6   const nx=406;    xmin=-Pi;    xmax=Pi;    {grid size and limits over x}
7         ny=406;    ymin=-Pi;    ymax=Pi;    {grid size and limits over y}
8   var FD: File of double;    x,y: double;    ix,iy: integer;
9   function Ineq(x,y:double): Boolean;
10  BEGIN
11    if Sqr(sin(x)+sin(y)) - (y*x+0.5) >= 0 then
12      Ineq:=TRUE
13    else
14      Ineq:=FALSE;
15  END; {..Ineq(x,y)}
16  BEGIN
17    ClrScr; Assign(FD,'F1_23.D2D'); Rewrite(FD);
18    for iy:=1 to ny do BEGIN
19      y:=ymin+(ymax-ymin)/(ny-1)*(iy-1);
20      for ix:=1 to nx do BEGIN
21        x:=xmin+(xmax-xmin)/(nx-1)*(ix-1);
22        if NOT Ineq(x,y) then Write(FD,x,y);
23      END;
24    END;
25    Close(FD);
26    Write('Data file generated successfully. Press <CR>..'); ReadLn;
27  END.
```

```pascal
1    Program P1_24A;
2    {==================================================================
3    Generate files F1_24A.DTA & F1_24A.D2D to plot Archimedean spiral:
4    x=Theta*cos(Theta)  & y=Theta*sin(Theta)  with Tmin<Theta<Tmax in rad.
5    ==================================================================}
6    uses CRT;
7    const nT = 91; {number of plot points (either 91 or 8*180+1=1441) }
8          Tmin=0.0;  Tmax=8*Pi; {limits over Theta }
9    var FD: File of double; {output file of double}
10       FT: Text;          {output ASCII file }
11       Theta,x,y: double;    iT: integer;
12   BEGIN
13     Assign(FD,'F1_24A.D2D'); Rewrite(FD);
14     Assign(FT,'F1_24A.DTA'); Rewrite(FT);
15     ClrScr;  {Write ASCII file header:}
16     WriteLn(FT,'x(Theta)  = Theta*cos(Theta)');
17     WriteLn(FT,'y(Theta)  = Theta*sin(Theta)');
18     WriteLn(FT,'  Theta  x(Theta)      y(Theta)');
19     for iT:=1 to nT do BEGIN
20       Theta:=Tmin+(Tmax-Tmin)/(nT-1)*(iT-1); {generate T}
21       x:=Theta*cos(Theta);  y:=Theta*sin(Theta);
22       WriteLn(FT,Theta:10:6,' ',x:14:10,' ',y:14:10);
23       Write(FD,x,y);
24     END;
25     Close(FT);  Close(FD);
26     Write('Data files generated successfully. Press <CR>..'); ReadLn;
27   END.
```

```pascal
1   Program P1_24B;
2   {==========================================================
3    Generate files F1_24B.DTA & F1_24B.D2D to plot Archimedean spiral:
4    x=Theta*cos(Theta)  &  y=Theta*sin(Theta)  with  Tmin<Theta<Tmax in rad.
5    Theta step is adjusted so that plot segment- lengths are DLavg ± DLtol.
6    NOTE 1:  If IncDT = 1/DecDT the 2nd repeat-until loop may become infinite.
7    NOTE 2:  Different IncDT or DecDT will reduce / increase FuncEvals.
8   ==========================================================}
9   uses CRT, LibMath;
10  const Tmin = 0.0;          Tmax = 8*Pi;          {limits of Theta}
11    nT = 101;                       {number of plot points to extract x & y bounds}
12    PlotDefx = 405;       PlotDefy = 405; {plot box resolution  }
13    DLavg = 31;            {approximate length of plot segment}
14    DLtol = 0.001;  {tolerance of segment length }
15    IncDT = 1.05;   {DT increaser coefficient i.e. DT:=IncDT*DT   }
16    DecDT = 0.95;   {DT decreaser coefficient i.e. DT:=DecDT*DT   }
17  var FD: File of double;     {output file of double }
18    FT: Text;  {output ASCII file      }
19    x, xmin,xmax,     {x point and x-range of plot in world units}
20    y, ymin,ymax,     {y point and y-range of plot in world units}
21    kx,ky,    {x,y scale factors}
22    T,DT,DL, x_1,y_1: double;
23    iT, DataPts, FuncEvals: LongInt;    OK: Boolean;
24  procedure FxFy(Theta: double; var Fx,Fy: double);
25  BEGIN
26    Fx:=Theta*cos(Theta);
27    Fy:=Theta*sin(Theta);
28    Inc(FuncEvals)
29  END; {..FxFy( )}
30  BEGIN
31    Assign(FD,'F1_24B.D2D');   Rewrite(FD);
```

```
32   Assign(FT, 'F1_24B.DTA'); Rewrite(FT);
33   ClrScr; {Next write the ASCII file header:}
34   WriteLn(FT,'x=Theta*cos(Theta)');
35   WriteLn(FT,'y=Theta*sin(Theta)');
36   WriteLn(FT,' DeltaL     Theta      x(Theta)      y(Theta)');
37   FuncEvals:=0; {reset function evaluation counter}
38   xmin:=InfD; xmax:=-InfD; ymin:=InfD; ymax:=-InfD;
39   for iT:=1 to nT do BEGIN{Estimate xmin,xmax,ymin,ymax and kx,ky:}
40     T:=Tmin+(Tmax-Tmin)/(nT-1)*(iT-1); {generate current T}
41     FxFy(T,x,y);
42     if (x < xmin) then xmin:=x;  {update xmin}
43     if (x > xmax) then xmax:=x;  {update xmax}
44     if (y < ymin) then ymin:=y;  {update ymin}
45     if (y > ymax) then ymax:=y;  {update ymax}
46   END;
47   kx:=PlotDefx/(xmax-xmin);   {x-axis scale factor}
48   ky:=PlotDefy/(ymax-ymin);   {y-axis scale factor}
49   T:=Tmin;  DL:=0.0;  DataPts:=0;
50   FxFy(T, x,y);  {compute first point}
51   DT:=(Tmax-Tmin)/nT;  {some initial T step size}
52   Repeat
53     Write(FD, x,y);  {..write data to file}
54     WriteLn(FT,DL:12:8,' ',T:12:8,' ',x:14:10,' ',y:14:10);
55     Inc(DataPts);
56     x_1:=x;  y_1:=y;  {..previous x & y}
57     Repeat
58       FxFy(T+DT, x,y);
59       DL:=Sqrt(Sqr(kx*(x-x_1))+Sqr(ky*(y-y_1)));
60       if (DL < DLavg-DLtol) then BEGIN
61         OK:=FALSE; DT:=IncDT*DT;  {increase T-step}
62       END
```

```pascal
63        else BEGIN
64          if (DL > DLavg+DLtol) then BEGIN
65            OK:=FALSE; DT:=DecDT*DT; {decrease T-step}
66          END
67          else OK:=TRUE;
68        END;
69      until OK; {..2nd repeat}
70      T:=T+DT;
71    until (T >= Tmax); {..1st repeat}
72    FxFy(Tmax, x,y); {calculate last point}
73    DL:=Sqrt(Sqr(kx*(x-x_1))+Sqr(ky*(y-y_1)));
74    WriteLn(FT,DL:12:8,'  ',Tmax:12:8,'  ',x:14:10,'  ',y:14:10);
75    Write(FD,x,y);
76    Close(FT); Close(FD);
77    Inc(DataPts); {Next write a short report on the screen:}
78    WriteLn('  DLavg = ',DLavg); WriteLn('  DLtol = ',DLtol:9);
79    WriteLn('  There were ',DataPts,' data points written to file.');
80    WriteLn('  There were ',FuncEvals,' FxFy(..) function calls.');
81    Write('  Press <CR>..'); ReadLn;
82  END.


1   Program P1_25;
2   {=================================================================
3   Generate files F1_25.DTA & F1_25.D2D to plot an Archimedean spiral:
4   x=Theta*cos(Theta) & y=Theta*sin(Theta) with Tmin<Theta<Tmax in rad.
5   Uses Coinc2Pts(..) and  Colin3Pts(..) to optimize the graph.
6   =================================================================}
7   uses CRT, LibMath, LibGe2D;
8   const Tmin = 0.0;    Tmax = 8*Pi;    {Theta limits}
9         nT0 = 100;     {number of plot points to extract x and y bounds}
10        nT  = 1000;    {number of actual plot points}
```

```pascal
11  var FD: File of double;    {output file of double}
12      FT: Text;              {output ASCII file       }
13      x, xmin,xmax,          {x point and x plot limits in world units}
14      y, ymin,ymax,          {y point and y plot limits in world units}
15      x2,y2,x1,y1,Eps2,Eps3,T: double;  iT, DataPts, FuncEvals: LongInt;
16  procedure FxFy(Theta: double; var Fx,Fy: double);
17  BEGIN
18      Inc(FuncEvals);
19      Fx:=Theta*cos(Theta);
20      Fy:=Theta*sin(Theta);
21  END;  {..FxFy()}
22  procedure Write2File(T,x,y:double);
23  BEGIN
24      WriteLn(FT,T:10:6,' ',x:14:10,' ',y:14:10);  {write T,x,y to ASCII}
25      Write(FD,x,y);  {write x,y to the file of doubles}
26      Inc(DataPts);
27  END;  {..Write2File}
28  label Label0,Label1,Label2;
29  BEGIN
30      Assign(FD,'F1_25.D2D');  Rewrite(FD);
31      Assign(FT,'F1_25.DTA');  Rewrite(FT);
32      ClrScr;   {Next write ASCII file header: }
33      WriteLn(FT,'x=Theta*cos(Theta)');
34      WriteLn(FT,'y=Theta*sin(Theta)');
35      WriteLn(FT,'  Theta  x(Theta)  y(Theta)');
36      FuncEvals:=0;   {reset function evaluation counter}
37      DataPts:=0;
38      xmin:=InfD; xmax:=-InfD; ymin:=InfD; ymax:=-InfD;
39      for iT:=1 to nT0 do BEGIN   {Estimate xmin,xmax,ymin,ymax and kx,ky:}
40          T:=Tmin+(Tmax-Tmin)/(nT0-1)*(iT-1);  {generate current T}
```

```
41    FxFy(T,x,y);
42    if (x < xmin) then xmin:=x;   {update xmin}
43    if (x > xmax) then xmax:=x;   {update xmax}
44    if (y < ymin) then ymin:=y;   {update ymin}
45    if (y > ymax) then ymax:=y;   {update ymax}
46    END; {..for}
47    Eps2:=0.000002*(Sqr(xmax-xmin)+Sqr(ymax-ymin));
48    Eps3:=0.000001*(Sqr(xmax-xmin)+Sqr(ymax-ymin));
49    iT:=1;
50    FxFy(Tmin,x,y);     {compute first point}
51    Write2File(Tmin,x,y);     {write first point to file}
52    x2:=InfD;   y2:=InfD;   x1:=InfD;   y1:=InfD;
53    repeat
54    Label2:
55    x2:=x1;   y2:=y1;   x1:=x;   y1:=y;
56    Label1:
57    Inc(iT);
58    T:=Tmin+(Tmax-Tmin)/(nT-1)*(iT-1);     {generate next T}
59    FxFy(T,x,y);     {compute next point}
60    if (iT = nT) then GoTo Label0;
61    if Coinc2Pts2D(x1,y1, x,y, Eps2) then     {x1,y1, x,y coincide ..}
62        GoTo Label1;     {.. eliminate x,y}
63    if (x2 = InfD) AND (y2 = InfD) then GoTo Label2;
64    if Colin3Pts2D(x2,y2, x1,y1, x,y, Eps3) then BEGIN
65        x1:=x; y1:=y;     {x2,y2, x1,y1, x,y collinear; eliminate x1,y1}
66        GoTo Label1;
67    END;
68    Write2File(T,x,y);
69    until FALSE; {..repeat}
```

```
70  Label0:
71  Write2File(T,x,y);
72  Close(FT);   Close(FD);   {Next write a short report on the screen:}
73  WriteLn('   Eps2 =   ',Eps2:8);   WriteLn('   Eps3 =   ',Eps3:8);
74  WriteLn('   There were ',DataPts,' data points written to file.');
75  WriteLn('   There were ',FuncEvals,' FxFy(..) function calls.');
76  Write('        Press <CR>..');   ReadLn;
77  END.
```

```
1   Program P1_26A;
2   {==================================================================
3   Generate files  F1_26A.DTA & F1_26A.D2D to plot n equally spaced
4   Archimedean spirals that grow one-at-a-time, of equations:
5   x=Theta*cos(Theta) & y=Theta*sin(Theta) with 0<Theta<2*Pi
6   ==================================================================}
7   uses CRT, LibMath;
8   const Tmin = 0.0;    Tmax = 2*Pi;         {limits of Theta }
9         nT   = 31;  {number of plot points i.e. of Theta values}
10        n    = 8;   {number of spirals}
11  var FD: File of double;    {output file of double}
12      FT: Text;              {output ASCII file }
13      Theta,Theta0,x,y: double;   iT,iC: integer;
14  BEGIN
15      Assign(FD,'F1_26A.D2D'); Rewrite(FD);
16      Assign(FT,'F1_26A.DTA'); Rewrite(FT);
17      ClrScr;   {Next write ASCII file header:}
18      WriteLn(FT,'Polar plot of ',n,' Archimedean spirals:');
19      WriteLn(FT,'        x                y');
20      for iC:=1 to n do BEGIN
21          Theta0:=(iC-1)*(2*Pi/n);   {generate angular offset}
```

```pascal
22        for iT:=1 to nT do BEGIN
23          Theta:=Tmin+(Tmax-Tmin)/(nT-1)*(iT-1);
24          x:=Theta*cos(Theta+Theta0);  y:=Theta*sin(Theta+Theta0);
25          WriteLn(FT,x:12:8,' ',y:12:8);  Write(FD,x,y);
26        END;
27        WriteLn(FT,'=====');  {ASCII line breaker}
28        Write(FD,InfD,InfD);  {D2D file line breaker}
29      END;
30      Close(FT);  Close(FD);
31      Write('Data files generated successfully. Press <CR>..'); ReadLn;
32    END;


1   Program P1_26B;
2   {=============================================================
3    Generate D_2D files F1_26B.DTA & F1_26B.D2D to plot n equally spaced,
4    simultaneously growing Archimedean spirals of equations:
5    x=Theta*cos(Theta) & y=Theta*sin(Theta) with  Tmin<Theta<Tmax in rad.
6    =============================================================}
7   uses CRT, LibMath;
8   const Tmin = 0.0; Tmax = 2*Pi; {Theta limits}
9         nT   = 31; {number of plot points i.e. Theta values}
10        n    = 8; {number of spirals}
11  var FD: File of double;    FT: Text;    {output files}
12      Theta,Theta0,x,y: double;    iT,iC: integer;
13  procedure Write2File(x,y:double);
14  BEGIN
15    WriteLn(FT,x:14:10,' ',y:14:10);  {write x,y to ASCII file}
16    Write(FD,x,y);                    {write x,y to the file of doubles}
17  END; {..Write2File}
18  BEGIN
19    Assign(FD,'F1_26B.D2D');Rewrite(FD);
```

```
20   Assign(FT,'F1_26B.DTA');Rewrite(FT);
21   ClrScr; {Next write ASCII file header:}
22   WriteLn(FT,'Polar plot of ',n,' Archimedean spirals:');
23   WriteLn(FT,'        x              y');
24   for iT:=2 to nT do BEGIN
25     for iC:=1 to n do BEGIN
26       Theta0:=(iC-1)*(2*Pi/n);  {angular offset}
27       Theta:=Tmin+(Tmax-Tmin)/(nT-1)*(iT-2);  {parameter value}
28       x:=Theta*cos(Theta+Theta0);  {x of 1st point of a new segment}
29       y:=Theta*sin(Theta+Theta0);  {y of 1st point of a new segment}
30       Write2File(x,y);
31       Theta:=Tmin+(Tmax-Tmin)/(nT-1)*(iT-1);
32       x:=Theta*cos(Theta+Theta0);  {x of 2nd point of a new segment}
33       y:=Theta*sin(Theta+Theta0);  {y of 2nd point of a new segment}
34       Write2File(x,y);
35       Write(FD,InfD,InfD);  {D2D line breaker}
36       WriteLn(FT,'======');  {ASCII line breaker}
37     END;
38   END;
39   Close(FT);  Close(FD);
40   Write('Data files generated successfully. Press <CR>..'); ReadLn;
41   END.
```

```
1   Program P1_31;
2   {===============================================================
3   Generate files  F1_31.DTA & F1_31.D2D  to plot spiraling polygons
4   ===============================================================}
5   uses CRT, LibMath;
6   const Tmin = 0.0;      Tmax = 2*Pi;      {Theta limits}
```

```
7      nT    = 31;     {number of plot points i.e. of Theta values}
8      n     = 8;      {number of spirals}
9   var FD: File of double;    {output file of double}
10      FT: Text;              {output ASCII f-le    }
11      Theta,Theta0,x,y: double;   iT,i: integer;
12  BEGIN
13    Assign(FD,'F1_31.D2D'); Rewrite(FD);
14    Assign(FT,'F1_31.DTA'); Rewrite(FT);
15    ClrScr;    {Write ASCII file header}
16    WriteLn(FT,'Spiraling Polygons:');
17    WriteLn(FT,'          x              y');
18    for iT:=1 to nT do BEGIN
19      Write(FD,InfD,InfD);   {D2D line breaker}
20      WriteLn(FT,'======');   {ASCII line breaker}
21      Theta:=Tmin+(Tmax-Tmin)/(nT-1)*(iT-1);
22      for i:=1 to n do BEGIN
23        Theta0:=(i-1)*(2*Pi/n);    {angular offset}
24        x:=Theta*cos(Theta+Theta0);
25        y:=Theta*sin(Theta+Theta0);
26        WriteLn(FT,x:12:8,' ',y:12:8);
27        Write(FD,x,y);
28      END;
29      x:=Theta*cos(Theta);   y:=Theta*sin(Theta);
30      WriteLn(FT,x:12:8,' ',y:12:8);    Write(FD,x,y);
31    END;
32    Close(FT);  Close(FD);
33    Write('Data files generated successfully. Press <CR>..'); ReadLn;
34  END.
```

```pascal
1  Program P1_33;
2  {=========================================================================
3  Generate files F1_33-#.DTA  &  F1_33-#.D2D to plot spiraling polygons
4  colored randomly in groups of 10 (# should be equal to n on line 9).
5  =========================================================================}
6  uses CRT, LibMath;
7  const Tmin =-2*Pi; Tmax = 2*Pi; {Theta limits}
8        nT   = 1441;     {number of rotated polygons i.e. of Theta values }
9        n    = 5;        {number of sides of the polygon i.e. 2,3 etc.    }
10       Fnme = 'F1_33-#'; {output file name i.e. 'F1_33-2' 'F1_33-3' etc. }
11 var FD: File of double; {output file of double }
12     FT: Text;           {output ASCII file     }
13     Theta,Theta0,x,y,Color: double;   iT,i: integer;
14 BEGIN
15 Assign(FD,Fnme+'.D2D');   Rewrite(FD);
16 Assign(FT,Fnme+'.DTA');   Rewrite(FT);
17 ClrScr; {Write ASCII file header}
18 WriteLn(FT,'Spiral polygon with ',n:1,' sides');
19 WriteLn(FT,'        x            y');
20 Randomize;   {Initialize the built-in random number generator}
21 Color:=(2+0.01*(Random(7)+1))*InfD; {assign random colors}
22 for iT:=1 to nT do BEGIN
23   if (iT MOD 10 = 0) then {groups of 10 have the same color}
24     Color:=(2+0.01*(Random(7)+1))*InfD;
25   WriteLn(FT,Color:11,' ',Color:11);  {new color in ASCII file}
26   Write(FD,Color,Color);               {new color in D2D file}
27   Write(FD,InfD,InfD);    {D2D line breaker}
28   WriteLn(FT,'======');   {ASCII line breaker}
29   Theta:=Tmin+(Tmax-Tmin)/(nT-1)*(iT-1);
30   for i:=1 to n do BEGIN
```

```
31          Theta0:=(i-1)*(2*Pi/n);      {angular offset}
32          x:=Theta*cos(Theta+Theta0);
33          y:=Theta*sin(Theta+Theta0);
34          WriteLn(FT,x:12:8,' ',y:12:8);
35          Write(FD,x,y);
36        END;
37        x:=Theta*cos(Theta);   y:=Theta*sin(Theta);
38        WriteLn(FT,x:12:8,' ',y:12:8);      Write(FD,x,y);
39      END;
40      Close(FT);   Close(FD);
41      Write('Data files generated successfully. Press <CR>..'); ReadLn;
42    END.
```

***

```
1   Program P2_123;
2   {====================================================================
3    Generate files F2_1.D3D, F2_2.D3D & F2_3.D3D to plot the functions:
4    F1 = 1/Sqrt(Sqr(1-Sqr(y))+Sqr(2*x*y))
5    F2 = 2/Exp(Sqr(x*x+y*y)-1.5))-1;
6    F3 = 10*(Exp(-Sqr(2*x+1)-Sqr(y+1))-Exp(-Sqr(x-1)-Sqr(y+1)))
7      +15*(Exp(-Sqr(x-1)-Sqr(y-1))-Exp(-Sqr(2*x+1)-Sqr(2*y-1)));
8    ====================================================================}
9   uses CRT, LibMath;
10  const Fxy='F1';   {'F1', 'F2' or 'F3' -- choose one}
11  var   FD: File of double;
12        F: argF2;   {argF2 is defined in LibMath}
13        nx,ny, xmin,xmax, ymin,ymax, x,y, Fv: double;
14        i,j: integer;
15  {$F+}
16  function F1(x,y: double): double;
```

```
17  BEGIN
18    F1:=1/Sqrt(Sqr(1-Sqr(y))+Sqr(2*x*y));
19  END;
20  function F2(x,y: double): double;
21  BEGIN
22    F2:=2/Exp(Sqr(Sqrt(x*x+y*y)-1.5))-1;
23  END;
24  function F3(x,y: double): double;
25  var T1,T2: double;
26  BEGIN
27    T1:=Exp(-Sqr(2*x+1)-Sqr(y+1))  - Exp(-Sqr(x-1)-Sqr(y+1));
28    T2:=Exp(-Sqr(x-1)-Sqr(y-1))  - Exp(-Sqr(2*x+1)-Sqr(2*y-1));
29    F3:=10*T1+15*T2;
30  END;
31  {$F-}
32  BEGIN
33    if (Fxy = 'F1') then BEGIN
34      F:=F1;
35      nx:=10;      ny:=261;
36      xmin:=0.1;  xmax:=1.0;  ymin:=0.0;   ymax:=2.5;
37      Assign(FD,'F2_1.D3D');
38    END;
39    if (Fxy = 'F2') then BEGIN
40      F:=F2;
41      nx:=481;     ny:=481;
42      xmin:=-Pi;  xmax:=Pi;  ymin:=-Pi;  ymax:= Pi;
43      Assign(FD,'F2_2.D3D');
44    END;
45    if (Fxy = 'F3') then BEGIN
46      F:=F3;
47      nx:=481;     ny:=481;
48      xmin:=-1.5; xmax:=2.5; ymin:=-2.5; ymax= 2.5;
```

```
49        Assign(FD,'F2_3.D3D');
50     END;
51     Rewrite(FD); ClrScr;
52     Write(FD, nx,ny, xmin,xmax, ymin,ymax);
53     for i:=1 to Round(nx) do BEGIN
54        x:=xmin+(xmax-xmin)/(nx-1)*(i-1);
55        for j:=1 to Round(ny) do BEGIN
56           y:=ymin+(ymax-ymin)/(ny-1)*(j-1);
57           Fv:=F(x,y);
58           Write(FD,Fv);
59        END;
60     END;
61     Close(FD);
62     Write('Output file generated successfully. <CR>..'); ReadLn;
63  END.
```

```
1   Program P2_3;
2   {============================================================================
3   Generate files F2_3.D3D, F2_3.R3D, F2_3.T3D, F2_3.G3D to plot:
4   F3(x,y)=10*(Exp(-Sqr(2*x+1)-Sqr(y+1))-Exp(-Sqr(x-1)-Sqr(y+1)))
5   +15*(Exp(-Sqr(x-1)-Sqr(y-1))-Exp(-Sqr(2*x+1)-Sqr(2*y-1)));
6   ============================================================================}
7   uses CRT,LibMath;
8   const nx  : double = 481;  ny  : double = 481;
9         xmin: double =-1.5;  xmax: double = 2.5;
10        ymin: double =-2.5;  ymax: double = 2.5;
11  var   FD: File of double;  {output file of double}
12        FR: File of real;    {output file of real - same format as D3D}
13        FT3D: Text;          {output ASCII file - same format as D3D}
14        FG3D: Text;          {output ASCII file - (xi,yj,zij) format }
```

```
15        ix,iy: integer;       x, y, z: double;    aReal: real;
16 function F3(x,y: double): double;
17 var T1,T2: double;
18 BEGIN
19    T1:=Exp(-Sqr(2*x+1)-Sqr(y+1)) - Exp(-Sqr(x-1)-Sqr(y+1));
20    T2:=Exp(-Sqr(x-1)-Sqr(y-1)) - Exp(-Sqr(2*x+1)-Sqr(2*y-1));
21    F3:=10*T1 + 15*T2;
22 END;      {.. F(x,y)}
23 BEGIN
24    ClrScr;
25    Assign(FD  ,'F2_3.D3D');    Rewrite(FD);
26    Assign(FR  ,'F2_3.R3D');    Rewrite(FR);
27    Assign(FT3D,'F2_3.T3D');    Rewrite(FT3D);
28    Assign(FG3D,'F2_3.G3D');    Rewrite(FG3D);
29    Write(FD,nx,ny, xmin,xmax, ymin,ymax);
30    aReal:=nx;   Write(FR,aReal);
31    aReal:=ny;   Write(FR,aReal);
32    aReal:=xmin;  Write(FR,aReal);
33    aReal:=xmax;  Write(FR,aReal);
34    aReal:=ymin;  Write(FR,aReal);
35    aReal:=ymax;  Write(FR,aReal);
36    WriteLn(FT3D,nx:16:6);
37    WriteLn(FT3D,ny:16:6);
38    WriteLn(FT3D,xmin:16:6);
39    WriteLn(FT3D,xmax:16:6);
40    WriteLn(FT3D,ymin:16:6);
41    WriteLn(FT3D,ymax:16:6);
42    for ix:=1 to Round(nx) do BEGIN
43      x:=xmin+(xmax-xmin)/(nx-1)*(ix-1);
44      for iy:=1 to Round(ny) do BEGIN
45        y:=ymin+(ymax-ymin)/(ny-1)*(iy-1);
```

```
46      Z:=F3(x,y);
47      Write(FD,Z);
48      aReal:=Z;  Write(FR,aReal);
49      WriteLn(FT3D,Z:16:6);
50      WriteLn(FG3D,'(',x:12:6,' ',y:12:6,' ',Z:12:6,')');
51      END;
52    END;
53    Close(FD);  Close(FR);  Close(FT3D);  Close(FG3D);
54    Write('Output files generated successfully! <CR>..'); ReadLn;
55  END.
```

```
1   Program P2_ZLCS;
2   {=========================================================
3   Generates ASCII file Z.LCS with level curve heights Log spaced from z0
4   towards zmin and zmax, to be appended to a CF3 file.
5   =========================================================}
6   const z0=0.265;       zmin=-12.8778;     zmax=14.8243;
7         NrLC = 28;
8   var   FT: Text;        {output ASCII file}
9         z: double;       NrLCdn, NrLCup, k: integer;
10  BEGIN
11    Assign(FT,'z.LCS');    Rewrite(FT);
12    if (z0-zmin) > (zmax-z0) then BEGIN
13      NrLCdn:=Round(NrLC*(z0-zmin)/(zmax-zmin));
14      NrLCup:=NrLC-NrLCdn;
15    END
16    else BEGIN
17      NrLCup:=Round(NrLC*(zmax-z0)/(zmax-zmin));
18      NrLCdn:=NrLC-NrLCup;
19    END;
```

```
20      for k:=NrLCdn downto 1 do BEGIN
21        z:=z0 - Exp((k-1)*Ln(z0-zmin+1)/(NrLCdn-1))+1;
22        WriteLn(FT,z:18:8);
23      END;
24      for k:=1 to NrLCup do BEGIN
25        z:=z0 + Exp((k-1)*Ln(zmax-z0+1)/(NrLCup-1))-1;
26        WriteLn(FT,z:18:8);
27      END;
28      Close(FT);
29    END.
```

```
1  Program P2_2T;
2  {=================================================================
3   Generate file F2_2.TXT with 36 x 36 pts. to plot the function:
4   F2=2/Exp(Sqr(Sqrt(x*x+y*y)-1.5))-1  using   Office Excel.
5   =================================================================}
6  uses CRT;
7  const nx: double = 36;   xmin: double =-Pi;   xmax: double = Pi;
8        ny: double = 36;   ymin: double =-Pi;   ymax: double = Pi;
9  var FT: Text;  {output ASCII file}
10     x, y, F2: double;   ix,iy: integer;
11 BEGIN
12   ClrScr;
13   Assign(FT,'F2_2.TXT');   Rewrite(FT);
14   Write(FT,'  x\y ');
15   for iy:=1 to Round(ny) do
16     Write(FT,ymin+(ymax-ymin)/(ny-1)*(iy-1):9:4,' ');
17   WriteLn(FT);
18   for ix:=1 to Round(nx) do BEGIN
19     x:=xmin+(xmax-xmin)/(nx-1)*(ix-1);
```

```
20      Write(FT,x:9:4,' ');
21      for iy:=1 to Round(ny) do BEGIN
22        y:=ymin+(ymax-ymin)/(ny-1)*(iy-1);
23        F2:=2.0/Exp(Sqr(Sqrt(x*x+y*y)-1.5))-1.0;
24        Write(FT,F2:9:5,' ');
25      END;
26      WriteLn(FT);
27    END;
28    Close(FT);
29    Write('Output ASCII file generated successfully. <CR>..'); ReadLn;
30 END.
```

```
1  Program P2_4;
2  {=====================================================================
3  Generate file F2_4.D3D to plot the function   F4(x,y)=0.1*x*y
4  =====================================================================}
5  uses CRT;
6  const nx: double = 496;  xmin: double =-Pi;  xmax: double = Pi;
7        ny: double = 496;  ymin: double =-Pi;  ymax: double = Pi;
8  var FD: File of double;  {output file of double}
9      ix,iy: integer; x,y, F4: double;
10 BEGIN
11   ClrScr;
12   Assign(FD,'F2_4.D3D');  Rewrite(FD);
13   Write(FD, ny,nx, xmin,xmax,ymin,ymax);
14   for ix:=1 to Round(nx) do BEGIN
15     x:=xmin+(xmax-xmin)/(nx-1)*(ix-1);
16     for iy:=1 to Round(ny) do BEGIN
17       y:=ymin+(ymax-ymin)/(ny-1)*(iy-1);
18       F4:=0.1*x*y;
```

```
19       Write(FD,F4);
20     END;
21   END;
22   Close(FD);
23   Write('Output file generated successfully. <CR>..'); ReadLn;
24 END.
```

```
1  Program P2_5;
2  {================================================================
3   Generate file F2_5P.D3D and F2_5N.D3D to plot the piecewise function:
4   F5(x,y)=0.1*x*y   for   x*x + y*y > 1
5   F5(x,y)=+/-1E30   for   x*x + y*y < 1
6   ================================================================}
7  uses CRT;
8  const nx: double = 481;   xmin: double =-Pi;   xmax: double = Pi;
9        ny: double = 481;   ymin: double =-Pi;   ymax: double = Pi;
10
11 var FD: File of double;   ix,iy: integer;   Inf, x,y, F: double;
12 BEGIN
13 { Decomment one of the lines below!     }
14 { Assign(FD,'F2_5N.D3D');  Inf:=-1.0E30; }
15   Assign(FD,'F2_5P.D3D');  Inf:= 1.0E30;
16   Rewrite(FD);   ClrScr;
17   Write(FD, ny,nx, xmin,xmax,ymin,ymax);
18   for ix:=1 to Round(nx) do BEGIN
19     x:=xmin+(xmax-xmin)/(nx-1)*(ix-1);
20     for iy:=1 to Round(ny) do BEGIN
21       y:=ymin+(ymax-ymin)/(ny-1)*(iy-1);
22       if (x*x+y*y < 2.25) then F:=Inf else F:=0.1*x*y;
23       Write(FD,F);
24     END;
```

```
25    END;
26    Close(FD);
27    Write('Output file generated successfully. <CR>..'); ReadLn;
28  END.


1   Program P2_6;
2   {=================================================================
3    Generate file F2_6.D3D (501 x 501 points) to plot the inequality:
4    Sqr(sin(x)+sin(y)) - (y*x+0.5) ò 0
5    =================================================================}
6   uses CRT;
7   const nxy: double = 501;
8         xmin: double =-Pi;      xmax: double = Pi;
9         ymin: double =-Pi;      ymax: double = Pi;
10  var FD: File of double;        ix,iy: integer; x,y, F: double;
11  function Ineq(x,y:double): double;
12  BEGIN
13    if Sqr(sin(x)+sin(y)) - (y*x+0.5)  >= 0 then
14      Ineq:=1.0E30
15    else
16      Ineq:=0.0;
17  END;  {.. Ineq(x,y) }
18  BEGIN
19    ClrScr;
20    Assign(FD,'F2_6.D3D'); Rewrite(FD);
21    Write(FD, nxy,nxy, xmin,xmax,ymin,ymax);
22    for ix:=1 to Round(nxy) do BEGIN
23      x:=xmin+(xmax-xmin)/(nxy-1)*(ix-1);
24      for iy:=1 to Round(nxy) do BEGIN
25        y:=ymin+(ymax-ymin)/(nxy-1)*(iy-1);
26        F:=Ineq(x,y);
```

```
27        Write(FD,F);
28      END;
29    END;
30  Close(FD);
31  Write('Output file generated successfully. <CR>..'); ReadLn;
32  END.


1  Program P2_7;
2  {==================================================================
3   Generate ASCII file F2_7.T3D with 181x181 pts. to plot using D_3D
4   two hemispheres centered at (0,0,0)
5  ==================================================================}
6  uses CRT,LibInOut;
7  const nx: double = 181; xmin: double = -PI; xmax: double = PI;
8        ny: double = 181; ymin: double = -PI; ymax: double = PI;
9  var FT: Text;   ix,iy: integer;  x, y, z1,z2: double;
10 function F7_up(x,y: double): double;
11 BEGIN
12   if (2.89-x*x-y*y) >= 0 then
13     F7_up:=Sqrt(2.89-x*x-y*y)
14   else
15     F7_up:=-1.0E30;
16 END; {.. F7_up(x,y)}
17 function F7_dn(x,y: double): double;
18 BEGIN
19   if (2.89-x*x-y*y) >= 0 then
20     F7_dn:=-Sqrt(2.89-x*x-y*y)
21   else
22     F7_dn:=+1.0E30;
23 END; {.. Dn(x,y)}
24 BEGIN
```

```
25    ClrScr;
26    Assign(FT,'F2_7.T3D');   Rewrite(FT);
27    WriteLn(FT,'   F7_up(x,y)    F7_dn(x,y)');
28    WriteLn(FT,     nX:16:9,'   ',nX:16:9);
29    WriteLn(FT,     nY:16:9,'   ',nY:16:9);
30    WriteLn(FT,Xmin:16:9,'   ',Xmin:16:9);
31    WriteLn(FT,Xmax:16:9,'   ',Xmax:16:9);
32    WriteLn(FT,Ymin:16:9,'   ',Ymin:16:9);
33    WriteLn(FT,Ymax:16:9,'   ',Ymax:16:9);
34    for ix:=1 to Round(nx) do BEGIN
35      x:=xmin+(xmax-xmin)/(nx-1)*(ix-1);
36      for iy:=1 to Round(ny) do BEGIN
37        y:=ymin+(ymax-ymin)/(ny-1)*(iy-1);
38        z1:=F7_up(x,y);  z2:=F7_dn(x,y);
39        WriteLn(FT,MyStr(z1,16),'  ',MyStr(z2,16));
40      END;
41    END;
42    Close(FT);
43    Write('Output file generated successfully. <CR>..'); ReadLn;
44    END.
```

```
1    Program P2_8;
2    {=========================================================================
3    Generate file   F2_8.T3D (251 x 251 pts.) to plot on the same graph:
4    F2(x,y)=2/Exp(Sqr(Sqrt(x*x+y*y)-1.5))-1 & F4(x,y) = 0.1*x*y
5    =========================================================================}
6    uses CRT,LibInOut;
7    const nx: double = 251;  xmin: double = -PI; xmax: double = PI;
8          ny: double = 251;  ymin: double = -PI; ymax: double = PI;
9    var  FT: Text;
```

```
10   x,y, F4,F2, Fu,Fd, F2u,F2d, F4u,F4d: double; i, ix,iy: LongInt;
11   function F_2(x,y: double): double;
12   BEGIN
13     F_2:=2.0/Exp(Sqr(Sqrt(x*x+y*y)-1.5))-1.0;
14   END; {.. F_2(x,y)}
15   function F_4(x,y: double): double;
16   BEGIN
17     F_4:=0.1*x*y;
18   END; {.. F_4(x,y)}
19   BEGIN
20     ClrScr;
21     Assign(FT,'F2_8.T3D');Rewrite(FT);
22     WriteLn(FT,'F_Up F_Dn F4_Up F4_Dn F2_Up F2_Dn');
23     for i:=1 to 6 do Write(FT,MyStr(nx,10),' ');   WriteLn(FT);
24     for i:=1 to 6 do Write(FT,MyStr(ny,10),' ');   WriteLn(FT);
25     for i:=1 to 6 do Write(FT,MyStr(xmin,10),' ');   WriteLn(FT);
26     for i:=1 to 6 do Write(FT,MyStr(xmax,10),' ');   WriteLn(FT);
27     for i:=1 to 6 do Write(FT,MyStr(ymin,10),' ');   WriteLn(FT);
28     for i:=1 to 6 do Write(FT,MyStr(ymax,10),' ');   WriteLn(FT);
29     for ix:=1 to Round(nx) do BEGIN
30       x:=xmin+(xmax-xmin)/(nx-1)*(ix-1);
31       for iy:=1 to Round(ny) do BEGIN
32         y:=ymin+(ymax-ymin)/(ny-1)*(iy-1);
33         F2u:=1E30;    F2d:=1E30;
34         F4u:=1E30;    F4d:=1E30;
35         F2:=F_2(x,y);
36         F4:=F_4(x,y);
37         if (F4 > F2) then BEGIN
38           Fu:=F4;    Fd:=F2;    F4u:=F4;    F2d:=F2;
39           END
40         else BEGIN
41           Fu:=F2;    Fd:=F4;    F4d:=F4;    F2u:=F2;
```

```
42        END;
43        WriteLn(FT,MyStr(Fd,10),' ',MyStr(Fu,10),' ',MyStr(F4d,10)
44        ,' ',MyStr(F4u,10),' ',MyStr(F2d,10),' ',MyStr(F2u,10));
45      END;
46    END;
47    Close(FT);
48    Write('Output file generated successfully. <CR>..');   ReadLn;
49  END.
```

***

```
1  program P3_01A;
2  {============================================================
3   Generate vectors t[..] and Y[..], plot them on the screen and
4   copy the screen to files  F3_01A.PCX  and  F3_01A.DXF
5   ============================================================}
6  uses  Graph,        {Red,CloseGraph }
7        LibMath,      {VDp,Pmax}
8        LibGraph,     {InitGr}
9        LibDXF,       {InitDXFfile,CloseDXFfile}
10       Unit_PCX,     {WritePCX}
11       LibPlots;     {PlotCurve,PlotXaxis,PlotYaxis}
12 const FileName = 'F3_01A';  {FileName.DXF or FileName.PCX }
13       nPts = 250;   {should not exceed 502 points i.e. Pmax }
14       On = Pi/2.5;  {natural circular frequency [rad/s] }
15       Z  = 0.2;     {damping ratio}
16       t0 = 0.0;     tmax = 15.0;   {start and end time [s]}
17 var t, Y: VDp;    i: Integer;    OK: Boolean;
18 BEGIN
19   for i:=1 to nPts do BEGIN    {generate t[..] and Y[..]}
20     t[i]:=(i-1)*(tmax-t0)/(nPts-1);
21     Y[i]:=2*exp(-Z*On*t[i])*sin(Sqrt(1-Sqr(Z))*On*t[i]+Pi/9);
```

```
22  END;
23  InitGr(0);                    {switch to graphic mode ..}
24  InitDXFfile(FileName+'.DXF');  {prepare to copy the screen to DXF}
25  PlotCurve(1, t,Y,nPts,  Red);
26  PlotYaxis(1, 0,  5,2,  'Y(t)    ');
27  PlotXaxis(1, 1,  6,3,  't [sec]    ');
28  WritePCX(FileName+'.PCX',  OK);   {copy the screen to PCX}
29  CloseDXFfile;  ReadLn;      {press <CR> to finish}
30  CloseGraph;
31  END.

1   program P3_01B;
2   {=======================================================================
3   Generate vectors t[..] and Y[..], plot them on the screen and
4   copy the screen to files  F3_01B.PCX   and   F3_01B.DXF
5   =======================================================================}
6   uses  Graph,      {Red,SolidLn,ThickWidth,SetLineStyle,CloseGraph}
7         LibMath,    {VDp,Pmax}
8         LibInOut,   {WaitToGo}
9         LibGraph,   {InitGr}
10        LibGIntf,   {DrawBorder}
11        LibDXF,     {InitDXFfile,CloseDXFfile}
12        Unit_PCX,   {WritePCX}
13        LibPlots;   {NewPlot,FitBox,PlotCurve,PlotYaxis,PlotXaxis, ..
14                     SetDivLine,UpdateLimitsY,NewLimitsX,NewLimitsY,
15                     GetXmax,GetYmax}
16  const FileName = 'F3_01B';   {FileName.DXF or FileName.PCX}
17        nPts = 250;      {should not exceed Pmax = 502}
18        On  = Pi/2.5;    {natural circular frequency [rad/s]}
19        Z   = 0.2;       {damping ratio}
20        t0  = 0.0;       tmax = 15.0;    {start and end time [s]}
```

```pascal
21  var t,Y: VDp; i: Integer; Ch: char;   OK: Boolean;
22  BEGIN
23    for i:=1 to nPts do BEGIN
24      t[i]:=(i-1)*(tmax-t0)/(nPts-1);
25      Y[i]:=2*exp(-Z*On*t[i])*sin(Sqr(1-Sqr(Z))*On*t[i]+Pi/9);
26    END;
27    InitGr(0);  {switch to graphic mode}
28    InitDXFfile(FileName+'.DXF');  {prepare to copy the screen to DXF}
29    NewPlot(1, FitBox, 150,80,500,430, 'Damped oscillation');
30    DrawBorder;
31    UpdateLimitsY(1, Y, nPts);
32    NewLimitsY(1, 1.1*GetYmax(1),1.15*GetYmin(1));
33    SetLineStyle(SolidLn, 0, ThickWidth);
34    PlotCurve(1, t,Y,nPts, Red);
35    SetDivLine(4, 10, 0.75);  {value 10 will cause the grid line}
36    PlotYaxis(1, 0, 6,2, 'Y(t)    ');
37    PlotXaxis(1, 0, 4,5, 't [sec]    ');
38    WritePCX(FileName+'.PCX', OK);  {copy the screen to PCX}
39    CloseDXFfile;  WaitToGo(Ch);  {press any key to finish}
40    CloseGraph;
41  END.
```

```pascal
1  program P3_02;
2  {======================================================================
3   Generate vectors t[..], Y1[..] and Y2[..], plot them on the screen and
4   copy the screen to files  F3_02.PCX  and  F3_02.DXF
5  ======================================================================}
6  uses  Graph,     {Blue,Red,CloseGraph}
7        LibInOut,  {EraseAll, ImplicitFileName}
8        LibGraph,  {InitGr}
9        LibMath,   {VDp}
```

```
10      LibDXF,         {InitDXFfile,CloseDXFfile}
11      Unit_PCX,       {WritePCX}
12      LibPlots;       {NewPlot,FitBox,UpdateLimitsX,UpdateLimitsY, ..
13                       ResizeY,SetMarker,PlotCurve,PlotYaxis,PlotXaxis}
14 const FileName = 'F3_02';  {.DXF or .PCX}
15      nPts = 250;           {should not exceed Pmax = 502}
16      On = Pi/2.5;          {Omega_n }
17      t0 = 0.0;  tmax = 15.0;  {start and end time [s]}
18 var t,Y1,Y2: VDp;  Z: double; i: Integer;  Ch: char;  OK: Boolean;
19 BEGIN
20   for i:=1 to nPts do BEGIN    {Generate vectors t,Y1 & Y2 ..}
21     t[i]:=(i-1)*(tmax-t0)/(nPts-1);
22     Z:=0.2;
23     Y1[i]:=2*exp(-Z*On*t[i])*sin(Sqrt(1-Sqr(Z))*On*t[i]+Pi/9);
24     Z:=0.3;
25     Y2[i]:=2*exp(-Z*On*t[i])*sin(Sqrt(1-Sqr(Z))*On*t[i]+Pi/9);
26   END;
27   InitGr(0);
28   InitDXFfile(FileName+'.DXF'); {prepare to copy the screen to DXF}
29   NewPlot(1, FitBox, 60,60,620,420, 'Damped oscillations');
30   UpdateLimitsX(1,  t,  nPts);   {get tmin & tmax from t[]}
31   UpdateLimitsY(1, Y1, nPts);    {get Ymin & Ymax from Y1[]}
32   UpdateLimitsY(1, Y2, nPts);    {update Ymin & Ymax using Y2[]}
33   ResizeY(1, 0.2);   {expand Y-axis nicely by about 10% both directions}
34   PlotCurve(1, t,Y1,nPts, Blue);
35   SetMarker(4, ':<>');
36   PlotCurve(1, t,Y1,nPts, -Blue);
37   PlotYaxis(1, 0, 7,2, ' Y1(t)          ');
38   PlotXaxis(1, 4,5, 't [sec]         ');
39   SetMarker(2, '|o');
40   PlotCurve(1, t,Y2,nPts, -Red);
```

```
41       PlotYaxis(1, 0, 7,2, ' Y2(t)');
42       WritePCX(FileName+'.PCX', OK);      {copy the screen to PCX}
43       CloseDXFfile;    WaitToGo(Ch);      {Press any key to finish}
44       CloseGraph;
45   END.


1    program P3_03A;
2    {=====================================================================
3    Read vectors X[1..nPts] and Y[1..nPts] from ASCII file F3_03.DTA, plot
4    them on the screen and copy the screen to files F3_03.PCX and F3_03.DXF
5    =====================================================================}
6    uses   Graph,
7           LibMath,          {VDp, Pmax}
8           LibInOut,         {Extract_V,WaitToGo}
9           LibGraph,         {InitGr,CloseGraph}
10          LibGIntf,         {DrawBorder}
11          LibDXF,           {InitDXFfile,CloseDXFfile}
12          Unit_PCX,         {WritePCX}
13          LibPlots;         {NewPlot,PlotCurve,PlotXaxis,PlotYaxis ..}
14                            NewLimitsX,NewLimitsY}
15   const  FileName = 'F3_03A';          {name of DXF and PCX files}
16          RowStart = 3;
17          Xcol = 1;  Ycol = 5;          {column number for X and Y}
18          nPts = 502;    {should not exceed 502 i.e. Pmax }
19   var    FT: Text;    {input ASCII file }
20          X,Y: VDp;
21          OneX,OneY, Xmin,Xmax, Ymin,Ymax: double;
22          RowFinish, jRow,i: Word;
23          Row: string;   Ch: char;    OK: Boolean;
24   BEGIN
```

```
25    Xmin:=0.0;   Xmax:=0.5;        {predefined X limits}
26    Ymin:=1.5;   Ymax:=3.5;        {predefined Y limits}
27    Assign(FT,'F3_03.DTA');   Reset(FT);
28    jRow:=0;
29    while NOT Eof(FT) do BEGIN   {count lines in FT file}
30      ReadLn(FT,Row); Inc(jRow);
31    END;
32    RowFinish:=jRow-RowStart + 1;
33    InitGr(0);   {switch to graphic mode}
34    InitDXFfile(FileName+'.DXF');  {prepare to copy the screen to DXF}
35    NewPlot(1, FitBox, 40,120,600,360, 'Experimental Data');
36    DrawBorder;
37    NewLimitsX(1, Xmin,Xmax);      {set Xmin & Xmax}
38    NewLimitsY(1, Ymin,Ymax);      {set Ymin & Ymax}
39    Reset(FT);
40    jRow:=0;     i:=0;
41    Repeat    {read the input file again to do the actual plot ..}
42      ReadLn(FT,Row);
43      Inc(jRow);
44      Extract_V(Row,Xcol,OneX); {extract from Row the Xcol-th value}
45      Extract_V(Row,Ycol,OneY); {extract from Row the Ycol-th value}
46      if (jRow >= RowStart) then BEGIN
47        Inc(i);   X[i]:=OneX;   Y[i]:=OneY;
48      END;
49      if (i = nPts) OR Eof(FT) then BEGIN   {X,Y have nPts components}
50        PlotCurve(1, X,Y, i, 1+Random(13));   {color sections at random}
51        i:=1; X[i]:=OneX; Y[i]:=OneY;
52      END;
53    until Eof(FT);
54    PlotYaxis(1, 0, 5,5, ' Volts ');
55    PlotXaxis(1, 2, 6,4, 't [sec] ');
```

```pascal
56      WritePCX(FileName+'.PCX', OK);              {copy the screen to PCX}
57      CloseDXFfile;        WaitToGo(Ch);          {press any key to finish}
58      CloseGraph;
59      END.
```

```pascal
1  Program P3_12;
2  {===============================================================
3  Generate file F3_12.TXT with 'saw-tooth' angle values
4  ===============================================================}
5  uses CRT, LibMath; {InfD, Atan2}
6  const FileName = 'F3_12.TXT';
7        Theta_min = -2*Pi;  Theta_max =  2*Pi;  {lower and upper bownds}
8        n=401;  {number of data points}
9  var FT: Text;  Theta0, Theta1, Theta2: double;  i: integer;
10 BEGIN
11   Assign(FT,FileName);  Rewrite(FT);
12   ClrScr;  {Next will write ASCII file header: }
13   WriteLn(FT,'   Theta0     Theta1     Theta2');
14   WriteLn(FT,Theta_min:9:6,'   ',Theta_min:9:6,'   ',Theta_min:9:6);
15   for i:=1 to n do BEGIN
16     Theta0:=Theta_min+(Theta_max-Theta_min)/(n-1)*(i-1); {original angle}
17     Theta0:=Theta0;
18     Theta1:=ArcTan(sin(Theta0)/cos(Theta0));  {break Theta0 with ArcTan}
19     Theta2:=ATan2(sin(Theta0),cos(Theta0));  {break Theta0 with ATan2 }
20     WriteLn(FT,Theta0:9:6,'   ',Theta1:9:6,'   ',Theta2:9:6);
21   END;
22   Close(FT);
23   Write('File '+FileName+' generated successfully. <CR>..');  ReadLn;
24 END.
```

```pascal
1   Program P3_23;
2   {==============================================================
3   Generates data file F3_14.G3D to plot a 3D helix of variable radius
4   ==============================================================}
5   uses CRT;
6   const FileName = 'F3_23.G3D';
7         Rf = 0.5;      {final radius }
8         nC = 8.0;      {total number of coils}
9         nl = 36;       {number of vertices per coil}
10        p = 1/8;       {axial pitch }
11  var   FT: Text;   r, t, X,Y,Z: double;   i,n: integer;
12  procedure XYZ(Theta: double; var X,Y,Z: double);
13  BEGIN
14    r:=Rf*Sqr(Theta/(2.0*Pi*nC));
15    X:=r*cos(Theta);   Y:=r*sin(Theta);   Z:=p*Theta/(2*Pi);
16  END;  {.. XYZ()}
17  BEGIN
18    ClrScr;
19    Assign(FT,FileName);   Rewrite(FT);
20    n:=Round(nC*nl);
21    for i:=1 to n do BEGIN
22      t:=(2*Pi*nC/n)*(i-1);
23      XYZ(t,X,Y,Z);
24      WriteLn(FT,'(',X:16:6,' ',Y:16:6,' ',Z:16:6,')');
25    END;
26    Close(FT);
27    Write('File '+FileName+' generated successfully. <CR>..'); ReadLn;
28  END.
```

```
1    Program P3_25;
2    {===============================================================
3     Generate file F3_25.G3D to plot a toroidal helix:
4    ===============================================================}
5    uses CRT;
6    const FileName = 'F3_25.G3D';
7    var FT: Text;      t,tmin,tmax, X,Y,Z: double;      nt,it: integer;
8    procedure XYZ(Theta: double; var X,Y,Z: double);
9    const rS=2;   {helix radius}
10        rT=10;   {torus centroidal radius}
11        nS=24;   {number of coils}
12   BEGIN
13     X:=(rT+rS*cos(nS*Theta))*cos(Theta);
14     Y:=(rT+rS*cos(nS*Theta))*sin(Theta);
15     Z:=rS*sin(nS*Theta);
16   END;   {.. XYZ()}
17   BEGIN
18     nt:=90*24;   tmin :=0.0;   tmax :=2*Pi;
19     ClrScr;   Assign(FT,FileName);   Rewrite(FT);
20     for it:=1 to nt do BEGIN
21       t:=tmin+(tmax-tmin)/(nt-1)*(it-1);
22       XYZ(t,X,Y,Z);
23       WriteLn(FT,'(',X:16:6,' ',Y:16:6,' ',Z:16:6,')');
24     END;
25     Close(FT);
26     Write('File '+FileName+' generated successfully.   <CR>..');   ReadLn;
27   END.
```

```
 1   program P3_31;
 2   {============================================================
 3   Writes to F3_31SW.M3D the commands to insert block "S_wheel2" and to file
 4   F3_31UCS.M3D the commands to insert a XYZ frame attached to "S_wheel2".
 5   ============================================================}
 6   uses   LibMath,  {VDn}
 7          LibGe3D;  {vect3, mat33, RT}
 8   var Rotx,Rotz: mat33;
 9       T3,Zero3: vect3;
10       Phi: VDn;            {steering wheel rotation angles}
11       x01,y01,z01,         {local point along the X axis of the "S_wheel" block}
12       x02,y02,z02,         {local point along the Y axis of the "S_wheel" block}
13       x1,y1,z1,            {global point along the Y axis of the "S_wheel" block}
14       x2,y2,z2,            {global point along the Y axis of the "S_wheel" block}
15       x,y,z: double;    i,n: Byte;  M3D1, M3D2: text;
16   BEGIN
17     Assign(M3D1,'F3_31SW.M3D');    Rewrite(M3D1);
18     Assign(M3D2,'F3_31UCS.M3D');   Rewrite(M3D2);
19     n:=7;
20     Phi[1]:=-45.120*RAD;
21     Phi[2]:=-35.088*RAD;
22     Phi[3]:=-19.968*RAD;
23     Phi[4]:= 00.000*RAD;
24     Phi[5]:= 24.608*RAD;
25     Phi[6]:= 53.648*RAD;
26     Phi[7]:= 86.992*RAD;
27     {Point along the local X-axis of the steering wheel:  }
28     x01:=300.0;   y01:=0.0;     z01:=0.0;
29     {Point along the local Y-axis of the steering wheel:  }
30     x02:=0.0;     y02:=300.0;   z02:=0.0;
```

```
31      {Zero translation vector: }
32      Zero3[1]:=0.0;      Zero3[2]:=0.0;      Zero3[3]:=0.0;
33      {Translation to the end of steering column:       }
34      T3[1]:=0.0; T3[2]:=971.338;   T3[3]:=658.399;
35      {Rotation about the X-axis by -30 deg (steering column tilt): }
36      Rotx[1,1]:=1.0;   Rotx[1,2]:=0.0;        Rotx[1,3]:=0.0;
37      Rotx[2,1]:=0.0;   Rotx[2,2]:=cos(-30*RAD);   Rotx[2,3]:=-sin(-30*RAD);
38      Rotx[3,1]:=0.0;   Rotx[3,2]:=sin(-30*RAD);   Rotx[3,3]:=cos(-30*RAD);
39      for i:=1 to n do BEGIN
40        {Rotation about the Z-axis (steering wheel turn): }
41        Rotz[1,1]:=cos(Phi[i]);   Rotz[1,2]:=-sin(Phi[i]);   Rotz[1,3]:=0.0;
42        Rotz[2,1]:=sin(Phi[i]);   Rotz[2,2]:=cos(Phi[i]);    Rotz[2,3]:=0.0;
43        Rotz[3,1]:=0.0;           Rotz[3,2]:=0.0;            Rotz[3,3]:=1.0;
44        RT(Rotz,Zero3, x01,y01,z01, x ,y ,z );
45        RT(Rotx,T3  ,  x  ,y  ,z  , x1,y1,z1);
46        RT(Rotz,Zero3, x02,y02,z02, x ,y ,z );
47        RT(Rotx,T3  ,  x  ,y  ,z  , x2,y2,z2);
48        WriteLn(M3D1,' ',i:2,' ');
49        WriteLn(M3D2,' ',i:2,' ');
50        WriteLn(M3D1,' BK "S_wheel"',T3[1]:6:3,T3[2]:9:3,T3[3]:9:3
51              ,x1:9:3,y1:9:3,z1:9:3, x2:9:3,y2:9:3,z2:9:3,' ');
52        WriteLn(M3D2,'(CL "GREEN") change color');
53        WriteLn(M3D2,'(AR',T3[1]:6:3, T3[2]:9:3, T3[3]:9:3, x1:9:3, y1:9:3, z1:9:3,' ');
54        WriteLn(M3D2,'(AR ',T3[1]:6:3, T3[2]:9:3, T3[3]:9:3, x2:9:3, y2:9:3, z2:9:3,' ');
55        WriteLn(M3D2,'(CL "WHITE") back to regular color');
56      END;
57      Close(M3D1);   Close(M3D2);
58      END.
```

***

```pascal
1   program P4_01;
2   {=============================================================
3   Finds the root of function  F1(x)=1/(Sqr(x-1)+0.1)+1.0/(Sqr(x-3)+0.2)-3
4   situated in the interval [0,1].
5   ============================================================}
6   uses LibMath;
7   const a = 0.0;    b = 1.0;
8   var  x: double;
9   {$F+}
10  function F1(x: double): double;
11  BEGIN
12    F1:=1/(Sqr(x-1)+0.1)+1.0/(Sqr(x-3)+0.2)-3;
13  END;
14  {$F-}
15  BEGIN
16    Zero(F1, a,b, x);
17    WriteLn('x=',x,'   F1(x)=',F1(x));
18    WriteLn('Function calls=',NrFev0);   ReadLn;   {press <CR>}
19  END.
```

```pascal
1   program P4_02;
2   {=============================================================
3   Finds the root of function F2(x)=(x*x*x-3*x)/(x*x-4) closest to x=1
4   ============================================================}
5   uses LibMath;
6   const Step = 0.1;
7   var  x: double;
8   {$F+}
9   function F2(x: double): double;
10  BEGIN
11    if Abs(x*x-4.0) > EpsD then {check for division by zero}
```

```
12        F2:=(x*x*x-3*x)/(x*x-4)
13      Else
14        F2:=InfD;
15    END;
16    {$F-}
17    BEGIN
18      x:=1.0;
19      ZeroStart(F2, Step, x);
20      WriteLn('x=',x,'  F2(x)=',F2(x));
21      WriteLn('Function calls=',NrRev0); ReadLn; {press <CR>}
22    END.
```

```
1    program P4_03;
2    {=====================================================================
3     Finds the roots of functions
4     F1(x) = 1/(Sqr(x-1)+0.1)+1.0/(Sqr(x-3)+0.2)-3 over interval [0,4]
5     F2(x) = (x*x*x-3*x)/(x*x-4) over interval [-4,4]
6    =====================================================================}
7    uses CRT, LibMath;
8    var X: VDn;    a,b: double;
9    {$F+}
10   function F1(x: double): double;
11   BEGIN
12     F1:=1/(Sqr(x-1)+0.1)+1.0/(Sqr(x-3)+0.2)-3;
13   END; {.. F1(x)}
14   function F2(x: double): double;
15   BEGIN
16     if Abs(x*x-4.0) > EpsD then {check for division by zero}
17       F2:=(x*x*x-3*x)/(x*x-4)
18     Else
19       F2:=InfD;
```

```
20   END;    {.. F2(x) }
21   {$F-}
22   BEGIN
23     a:=0.0;   b:=4.0;
24     ZeroGrid(F1, a,b, 25, X);
25     WriteLn('x1=',X[1],'    F1(x1)=',F1(X[1]));
26     WriteLn('x2=',X[2],'    F1(x2)=',F1(X[2]));
27     WriteLn('x3=',X[3],'    F1(x3)=',F1(X[3]));
28     WriteLn('x4=',X[4],'    F1(x4)=',F1(X[4]));
29     WriteLn('Function calls=',NrFev0);
30     WriteLn;
31     NrFev0:=0;
32     a:=-4.0;  b:= 4.0;
33     ZeroGrid(F2, a,b, 25, X);
34     WriteLn('x1=',X[1],'    F2(x1)=',F2(X[1]));
35     WriteLn('x2=',X[2],'    F2(x2)=',F2(X[2]));
36     WriteLn('x3=',X[3],'    F2(x3)=',F2(X[3]));
37     WriteLn('Function calls=',NrFev0);  ReadLn; {press <CR>}
38   END.
```

```
1   program P4_04;
2   {=============================================================================
3   Finds the minimum of function F1(x)=1/(Sqr(x-1)+0.1)+1/(Sqr(x-3)+0.2)-3
4   situated in the interval [1,3].
5   =============================================================================}
6   uses  LibMin1;
7   const a = 1.0;  b = 3.0;
8   var   x,vF: double;
9   {$F+}
10  function F1(x: double): double;
11  BEGIN
```

```
12    F1:=1/(Sqr(x-1)+0.1)+1.0/(Sqr(x-3)+0.2)-3;
13  END;
14  {$F-}
15  BEGIN
16    Brent(F1, a,b, vF,x);
17    WriteLn('x=',x,'  F1(x)=',vF);
18    Write('Obj. function calls=',NrFevl); ReadLn; {press <CR>}
19  END.
```

```
1   program P4_05;
2   {=========================================================================
3   Finds the minimum of function  F1(x)=1/(Sqr(x-1)+0.1)+1/(Sqr(x-3)+0.2)-3
4   situated closest to x=0.75.
5   =========================================================================}
6   uses LibMin1;
7   var  x,vF: double;
8   {$F+}
9   function F1(X: double): double;
10  BEGIN
11    F1:=1/(Sqr(x-1)+0.1)+1/(Sqr(x-3)+0.2)-3;
12  END;
13  {$F-}
14  BEGIN
15    x:=0.75;
16    BrentStart(F1, 0.01, vF,x);
17    WriteLn('x=',x,'  F1(x)=',vF);
18    Write('Obj. function calls=',NrFevl); ReadLn; {press <CR>}
19  END.
```

```
1    program P4_06;
2    {================================================================
3    Finds all minima and maxima of functions
4    F1(x)=1/(Sqr(x-1)+0.1)+1/(Sqr(x-3)+0.2)-3   situated in the interval [0,4]
5    F2(x)=(x*x*x-3*x)/(x*x-4)   situated in the interval [-4,4].
6    ================================================================}
7    uses  LibMin1, LibMath;
8    var   x,vF:VDn;   a,b:double;
9    {$F+}
10   function F1(X: double): double;
11   BEGIN
12     F1:=1/(Sqr(x-1)+0.1)+1/(Sqr(x-3)+0.2)-3;
13   END;  {.. F1()}
14   function _F1(X: double): double;
15   BEGIN
16     _F1:=-(1/(Sqr(x-1)+0.1)+1/(Sqr(x-3)+0.2)-3);
17   END;  {.. _F1()}
18   function F2(x: double): double;
19   BEGIN
20     F2:=InfD;
21     if Abs(x*x-4) > EpsD then F2:=(x*x*x-3*x)/(x*x-4)
22   END;  {.. F2()}
23   function _F2(x: double): double;
24   BEGIN
25     _F2:=InfD;
26     if Abs(x*x-4) > EpsD then _F2:=-(x*x*x-3*x)/(x*x-4)
27   END;  {.. _F2()}
28   {$F-}
```

```
29   BEGIN
30     ClrScr;
31     a:=0.0;  b:=4.0;
32     BrentGrid(F1, a,b,20, vF,X);
33     WriteLn('x1=',X[1],'  F1(x1)=',vF[1]);     WriteLn;
34     WriteLn('Obj. function calls=',NrFev1);    WriteLn;
35     NrFev1:=0;
36     BrentGrid(_F1, a,b,20, vF,X);
37     WriteLn('x1=',X[1],'  F1(x1)=',-vF[1]);
38     WriteLn('x2=',X[2],'  F1(x2)=',-vF[2]);
39     WriteLn('Obj. function calls=',NrFev1);    WriteLn;
40     a:=-4.0;   b:=4.0;
41     NrFev1:=0;
42     BrentGrid(F2, a,b,20, vF,X);
43     WriteLn('x1=',X[1],'  F2(x1)=',vF[1]);
44     WriteLn('x2=',X[2],'  F2(x2)=',vF[2]);
45     WriteLn('x3=',X[3],'  F2(x3)=',vF[3]);
46     WriteLn('x4=',X[4],'  F2(x4)=',vF[4]);
47     WriteLn('Obj. function calls=',NrFev1);    WriteLn;
48     NrFev1:=0;
49     BrentGrid(_F2, a,b,20, vF,X);
50     WriteLn('x1=',X[1],'  F2(x1)=',-vF[1]);
51     WriteLn('x2=',X[2],'  F2(x2)=',-vF[2]);
52     WriteLn('x3=',X[3],'  F2(x3)=',-vF[3]);
53     WriteLn('x4=',X[4],'  F2(x4)=',-vF[4]);
54     WriteLn('Obj. function calls=',NrFev1);   ReadLn; {press <CR>}
55   END.
```

```
1   program P4_08;
2   {=========================================================================
3   Uses Nelder-Mead Simplex method to minimize function:
4   Fn(x,y)=10*(Exp(-Sqr(2*x+1)-Sqr(y+1))-Exp(-Sqr(x-1)-Sqr(y+1)))
5   +15*(Exp(-Sqr(x-1)-Sqr(y-1))-Exp(-Sqr(2*x+1)-Sqr(2*y-1)));
6   =========================================================================}
7   uses  DOS,CRT,
8         LibMath,    {VDn}
9         LibMinN;    {NelderMead}
10  const Nvar = 2;
11        LimAF = 5000; {maximum obj. function calls}
12  var   XX, XXmin, XXmax: VDn;   vF: double;  PlsMns,i,j: integer;
13  {$F+}
14  function Fn(vX: VDn): double;
15  var x,y,T1,T2,T3,T4: double;
16  BEGIN
17    x:=vX[1];
18    y:=vX[2];
19    T1:= Exp(-Sqr(2*x+1)-Sqr(  y+1));
20    T2:=-Exp(-Sqr(  x-1)-Sqr(  y+1));
21    T3:= Exp(-Sqr(  x-1)-Sqr(  y-1));
22    T4:=-Exp(-Sqr(2*x+1)-Sqr(2*y-1));
23    Fn:=PlsMns*(10*(T1+T2)+15*(T3+T4));
24  END;
25  {$F-}
26  BEGIN
27    ClrScr;
28    XXmin[1]:=-1.5;    XXmax[1]:=2.5;
29    XXmin[2]:=-2.5;    XXmax[2]:=2.5;
30    PlsMns:=+1;  {'+' for minimization, '-' for maximization }
31    for i:=1 to Nvar do XX[i]:=InfD;
```

```
32    NelderMead('P4_08-1.SPX',Fn,Nvar, LimAF, 1E-32, XXmin,XXmax, vF,XX);
33    for i:=1 to Nvar do WriteLn('x',i:1,' =',XX[i]);
34    WriteLn('F_opt=',PlsMns*vF);
35    WriteLn('Obj. function calls=',NrFevN);  WriteLn;
36    PlsMns:=-1;   {'-' for maximization}
37    for i:=1 to Nvar do XX[i]:=InfD;
38    NelderMead('P4_08-2.SPX',Fn,Nvar, LimAF, 1E-32, XXmin,XXmax, vF,XX);
39    for i:=1 to Nvar do WriteLn('x',i:1,' =',XX[i]);
40    WriteLn('F_opt=',PlsMns*vF);
41    WriteLn('Obj. function calls=',NrFevN);  ReadLn;   {press <CR>}
42    END.


1     program P4_09;
2     {========================================================================
3     Uses Nelder-Mead Simplex method to minimize function:
4     F3(x,y)=10*(Exp(-Sqr(2*x+1)-Sqr(y+1))-Exp(-Sqr(x-1)-Sqr(y+1)))
5     +15*(Exp(-Sqr(x-1)-Sqr(y-1))-Exp(-Sqr(2*x+1)-Sqr(2*y-1)));
6     ========================================================================}
7     uses  DOS,CRT,
8           LibMath,   {VDn}
9           LibMinN;   {NelderMead}
10    const   PlsMns= +1;  {'+' for minimization '-' for maximization}
11            Nvar  = 2;
12            LimAF = 1000;  {maximum obj. function calls}
13    var FileName: PathStr;
14        XX,XXmin,XXmax,XXbest: VDn;   vF,vFbest: double;   i,j,TotalFev: integer;
15    {$F+}
16    function Fn(vX: VDn): double;
17    var x,y, T1,T2,T3,T4: double;
18    BEGIN
```

```
19    x:=vX[1];
20    y:=vX[2];
21    T1:= Exp(-Sqr(2*x+1)-Sqr( y+1));
22    T2:=-Exp(-Sqr( x-1)-Sqr( y+1));
23    T3:= Exp(-Sqr( x-1)-Sqr( y-1));
24    T4:=-Exp(-Sqr(2*x+1)-Sqr(2*y-1));
25    Fn:=PlsMns*(10*(T1+T2)+15*(T3+T4));
26  END;  {.. Fn()}
27  {$F-}
28  BEGIN
29    ClrScr;
30    vFbest:=InfD;     TotalFev:=0;
31    XXmin[1]:=-1.5;   XXmax[1]:=2.5;
32    XXmin[2]:=-2.5;   XXmax[2]:=2.5;
33    for j:=1 to 100 do BEGIN
34      for i:=1 to Nvar do
35        XX[i]:=XXmin[i]+Random*(XXmax[i]-XXmin[i]);
36      NelderMead('',Fn,Nvar,LimAF, 1.0E-32, XXmin,XXmax, vF,XX);
37      TotalFev:=TotalFev+NrFevN;
38      if (vF < vFbest) then BEGIN
39        vFbest:=vF;
40        for i:=1 to Nvar do XXbest[i]:=XX[i];
41      END;
42    END;
43    for i:=1 to Nvar do WriteLn('x',i:1,' =',XXbest[i]);
44    WriteLn('F_opt=',PlsMns*vFbest);
45    WriteLn('Obj. function calls=',TotalFev); ReadLn; {Press <CR>}
46  END.
```

```pascal
1    Program F4_05A;
2    {=================================================================
3    Generate files F4_5A.D3D to plot the function  Fn(x,y)=0.1*x*y
4    subjected to (x*x+y*y) < Sqr(rT+rS*cos(n*Atan2(x,y)))
5    =================================================================}
6    uses LibMath; {InfD}
7    const nx: double = 501;      xmin: double =-1.25; xmax: double =  1.25;
8          ny: double = 501;      ymin: double =-1.25; ymax: double =  1.25;
9    var  FD: File of double;
10        ix,iy: integer; x, Y, z: double;
11   function Fn(x,y: double): double; {Function to be optimized}
12   const rT=1.0; rS=0.2; n=8;
13   var Theta: double;
14   BEGIN
15     Fn:=InfD;
16     Theta:=Atan2(y,x);
17     if (Theta < InfD) then BEGIN
18       if (x*x+y*y) > Sqr(rT+rS*cos(n*Theta)) then Exit;
19     END;
20     Fn:=0.1*x*y;
21   END; {.. Fn()}
22   BEGIN
23     Assign(FD,'F4_5A.D3D'); Rewrite(FD);
24     Write(FD,nx,ny, xmin,xmax, ymin,ymax);
25     for ix:=1 to Round(nx) do BEGIN
26       x:=xmin+(xmax-xmin)/(nx-1)*(ix-1);
27       for iy:=1 to Round(ny) do BEGIN
28         y:=ymin+(ymax-ymin)/(ny-1)*(iy-1);
29         z:=Fn(x,y);   Write(FD,z);
30       END;
31     END;
32     Close(FD);
33   END.
```

```
1   program P4_10;
2   {===========================================================================
3   Uses Nelder-Mead Simplex method to minimize function:
4   Fn(x,y)=0.1*x*y  subjected to  (x*x+y*y) < Sqr(rT+rS*cos(n*Theta))
5   ===========================================================================}
6   uses  DOS,CRT,
7         LibMath,   {VDn}
8         LibMinN;   {NelderMead}
9   const PlsMns=+1.0; {'+' for minimization '-' for maximization}
10        Nvar = 2;
11        LimAF = 5000;  {maximum obj. function calls}
12   var  XX, XXmin, XXmax, XXbest: VDn;
13        vF, vFbest: double;
14        i,j: Byte;
15        TotalFev: LongInt;
16   {$F+}
17   function Fn(vX: VDn): double;
18   const  rT=1.0; rS=0.2; n=8;
19   var x,y,Theta:double;  j:Byte;
20   BEGIN
21     Fn:=InfD;
22     x:=vX[1];  y:=vX[2];
23     Theta:=Atan2(y,x);
24     if (Theta < InfD) then BEGIN
25       if (x*x+y*y) > Sqr(rT+rS*cos(n*Theta)) then Exit;
26     END;
27     Fn:=PlsMns*0.1*x*y;
28   END;  {.. Fn()}
29   {$F-}
30   BEGIN
31     ClrScr;  WriteOutN:=FALSE;  {do not displays search status info}
```

```
32      vFbest:=InfD;   TotalFev:=0;
33      XXmin[1]:=-1.5;   XXmax[1]:=2.5;
34      XXmin[2]:=-2.5;   XXmax[2]:=2.5;
35      for j:=1 to 100 do BEGIN
36        for i:=1 to Nvar do
37          XX[i]:=XXmin[i]+Random*(XXmax[i]-XXmin[i]);
38        NelderMead('',Fn,Nvar, LimAF, 1.0E-32, XXmin,XXmax, vF,XX);
39        TotalFev:=TotalFev+NrFevN;
40        if (vF < vFbest) then BEGIN
41          vFbest:=vF;
42          for i:=1 to Nvar do XXbest[i]:=XX[i];
43        END;
44      END;
45      for i:=1 to Nvar do WriteLn('x',i:1,' =',XXbest[i]);
46      WriteLn('F_opt=',PlsMns*vFbest);
47      WriteLn('Obj. function calls=',TotalFev);  ReadLn; {Press <CR>}
48  END.
```

```
1   Program F4_5B;
2   {==============================================================
3    Generate file: F4_5B.D2D to plot the parametric curve:
4    x = (rT + rS*cos(n*Theta))*cos(Theta)
5    y = (rT + rS*cos(n*Theta))*sin(Theta)  with 0 < Theta < 2Pi
6    ==============================================================}
7   const Tmin = 0.0;    Tmax = 2*Pi;  {limits of Theta }
8         nT   = 361;    {number of plot points}
9   var FD: File of double;
10      Theta,rT,rS,x,y: double;    n,iT: integer;
11  BEGIN
12    Assign(FD,'F4_5B.D2D'); Rewrite(FD);
```

```
13    rT:=1.0;        {median circle radius}
14    rS:=0.2;        {radial oscillation amplitude}
15    n:=8;           {number of radial oscillations}
16    for iT:=1 to nT do BEGIN
17      Theta:=Tmin+(Tmax-Tmin)/(nT-1)*(iT-1);
18      x:=(rT+rS*cos(n*Theta))*cos(Theta);
19      y:=(rT+rS*cos(n*Theta))*sin(Theta);
20      Write(FD,x,y);
21    END;
22    Close(FD);
23  END.
```

```
1   program P4_12;
2   {=====================================================================
3   Uses Nelder-Mead Simplex method to minimize the function
4   F5 = 0.7854*x1*Sqr(x2)*(3.3333*Sqr(x3)+14.9334*x3-43.0934)
5   -1.508*x1*(Sqr(x6)+Sqr(x7))+7.4777*(x6*x6*x6+x7*x7*x7);
6   subjected to:
7    2.6 <= x[1]  <= 3.6;      0.7 <= x[2]  <= 0.8;
8   17.0 <= x[3]  <= 28.0;     7.3 <= x[4]  <= 8.3;
9    7.3 <= x[5]  <= 8.3;      2.9 <= x[6]  <= 3.9;
10   5.0 <= x[7]  <= 5.5;
11  =====================================================================}
12  uses   DOS,CRT,
13         LibMath,   {VDn}
14         LibInOut,  {MyVal, MySt, BackUpFile}
15         LibMinN;   {NelderMead}
16  const Nvar = 7;    {number of variables}
17        LimAF = 20000;{max function calls per iterations}
18  var XX, XXmin, XXmax, XXbest: VDn;
19      vF, vFbest: double;  i,j: Word;   TotalFev: LongInt;
```

```pascal
20        FT:Text;
21   {$F+}
22   function F5(vX: VDn): double;
23   var  x1,x2,x3,x4,x5,x6,x7,TT:double; i:Word;
24   BEGIN
25     F5:=InfD;
26     for i:=1 to Nvar do
27       if (vX[i] < XXmin[i]) OR (vX[i] > XXmax[i]) then EXIT;
28     x1:=vX[1];  x2:=vX[2];  x3:=vX[3];  x4:=vX[4];
29     x5:=vX[5];  x6:=vX[6];  x7:=vX[7];
30     TT:=3.3333*Sqr(x3) + 14.9334*x3 - 43.0934;
31     TT:=0.7854*x1*Sqr(x2)*TT;
32     TT:=TT - 1.508*x1*(Sqr(x6) + Sqr(x7));
33     TT:=TT + 7.4777*(x6*x6*x6 + x7*x7*x7);
34     F5:=TT + 0.7854*(x4*Sqr(x6) + x5*Sqr(x7));
35   END; {.. F5() }
36   {$F-}
37   BEGIN
38     ClrScr;
39     XXmin[1]:= 2.6;    XXmax[1]:= 3.6;
40     XXmin[2]:= 0.7;    XXmax[2]:= 0.8;
41     XXmin[3]:=17.0;    XXmax[3]:=28.0;
42     XXmin[4]:= 7.3;    XXmax[4]:= 8.3;
43     XXmin[5]:= 7.3;    XXmax[5]:= 8.3;
44     XXmin[6]:= 2.9;    XXmax[6]:= 3.9;
45     XXmin[7]:= 5.0;    XXmax[7]:= 5.5;
46     vFbest:=InfD;  TotalFev:=0;
47     WriteOutN:=FALSE;    {do not displays search status info}
48     for i:=1 to Nvar do {first initial guess ..}
49       XXbest[i]:=XXmin[i]+0.5*(XXmax[i]-XXmin[i]);
50     for j:=1 to 500 do BEGIN
```

```pascal
51      for i:=1 to Nvar do BEGIN {1st guess base on the previous XXbest}
52        repeat
53          XX[i]:=XXbest[i]+(Random-0.5)/j*(XXmax[i]-XXmin[i]);
54        until (XXmin[i] <= XX[i]) AND (XX[i] <= XXmax[i]);
55      END;
56      NelderMead('',F5, Nvar, LimAF, 1.0E-64, XXmin,XXmax, vF,XX);
57      TotalFev:=TotalFev + NrFevN;
58      if (vF < vFbest) then BEGIN
59        for i:=1 to Nvar do XXbest[i]:=MyVal(MySt(XX[i],6));
60        vFbest:=F5(XXbest);
61        WriteLn(j:4,'') F(X)=',vFbest);
62      END;
63    END;
64    WriteLn('^j'F_opt=',vFbest);
65    for i:=1 to Nvar do WriteLn('x',i:1,'=',XXbest[i]);
66    Write('Obj. function calls = ',TotalFev,'   <CR>..');
67    BackUpFile('Results.TXT');
68    Assign(FT,'Results.TXT'); Rewrite(FT);
69    WriteLn(FT,'Obj. function calls=',TotalFev);
70    WriteLn(FT,'F_opt=',vFbest);
71    for i:=1 to Nvar do WriteLn(FT,'x',i:1,'=',XXbest[i]);
72    Close(FT);   ReadLn; {Press <CR>}
73  END.
```

```pascal
1   Program P4_14;
2   {=================================================================
3    Generates data files to plot using D_2D the design space and
4    performance space of the bicriterion minimization problem:
5    F1=pi/4*((L-x1)*(D1^2-x2^2)+x1*(D2^2-x2^2)) and
6    F2=64*F/(3*pi*E)*((L^3+x1^3)/(D1^4-x2^4)+x1^3/(D2^4-x2^4))
7    subjected to
```

```
 8      32*D1*F*L /(pi*(D1^4-x2^4))  <= Sigma_Y
 9      32*D2*F*x1/(pi*(D2^4-x2^4))  <= Sigma_Y
10    x1 >= 0;   x1 <= L;  x2 >= 40;   x2 <= 75;
11    ===============================================================}
12    uses CRT, LibMath;
13    var FD1,FD2: File of double;
14        x1, x1min,x1max,  x2, x2min,x2max,  F1,F2: double;
15        nx1,nx2, i,j: word;
16    function F12(x1,x2: double; var F1,F2: double): Boolean;
17    const L=1000; D1=100; D2=80; E=206E3; F=15000; Sigma_Y=220;
18    BEGIN
19      F12:=FALSE;
20      if (x1 <   0) then Exit;
21      if (x1 >   L) then Exit;
22      if (x2 <  40) then Exit;
23      if (x2 >  75) then Exit;
24      if 32*D1*F* L/(pi*(Pow(D1,4)-Pow(x2,4))) > Sigma_Y then Exit;
25      if 32*D2*F*x1/(pi*(Pow(D2,4)-Pow(x2,4))) > Sigma_Y then Exit;
26      F1:=pi/4*((L-x1)*(D1*D1-x2*x2) + x1*(D2*D2-x2*x2));
27      F2:=(Pow(L,3)-Pow(x1,3))/(Pow(D1,4)-Pow(x2,4));
28      F2:=Pow(x1,3)/(Pow(D2,4)-Pow(x2,4)) + F2;
29      F2:=64*F/(3*pi*E)*F2;
30      F12:=TRUE;
31    END; {.. F12()}
32    BEGIN
33      nx1:=250;   x1min:=0.0;   x1max:=1000;
34      nx2:=250;   x2min:=40;    x2max:=100; ClrScr;
35      Assign(FD1,'F4_10A.D2D'); Rewrite(FD1);
36      Assign(FD2,'F4_10B.D2D'); Rewrite(FD2);
37      for i:=1 to nx1 do BEGIN
38        x1:=x1min+(x1max-x1min)/(nx1-1)*(i-1);
```

```
39      for j:=1 to nx2 do BEGIN
40        x2:=x2min+(x2max-x2min)/(nx2-1)*(j-1);
41        if F12(x1,x2, F1,F2) then BEGIN
42          Write(FD1, x1,x2);
43          Write(FD2, F1,F2);
44        END;
45      END;
46      WriteIn(i:4,' /',nx2:4);
47    END;
48    Close(FD1); Close(FD2);
49  END.


1   Program P4_15;
2   {=============================================================
3    Generates data files to plot using D_2D the design space and
4    performance space of the bicriterion minimization problem:
5    F1=0.4*x1+x2  &  F2=1.0+x1*x1-x2+0.2*cos(4.75*x2)  subjected to:
6    0.8*x1*x1+x2*x2 <= 1.0
7   =============================================================}
8   uses CRT,LibMath;
9   var FD1, FD2: File of double;
10      x1, x1min,x1max, x2, x2min,x2max, F1,F2: double;
11      n,i: word;
12  function F12(x1,x2: double; var F1,F2: double): Boolean;
13  BEGIN
14    F12:=FALSE;
15    if (0.8*x1*x1+x2*x2 > 1.0) then Exit;
16    F1:=0.4* x1+x2;
17    F2:=1.0+x1*x1-x2+0.2*cos(4.75*x2);
18    F12:=TRUE;
```

```
19  END; {.. F12()}
20  BEGIN
21    Assign(FD1,'F4_12a.D2D'); Rewrite(FD1);
22    Assign(FD2,'F4_12b.D2D'); Rewrite(FD2);
23    n:=62500;   {total number of random points}
24    x1min:=-1.2; x1max:=1.2; x2min:=-1.2; x2max:=1.2;
25    Randomize; ClrScr;
26    for i:=1 to n do BEGIN
27      x1:=x1min+Random*(x1max-x1min);
28      x2:=x2min+Random*(x2max-x2min);
29      if F12(x1,x2, F1,F2) then BEGIN
30        Write(FD1,x1,x2);
31        Write(FD2,F1,F2);
32      END;
33      if (i MOD 500 = 0) then WriteLn(i:6,' ',' ',n:6);
34    END;
35    Close(FD1); Close(FD2);
36  END.


1   program P4_16;
2   {=================================================================
3   Uses Nelder-Mead Simplex method to simultaneously minimize functions:
4   F1=0.4*x1+x2 and F2=1.0+x1*x1-x2+0.2*cos(4.75*x2)
5   subjected to 0.8*x1*x1+x2*x2 <= 1.0
6   =================================================================}
7   uses DOS,CRT,
8        LibMath,   {VDn}
9        LibInOut,  {BackUpFile}
10       LibMinN;   {NelderMead}
11  const Nvar = 2;      {number of design variables}
```

```
12        LimAF = 25000; {max function calls per iterations}
13  var XX, XXmin, XXmax, XXbest: VDn;
14      MiMax, Fmin, Fmax, F1min, F1max, F2min, F2max,
15      W1, W2, F1, F2, vF, vFbest: double;
16      FT: Text;  i,j: Word;  TotalFev: LongInt;
17  {$F+}
18  function F12(vX: VDn): double;
19  var x1,x2:double;
20  BEGIN
21    F12:=InfD;
22    x1:=vX[1]; x2:=vX[2];
23    if (0.8*x1*x1+x2*x2 > 1.0) then Exit;
24    F1:=0.4*x1+x2;
25    F2:=1.0+x1*x1-x2+0.2*cos(4.75*x2);
26    F12:=MiMax*(W1*(F1-F1min)/(F1max-F1min)+W2*(F2-F2min)/(F2max-F2min));
27  END;
28  {$F-}
29  BEGIN
30    XXmin[1]:=-1.2;      XXmax[1]:=1.2;
31    XXmin[2]:=-1.2;      XXmax[2]:=1.2;
32    Assign(FT,'P4_16.TXT'); Rewrite(FT);
33    WriteLn(FT,' W1   F1   F2    x1    x2');
34    ClrScr; WriteOutN:=FALSE;     {NelderMead search status info OFF}
35    F1min:=0.0; F1max:=1.0; F2min:=0.0; F2max:=1.0;
36    W1:=+1.0; W2:=0.0;
37    MiMax:=-1;   {+1 for minimization and -1 for maximization}
38    for i:=1 to Nvar do XX[i]:=XXmin[i]+Random*(XXmax[i]-XXmin[i]);
39    NelderMead('',F12,Nvar,LimAF, 1.0E-19, XXmin,XXmax, Fmax,XX);
40    MiMax:=+1;   {+1 for minimization and -1 for maximization}
41    for i:=1 to Nvar do XX[i]:=XXmin[i]+Random*(XXmax[i]-XXmin[i]);
42    NelderMead('',F12,Nvar, LimAF, 1.0E-19, XXmin,XXmax, Fmin,XX);
```

```
43    Flmin:=Fmin; Flmax:=-Fmax;
44    W1:=0.0;W2:=+1.0;
45    MiMax:=-1;     {+1 for minimization and -1 for maximization}
46    for i:=1 to Nvar do XX[i]:=XXmin[i]+Random*(XXmax[i]-XXmin[i]);
47    NelderMead('',F12,Nvar, LimAF, 1.0E-19, XXmin,XXmax, Fmax,XX);
48    MiMax:=+1;     {+1 for minimization and -1 for maximization}
49    for i:=1 to Nvar do XX[i]:=XXmin[i]+Random*(XXmax[i]-XXmin[i]);
50    NelderMead('',F12,Nvar, LimAF, 1.0E-19, XXmin,XXmax, Fmin,XX);
51    F2min:=Fmin; F2max:=-Fmax;
52    MiMax:=+1;     {+1 for minimization and -1 for maximization}
53    W2:=-0.01;
54    Repeat
55    W2:=W2+0.01;    W1:=1.0-W2;
56    WriteLn('Pareto point ',W1:1:5,'/',W2:1:5);
57    vFbest:=InfD;
58    TotalFev:=0;
59    for i:=1 to Nvar do {first initial guess}
60    XX[i]:=XXmin[i]+Random*(XXmax[i]-XXmin[i]);
61    for j:=1 to 10 do BEGIN
62    for i:=1 to Nvar do BEGIN {initial guess based on previous XX}
63    Repeat
64    XX[i]:=XX[i]+(Random-0.5)/j*(XXmax[i]-XXmin[i]);
65    Until
66    (XXmin[i] <= XX[i]) AND (XX[i] <= XXmax[i]);
67    END;
68    NelderMead('',F12,Nvar, LimAF, 1.0E-19, XXmin,XXmax, vF,XX);
69    TotalFev:=TotalFev+NrFevN;
70    if (vF < vFbest) then BEGIN
71    vFbest:=vF; for i:=1 to Nvar do XXbest[i]:=XX[i];
72    END;
73    END;
```

```pascal
74      for i:=1 to Nvar do XX[i]:=XXbest[i];    vF:=F12(XX);
75      WriteLn(FT,  W1:3,'  ',F1:16,'  ',F2:16,'  ',XX[1]:16,'  ',XX[2]:16);
76    unt-l W2 > 1.0;
77    Close(FT);
78  END.


1  program P4_17;
2  {=============================================================
3   Uses Nelder-Mead Simplex method to simultaneously minimize:
4   F1=pi/4*((L-x1)*(D1^2-x2^2)+x1*(D2^2-x2^2))
5   F2=64*F/(3*pi*E)*((L^3+x1^3)/(D1^4-x2^4)+x1^3/(D2^4-x2^4))
6   subjected to
7   32*D1*F*L /(pi*(D1^4-x2^4))  <= Sigma_Y
8   32*D2*F*x1/(pi*(D2^4-x2^4))  <= Sigma_Y
9   x1 >= 0;  x1 <= L;  x2 >= 40;  x2 <= 75;
10  =============================================================}
11  uses  DOS,CRT,
12        LibMath,     {VDn}
13        LibInOut,    {BackUpFile}
14        LibMinN;     {NelderMead}
15  const Nvar  = 2;       {number of design variables}
16        LimAF = 25000; {max function calls per iterations}
17  var XX, XXmin,XXmax, XXbest: VDn;
18      MiMax,Fmin,Fmax,F1min,F1max,F2min,F2max,W1,W2,F1,F2,vF,vFbest:double;
19      FT: Text;  i,j: Word;  TotalFev: LongInt;
20  {$F+}
21  function F12(vX: VDn): double;
22  const L=1000; D1=100; D2=80; E=206E3; F=15000; Sigma_Y=220;
23  var x1,x2:double;
24  BEGIN
```

```
25        F12:=InfD;    x1:=vX[1];    x2:=vX[2];
26        if (x1 <  0) then Exit;
27        if (x1 >  L) then Exit;
28        if (x2 < 40) then Exit;
29        if (x2 > 75) then Exit;
30        if 32*D1*F* L/(pi*(Pow(D1,4)-Pow(x2,4)))  > Sigma_Y then Exit;
31        if 32*D2*F*x1/(pi*(Pow(D2,4)-Pow(x2,4)))  > Sigma_Y then Exit;
32        F1:=pi/4*((L-x1)*(D1*D1-x2*x2) + x1*(D2*D2-x2*x2));
33        F2:=(Pow(L,3)-Pow(x1,3))/(Pow(D1,4)-Pow(x2,4));
34        F2:=Pow(x1,3)/(Pow(D2,4)-Pow(x2,4)) + F2;
35        F2:=64*F/(3*pi*E)*F2;
36        F12:=MiMax*(W1*(F1-F1min)/(F1max-F1min)+W2*(F2-F2min)/(F2max-F2min));
37     END; {.. F12()}
38     {$F-}
39     BEGIN
40        XXmin[1]:=0;    XXmax[1]:=1000; XXmin[2]:=40; XXmax[2]:=100;
41        Assign(FT,'P4_17.TXT'); Rewrite(FT);
42        WriteLn(FT,'  W1    F2    F1    F2     x1      x2');
43        ClrScr; WriteOutN:=FALSE;  {NelderMead search status OFF }
44        F1min:=0.0;  F1max:=1.0; F2min:=0.0; F2max:=1.0; W1:=+1.0;      W2:=0.0;
45        MiMax:=-1;    {+1 for minimization and -1 for maximization}
46        for i:=1 to Nvar do XX[i]:=XXmin[i]+Random*(XXmax[i]-XXmin[i]);
47        NelderMead('',F12,Nvar, LimAF,  1.0E-19, XXmin,XXmax, Fmax,XX);
48        MiMax:=+1;   {+1 for minimization and -1 for maximization}
49        for i:=1 to Nvar do XX[i]:=XXmin[i]+Random*(XXmax[i]-XXmin[i]);
50        NelderMead('',F12,Nvar, LimAF,  1.0E-19, XXmin,XXmax,  Fmin,XX);
51        F1min:=Fmin;  F1max:=-Fmax;
52        W1:=0.0;  W2:=+1.0;
53        MiMax:=-1;   {+1 for minimization and -1 for maximization}
54        for i:=1 to Nvar do XX[i]:=XXmin[i]+Random*(XXmax[i]-XXmin[i]);
55        NelderMead('',F12,Nvar, LimAF,  1.0E-19, XXmin,XXmax,  Fmax,XX);
```

```
56    MiMax:=+1;    {+1 for minimization and -1 for maximization}
57    for i:=1 to Nvar do XX[i]:=XXmin[i]+Random*(XXmax[i]-XXmin[i]);
58    NelderMead('',F12,Nvar, LimAF, 1.0E-19, XXmin,XXmax, Fmin,XX);
59    F2min:=Fmin; F2max:=-Fmax;
60    MiMax:=+1;    {+1 for minimization and -1 for maximization}
61    W2:=-0.01;
62    repeat
63      W2:=W2+0.01;    W1:=1.0-W2;
64      WriteLn('Pareto point ',W1:1:5,'/',W2:1:5);
65      vFbest:=InfD;    TotalFev:=0;
66      for i:=1 to Nvar do {first initial guess}
67        XX[i]:=XXmin[i]+Random*(XXmax[i]-XXmin[i]);
68      for j:=1 to 10 do BEGIN
69        for i:=1 to Nvar do BEGIN {initial guess based on previous XX}
70          repeat
71            XX[i]:=XX[i]+(Random-0.5)/j*(XXmax[i]-XXmin[i]);
72          until
73            (XXmin[i] <= XX[i]) AND (XX[i] <= XXmax[i]);
74        END;
75        NelderMead('',F12,Nvar, LimAF, 1.0E-19, XXmin,XXmax, vF, XX);
76        TotalFev:=TotalFev+NrFevN;
77        if (vF < vFbest) then BEGIN
78          vFbest:=vF; for i:=1 to Nvar do XXbest[i]:=XX[i];
79        END;
80      END;
81      for i:=1 to Nvar do XX[i]:=XXbest[i];    vF:=F12(XX);
82      WriteLn(FT, W1:3,' ',F1:16,' ',F2:16,' ',XX[1]:16,' ',XX[2]:16);
83    until W2 > 1.0;
84    Close(FT);
85  END.
```

***

```
1   program P5_01;
2   {====================================================================
3    Locus/CometLocus animation of n Archimedes spirals of equations:
4    x(Theta)=Theta*cos(Theta+2*Pi/n*(i-1))
5    y(Theta)=Theta*sin(Theta+2*Pi/n*(i-1))   whith***  Tmin <= Theta <= Tmax.
6   ====================================================================}
7   uses LibDXF,     {InitDXFfile}
8        LibInOut,   {MySt, IsKeyPressed}
9        LibMec2D;   {OpenMecGraph,NewFrame,SetJointSize,Locus,CometLocus,}
10                   {SetTitle,PutPoint, MecOut,CloseMecDXF,CloseMecGraph}
11  const Tmin = 0.0;      {Theta lower bound}
12        Tmax = 2*Pi;     {Theta upper bound}
13        nFr  = 31;       {number of plot points i.e. Theta values}
14        n    = 8;        {number of spirals}
15  var i,iFr:Integer;     Theta0,Theta,x,y:double;
16  BEGIN
17    OpenMecGraph(-8,8,-8,8);
18    InitDXFfile('F5_01.DXF');
19    iFr:=0;     SetTitle('Archimedean Spirals');
20    Repeat
21      if (iFr > nFr) then BEGIN
22        iFr:=0;  CloseMecDXF;
23      END;
24      NewFrame(500);
25      SetJointSize(Round(2.0 + 4*iFr/nFr));
26      for i:=1 to n do BEGIN
27        Theta0:=(i-1)*(2*Pi/n);           {..angular spacing between curves}
28        Theta:=Tmin+(Tmax-Tmin)/(nFr-1)*iFr;  {..current angle Theta}
29        x:=Theta*cos(Theta+Theta0);
30        y:=Theta*sin(Theta+Theta0);
31        Locus(i MOD 15 +1,x,y,'p'+MySt(i,2));   {..Locus or CometLocus}
32        PutPoint(i MOD 15 +1,'O',x,y,'');
```

```
33      END;
34      Inc(iFr);
35    until IsKeyPressed(27);    {..<Esc> stops animation}
36    {until (Not MecOut) AND IsKeyPressed(27); }
37    CloseMecGraph(TRUE);       {..save .$2D files as .D2D}
38  END.
```

```
1   program P5_04;
2   {===============================================================}
3   Simulation of two cranks in series tracing an epicycloidal curve
4   {===============================================================}
5   uses Graph,          {Magenta,Red,White}
6        LibMath,        {_}
7        LibDXF,         {InitDXFfile}
8        LibInOut,       {IsKeyPressed}
9        LibMecIn,       {gCrank,Crank}
10       LibMec2D;       {SetJointSize,Locus,NewFrame,CloseMecDXF,CloseMecGraph}
11  const nPoz = 144;  {number of positions}
12  var i:Integer;  t,AB,BC,xA,yA,xB,yB,xC,yC,Phi1,Phi2:double;
13  BEGIN
14    InitDXFfile('F5_04A.DXF');  {..'F5_04A.DXF' or 'F5_04B.DXF'}
15    SetJointSize(6);    {+6 with 'F5_04A.DXF and -6 with 'F5_04B.DXF}
16    xA:=0; yA:=0;       {ground joint}
17    AB:=40;             {first crank length}
18    BC:=30;             {second crank length}
19    OpenMecGraph(-75,75,-75,75);
20    i:=0;
21    repeat
22      if (i > nPoz) then BEGIN
23        i:=0;   CloseMecDXF;
24      END;
```

```
25      NewFrame(5000);
26      t:=i/nPoz; {t = time}
27      Phi1:=Pi/4+2*Pi*t;
28      Phi2:=-8*Pi*t;
29      gCrank(Magenta,xA,yA,Phi1,_,_,AB,xB,yB,_,_,_);
30      Crank(Red,xA,yA,0,0,0,xB,yB,_,_,_,_,Phi2,_,_,BC,xC,yC,_,_,_);
31      Locus(White,xC,yC,'C');    {replace with CometLocus for F5_04B.DXF}
32      Inc(i);
33      until IsKeyPressed(27);
34      CloseMecGraph(FALSE); {..do not retain.$2D files}
35   END.
```

```
1    program P5_06A;
2    {=========================================================================
3    Simulation of two pistons in series tracing a Lissajous curve
4    =========================================================================}
5    uses Graph,       {Black,Cyan,Green,Magenta,Red}
6         LibMath,     {_}
7         LibInOut,    {IsKeyPressed}
8         LibDXF,      {InitDXFfile, SetDXFlayer}
9         LibMecIn,    {gSlider, Slider}
10        LibMec2D;    {OpenMecGraph, SetJointSize, Offset, PutPoint,}
11                     {Locus, NewFrame, CloseMecDXF, CloseMecGraph}
12   const nPoz = 90;      {number of positions}
13   var  xP1,yP1, xQ1,yQ1, xA1,yA1, xB1,yB1, xA_2,yA_2, xQ_2,yQ_2, xP2,yP2,
14        xQ2,yQ2, xA2,yA2, xB2,yB2, A1B1,A2B2, s1,s2, t:double;  i:integer;
15   BEGIN
16     A1B1:=60;    A2B2:=50;
17     xP1:=-27.0;  yP1 :=0.0;
18     xQ1:= 27.0;  yQ1 :=0.0;
```

```
19   xA_2:=0.0;    yA_2:= 22.5;
20   xQ_2:=0.0;    yQ_2:=-22.5;
21   InitDXFfile('F5_06A.DXF');
22   OpenMecGraph(-35,85,-35,75);
23   SetJointSize(6);
24   i:=0;
25   Repeat
26   if (i > nPoz) then BEGIN
27     i:=0;   CloseMecDXF;
28   END;
29   NewFrame(3000);
30   t:=i/nPoz; {t = time}
31   s1:=80.0 + 20.0*sin(2*Pi*t-Pi/4);
32   s2:=70.0 + 15.0*sin(4*Pi*t+Pi/4);
33   gSlider(Magenta,xP1,yP1, xQ1,yQ1
34   ,A1B1, s1,_,_, xB1,yB1,_,_,_, xA1,yA1,_,_,_,_);
35   Offset(Black,' ',xB1,yB1,_,_,_, xA1,yA1,_,_,_,_
36   ,xA_2,yA_2, xP2,yP2,_,_,_,_);
37   Offset(Black,' ',xB1,yB1,_,_,_, xA1,yA1,_,_,_,_
38   ,xQ_2,yQ_2, xQ2,yQ2,_,_,_,_);
39   Slider(Cyan,xP2,yP2,_,_,_,_, xQ2,yQ2,_,_,_,_
40   ,A2B2, s2,_,_, xB2,yB2,_,_,_, xA2,yA2,_,_,_,_);
41   PutPoint(Magenta,'x', xB1,yB1,'');   {..make joint B1 appear locked}
42   PutPoint(White,' ',xB1,yB1,' B_1');
43   PutPoint(White,' ', xA1,yA1,'A_1 ');
44   LabelJoint(White,xB2,yB2, xA2,yA2,' A_2');
45   LabelJoint(White,xB2,yB2, xP2,yP2,' P_2');
46   PutPoint(White,'o',xB2,yB2,' B_2');
47   PutPoint(White,' ',xP1,yP1,'P_1 ');
48   PutPoint(White,' ',xQ1,yQ1,'Q_1');
49   PutPoint(White,' ',xQ2,yQ2,' Q_2');
```

```
50    Locus(Green,xB2,yB2,'B_2');
51    Inc(i);
52    until IsKeyPressed(27);
53    CloseMecGraph(FALSE);   {..do not retain .$2D files }
54 END.


 1 program P5_08A;
 2 {=============================================================
 3 A crank rotating about a base with an offset point.
 4 =============================================================}
 5 uses Graph,           {Red}
 6      LibMath,         {_}
 7      LibInOut,        {IsKeyPressed}
 8      LibDXF,          {InitDXFfile}
 9      LibMecIn,        {gCrank}
10      LibMec2D;  {OpenMecGraph,NewFrame,SetJoin=Size,Offset,CloseMecDXF}
11 const nPoz = 36;      {..number of positions}
12 var i:Integer; Style:Char; Phi,AB,xA,yA,xB,y3,x1P,y1P,xP,yP:double;
13 BEGIN
14    InitDXFfile('F5_08A.DXF');    {..DXF file for M3D animation}
15    Style:='T';  {..'T','I','/','\','V','A'}
16    AB:=1.0;     {..crank length}
17    xA:=0.0;     yA:=0.0;   {..motor location}
18    x1P:=0.65;   y1P:=0.5;  {..relative coordinates of point P}
19    OpenMecGraph(-1.5,1.5, -1.5,1.5);    SetJointSize(5);
20    i:=0;
21    Repeat
22      if (i > nPoz) then BEGIN
23        CloseMecDXF;  i:=0;
24      END;
```

```
25    NewFrame(50);
26    Phi:=2*Pi*i/nPoz;
27    gCrank(Blue, xA,yA, Phi,_,_, AB, xB,yB,_,_,_,_);
28    Offset(Red,Style,xA,yA,0,0,0,xB,yB,_,_,_,x1P,y1P,xP,yP,_,_,_,_);
29    PutGPoint(White,' ',xA,yA,'A ');
30    LabelJoint(White,xA,yA, xB,yB,' B');
31    LabelJoint(White,0.5*(xA+xB),0.5*(yA+yB),xP,yP,' P');
32    Inc(i);
33  until IsKeyPressed(27);
34  CloseMecGraph(FALSE);  {..do not retain .$2D files}
35 END.
```

```
1  program P5_10;
2  {===================================================================
3   A crank driving a slider tracing a polar curve.  Also shown is the
4   Coriolis acceleration vector of the slider relative to its guide Q-Q'.
5   ===================================================================}
6  uses  Graph,      {Red,Black,Blue,Cyan,White,LightRed}
7        LibMath,    {_,DEG}
8        LibInOut,   {IsKeyPressed}
9        LibDXF,     {InitDXFfile}
10       LibMecIn,   {gCrank,Slider}
11       LibMecOut,  {Offset,Locus,VarDist,NewFrame,MecOut,OpenMecGraph}
12       LibMec2D;   {PutPoint,PutVector,CloseMecDXF,CloseMecGraph}
13 const nPoz = 90;      {..number of positions}
14 var FT1,FT2:text;   i:Integer;  t;  xO,yO, OP,Q_Q, axCor,ayCor,
15     xA,yA, vxA,vyA, axA,ayA, xB,yB, vxB,vyB, axB,ayB, xP,yP, vxP,vyP,
16     axP,ayP, xQ,yQ, vxQ,vyQ, axQ,ayQ,   x_Q,y_Q, vx_Q,vy_Q, ax_Q,ay_Q,
17     Phi,dPhi,ddPhi, s,ds,dds, Theta,dTheta,ddTheta, r,dr,ddr:double;
18 BEGIN
```

```
19   xO:=0.0;     yO:=0.0;
20   OP:=35;     Q_Q:=80;     {..distance QQ' - note that _Q is Q'}
21   Assign(FT1,'P5_10A.TXT');     Rewrite(FT1);
22   WriteLn(FT1,' t     s     ds     dds     r     dr     ddr');
23   Assign(FT2,'P5_10B.TXT');     Rewrite(FT2);
24   WriteLn(FT2,' t     Phi     dPhi     ddPhi     Theta     dTheta     ddTheta');
25   InitDXFfile('F5_10.DXF');
26   OpenMecGraph(-70,70, -20,70);
27   Theta:=Pi/2;
28   i:=0;
29   Repeat
30       if (i > nPoz) then BEGIN
31           i:=0;  CloseMecDXF;
32       END;
33       NewFrame(5000);
34       t:=i/nPoz;          {t = time}
35       Phi:=Pi/2  +  Pi/4*sin(2*Pi*t);
36       dPhi :=  2*Pi *Pi/4*cos(2*Pi*t);
37       ddPhi:=-Sqr(2*Pi)*Pi/4*sin(2*Pi*t);
38       s  :=  0.45*Q_Q*cos(2*Pi*t);
39       ds :=  -2*Pi *0.45*Q_Q*sin(2*Pi*t);
40       dds:=-Sqr(2*Pi)*0.45*Q_Q*cos(2*Pi*t);
41       gCrank(Magenta,xO,yO,  Phi,dPhi,ddPhi,  CP,  xP,yP,vxP,vyP,axP,ayP);
42       Offset(Blue,'I',xP,yP,vxP,vyP,axP,ayP,  xO,yO,0,0,0,0
43           ,0,-0.5*Q_Q, xO,yO,vxO,vyO,axO,ayO);
44       Offset(Blue,'I',xP,yP,vxP,vyP,axP,ayP,  xO,yO,0,0,0,0
45           ,0, 0.5*Q_Q, x_Q,Y_Q,vx_Q,vy_Q,ax_Q,ay_Q);
46       Slider(-Red,  xP,yP,vxP,vyP,axP,ayP,  xQ,yQ,vxQ,vyQ,axQ,ayQ
47           ,0, s,ds,dds, xB,yB,vxB,vyB,axB,ayB,  xA,yA,vxA,vyA,axA,ayA);
48       Locus(Cyan, xB,yB, 'B');
49       PutPoint(White,'o',xB ,yB ,'B     ');
```

```
50    LabelJoint(White,xO,yO,xP ,yP ,' P');
51    LabelJoint(White,x_Q,y_Q,xQ,yQ,' Q');
52    LabelJoint(White,xQ,yQ,x_Q,y_Q,' Q''');
53    PutGPoint(White,' ',xO ,yO ,'O   ');
54    VarDist(x_Q,y_Q,vx_Q,vy_Q,ax_Q,ay_Q,    xB,yB,vxB,vyB,axB,ayB
55         ,r,dr,ddr, Theta,dTheta,ddTheta);
56    axCor:=2*dTheta*dr*cos(Theta+Pi/2);    {x comp. of Coriolis accel.}
57    ayCor:=2*dTheta*dr*sin(Theta+Pi/2);    {y comp. of Coriolis accel.}
58    PutVector(LightRed, '|' , xB,yB, axCor,ayCor, 0.03,' a_c');
59    if MecOut then BEGIN
60      WriteLn(FT1,t:7:4,' ',s:8:4,' ',ds:9:4
61         ,' ',dds:9:3,' ',r-0.5*Q_Q:8:4,' ',dr:9:4,' ',ddr:9:3);
62      WriteLn(FT2,t:7:4,' ',Phi*DEG:7:3,' ',dPhi:7:4,' ',ddPhi:8:4
63         ,' ',(Theta-Pi/2)*DEG:7:3,' ',dTheta:7:4,' ',ddTheta:8:4);
64      END;
65    Inc(i);
66    until IsKeyPressed(27);
67    Close(FT1);   Close(FT2);
68    CloseMecGraph(FALSE);   {..do not retain .$2D files}
69    END.


1    program P5_11;
2    {=========================================================================
3    A crank driving a slider tracing a polar curve.  Also shown as vectors
4    are the velocity and the normal & tangential acceleration of the slider.
5    =========================================================================}
6    uses Graph,    {Red,Black,Blue,Cyan,White,LightBlue,LightRed}
7         LibMath,    {_,DEG}
8         LibInOut,   {IsKeyPressed}
9         LibDXF,     {InitDXFfile}
10        LibMecIn,   {gCrank,Slider}
```

```pascal
11    LibMec2D; {OpenMecGraph,NewFrame,XminWS,XmaxWS,YminWS,YmaxWS,Offset}
12    {PutGPoint,PutPoint,Locus,PutVector,ntAccel,CloseMecDXF,CloseMecGraph}
13    const nPoz = 90;       {..number of positions}
14    var FT:text;  i:Integer;  t,xO,yO, OP,Q_Q, x_Q,y_Q, xA,yA,vxA,vyA,axA,ayA,
15        xB,yB, vxB,vyB, axB,ayB, xP,yP, vxP,vyP, axP,ayP, xQ,yQ, vxQ,vyQ,
16        axQ,ayQ, Phi,dPhi,ddPhi, s,ds,dds, axBt,ayBt, axBn,ayBn:double;
17    BEGIN
18      OP:=35;     Q_Q:=80;       {..distances OP and QQ' - note that _Q is Q'}
19      xO:=0.0;    yO:=0.0;
20      Assign(FT,'P5_11.TXT'); Rewrite(FT);
21      WriteLn(FT,' t    xB    yB    vxB    vyB    axB    ayB');
22      InitDXFfile('F5_11.DXF');
23      i:=0;
24      Repeat
25        if (i > nPoz) then BEGIN        {..no effect until OpenMecGraph is called}
26          i:=0; CloseMecDXF;
27          xB:=0.1*(XmaxWS-XminWS); {..expand by 10% - multiple uses of xB}
28          yB:=0.1*(YmaxWS-YminWS); {..expand by 10% - multiple uses of yB}
29          OpenMecGraph(XminWS-xB,XmaxWS+xB, YminWS-yB,YmaxWS+yB);
30        END;
31        NewFrame(5000);
32        PutGPoint(White,' ',0,60,'Simulation with ntAccel');
33        t:=i/nPoz; {t = time}
34        Phi  :=Pi/2 +  Pi/4*sin(2*Pi*t);
35        dPhi := 2*Pi *Pi/4*cos(2*Pi*t);
36        ddPhi:=-Sqr(2*Pi)*Pi/4*sin(2*Pi*t);
37        s:= 0.45*Q_Q*cos(2*Pi*t);
38        ds :=  -2*Pi *0.45*Q_Q*sin(2*Pi*t);
39        dds:=-Sqr(2*Pi)*0.45*Q_Q*cos(2*Pi*t);
40        gCrank(Red,xO,yO, Phi,dPhi,ddPhi, OP, xP,yP,vxP,vyP,axP,ayP);
41        PutGPoint(White,' ',xO ,yO ,'O    ');
42        LabelJoint(White,xO,yO,xP ,yP ,' P');
```

```
43    Offset(Blue,'I', xP,yP,vxP,vyP,axP,ayP,  xO,yO,0,0,0,0
44    ,0,-0.5*Q_Q, xQ,yQ,vxQ,vyQ,axQ,ayQ);
45    LabelJoint(White,x_Q,y_Q,xQ,yQ,' Q');
46    Offset(Blue,'I', xP,yP,vxP,vyP,axP,ayP,  xO,yO,0,0,0,0
47    ,0, 0.5*Q_Q, x_Q,y_Q,_,_,_,_);
48    LabelJoint(White,xQ,yQ,x_Q,y_Q,' Q''');
49    Slider(Blue, xP,yP,vxP,vyP,axP,ayP, xQ,yQ,vxQ,vyQ,axQ,ayQ
50    ,0, s,ds,dds, xB,yB,vxB,vyB,axB,ayB, _,_,_,_,_,_);
51    PutPoint(White,'o',xB ,yB ,'B           ');
52    Locus(Cyan, xB,yB, 'B');
53    PutVector(LightBlue, '|' ,xB,yB, vxB,vyB, 0.1,'v_B');
54    ntAccel(vxB,vyB,axB,ayB, axBn,ayBn,axBt,ayBt);
55    PutVector(Magenta, ' ' ,xB,yB, axBt,ayBt, 0.01,'at_B');
56    PutVector(Magenta, ' ' ,xB,yB, axBn,ayBn, 0.01,'an_B');
57    PutPoint(White,' ',15,0,'Phi='+MyStr(Phi*DEG,5)+'°,   s='+MyStr(s,5));
58    if MecOut then WriteLn(FT,t:8:6,' ',xB:9:6,' ',yB:9:6,' `
59    ,vxB:9:6,' ',vyB:9:6,' ',axB:10:6,' ',ayB:10:6);
60    Inc(i);
61    until IsKeyPressed(27);
62    Close(FT);
63    CloseMecGraph(TRUE);   {..retain the B.$2D locus file}
64    WriteLn(XminWS:6:3,' < x < ',XmaxWS:6:3);   {report workspace limits..}
65    Write(YminWS:6:3,' < y < ',YmaxWS:6:3,' <CR>..');  ReadLn;
66    END.
```

```
1    program P5_15A;
2    {==================================================================
3    Two cranks in series with PutAng, PutDist, PutText & PutGtext
4    ==================================================================}
5    uses Graph,      {Magenta,Red,White}
```

```
6    LibMath,   {_,DEG}
7    LibGe2D,   {Dist2Pts2D,S123}
8    LibDXF,    {InitDXFfile}
9    LibInOut,  {IsKeyPressed}
10   LibMecIn,  {gCrank,Crank}
11   LibMec2D;  {SetJointSize,NewFrame,CloseMecDXF,CloseMecGraph,}
12              {PutAng,PutDist,PutText,PutGText}
13   const nPoz = 144;   {..number of positions}
14   var i:Integer; t,AB,BC,xA,yA,xB,yB,xC,yC,Phi1,Phi2,ExtLlgt,AC:double;
15   BEGIN
16     InitDXFfile('F5_14.DXF');
17     xA:=0; yA:=0;      {..ground joint location}
18     AB:=40; BC:=30;    {..crank lengths}
19     OpenMecGraph(-60,60,-60,65);
20     SetJointSize(4);
21     i:=0;
22     Repeat
23       if (i > nPoz) then BEGIN
24         CloseMecDXF;  i:=0;
25       END;
26       NewFrame(5000);
27       PutGText(White,0,_,'Simulation with PutAng and PutDist');
28       t:=i/nPoz; {t = time}
29       Phi1:=Pi/4+2*Pi*t;
30       Phi2:=   -8*Pi*t;
31       gCrank(Magenta,xA,yA, Phi1,_,_, AB,xB,yB,_,_,_);
32       Crank(Red,xA,yA,0,0,0,xB,yB,_,_,_,Phi2,_,_,BC,xC,yC,_,_,_);
33       AC:=Dist2Pts2D(xA,yA,xC,yC);
34       ExtLlgt:=8*Sgn(S123(xA,yA,xB,yB,xC,yC));
35       PutDist(White,xA,yA, xC,yC, ExtLlgt,' >|'); {..put distance AC}
36       PutAng(White,xA,yA, xA,yA, xB,yB, 4,'| >'); {..put Phil}
```

```
37      PutAng(White,2*xB-xA,2*yB-yA, xB,yB, xC,yC, 4,'|  >'); {..put Phi2}
38      PutText(White,_,_,'AC ='+MyStr(AC,4)+'n\Phi1='+MyStr(Phi1*DEG,5));
39      +'n\Phi2='+MyStr(Phi2*DEG,5));
40      PutGPoint(White,'  ',xA ,yA ,'A   ');
41      PutPoint(White,' ',xB ,yB ,'B   ');
42      PutPoint(White,'  ',xC ,yC ,'C');
43      Inc(i);
44    until IsKeyPressed(27);
45    CloseMecGraph(FALSE);  {..do not retain.$2D files}
46  END.


1   program P5_16A;
2   {==============================================================================}
3   A crank rotating about a base.  Uses Link and Base subroutines.
4   {==============================================================================}
5   uses Graph,        {Blue,Cyan,Green,Red,Magenta}
6        LibMath,        {_}
7        LibInOut,       {IsKeyPressed}
8        LibDXF,         {InitDxFfile}
9        LibMecIn,       {gCrank}
10       LibMec2D;       {OpenMecGraph,NewFrame,SetJointSize,}
11                       {gShape,Shape,Link,Base,CloseMecDXF}
12  const nPoz = 36;  {number of positions }
13        OA   = 20;  {crank length }
14        w    = 6;   {crank width }
15        Col  = 2;   {gCrank color - either 2, -2 or 0}
16  var i:Word;    Phi,xA,yA:double;
17  BEGIN
18    InitDXFfile('F5_16A.DXF');       {..DXF file for M3D animation}
19    OpenMecGraph(-25,25, -25,25);
```

```
20    SetJointSize(6);
21    i:=0;
22    Repeat
23      if (i > nPoz) then BEGIN
24        i:=0;  CloseMecDXF;
25      END;
26      NewFrame(5000);
27      Phi:=2*Pi*i/nPoz;
28      gCrank(Col, 0,0, Phi,_,_, OA, xA,yA,_,_,_);
29      Base(White, 0,0.1*OA, 0,-0.25*OA, 4*w, 0.5*w,  0.0);
30      Link(Red, 0,0, xA,yA, w, w, -0.25*w);
31      gShape('1.6',Red, 0,0);   {..draw a circle of radius 1.6 at (0,0)}
32      Shape('0.8',Red, xA,yA, xA,yA);   {..draw a circle of radius 0.8 at A}
33      Inc(i);
34    until IsKeyPressed(27);
35    CloseMecGraph(FALSE);
36  END.
```

```
1  program P5_16B;
2  {=========================================================================
3  Gear reducer animation with Shapes read from ASCII files
4  Housing.XY, Pinion.XY & Gear.XY
5  =========================================================================}
6  uses Graph,     {LightBlue,Red,White}
7       LibMath,   {_}
8       LibInOut,  {IsKeyPressed}
9       LibDXF,    {InitDXFfile}
10      LibMecIn,  {gCrank}
11      LibMec2D;  {Shape,OpenMecGraph,CloseMecGraph,CloseMecDXF}
12  const nPoz = 108;  {number of positions}
```

```
13   var i,jh,nh:Integer; t, il2, Phi1,Phi2,Phi20,
14       xO1,yO1, xO2,yO2,    xA,yA, xB,yB, r1,r2, rh:double;
15   BEGIN
16     InitDXFfile('F5_16B.DXF');
17     xO1:=0.0;    yO1:=0.0;        {pinion axis location}
18     xO2:=19.5;   yO2:=0;          {gear axis location}
19     r1:=5;
20     Phi20:=Pi;
21     r2:=5; {peripheral hole eccentricity of gear}
22     rh:=1; {peripheral hole radius}
23     nh:=5; {number of peripheral holes of gear}
24     il2:=-2; {gear ratio}
25     OpenMecGraph(-20,50,-15,15);
26     i:=0;
27   Repeat
28     if (i > nPoz) then BEGIN
29       i:=0; CloseMecDXF;
30     END;
31     NewFrame(5000);
32     t:=i/nPoz; {t = time}
33     Phi1:=il2*(2*Pi*t);
34     Phi2:=Phi20 + Phi1/il2;
35     gShape('Housing.xy',White, xO1,yO1); {..transmission housing}
36     gCrank(Black,xO1,yO1,Phi1,_,_, r1, xA,yA,_,_,_);
37     Shape('Pinion.xy',Red, xO1,yO1, xA,yA); {..involute & hole + keyway}
38     gCrank(Black, xO2,yO2,Phi2,_,_, r2, xB,yB,_,_,_);
39     Shape('Gear.xy',LightBlue, xO2,yO2, xB,yB);
40     Shape('',LightBlue, xO2,yO2, xO2+2*rh,0); {..center hole}
41     Shape('',LightBlue, xO2,yO2, xO2+9*rh,0); {..rim circle}
42     for jh:=0 to nh-1 do BEGIN
43       gCrank(Black,xO2,yO2,Phi2+2*Pi/nh*jh,_,_, r2, xB,yB,_,_,_);
```

```
44        Shape('',LightBlue, xB,yB, xB+rh,yB);  {..the peripheral holes}
45      END;
46      Inc(i);
47    until IsKeyPressed(27);
48    CloseMecGraph(FALSE);  {..do not retain *.$2D files}
49  END.


1   program P5_17A;
2   {===========================================================================
3   A rotating vector OA and plot of yA(Theta) with scan line and point
4   ===========================================================================}
5   uses CRT,          {Delay}
6        Graph,        {SetViewPort,SetColor,Red}
7        LibMath,      {_, VDp}
8        LibGIntf,     {DrawBorder}
9        LibInOut,     {EraseAll,ImplicitFileName,IsKeyPressed}
10       LibGraph,     {MaxX,MaxY, Obj2Scr, X_p,Y_p,R_p}
11       LibDXF,       {InitDXFfile,PDcircle,CloseDXFfile}
12       Unit_PCX,     {WritePCX}
13       LibMec2D,     {PutRefSystem,CloseMechGraph}
14       LibPlots;     {PlotTitle,SetDigitsDivs,SetDivLine,NewPlot,PlotCurve,}
15                     {PlotYaxis,PlotXaxis,ResizeY,UpdateLimitsX,UpdateLimitsY,}
16                     {PlotScanLine,PlotScanPoint,DXFScanLine,DXFScanPoint}
17  const N = 37;      {..number of animation points}
18     OA= 294/2;      {phasor magnitude = 1/2 of plot height}
19     Theta0=0;       {..initial Theta }
20     ThetaN=2*Pi;    {..final Theta }
21  var _Theta, _yA: VDp;  {..vectors to be graphed}
22     i, x1L,y1L,x2L,y2L, x1R,y1R,x2R,y2R: Integer;
23     Theta,    xA,yA: double;
```

```
24      FirstTime, SecondTime, OK: Boolean;
25 BEGIN
26      InitDXFfile('F5_17-1.DXF');      {..DXF output of phasor}
27      EraseAll('F5_17*.PCX');   {..erase old PCX files}
28      InitGr(0);
29      x1L:=0;        y1L:=15; x2L:=MaxX DIV 2; y2L:=MaxY;  {..left window }
30      x1R:=MaxX DIV 2; y1R:=15; x2R:=MaxX; y2R:=MaxY;  {..right window}
31      FirstTime :=TRUE;
32      SecondTime:=FALSE;
33      PlotTitle('Phasor Diagram');
34      MecOut:=TRUE;    {..this is because we use Obj2Scr and not OpenMecGraph!}
35      Repeat
36      for i:=1 to N do BEGIN
37      SetViewPort(x1R,y1R,x2R,y2R, ClipOn);    {..right window}
38      Obj2Scr(TRUE, -1.05*OA,1.15*OA,-1.1*OA,1.1*OA);  {..w-space limits}
39      NewFrame(5000);
40      Theta:=Theta0+(ThetaN-Theta0)/(N-1)*(i-1);
41      xA:=OA*cos(Theta);
42      yA:=OA*sin(Theta);
43      _Theta[i]:=Theta;
44      _yA[i]:=yA/OA;        {..plot the normalized yA}
45      PutRefSystem(4.1,4.1,'x','y');   {..reference frame at (0,0)}
46      PDcircle('0',X_p(0),Y_p(0),R_p(OA));   {..unit circle at (0,0)}
47      PutVector(Red,'-', 0,0, xA,yA, 1.0, '');
48      if NOT FirstTime then BEGIN
49      PlotScanLine(1,_Theta[i],-8000);
50      PlotScanPoint(1,_Theta[i],_yA[i],-8000);
51      if NOT (FirstTime OR SecondTime) then Delay(50000);
52      END;
```

```
53      if SecondTime then WritePCX(ImplicitFileName('F5_17.PCX'),OK);
54      if NOT FirstTime then BEGIN
55          PlotScanLine(1,_Theta[il,-8000);
56          PlotScanPoint(1,_Theta[il,_yA[il,-8000);
57      END;
58   END;
59   CloseMecDXF;  {..MecOut becomes FALSE}
60   if FirstTime then BEGIN  {plot the graph yA(Theta)..}
61      SetDigitsDivs(5);
62      SetDivLine(2,1,1.0);
63      SetViewPort(x1L,y1L,x2L,y2L, ClipOn);
64      InitDXFfile('F5_17-2.DXF');        {..DXF output of yA(Theta) }
65      NewPlot(1,FitBox,25,y1L+85, x2L-10,y2L-85,'');
66      DrawBorder;           {..draw a box around the plot}
67      UpdateLimitsX(1, _Theta, N);
68      UpdateLimitsY(1, _yA, N);
69      ResizeY(1,0.1);  {..expand y-range by 10%}
70      DXFScanLine(1);
71      DXFScanPoint(1);
72      PlotCurve(1, _Theta, _yA, N, Red);      {..plot yA(Theta) }
73      PlotXaxis(1, 1, 9,2,'ê     ');    {..horizontal axis}
74      PlotYaxis(1, 0, 3,5,'Y     ');    {..vertical axis    }
75      CloseMecDXF;
76   END;
77   if FirstTime then SecondTime:=TRUE else SecondTime:=FALSE;
78   FirstTime:=FALSE;
79   until IsKeyPressed(27);
80   CloseMecGraph(FALSE);   {..do not retain .$2D files}
81 END.
```

```pascal
 1    program P5_17B;
 2    {=============================================================
 3    A rotating vector OA and data output yA(Theta)
 4    =============================================================}
 5    uses Graph, {Red}
 6         LibInOut,     {IsKeyPressed}
 7         LibGraph,     {Obj2Scr}
 8         LibDXF,       {InitDXFfile,PDcircle,CloseDXFfile}
 9         LibMec2D;     {MecOut,PutRefSystem,PutVector,CloseMechGraph}
10    const N = 37;      {number of animation points}
11          OA=250/2;    {phasor magnitude i.e. 1/2 of plot box height}
12          Theta0=0;    {initial Theta    }
13          ThetaN=2*Pi; {final Theta       }
14    var   FD: file of double;      i:Integer;  Theta,xA,yA:double;
15    BEGIN
16      InitDXFfile('F5_17B.DXF');  {..DXF file for M3D animation}
17      Assign(FD,'F5_17B.D2D');   Rewrite(FD);
18      OpenMecGraph(-2*OA,2*OA,  -2*OA,2*OA);
19      i:=0;
20      repeat
21        if (i > N) then BEGIN
22          i:=0;  CloseMecDXF;
23        END;
24        NewFrame(5000);
25        Theta:=Theta0+(ThetaN-Theta0)/N*i;
26        xA:=OA*cos(Theta);
27        yA:=OA*sin(Theta);
28        gShape(MySt(OA,7),White, 0,0);  {..draw circle of radius OA at (0,0)}
29        PutVector(Red,'-', 0,0, xA,yA, 1, '');
30        PutRefSystem(0.125*OA/RJtSz,0.125*OA/RJtSz,'x','y');
31        yA:=yA/OA;
```

```
32         if MecOut then Write(FD,Theta,yA);
33         Inc(i);
34      until IsKeyPressed(27);
35      Close(FD);
36      CloseMecGraph(FALSE);  {..do not retain.$2D files }
37   END.
```

***

```
1   program P6_01;
2   {=================================================================
3    Kinematic simulation of the needle drive of a sewing machine
4   =================================================================}
5   uses Graph,        {Blue,Brown,Green,Red,Magenta,White}
6        LibMath,       {_}
7        LibInOut,      {IsKeyPressed}
8        LibDXF,        {InitDXFfile}
9        LibMecIn,      {gCrank}
10        LibAssur,      {RRR,RR_T,RRT_}
11        LibMec2D;      {OpenMecGraph,SetJointSize,Offset,Locus,Left,Right,}
12                       {NewFrame,PutGPoint,PutPoint,LabelJoint,CloseMecGraph,}
13                       {CloseMecDXF,MecOut,XminWS,XmaxWS,YminWS,YmaxWS}
14   const DXFname = 'F6_01A'  {either 'F6_01A' or 'F6_01B'}
15         nPoz    = 72;        {number of animation frames}
16   var i:Integer; t,Phi,xO,yO,xA,yA,OA,xB,yB,AB,xP,yP,xQ,yQ,PQ,
17        x_C,y_C,xC,yC,xD,yD,CD,xE,yE,DE,x_F,y_F,xF,yF:double;
18   BEGIN
19     InitDXFfile(DXFname+'.DXF');
20     OA:=15;   AB:=43;
21     CD:=23;   DE:=25;
22     xO:=0;    yO:=0;
23     xP:=3;    yP:=-63.0;
```

```
24    xQ:=3;     yQ:=-23.0;
25    x_C:=10;   y_C:=12;
26    xE:=-15;   yE:=23;
27    x_F:=-8;   y_F:=25;
28    PQ:=Abs(yP-yQ);
29    OpenMecGraph(-16,33,-98,46);
30    SetJointSize(3);
31    i:=0;
32    repeat
33      if (i > nPoz) then BEGIN
34        i:=0;   CloseMecDXF;
35      END;
36      NewFrame(5000);
37      t:=i/nPoz; {t = time}
38      Phi:=2*Pi*t;
39      gCrank(Green,xO,yO,Phi,_,_,OA, xA,yA,_,_,_,_);
40      if (Pos('A',DXFname) > 0) then BEGIN   {mechanism version A..}
41        RR_T(Red,xA,yA,_,_,_,_, xP,yP,0,0,0, 0.5*Pi,0,0, AB,0,PQ
42        ,Left, xB,yB,_,_,_,_, xB,yB,_,_,_,_, xQ,yQ,_,_,_,_ );
43        PutPoint (White,' ', xQ,yQ ,'Q ');
44      END;
45      if (Pos('B',DXFname) > 0) then BEGIN   {mechanism version B..}
46        RRT_(Red,xA,yA,_,_,_,_, xP,yP,0,0,0,0, xQ,yQ,0,0,0, AB,0
47        ,Left ,xB,yB,_,_,_,_, xB,yB,_,_,_,_ );
48        PutGPoint(White,' ', xQ,yQ ,'Q ');
49      END;
50      Offset(Blue,'^',xO,yO,_,_,_,_, xA,yA,_,_,_,_,x_C,y_C, xC,yC,_,_,_,_);
51      RRR(Magenta,xC,yC,_,_,_,_ xE,yE,0,0,0,0
52        ,CD,DE, Right, xD,yD,_,_,_,_ );
53      Offset(Brown,'/',xD,yD,_,_,_,_ xC,yC,_,_,_,_,x_F,y_F, xF,yF,_,_,_,_);
54      Locus(Cyan, xF,yF,'F');
```

```
55        LabelJoint(White,xC,yC,xD,yD,' D');
56        PutGPoint(White,' ',xO,yO,'O');
57        PutPoint (White,' ',xA,yA,' A');
58        PutPoint (White,' ',xB,yB,' B');
59        PutPoint (White,' ',xC,yC,'C ');
60        PutGPoint(White,' ',xE,yE,'E ');
61        PutPoint (White,' ',xF,yF,' F');
62        PutGPoint(White,' ',xP,yP,'P ');
63        Inc(i);
64      until IsKeyPressed(27);
65      CloseMecGraph(FALSE);        {..do not save _ocus files}
66      WriteLn(XminWS:6:3,' < x < ',XmaxWS:6:3);  {report workspace limits ..}
67      Write(YminWS:6:3,' < y < ',YmaxWS:6:3,' <CR>..'); ReadLn;
68  END.
```

```
1  program P6_06;
2  {==========================================================================
3  RTRTR double oscillating slide driven by two cranks. Uses RTRTR(..)
4  ==========================================================================}
5  uses Graph,        {Red,White,Green}
6       LibMath,      {_}
7       LibInOut,     {IsKeyPressed}
8       LibDXF,       {InitDXFfile}
9       LibMecIn,     {gCrank,RTRTRc}
10      LibMec2D;     {OpenMecGraph,SetJointSize,BumpPiston,NewFrame..}
11                    {Locus,CloseMecDXF,CloseMecGraph}
12  const nPoz = 90;  {..number of positions}
13  var i:Word; LftRgt:shortint; t,xO1,yO1,O1A,xA,yA,A0A,A0Q1,P1C,
14      Phi1,s1,xO2,yO2,O2B,xB,yB,B0B,B0Q2,P2C,Phi2,s2,xC,yC:double;
15  BEGIN
```

```
16   InitDXFfile('F6_06.DXF');
17   BumpPiston:=FALSE; {TRUE constrains pistons inside cylinders}
18   LftRgt:=-1;   {orientation of the A0-C-B0 loop    }
19   xO1 :=-0.4;  yO1:=0;  {ground joint of crank 1      }
20   xO2 := 0.4;  yO2:=0;  {ground joint of crank 2      }
21   O1A  := 0.12;   {length of crank 1                  }
22   O2B  := 0.10;   {length of crank 2                  }
23   A0A  := 0.05;   {eccentricity of cylinder 1         }
24   B0B  :=-0.05;   {eccentricity of cylinder 2         }
25   A0Q1:= 0.35;    {length of cylinder 1               }
26   B0Q2:= 0.35;    {length of cylinder 2               }
27   P1C  := 0.40;   {length of piston 1                 }
28   P2C  := 0.45;   {length of piston 2                 }
29   OpenMecGraph(-0.5,0.5, -0.25,1.0);
30   SetJointSize(4);
31   i:=0;
32   repeat
33     if (i > nPoz) then BEGIN
34       i:=0;  CloseMecDXF;
35     END;
36     NewFrame(3000);
37     t:=i/nPoz;   {t = time}
38     Phi1:=     Pi/3*sin(2*Pi*t);
39     Phi2:=Pi + Pi/3*sin(2*Pi*t);
40     gCrank(Red,xO1,yO1, Phi1,_,_, O1A,  xA,yA,_,_,_,_);
41     gCrank(Red,xO2,yO2, Phi2,_,_, O2B,  xB,yB,_,_,_,_);
42     s1:=0.65 + 0.15*cos(2*Pi*t);
43     s2:=0.60 + 0.12*cos(4*Pi*t);
44     RTRTRc(White, xA,yA,_,_,_,_,  xB,yB,_,_,_,_,  A0A,A0Q1,P1C
45     ,B0B,B0Q2,P2C, s1,_,_, s2,_,_, LftRgt, xC,yC,_,_,_,_, _);
46     Locus(Green, xC,yC, 'C');
```

```
47    Inc(i);
48  until (IsKeyPressed(27));
49  CloseMecGraph(FALSE);   {..do not save locus files}
50  END.
```

```
1  program P6_11;
2  {==============================================================
3  RTRTR actuator driven by two cranks with ASCII output.
4  Uses both the RTRTR and RTRTRc subroutines.
5  ==============================================================}
6  uses Graph,        {White,Green,LightBlue}
7    LibMath,         {DEG}
8    LibInOut,        {MySt,MyStr}
9    LibGe2D,         {Dist2P_2D}
10   LibDXF,          {InitDXFfile,SetDXFlayer}
11   LibMecIn,        {gCrank,RTRTR,RTRTRc}
12   LibMec2D;        {NewAnimFrame,SetJointSize,BumpPiston,Ang3PVA, }
13                    {OpenMecGraph,MecOutp,PutPoint,CloseMecDXF,CloseMecGraph}
14 const FName ='P6_11A';  {'P6_11A' for (a); 'P6_11B' for (b) - file name }
15   DPhi = Pi/3;    {Pi/3 for (a); Pi  for (b)  - rocker amplitude }
16   O1A  = 0.12;    {0.12 for (a); 0.45 for (b) - length of crank 1 }
17   O2B  = 0.10;    {0.10 for (a); 0.35 for (b) - length of crank 2 }
18   nPoz = 120;     {120 for (a); 600 for (b) - positions for 0<t<1}
19 var FT:text;  i:Word;      LftRgt:shortint;  t,xO1,yO1,xA,yA,xO2,yO2,
20 xB,yB,xC,yC,A0A,A0Q1,P1C,B0B,B0Q2,P2C,Delta,Deltav,ACB,kACB:double;
21 function Phi1(t:double):double; BEGIN
22   Phi1:=DPhi*sin(2*Pi*t);
23 END;
24 function Phi2(t:double):double; BEGIN
25   Phi2:=Pi + DPhi*sin(2*Pi*t);
26 END;
```

```
27  function s1(t:double):double; BEGIN
28    s1:=0.65 + 0.15*cos(2*Pi*t);
29  END;
30  function s2(t:double):double; BEGIN
31    s2:=0.60 + 0.12*cos(4*Pi*t);
32  END;
33  BEGIN
34    InitDXFfile(FName+'.DXF');      {DXF file for M3D animation}
35    Assign(FT,FName+'.TXT');  Rewrite(FT);
36    WriteLn(FT,'  t    Delta   Deltav (AC+BC)/AB   <ACB');
37    BumpPiston:=FALSE;      {TRUE -> constrains pistons inside cylinders}
38    LftRgt:=-1;             {orientation of A0-C-B0 loop }
39    xO1 :=-0.4;  yO1:=0;    {ground joint of crank 1 }
40    xO2 := 0.4;  yO2:=0;    {ground joint of crank 2 }
41    A0A := 0.05;            {eccentricity of cylinder 1 }
42    B0B :=-0.05;            {eccentricity of cylinder 2 }
43    SizeLinMotor(_,A0Q1,P1C);   {..makes A0Q1=-InfD & P1C=InfD}
44    SizeLinMotor(_,B0Q2,P2C);   {..makes B0Q2=-InfD & P2C=InfD}
45    for i:=0 to nPoz do BEGIN {generate a full set of s1 and s2..}
46      t:=i/nPoz; {t = time}
47      SizeLinMotor(s1(t),A0Q1,P1C);  {..update A0Q1,P1C}
48      SizeLinMotor(s2(t),B0Q2,P2C);  {..update B0Q2,P2C}
49    END;
50    Delta:=0; Deltav:=0; {Delta's may accidentally become InfD}
51    OpenMecGraph(-0.85,0.8, -0.83,0.75);
52    SetJointSize(4);
53    i:=0;
54    Repeat
55      if (i > nPoz) then BEGIN
56        i:=0;  CloseMecDXF;
57      END;
```

```
58      NewFrame(500);
59      t:=i/nPoz;      {t = time}
60      gCrank(Red,xO1,yO1,Phi1(t),_,_,  O1A,  xA,yA,_,_,_,_);
61      gCrank(Red,xO2,yO2,Phi2(t),_,_,  O2B,  xB,yB,_,_,_,_);
62      RTRTRc(White,xA,yA,_,_,_,_,xB,yB,_,_,_,_,AOA,AOQ1,P1C
63          ,BOB,BOQ2,P2C,s1(t),_,_,s2(t),_,_,LftRgt,xC,yC,_,_,_,_,Delta);
64      RTRTR(White,xA,yA,_,_,_,_,xB,yB,_,_,_,_,AOA,AOQ1,P1C
65          ,BOB,BOQ2,P2C,s1(t),_,_,s2(t),_,_,LftRgt,xC,yC,_,_,_,_,Deltav);
66      ACB:=0.0;
67      Ang3PVA(xA,yA,_,_,_,_,xC,yC,_,_,_,_,xB,yB,_,_,_,_,ACB,_,_);
68      kACB:=Dist2Pts2D(xA,yA,xC,yC)+Dist2Pts2D(xB,yB,xC,yC);
69      kACB:=kACB/Dist2Pts2D(xA,yA,xB,yB);
70      if MecOut then WriteLn(FT,t:8:6,' ',Delta:8:6
71          ,' ',Deltav:8:6,' ',kACB:8:6,' ',ACB*DEG:7:4);
72      Inc(i);
73      until IsKeyPressed(27);
74      CloseMecGraph(FALSE);      {..do not save locus files}
75      Close(FT);
76  END.


1   program P6_17;
2   {=============================================================
3    Simulation of an RRT_ dyadic isomer driven by two cranks
4    =============================================================}
5   uses Graph,        {Cyan,Red,White}
6       Libmath,       {_}
7       LibInOut,      {IsKeyPressed}
8       LibDXF,        {InitDXFfile}
9       LibMecIn,      {gCrank}
10      LibAssur,      {RRT_}
```

```pascal
11    LibMec2D;          {OpenMecGraph,NewFrame,SetJointSize}
12                       {MecOut,PutGPoint,PutPoint,CloseMecDXF}
13  const nPoz = 90;     {number of positions}
14  var  i:Word;    PlsMns:shortint;  t,xO1,yO1,xO2,yO2,Phi1,Phi2,
15       O1A,O2P,xA,yA,xB,yB,xP,yP,xQ,yQ,AC,BC,BQ,xC,yC:double;
16  BEGIN
17    InitDXFfile('F6_17.DXF');
18    PlsMns:=-1;              {orientation of the RRT dyad }
19    xO1:=-0.50; yO1:=0;      {ground joint of crank #1  }
20    xO2:= 0.35; yO2:=0;      {ground joint of crank #2  }
21    O1A:= 0.15; O2P:= 0.15;  {lengths of cranks #1 and #2 }
22    AC := 0.55;              {length of connecting rod  }
23    BC := 0.10;              {sliding rod offset }
24    BQ := 0.65;              {sliding rod length }
25    OpenMecGraph(-0.8,1.1, -0.2,0.4);
26    SetJointSize(4);
27    i:=0;
28    repeat
29      if (i > nPoz) then BEGIN
30        i:=0; CloseMecDXF;
31      END;
32      NewFrame(500);
33      t:=i/nPoz; {t = time}
34      Phi1:=2*Pi*t;
35      Phi2:=Pi/2+Pi/9*sin(2*Pi*t);
36      gCrank(Red,xO1,yO1,Phi1,_,_,O1A,xA,yA,_,_,_);
37      gCrank(Red,xO2,yO2,Phi2,_,_,O2P,xP,yP,_,_,_);
38      RR_T(Cyan, xA,yA,_,_,_,xP,yP,_,_,_,Phi2+Pi/2,_,_,AC,BC,BQ
39      ,PlsMns, xB,yB,_,_,_,xC,yC,_,_,_,xQ,yQ,_,_,_, _);
40      PutGPoint(White,' ', xO1,yO1 ,'O_1  ');
41      PutGPoint(White,' ', xO2,yO2 ,' O_2');
```

```
42      PutPoint(White,'  ',  xA,yA ,'A   ');
43      PutPoint(White,'  ',  xP,yP ,'P');
44      PutPoint(White,'  ',  xB,yB ,'B  ');
45      PutPoint(White,'  ',  xC,yC ,'   C');
46      PutPoint(White,'  ',  xQ,yQ ,'  Q');
47      Locus(Magenta,xQ,yQ,'Q');
48      Inc(i);
49    until IsKeyPressed(27);
50    CloseMecGraph(FALSE);       {..do not save locus files}
51 END.
```

```
1  program P6_21;
2  {=======================================================
3  Simulation of an RTR dyad driven by a crank and a rocker
4  =======================================================}
5  uses Graph,        {Cyan,Magenta,Red}
6       LibMath,      {_}
7       LibInOut,     {IsKeyPressed}
8       LibDXF,       {InitDXFfile}
9       LibMecIn,     {gCrank}
10      LibAssur,     {RT_R}
11      LibMec2D;     {OpenMecGraph,NewFrame,Locus,CloseMecDXF,MecOutp}
12 const nPoz = 90;   {number of positions}
13 var i:Word;  t,xO1,yO1,Phi1,O1A,xA,yA,xO2,yO2,
14     Phi2,O2B,xB,yB,xP,yP,xC,yC,xQ,yQ,AC,BP,PQ:double;
15 BEGIN
16   InitDXFfile('F6_21.DXF');
17   xO1:=-0.30; yO1:=0;    {ground joint of crank #1  }
18   xO2:= 0.35; yO2:=0;    {ground joint of crank #2  }
19   O1A:= 0.10; O2B=0.20;  {length of crank #1 and #2 }
```

```
20   AC := 0.25;              {length of slider offset }
21   BP := 0.15;              {sliding rod offset }
22   PQ := 1.00;              {sliding rod length }
23   OpenMecGraph(-0.9,0.6, -0.2, 0.4);
24   SetJointSize(4);
25   i:=0;
26   repeat
27   if (i > nPoz) then BEGIN
28       i:=0;  CloseMecDXF;
29   END;
30   NewFrame(500);
31   t:=i/nPoz; {t = time}
32   Phi1:=2*Pi*t*t*t;
33   Phi2:=Pi/2 + Pi/3*cos(2*Pi*t);
34   gCrank(Red, xO1,yO1, Phi1,_,_, O1A, xA,yA,_,_,_,_);
35   gCrank(Red, xO2,yO2, Phi2,_,_, O2B, xB,yB,_,_,_,_);
36   RT_R(Cyan, xA,yA,_,_,_,_, xB,yB,_,_,_,_, AC,BP,PQ,
37   xP,yP,_,_,_,_, xC,yC,_,_,_,_, xQ,yQ,_,_,_,_,_);
38   PutGPoint(White,' ', xO1,yO1 ,'O1 ');
39   PutGPoint(White,' ', xO2,yO2 ,'   O2');
40   PutPoint(White,' ', xA,yA ,' A');
41   PutPoint(White,' ', xP,yP ,' P');
42   PutPoint(White,' ', xB,yB ,' B');
43   PutPoint(White,' ', xC,yC ,'C');
44   PutPoint(White,' ', xQ,yQ ,'Q ');
45   Locus(Magenta,xQ,yQ,'Q');
46   Inc(i);
47   until IsKeyPressed(27);
48   CloseMecGraph(FALSE);       {..do not save locus files}
49   END.
```

```
1   program P6_24;
2   {===================================================================
3   Simulation of a T_R_T dyadic isomer driven by two rockers
4   ===================================================================

5   uses  Graph,        {Cyan,Red,White}
6         LibMath,       {_}
7         LibInOut,      {IsKeyPressed}
8         LibDXF,        {InitDXFfile}
9         LibMecIn,      {gCrank}
10        LibAssur,      {T_R_T}
11        LibMec2D;      {OpenMecGraph,NewFrame,Locus,}
12                       {CloseMecDXF,MecOutp,CloseMecGraph}

13  const nPoz = 90; {number of positions}
14  var i:Word;    OK:Boolean; t,Theta1,Theta2,xO1,yO1,O1A,
15        xO2,yO2,O2B,xA,yA,xB,yB,xC,yC,P1Q1,Q1C,P2Q2,Q2C,
16        xP1,yP1,xQ1,yQ1,xP2,yP2,xQ2,yQ2:double;

17  BEGIN
18    InitDXFfile('F6_24.DXF');
19    xO1 :=-0.20;  yO1:=0;              {ground joint of rocker #1  }
20    xO2 := 0.15;  yO2:=0;              {ground joint of rocker #2  }
21    O1A := 0.20;  O2B:= 0.20;          {lengths of rocker #1 and #2}
22    P1Q1:= 0.50;
23    Q1C := 0.10;
24    P2Q2:= 0.50;
25    Q2C := 0.15;
26    OpenMecGraph(-0.6,0.6, -0.2,0.5);
27    SetJointSize(4);
28    i:=0;
```

```
29    repeat
30      if (i > nPoz) then BEGIN
31        i:=0;  CloseMecDXF;
32      END;
33      NewFrame(500);
34      t:=i/nPoz; {t = time}
35      Theta1:=3*Pi/4 + Pi/8*sin(2*Pi*t);
36      Theta2:=  Pi/4 + Pi/9*sin(3*Pi*t);
37      gCrank(Red,xO1,yO1,Theta1,_,_, O1A,  xA,yA,_,_,_,_);
38      gCrank(Red,xO2,yO2,Theta2,_,_, O2B,  xB,yB,_,_,_,_);
39      T_R_T(Cyan, xA,yA,_,_,_,_, xB,yB,_,_,_,_, Theta1-Pi/2,_,_
40        ,Theta2+Pi/2,_,_,_,P1Q1,Q1C,P2Q2,Q2C,xC,yC,_,_,_,_,xP1,yP1,_,_,_,_
41        ,xQ1,yQ1,_,_,_,_, xP2,yP2,_,_,_,_, xQ2,yQ2,_,_,_,_, OK);
42      PutGPoint(White,' ', xO1,yO1 ,'O1 ');
43      PutGPoint(White,' ', xO2,yO2 ,'   O2');
44      PutPoint(White,' ', xA,yA ,'A');
45      PutPoint(White,' ', xB,yB ,'B');
46      LabelJoint(White,0.5*(xQ1+xQ2),0.5*(yQ1+yQ2),xC,yC,' C');
47      LabelJoint(White, xP1,yP1,xQ1,yQ1,' Q1');
48      LabelJoint(White, xQ1,yQ1,xP1,yP1,' P1');
49      LabelJoint(White, xP2,yP2,xQ2,yQ2,' Q2');
50      LabelJoint(White, xQ2,yQ2,xP2,yP2,' P2');
51      Locus(Magenta,xC,yC,'C');
52      Inc(i);
53    until IsKeyPressed(27);
54    CloseMecGraph(FALSE); {..do not save locus files}
55  END.
```

```
1   program P6_26;
2   {=================================================================
3   Simulation of an _TRT_ dyadic isomer driven by two rockers
4   =================================================================}
5   uses Graph,          {Cyan,Magenta,Red}
6        LibMath,        {_}
7        LibInOut,       {IsKeyPressed}
8        LibDXF,         {InitDXFfile}
9        LibMecIn,       {gCrank}
10       LibAssur,       {_TRT_}
11       LibMec2D;       {OpenMecGraph,NewFrame,Locus,}
12                       {CloseMecDXF,MecOutp,CloseMecGraph}
13  const nPoz = 90; {number of positions}
14  var i: Word; OK: Boolean;
15      t, Theta1, Theta2, P1Q1, P2Q2, xP1,yP1, xQ1,yQ1, xP2,yP2,
16      xQ2,yQ2, xA,yA, xB,yB, xC,yC, AC,BC: double;
17  BEGIN
18    InitDXFfile('F6_26.DXF');     {ground joint of rocker #1  }
19    xP1 :=-0.20; yP1:=0;          {ground joint of rocker #2  }
20    xP2 := 0.20; yP2:=0;          {lengths of rocker #1 and #2 }
21    P1Q1:= 1.15; P2Q2:=1.15;
22    AC  := 0.65;
23    BC  := 0.70;
24    OpenMecGraph(-1.50,1.50, -0.05, 1.4);
25    SetJointSize(4);
26    i:=0;
27    repeat
28      if (i > nPoz) then BEGIN
29        i:=0; CloseMecDXF;
30        END;
```

```
31      NewFrame(500);
32      t:=i/nPoz; {t = time}
33      Theta1:=3*Pi/4 + Pi/8*sin(2*Pi*t);
34      Theta2:=  Pi/4 + Pi/9*sin(3*Pi*t);
35      gCrank(Red, xP1,yP1, Theta1,_,_, P1Q1, xQ1,yQ1,_,_,_,_);
36      gCrank(Red, xP2,yP2, Theta2,_,_, P2Q2, xQ2,yQ2,_,_,_,_);
37      _TRT_(Cyan, xP1,yP1,0,0,0, xQ1,yQ1,_,_,_,_,xP2,yP2,0,0,0,xQ2,yQ2,
38      _,_,_,_, AC,BC, xA,yA,_,_,_,_, xB,yB,_,_,_,_, xC,yC,_,_,_,_, OK);
39      PutGPoint(White,' ', xP1,yP1 ,'P1 ');
40      PutGPoint(White,' ', xP2,yP2 ,' P2');
41      PutPoint(White,' ', xA,yA ,'A');
42      PutPoint(White,' ', xB,yB ,'B');
43      LabelJoint(White, xP1,yP1,xQ1,yQ1,' Q1');
44      LabelJoint(White, xP2,yP2,xQ2,yQ2,' Q2');
45      LabelJoint(White,0.5*(xA+xB),0.5*(yA+yB),xC,yC,'   C');
46      Locus(White,xC,yC,'C');
47      Inc(i);
48    until IsKeyPressed(27);
49    CloseMecGraph(FALSE); {..do not save locus files}
50  END.
```

```
1  program P6_28;
2  {=============================================================
3   Simulation of a T_RT_ dyadic isomer driven by two rockers
4  =============================================================}
5  uses Graph,    {Cyan,Magenta,Red}
6       LibMath,  {_}
7       LibInOut, {IsKeyPressed}
8       LibDXF,   {InitDXFfile}
9       LibMecIn, {gCrank}
```

```
10      LibAssur,      {T_RT_}
11      LibMec2D;      {OpenMecGraph,NewFrame,Locus,}
12                     {CloseMecDXF, MecOutp,CloseMecGraph}
13   const nPoz = 90;  {number of positions}
14   var i: Word;   OK: Boolean;
15      xO1,yO1,O1A,Theta1,xA,yA, Theta2, P1Q1,Q1C,P2Q2, xP1,yP1,
16      xQ1,yQ1, xP2,yP2, xQ2,yQ2, BC, xB,yB, xC,yC, t: double;
17   BEGIN
18      InitDXFfile('F6_28.DXF');
19      xO1 :=-0.05; yO1:=0;           {ground joint of rocker #1  }
20      xP2 := 0.05; yP2:=0;           {ground joint of rocker #2  }
21      O1A := 0.10; P2Q2:= 0.73;      {lengths of rocker #1 and #2 }
22      P1Q1:= 0.78;
23      Q1C := 0.35;
24      BC  := 0.30;
25      OpenMecGraph(-0.26,0.80, -0.57,0.72);
26      SetJointSize(4);
27      i:=0;
28      repeat
29         if (i > nPoz) then BEGIN
30            i:=0;   CloseMecDXF;
31         END;
32         NewFrame(500);
33         t:=i/nPoz;     {t = time}
34         Theta1:=Pi   +Pi/8*sin(2*Pi*t);
35         Theta2:=Pi/16+Pi/9*sin(3*Pi*t);
36         gCrank(Red, xO1,yO1, Theta1,_,_, O1A,xA,yA,_,_,_);
37         gCrank(Red, xP2,yP2, Theta2,_,_, P2Q2, xQ2,yQ2,_,_,_);
38         T_RT_(Cyan, xA,yA,_,_,_, Theta1-Pi/2,_,_, xP1,yP1,_,_,_,
39            xQ2,yQ2,_,_,_,_, P1Q1,Q1C,BC, xP2,yP2,0,0,0,
40            xQ1,yQ1,_,_,_,_, xB,yB,_,_,_,_, xC,yC,_,_,_,_, OK);
```

```
41       PutGPoint(White,' ', xO1,yO1 ,'O1  ');
42       PutGPoint(White,' ', xP2,yP2 ,'   P2');
43       PutPoint(White,' ', xA,yA ,'A');
44       PutPoint(White,' ', xB,yB ,'B');
45       LabelJoint(White, xP1,yP1,xQ1,yQ1,' Q1');
46       LabelJoint(White, xQ1,yQ1,xP1,yP1,' P1');
47       LabelJoint(White, xP2,yP2,xQ2,yQ2,' Q2');
48       LabelJoint(White,0.5*(xA+xB),0.5*(yA+yB),xC,yC,'   C');
49       Locus(White,xC,yC,'C');
50       Inc(i);
51    until IsKeyPressed(27);
52    CloseMecGraph(FALSE);      {..do not save locus files}
53 END.
```

```
1  program P6_30;
2  {==========================================================================
3  Simulation of an R_T_T dyadic isomer driven by a crank and a rocker
4  ==========================================================================}
5  uses Graph,    {Cyan,LightBlue,LightRed,Magenta,Red,White}
6       LibMath,     {_,RAD}
7       LibInOut,    {IsKeyPressed}
8       LibDXF,      {InitDXFfile}
9       LibMecIn,    {gCrank}
10      LibAssur,    {R_T_T}
11      LibMec2D; {OpenMecGraph,NewFrame,Locus,}
12                {SetJointSize,CloseMecDXF,MecOutp}
13 const nPoz = 90; {number of positions}
14 var i: Word;  OK: Boolean; t, O1A, O2B, AD,DK,PQ,QC, Phi1,Phi2,
15     Alph1,Alph2, xO1,yO1, xO2,yO2, xA,yA, xB,yB, xC,yC, xD,yD,
16     xK,yK, xP,yP, xQ,yQ: double;
```

```
17  BEGIN
18    xO1:=-0.05; yO1:=0;        {ground joint of crank #1  }
19    xO2:= 0.05; yO2:=0;        {ground joint of crank #2  }
20    O1A:= 0.05; O2B:= 0.10;    {lengths of crank #1 and #2}
21    AD := 0.05; DK:=0.30;      {L-rod of sliding block C  }
22    QC := 0.25;                {length of spacer QC       }
23    PQ := 0.45;                {rod of sliding block B    }
24    Alph1:= 60*RAD;
25    Alph2:=110*RAD;
26    InitDXFfile('F6_30.DXF');
27    OpenMecGraph(-0.4,0.3, -0.14,0.23);
28    SetJointSize(4);
29    i:=0;
30    repeat
31    if (i > nPoz) then BEGIN
32       i:=0;  CloseMecDXF;
33    END;
34    NewFrame(5000);
35    t:=i/nPoz;  {t = time}
36    Phi1:=2*Pi*t*t*t;
37    Phi2:=Pi/3+Pi/9*sin(Pi*t);
38    gCrank(Red, xO1,yO1, Phi1,_,_, O1A, xA,yA,_,_,_,_);
39    gCrank(Red, xO2,yO2, Phi2,_,_, O2B, xB,yB,_,_,_,_);
40    R_T_T(Cyan, xA,yA,_,_,_, xB,yB,_,_,_, Phi2+0.5*Pi,_,_
41     ,AD,DK,PQ,QC, Alph1,Alph2, xC,yC,_,_,_, xD,yD,_,_,_,
42     ,xK,yK,_,_,_, xP,yP,_,_,_, xQ,yQ,_,_,_, OK);
43    PutGPoint(White,' ', xO1,yO1 ,'O_1');
44    PutGPoint(White,' ', xO2,yO2 ,' O_2');
45    PutPoint(White,' ', xA,yA ,' A');
46    PutPoint(White,' ', xB,yB ,'B');
47    PutPoint(White,' ', xC,yC ,' C');
```

```
48    LabelJoint(White,  xD,yD,xK,yK,' K');
49    LabelJoint(White,  xQ,yQ,xP,yP,' P');
50    LabelJoint(White,xC,yC,xQ,yQ,'  Q');
51    LabelJoint(White,0.5*(xA+xK),0.5*(yA+yK),xD,yD,' D');
52    PutAng(White, xK,yK, xC,yC, 2*xC-xQ,2*yC-yQ,  -8, '|<α_1');
53    PutAng(White, 2*xQ-xP,2*yQ-yP, xQ,yQ, xC,yC,   8, '<α_2|');
54    Locus(Magenta,xQ,yQ,'Q');
55    Inc(i);
56    until IsKeyPressed(27);
57    CloseMecGraph(FALSE);      {..do not save locus files}
58  END.

1   program P6_32;
2   {==================================================================
3   Simulation of an RT_T_ dyadic isomer driven by a crank and a rocker
4   ==================================================================}
5   uses Graph,        {Cyan,Red,White}
6        LibInOut,     {IsKeyPressed}
7        LibMath,      {_,RAD}
8        LibDXF,       {InitDXfile}
9        LibMecIn,     {gCrank}
10       LibAssur,     {RT_T_}
11       LibMec2D;     {OpenMecGraph,NewFrame,Locus,}
12                     {CloseMecDXF,MecOutp,CloseMecGraph}
13  const nPoz = 90;   {number of positions}
14  var i: Word;    OK: Boolean;    t, Phi1,Phi2,Alpha2, OA,PQ,AC,BD,
15      xO,yO, xP,yP, xQ,yQ, xA,yA, xB,yB, xC,yC, xD,yD: double;
16  BEGIN
17    xO:=-0.20;  yO:=0;    {ground joint of rocker #1 }
18    xP:= 0.20;  yP:=0;    {ground joint of rocker #2 }
```

```
19   OA:= 0.20;   PQ:=0.7; {length of rocker #1 and #2 }
20   AC:= 0.35;
21   BD:= 0.85;
22   Alpha2:=110*RAD;
23   InitDXFfile('F6_32.DXF');
24   OpenMecGraph(-0.8,1.0,  -0.25,1.0);
25   SetJointSize(4);
26   i:=0;
27   repeat
28     if (i > nPoz) then BEGIN
29       i:=0;   CloseMecDXF;
30     END;
31     NewFrame(500);
32     t:=i/nPoz;  {t = time}
33     Phil:=2*Pi*t*t;
34     Phi2:=Pi/4+Pi/6*sin(Pi*t);
35     gCrank(Red,xO,yO, Phil,_,_, OA, xA,yA,_,_,_,_);
36     gCrank(-Red,xP,yP, Phi2,_,_, PQ, xQ,yQ,_,_,_,_);
37     RT_T (Cyan, xA,yA,_,_,_,_, xP,yP,0,0,0,0, xQ,yQ,_,_,_,_,
38       ,AC,BD,Alpha2,xB,yB,_,_,_,_,_, xC,yC,_,_,_,_,  xD,yD,_,_,_,_,  OK);
39     PutGPoint(White,' ',  xO,yO ,'O    ');
40     PutGPoint(White,' ',  xP,yP ,'    P');
41     PutPoint(White,'  ',  xA,yA ,'  A');
42     PutPoint(White,'  ',  xQ,yQ ,'  Q');
43     PutPoint(White,'  ',  xD,yD ,'D  ');
44     LabelJoint(White,xC,yC,xB,yB,'      B');
45     LabelJoint(White,xA,yA,xC,yC,'      C');
46     PutAng(White, xQ,yQ, xB,yB, xC,yC,  8,  '<a_2');
47     Locus(White,xC,yC,'C');
48     Inc(i);
```

```
49      until IsKeyPressed(27);
50      CloseMecGraph(FALSE);        {..do not save locus files}
51  END.

 1  program P6_34;
 2  {===============================================================
 3   Simulation of an R_TT_ dyad driven by a crank and a rocker
 4   ===============================================================}
 5  uses Graph,        {Cyan,Magenta,Red}
 6       LibMath,      {_,RAD}
 7       LibInOut,     {IsKeyPressed}
 8       LibDXF,       {InitDXFfile}
 9       LibMecIn,     {gCrank}
10       LibAssur,     {R_TT_}
11       LibMec2D;     {OpenMecGraph,NewFrame,Locus,}
12                     {CloseMecDXF,MecOutp,CloseMecGraph}
13  const nPoz = 90; {number of positions}
14  var i:Word;    OK:Boolean;  t,Phi1,Phi2,Alpha1,Alpha2,  OA,PQ,AD,DK,BC,
15      xO,yO,xK,yK,xB,yB,xA,yA,xP,yP,  xD,yD,xQ,yQ,xC,yC:double;
16  BEGIN
17    InitDXFfile('F6_34.DXF');
18    xO:=-0.10; yO:=0;        {ground joint of crank   }
19    xP:= 0.10; yP:=0;        {ground joint of rocker  }
20    OA:= 0.07; PQ:= 0.35;    {lengths of crank and rocker}
21    AD:= 0.05; DK:=0.55;     {L-rod of sliding block C  }
22    BC:=0.28;                {length of spacer BC   }
23    Alpha1:= 60*RAD;
24    Alpha2:=110*RAD;
25    OpenMecGraph(-0.20,0.5,-0.2,0.7);
26    SetJointSize(4);
```

```
27   i:=0;
28   repeat
29     if (i > nPoz) then BEGIN
30       i:=0;    CloseMecDXF;
31     END;
32     NewFrame(500);
33     t:=i/nPoz; {t = time}
34     Phi1:=2*Pi*t*t;
35     Phi2:=Pi/9 + Pi/6*sin(2*Pi*t);
36     gCrank(Red,xO,yO, Phi1,_,_, OA, xA,yA,_,_,_,_) ;
37     gCrank(-Red,xP,yP, Phi2,_,_, PQ, xQ,yQ,_,_,_,_);
38     R_TT_(Cyan, xA,yA,_,_,_,_, xP,yP,0,0,0,0,
39          xQ,yQ,_,_,_,_, AD,DK,BC, Alpha1,Alpha2, xB,yB,_,_,_,_,
40          xC,yC,_,_,_,_, xD,yD,_,_,_,_, xK,yK,_,_,_,_, OK) ;
41     PutGPoint(White,'   ',  xO,yO ,'O  '   );
42     PutGPoint(White,'   ',  xP,yP ,'    P');
43     PutPoint(White,'   ', xA,yA ,' A') ;
44     PutPoint(White,'   ', xQ,yQ ,' Q') ;
45     PutPoint(White,'   ', xD,yD ,' D') ;
46     PutPoint(White,'   ', xK,yK ,' K') ;
47     LabelJoint(White,xC,yC,xB,yB,'    B') ;
48     LabelJoint(White,xA,yA,xC,yC,'    C') ;
49     PutAng(White, 2*xC-xD,2*yC-yD,xC,yC,2*xC-xB,2*yC-yB,-9,' |<α_1') ;
50     PutAng(White, xQ,yQ, xB,yB, xC,yC, 9, '<α_2') ;
51     Locus(White,xC,yC,'C') ;
52     Inc(i);
53   until IsKeyPressed(27);
54   CloseMecGraph(FALSE);    {..do not save locus files}
55 END.
```

```pascal
1    program P6_36;
2    {========================================================================
3    Simulation of a RT_T dyadic isomer driven by a crank and a rocker
4    ========================================================================}
5    uses Graph,              {Cyan,Magenta,Red}
6         LibMath,            {_,RAD}
7         LibInOut,           {IsKeyPressed}
8         LibDXF,             {InitDXFfile}
9         LibMecIn,           {gCrank}
10        LibAssur,           {RT:T}
11        LibMec2D;           {OpenMecGraph,NewFrame,Locus,CloseMecDXF,}
12                            {MecOutp,CloseMecGraph}
13   const nPoz = 90;  {number of positions}
14   var i:Word;       OK:Boolean;   t,Phi1,O1A,Phi2,O2B,PQ,QD,AC,Alpha2,
15        xO1,yO1,xO2,yO2,xA,yA,xB,yB,xC,yC,xD,yD,xP,yP,xQ,yQ:double;
16   BEGIN
17     InitDXFfile('F6_36.DXF');
18     xO1:=-0.20; yO1:=0;              {ground joint of crank     }
19     xO2:= 0.20; yO2:=0;              {ground joint of rocker    }
20     O1A:= 0.20; O2B:=0.40;           {lengths of crank and rocker}
21     PQ := 0.90; QD:=1.30;            {L-rod dimensions          }
22     AC := 0.35;                      {length of offset AC       }
23     Alpha2:=110*RAD;
24     OpenMecGraph(-0.67,0.86,-0.63,0.88);
25     SetJointSize(4);
26     i:=0;
27     Repeat
28       if (i > nPoz) then BEGIN
29         i:=0; CloseMecDXF;
30       END;
```

```
31      NewFrame(500);
32      t:=i/nPoz+EpsD;      {t = time}
33      Phi1:=2*Pi*t*t;
34      Phi2:=Pi/8*sin(2*Pi*t);
35      gCrank(Red,xO1,yO1,Phi1,_,_,  O1A,  xA,yA,_,_,_,_);
36      gCrank(Red,xO2,yO2,Phi2,_,_,  O2B,  xB,yB,_,_,_,_);
37      RT_T(Cyan,xA,yA,_,_,_,_,xB,yB,_,_,_,_,Phi2+Pi/2,_,_,AC,PQ,QD,
38      Alpha2,xC,yC,_,_,_,_,xD,yD,_,_,_,_,xP,yP,_,_,_,_,xQ,yQ,_,_,_,_,OK);
39      PutGPoint(White,'  ',  xO1,yO1 ,'O1 ');
40      PutGPoint(White,'  ',  xO2,yO2 ,'   O2');
41      PutPoint(White,'  ',  xA,yA ,'  A');
42      PutPoint(White,'  ',  xB,yB ,'B');
43      PutPoint(White,'  ',  xC,yC ,'C');
44      PutPoint(White,'  ',  xD,yD ,' D');
45      PutPoint(White,'  ',  xP,yP ,'P ');
46      PutPoint(White,'  ',  xQ,yQ ,' Q');
47      PutAng(White, 2*xQ-xP,2*yQ-yP, xQ,yQ, xD,yD,   8,  '<α_2|');
48      Locus(White,xQ,yQ,'Q');
49      Inc(i);
50      until IsKeyPressed(27);
51      CloseMecGraph(FALSE);        {..do not save locus files}
52      END.
```

***

```
1      program P7_06;
2      {======================================================================
3      Synthesis of disc cams w/ translating follower knife edge and w/ roller
4      ======================================================================}
5      uses Graph,    {Green,Red,White}
6           LibMath,  {_}
```

```
7        LibDXF,          {InitDXFfile}
8        LibMecIn,        {gCrank}
9        LibInOut;        {IsKeyPressed}
10       LibMec2D; {OpenMecGraph,Slider,Offset,PutGPoint,PutPoint,LabelJoint}
11                 {PutAng,PutDist,Locus,NewFrame,CloseMecDXF,CloseMecGraph}
12 const Color = 0;     {0 follower & pitch cam, 1 entire mechanism }
13       CamXY ='Cam06'; {pitch cam output to Cam06.D2D and to layer Cam06}
14       DTheta= 2*Pi;   {cam rotational amplitude - cannot be 0}
15       Ds    = 1.0;    {follower amplitude - cannot be 0}
16       s0    = 0.25;   {follower bias - cannot be 0}
17       OP    = 0.20;   {follower offset eF - positive or negative}
18       r     = 0.15;   {roller radius; r=0 for knife edge}
19 var FT:text;           Ch:char;
20       rb,rt, dF,dC,Theta, s,xA,yA, xP,yP,xQ,yQ, xB,yB,xC,yC:double;
21 BEGIN
22    if (Color <> 0) then BEGIN
23       Assign(FT,'dFvdC_L.XY');   {..follower motion file - full size}
24       InitDXFfile('F7_06A.DXF'); {..output M3D-DXF file}
25    END
26    else BEGIN
27       InitDXFfile('F7_06B.DXF'); {..output M3D-DXF file}
28       Assign(FT,'dFvdC.XY');     {..follower motion file - reduced size}
29    END;
30    Reset(FT);
31    rb:=Sqrt(OP*OP+s0*s0);       {..base circle radius}
32    rt:=Sqrt(OP*OP+Sqr(s0+Ds));  {..top circle radius}
33    Write('rb=',rb:9:5,'  rt=',rt:9:5,'  <CR>..');  ReadLn;
34    OpenMecGraph(-rt-r,rt+r, -rt-r,rt+r);
35    repeat
36       if (Color > 0) then NewFrame(50);
37       ReadLn(FT,dC,dF);
38       Theta:=dC*DTheta;         {..cam angle}
```

```
39     s:=s0+dF*Ds;              {..follower displacement s(Theta) }
40     gCrank(Color,0,0,Theta,_,_,OP+20.5*RJtSz,xA,yA,_,_,_,_);
41     Offset(Black,'T',0,0,_,_,_,xA,yA,_,_,_,_,OP, 0, xP,yP,_,_,_,_);
42     Offset(Black,'/',0,0,_,_,_,xA,yA,_,_,_,_,OP,s0, xQ,yQ,_,_,_,_);
43     Slider(Color,xP,yP,_,_,_,xQ,yQ,_,_,_,_,s0,s,_,_
44     ,xC,yC,_,_,_,_, xB,yB,_,_,_,_);
45     Locus(Green,xC,yC,CamXY);    {..locus of roller center}
46     Shape('',Red,xC,yC,xC+r,yC);  {..draw roller}
47     if (Color = 0) then PutGPoint(Green,'+', 0,0,'')
48     else BEGIN
49       PutGPoint(White,' ', 0,0 ,'O       ');
50       LabelJoint(White, xC,yC, xA,yA,' A');
51       LabelJoint(White, xC,yC, xB,yB,' B');
52       PutPoint(Red,'o',xC,yC ,'');
53       LabelJoint(White, xP,yP, xC,yC,' C');
54       LabelJoint(White, xC,yC, xP,yP,'P ');
55       PutPoint(White,' ', xQ,yQ,'Q');
56       PutAng(White, 0,0,0,xA,yA, 5, '<'+#233+'|'); {..#233 = Theta }
57       PutDist(White, xC,yC,xP,yP, 20, '|<s'); {..note the 20 in gCrank}
58     END;
59     until Eof(FT);
60     CloseMecDXF;
61     WaitToGo(Ch);
62     CloseMecGraph(Color = 0);
63   END.
```

```
1    program P7_07;
```
```
5    uses Graph,     {Blue,Cyan,Green,Magenta,White}
```

```pascal
6        LibDXF,        {InitDXFfile}
7        LibMath,       {_,DEG}
8        LibInOut,      {IsKeyPressed}
9        LibMecIn,      {gCrank}
10       LibGe2D,       {U2dirs2D90,RT2D}
11       LibCams,       {TransFolRotCam,DoubleOffset}
12       LibMec2D;      {OpenMecGraph,NewFrame,Slider,Offset,PutPoint,}
13                      {PutVector,PutText,Shape,CloseMecDXF,CloseMecGraph}
14   const nPoz   = 360;      {number of cam positions (90 with Anim = 1) }
15         Anim   = 0;        {0 = animation OFF, 1 = animation ON}
16         CamXY  = 'Cam06';  {.D2D input file with cam profile points}
17         DTheta = 2*Pi;     {cam rotational cycle - cannot be 0}
18         OP     = 0.20;     {follower offset eF - positive or negative }
19         r      = 0.15;     {roller radius; r=0 for knife edge}
20   var FT:text;    Ch:char;    i,Skip:Integer;    s,rb,rt,xA,yA,Theta
21       ,DnX,DnY,Rho,Gamma,xC,yC,xCi,yCi,xCo,yCo:double;
22   BEGIN
23     if (Anim <> 0) then InitDXFfile('F7_07.DXF');  {..output M3D-DXF file}
24     Assign(FT,'F7_07.TXT'); Rewrite(FT);           {..output ASCII data file}
25     GetProfileRadii(CamXY+'.D2D',rb,rt);           {..extract base & top circ. radii}
26     WriteLn(FT,CamXY+'.D2D cam with knife-edge translating follower');
27     WriteLn(FT,'DTheta=',DTheta*DEG:6:2,';   OP=',OP:6:2,';   r=',r:6:2);
28     WriteLn(FT,'  Theta s     Gamma Rho    rb     xC     yC'
29     ,'  xCi yCi   xCo   yCo   rb    rt');
30     Write('rb=',rb:9:5,',  rt=',rt:9:5,'  <CR>..');   ReadLn;
31     If (Abs(OP) > rt) then BEGIN
32       Write('Eccentricity OP too big!    <CR>..');   ReadLn;
33       Halt;
34     END;
35     OpenMecGraph(-rt,rt, -rt,1.5*rt);
```

```
36    if (Anim > 0) then Skip:=1 else Skip:=nPoz DIV 10;
37    i:=0;
38    repeat
39      if (i > nPoz) then BEGIN
40        i:=0;   CloseMecDXF;
41      END;
42      if (Anim = 1) OR (i MOD Skip = 0) then NewFrame(0);
43      Theta:=-DTheta*i/nPoz;
44      gCrank(Anim*Magenta,0,0,Theta,_,_,rb,xA,yA,_,_,_,_);
45      Shape('',Anim*Magenta,0,0,xA,yA);    {..draw cam base circle}
46      Shape(CamXY+'.D2D',Anim*Green,0,0,xA,yA);    {..draw cam from file}
47      RotCamTransPointed(CamXY+'.D2D',OP,Theta,s,xC,yC,DnX,DnY,Rho);
48      DoubleOffset(xC,yC, DnX,DnY, Rho, r, xCi,yCi,xCo,yCo);
49      Gamma:=U2dirs2D90(DnX,DnY,0,1.0)*DEG;
50      PutVector(Anim*Cyan,'-',xC,yC, 0,1.0, 0.5,'Vc'); {..velocity}
51      PutVector(Anim*Blue,'-',xC,yC, DnX,DnY, 0.4,'Fc'); {..normal force}
52      PutPoint(Anim*White,'o', xC,yC,   C'); {..contact point}
53      if (i MOD Skip = 0) then PutText(White,_,_,'θ ='+ MySt2(Theta*DEG,7,2)
54        +'°'+'\n\γ ='+MySt2(Gamma,7,2)+'°\n\Rho='+ MyStr(Rho,6));
55      RT2D(xC , yC,-Theta,0,0,xC ,yC ); {..rotate backwards point C }
56      RT2D(xCi,yCi,-Theta,0,0,xCi,yCi); {..rotate backwards point Ci}
57      RT2D(xCo,yCo,-Theta,0,0,xCo,yCo); {..rotate backwards point C }
58      if MecOut then WriteLn(FT,Theta:8:4,' ',s:10:6,' ',Gamma:8:4
59        ,' ',Rho:10:6,' ',xC:10:6,' ',xCi:10:6,' ',yCi:10:6
60        ,' ',xCo:10:6,' ',yCo:10:6,' ',rb:9:6,' ',r:9:6);
61      Inc(i);
62    until (NOT MecOut) AND (IsKeyPressed(27) OR (Anim = 0));
63    Close(FT);
64    CloseMecGraph(FALSE);
65  END.
```

```pascal
1   program P7_10;
2   {==============================================================
3    Synthesis of disc cams with oscillating follower, pointed or with roller
4   ==============================================================}
5   uses Graph,        {Red,White}
6        LibMath,      {_,RAD}
7        LibDXF,       {InitDXFfile}
8        LibMecIn,     {gCrank}
9        LibInOut,     {WaitToGo}
10       LibMec2D;     {OpenMecGraph,Slider,Offset,PutGPoint,LabelJoint,}
11                     {Shape,NewFrame,CloseMecDXF,CloseMecGraph}
12  const Color = 1;      {0 follower & pitch cam, 1 entire mechanism}
13        CamXY ='Cam10'; {.D2D output cam profile name}
14        DTheta= 2*Pi;   {cam rotational amplitude - cannot be 0}
15        DPhi  = 25*RAD; {follower amplitude - cannot be 0}
16        Phi0  = 20*RAD; {follower bias - cannot be 0}
17        OO1   = 2.0;    {cam-follower pin-joint distance}
18        O1C   = 2.0;    {follower arm length}
19        r     = 0.2;    {roller radius}
20  var FT:text;  Ch:char;  rb,rt,  dF,dC,Theta,Phi,xO1,yO1,xC,yC:double;
21  BEGIN
22    if (Color <> 0) then BEGIN
23      Assign(FT,'dFvdC_L.XY');  {..follower motion file - full size}
24      InitDXFfile('F7_10A.DXF');  {..output DXF file}
25    END
26    else BEGIN
27      Assign(FT,'dFvdC.XY');  {..follower motion file - reduced size}
28      InitDXFfile('F7_10B.DXF');  {..output DXF file}
29    END;
```

```
30   Reset(FT);
31   rb:=Sqrt(OO1*OO1+O1C*O1C-2*OO1*O1C*cos(Phi0));       {..base circle radius}
32   rt:=Sqrt(OO1*OO1+O1C*O1C-2*OO1*O1C*cos(Phi0+DPhi));  {..top circle radius}
33   Write('rb=',rb:9:5,'', rt=',rt:9:5,' <CR>..');  ReadLn;
34   OpenMecGraph(-1.25*OO1,1.25*OO1,-1.25*OO1,1.25*OO1);
35   repeat
36     if (Color > 0) then   NewFrame(50);
37     ReadLn(FT,dC,dF);
38     Theta:=dC*2*Pi;       {..cam rotation }
39     Phi:=Phi0+dF*DPhi;    {..follower displacement}
40     gCrank(Color,0,0,Theta,_,_,OO1,xO1,yO1,_,_,_);
41     Crank(Color,0,0,_,_,_,xO1,yO1,_,_,_,Pi-Phi,_,_,O1C,xC,yC,_,_,_);
42     Locus(Green,xC,yC,CamXY);       {..draw knife edge cam profile}
43     Shape('',Red,xC,yC,xC+r,yC);    {..draw rcller}
44     if (Color = 0) then   PutGPoint(Green,'+', 0,0,'')
45     else BEGIN
46       PutGPoint(White,' ', 0 , 0,'O      ');
47       PutPoint(Red,'o',xC,yC,'');
48       LabelJoint(White,xC,yC, xO1,yO1,'       O1');
49       LabelJoint(White,0,0,xC,yC,'   C');
50       PutAng(White, 0,0,0,xO1,yO1, 5, '|'+#233+'>');     {..#233 = Theta}
51       PutAng(-White, xC,yC,xO1,yO1,0,0, 16, #237+'>');   {..#237 = Phi}
52     END;
53   until Eof(FT);
54   CloseMecDXF;
55   WaitToGo(Ch);
56   CloseMecGraph(Color = 0);
57 END.
```

```pascal
1   program P7_11;
2   {================================================
3    Kinematic analysis of cams with knife edged oscillating follower and
4    inner and outer offset cam profiles calculation
5   ================================================}
6   uses Graph,
7        LibMath,     {_,DEG,RAD}
8        LibGraph,    {X_p,Y_p,p_X,p_Y}
9        LibDXF,      {InitDXFfile}
10       LibInOut,    {IsKeyPressed}
11       LibGe2D,     {U2dirs2D90}
12       LibMecIn,    {gCrank}
13       LibCams,     {RotCamOscilPointed,DoubleOffset}
14       LibMec2D;    {OpenMecGraph,Slider,Offset,}
15                    {Locus,NewFrame,CloseMecDXF,CloseMecGraph}
16   const nPoz = 360;     {number of cam positions (90 with Anim = 1)}
17         Anim = 1;       {0 = animation OFF, 1 = animation ON}
18         CamXY = 'Cam10'; {.D2D input file with cam profile points  }
19         DTheta= 2*Pi;   {cam rotational amplitude - cannot be 0}
20         OO1  = 2.0;     {cam-joint to follower-joint distance}
21         O1C  = 2.0;     {follower arm length}
22         r    = 0.2;     {roller radius}
23   var FT:text; Ch:char; i,Skip:Integer; Theta,rb,rt, xA,yA, Phi,DnX,DnY
24       ,Rho,O1Ci,O1Co, xC,yC, xCi,yCi, xCo,yCo, Gamma,Gam_i,Gam_o:double;
25   BEGIN
26   if (Anim <> 0) then InitDXFfile('F7_11.DXF'); {..output M3D-DXF file}
27   Assign(FT,'F7_11.TXT'); {..output ASCII data file  }
28   Rewrite(FT);
29   WriteLn(FT,CamXY+'.D2D cam with knife-edge oscillating follower.');
30   GetProfileRadii(CamXY+'.D2D',rb,rt); {..get base & top circle radii}
31   Write('rb=',rb:9:5,'   rt=',rt:9:5,' <CR>..');  ReadLn;
```

```
32   If (rt > OO1) OR (OO1-O1C < -rt) then BEGIN
33     Write('Inproper OO1 and O1C values!    <CR>..'); ReadLn;
34     Halt;
35   END;
36   WriteLn(FT,'DTheta=',DTheta*DEG:6:2,';  OO1=',OO1:6:2
37   ,';  O1C=',O1C:6:2,';  r=',r:6:2);
38   WriteLn(FT,' Theta  Phi    Gamma  Gam_i  Gam_o  Rho   ',
39   'xC   yC   xCi  yCi   xCo   yCo   rb    rt');
40   OpenMecGraph(-rt,OO1, -rt-r,rt+3*r);
41   if (Anim > 0) then Skip:=1 else Skip:=nPoz DIV 10;
42   i:=0;
43   repeat
44     if (i > nPoz) then BEGIN
45       i:=0;  CloseMecDXF;
46     END;
47     if (Anim <> 0) OR (i MOD Skip = 0) then NewFrame(0);
48     Theta:=-DTheta*i/nPoz;
49     gCrank(Anim*Magenta,0,0,Theta,_,_,rb,xA,yA,_,_,_);
50     Shape('',Anim*Magenta,0,0,xA,yA);   {..draw cam base circle}
51     Shape(CamXY+'.D2D',Anim*Green,0,0,xA,yA);   {..draw from file}
52     RotCamOscilPointed(CamXY+'.D2D',OO1,O1C,Theta, Phi,xC,yC,DnX,DnY,Rho);
53     DoubleOffset(xC,yC, DnX,DnY, Rho, r, xCi,yCi,xCo,yCo);
54     Gamma:=U2dirs2D90(DnX,DnY,yC/O1C,(OO1-xC)/O1C)*DEG;
55     O1Ci:=Dist2Pts2D(xCi,yCi,OO1,0);
56     Gam_i:=U2dirs2D90(DnX,DnY,yCi/O1Ci,(OO1-xCi)/O1Ci)*DEG;
57     O1Co:=Dist2Pts2D(xCo,yCo,OO1,0);
58     Gam_o:=U2dirs2D90(DnX,DnY,yCo/O1Co,(OO1-xCo)/O1Co)*DEG;
59     PutGPoint(Anim*White,'/', OO1,0,'');   {..ground pin joint}
60     if (Anim <> 0) then
61       SkLine(X_p(OO1),Y_p(0),X_p(xC),Y_p(yC),JtSz,JtSz-1);
```

```
62     PutVector(Anim*Blue,'-',xC,yC, DnX,DnY, 0.6,' Fc'); {..normal force}
63     PutVector(Anim*Cyan,'-',xC,yC, yC/O1C, (OO1-xC)/O1C, 0.5,'Vc');
64     PutPoint(Anim*White,'o', xC,yC,'C ');
65     if (i MOD Skip = 0) then PutText(White,_,_,'θ ='+MySt2(Theta*DEG,7,2)
66     +'°'+'n\γ ='+MySt2(Gamma,7,2)+'°n\Rho='+MyStr(Rho,6));
67     RT2D(xC ,yC ,-Theta,0,0,xC ,yC );  {..rotate backwards point C }
68     RT2D(xCi,yCi,-Theta,0,0,xCi,yCi);  {..rotate backwards point Ci}
69     RT2D(xCo,yCo,-Theta,0,0,xCo,yCo);  {..rotate backwards point Co}
70     if MecOut then WriteLn(FT,Theta:8:4,' ',Phi*DEG:10:6,' '
71     ,Gamma:8:4,' ',Gam_i:8:4,' ',Gam_o:8:4,' ',Rho:10:6,' ',xC:10:6
72     ,' ',yC:10:6,' ',xCi:10:6,' ',yCi:10:6,' ',xCo:10:6,' ',yCo:10:6
73     ,' ',rb:9:6,' ',rt:9:6);
74     Inc(i);
75     until (NOT MecOut) AND (IsKeyPressed(27) OR (Anim = 0));
76     Close(FT);
77     CloseMecGraph(FALSE);
78   END.
```

```
1  program P7_14;
2  {=================================================================
3  Synthesis of disc cams with flat-faced translating follower
4  =================================================================}
5  uses Graph,        {Red,White}
6       LibMath,      {_,RAD}
7       LibDXF,       {InitDXFfile}
8       LibInOut,     {IsKeyPressed}
9       LibMecIn,     {gCrank}
10      LibCams,      {EnvelOfLines, EndEnvelopes}
11      LibMec2D;     {OpenMecGraph,Slider,Offset,}
12                    {Locus,NewFrame,CloseMecDXF,CloseMecGraph}
13  const Color = 1;    {0 follower face only, 1 more details  }
```

```
14      Anim    = 1;        {0 = accumulate frames, 1 = animate        }
15      nPL     = 1000;     {polar lines in EnvelOfLines - maximum 1000 }
16      CamXY   ='Cam14';   {D2D file name for cam profile              }
17      DTheta  = 2*Pi;     {cam rotational amplitude - cannot be 0     }
18      Gamma   = 20*RAD;   {follower face angle - positive or negative }
19      Ds      = 1.0;      {follower amplitude - cannot be 0           }
20      s0      = 1.5;      {follower bias - cannot be 0                }
21      xPC     = 0.0;      {x coord. of polar center in EnvelOfLines   }
22      yPC     = 0.0;      {y coord. of polar center in EnvelOfLines   }
23      PLL     = 1.2*s0+Ds; {initial polar line length in EnvelOfLines }
24    var FT:text;   Ch:char;    rb,rt, dF,dC,Theta, s,xA,yA,xB,yB
25        ,xD,yD, xP,yP, xQ,yQ, PQ, sG,cG:double;
26    BEGIN
27      if (Anim = 1) OR (Color = 1) then BEGIN
28        Assign(FT,'dFvdC_L.XY');{..follower motion file - reduced size}
29        InitDXFfile('F7_14A.DXF');    {..output M3D-DXF file}
30      END
31      else BEGIN
32        Assign(FT,'dFvdC.XY'); {..follower motion file - full size}
33        InitDXFfile('F7_14B.DXF');    {..output M3D-DXF file}
34      END;
35      Reset(FT);
36      sG:=sin(Gamma);    cG:=cos(Gamma);
37      rb:=s0+cG;          {..base circle radius}
38      rt:=(s0+Ds)*cG;     {..top circle radius}
39      Write('rb=',rb:9:5,'    rt=',rt:9:5,'   <CR>..');    ReadLn;
40      PQ:=rt;            {..half of follower face length}
41      OpenMecGraph(-rt,rt, -1.75*rt,1.5*rt);
42      repeat
43        if (Anim > 0) then NewFrame(50);
44        ReadLn(FT,dC,dF);
```

```
45    Theta:=dC*DTheta+Pi/2;      {..cam rotation angle}
46    s:=s0+dF*Ds;                {..follower displacement}
47    gCrank(-Anim*Color,0,0,Theta,_,_,s0,xA,yA,_,_,_,_);
48    Slider(Anim*Color,0,0,0,0,0,0, xA,yA,_,_,_,_, s0, s,_,_
49    ,xD,yD,_,_,_,_, xB,yB,_,_,_,_);
50    Offset(0,'T',xD,yD,_,_,_,_,0,0,_,_,_,_,-PQ*sG, PQ*cG,xP,yP,_,_,_,_);
51    Offset(0,'T',xD,yD,_,_,_,_,0,0,_,_,_,_, PQ*sG, -PQ*cG,xQ,yQ,_,_,_,_);
52    Offset(Red,'/',xP,yP,_,_,_,_,xQ,yQ,_,_,_,_,2*PQ,0,_,_,_,_,_);
53    if (Color = 0) then
54      EnvelofLines(-Anim*White,nPL,xPC,yPC, PLL, xP,yP,xQ,yQ);
55    else BEGIN
56      EnveloflLines(0,nPL,xPC,yPC, PLL, xP,yP,xQ,yQ);
57      PutGPoint(Anim*White,' ','0,0,'o ');
58      PutPoint(Anim*White,' ',xA,yA,' A');
59      LabelJoint(Anim*White,xD,yD,xB,yB,' B');
60      LabelJoint(Anim*White,xB,yB,xD,yD,' D');
61      LabelJoint(Anim*White,xQ,yQ,xP,yP,'P');
62      LabelJoint(Anim*White,xP,yP,xQ,yQ, Q');
63      PutAng(Anim*White,0,0,0,xA,yA,6, '<'+#233+'|'); {..#233=Theta}
64      PutDist(Anim*White,0,0,xD,yD,-10, '|s>|');
65    END;
66    until Eof(FT);
67    PutGPoint(Green,'+', 0,0,''); {..put cam center}
68    if (Color = 0) AND (Anim = 0) then
69      EndEnvelopes(CamXY,Green)  {..write cam to file CamXY.D2D}
70    else
71      EndEnvelopes(MySt(LastNrLayer,3),Green); {..write cam to last layer}
72    CloseMecDXF;
73    WaitToGo(Ch);
74    CloseMecGraph(Anim = 0);
75    END.
```

```pascal
1  program P7_15;
2  {=================================================================
3   Kinematic analysis of disc cams with flat-faced translating follower
4   =================================================================}
5  uses Graph,       {Magenta,Blue,Cyan,White}
6       LibMath,     {_,DEG}
7       LibDXF,      {InitDXFfile}
8       LibInOut,    {IsKeyPressed}
9       LibGraph,    {p_X,p_Y}
10      LibGe2D,     {RT2D}
11      LibMecIn,    {gCrank}
12      LibCams,     {GetProfileRadii,RotCamTransF_at}
13      LibMec2D;    {OpenMecGraph,Slider,Offset,}
14                   {Locus,NewFrame,CloseMecDXF,C oseMecGraph}
15  const nPoz = 90;      {number of cam positions (90 with Anim = 1)}
16        Anim = 1;       {0 = animation OFF, 1 = animation ON     }
17        CamXY ='Cam14'; {.D2D input file with cam profile points }
18        DTheta= 2*Pi;   {cam rotational amplitude - cannot be 0   }
19        Gamma = 20*RAD; {follower angle in degrees                }
20  var FT:text;  Ch:char;   i,Skip:Integer;   rb,rt,  s,  sG,cG
21      ,xA,yA, Theta,Rho, xC,yC,xP,yP,xQ,yQ,PQ: double;
22  BEGIN
23  if (Anim <> 0) then InitDXFfile('F7_15.DXF'); {..output M3D-DXF file}
24  Assign(FT,'F7_15.TXT'); Rewrite(FT); {..output ASCII data file}
25  GetProfileRadii(CamXY+'.D2D',rb,rt); {..get base & top circle radii}
26  Write('rb=',rb:9:5,'', rt=',rt:9:5,'    <CR>..'); ReadLn;
27  WriteLn(FT,CamXY+'.D2D cam with flat-face translating follower');
28  WriteLn(FT,'DTheta=',DTheta*DEG:6:2,' Gamma=',Gamma*DEG:8:3);
29  WriteLn(FT,'  Theta     s       Rho      xC      yC'
30      ,'    rb   rt');
31  sG:=sin(Gamma);     cG:=cos(Gamma);
```

```
32  PQ:=0.85*rt;
33  OpenMecGraph(-1.2*rt,1.2*rt,1.2*rt, -1.2*rt,1.5*rt);
34  if (Anim > 0) then Skip:=1 else Skip:=nPoz DIV 10;
35  i:=0;
36  repeat
37    if (i > nPoz) then BEGIN
38      i:=0;  CloseMecDXF;
39    END;
40    if (Anim = 1) OR (i MOD Skip = 0) then NewFrame(0);
41    Theta:=-DTheta*i/nPoz;
42    gCrank(Anim*Magenta,0,0,Theta,_,_,rb,xA,yA,_,_,_,_);
43    Shape('',Anim*Magenta,0,0,xA,yA);      {..draw cam base circle}
44    Shape(CamXY+'.D2D',Anim*Red,0,0,xA,yA);  {..draw cam from file}
45    RotCamTransFlat(CamXY+'.D2D',Theta,Gamma,s,xC,yC,Rho);
46    xP:= PQ*cG;   yP:=s+PQ*sG;
47    xQ:=-PQ*cG;   yQ:=s-PQ*sG;
48    if (Anim <> 0) then
49      PDline('',X_p(xP),Y_p(yP),X_p(xQ),Y_p(yQ));  {..draw follower face}
50    PutVector(Anim*Cyan,'-',xC,yC, 0,1.0, 1.0,'Vc');
51    PutVector(Anim*Blue,'-',xC,yC,-(yP-yQ)/PQ,(xP-xQ)/PQ, 0.55,'Fc');
52    PutPoint(Anim*White,'o',xC,yC, C');     {..contact point C}
53    if (i MOD Skip = 0) then PutText(White,_,_,'θ ='+MySt2(Theta*DEG,7,2)
54      +'°'+'n\γ  ='+MySt2(Gamma*DEG,7,2)+'°n\Rho='+MyStr(Rho,6));
55    RT2D(xC,yC,-Theta,0,0,xC,yC);  {..rotate backwards point C}
56    if MecOut then WriteLn(FT,Theta:8:4,' ',s:10:6,' ',Rho:10:6
57      ,' ',xC:10:6,' ',yC:10:6,' ',rb:9:5,' ',rt:9:5);
58    Inc(i);
59  until (NOT MecOut) AND (IsKeyPressed(27) OR (Anim = 0));
60  Close(FT);
61  CloseMecGraph(FALSE);   {..erase all .$2D files}
62 END.
```

```
1   program P7_18;
2   {===============================================================
3   Synthesis of disc cams with flat-face oscillating follower
4   ===============================================================}
5   uses Graph,          {Red,White}
6        LibMath,        {_}
7        LibGraph,       {p2R121}
8        LibDXF,         {InitDXFfile}
9        LibInOut,       {IsKeyPressed}
10       LibMecIn,       {gCrank}
11       LibCams,        {EnveloFlines,EndEnvelopes}
12       LibMec2D;       {OpenMecGraph,Slider,Offset,}
13                       {Shape,NewFrame,CloseMecDXF,CloseMecGraph}
14  const Color = 0;     {0 follower face only, 1 entire mechanism }
15        Anim  = 0;     {0 = accumulate frames, 1 = animate}
16        nPL   = 1000;  {polar lines in EnveloFlines - maximum 1000 }
17        CamXY ='Cam18'; {cam profile name - implicit file ext. D2D }
18        DTheta= 2*Pi;  {cam rotational amplitude - cannot be 0 }
19        Phi0  = 35*RAD; {follower byas in degrees - cannot be 0 }
20        DPhi  = 15*RAD; {follower amplitude - cannot be 0 }
21        OO1   = 3.5;   {cam-follower center distance - positive only }
22        O1P   = 0.4;   {follower offset - positive or negative }
23        xPC   = 0.0;   {x coord. of polar center in EnveloFlines }
24        yPC   = 0.0;   {y coord. of polar center in EnveloFlines }
25  var FT:text;  Ch:char;      rb,rt, PQ, dC,dF, Theta,Phi,xO1,yO1
26      ,xA,yA, xP,yP, xQ,yQ, xPP,yPP: double;
27  BEGIN
28    if (Anim = 1) OR (Color = 1) then BEGIN
29      Assign(FT,'dFvdC_L.XY'); {..follower motion file - reduced size}
30      InitDXFfile('F7_18A.DXF');
31    END
```

```
32   else BEGIN
33     Assign(FT,'dFvdC.XY');  {..follower motion file - full size}
34     InitDXFfile('F7_18B.DXF');
35   END;
36   Reset(FT);
37   rb:=O1P+O01*sin(Phi0);          {..base circle radius}
38   rt:=O1P+O01*sin(Phi0+DPhi);       {..top circle radius}
39   Write('rb=',rb:9:5,'  rt=',rt:9:5,'  <CR>..'); ReadLn;
40   PQ:=1.5*O01;  {..follower face length}
41   OpenMecGraph(-1.1*O01,1.1*O01,-1.5*O01,1.5*O01);
42   repeat
43     if (Anim > 0) then NewFrame(50);
44     ReadLn(FT,dC,dF);
45     Theta:=dC*DTheta;         {..cam rotation angle}
46     Phi:=Phi0+dF*DPhi;         {..follower displacement}
47     gCrank(Anim*Color,0,0,Theta,_,_,O01,xO1,yO1,_,_,_,_);
48     Crank(Anim*Color,0,0,_,_,_,_,xO1,yO1,_,_,_
49       ,0.5*Pi-Phi,_,_,p2R121(10.5*JtSz),xPP,yPP,_,_,_,_);
50     Offset(Color,' ',xO1,yO1,_,_,_,xPP,yPP,_,_,_,O1P,0,xP,yP,_,_,_,_);
51     Offset(Red,'I',xP,yP,_,_,_,xO1,yO1,_,_,_
52       ,0,-Sgn(O1P)*PQ,xQ,yQ,_,_,_,_);
53     if (Color = 0) then
54       EnvelofLines(-Anim*White,nPL,xPC,yPC, O01, xP,yP,xQ,yQ)
55     else BEGIN
56       EnvelofLines(0,nPL,xPC,yPC, O01, xP,yP,xQ,yQ)
57       PutGPoint(Anim*White,' ',0,0,'o ');
58       LabelJoint(Anim*White,xP,yP,xO1,yO1,' O1');
59       LabelJoint(Anim*White,xQ,yQ,xP,yP,' P');
60       LabelJoint(Anim*White,xP,yP,xQ,yQ,' Q');
61       PutAng(Anim*White,0,0,0,xO1,yO1, 5, '<'+#233+'|');   {..#233=Theta}
62       PutAng(Anim*White,2*xO1,2*yO1,xO1,yO1,xPP,yPP, 10
```

```
63              ,'<'+#227+'/2-'+#237+'|');  {..#227=Pi and #237=Phi }
64          END;
65      until Eof(FT);
66      PutGPoint(Green,'+', 0,0,'');   {..mark cam center}
67      if (Color = 0) AND (Anim = 0) then
68          EndEnvelopes(CamXY,Green)          {..write cam to file CamXY.D2D}
69      else
70          EndEnvelopes(MySt(LastNrLayer,3),Green); {..write cam to last layer}
71      CloseMecDXF;
72      WaitToGo(Ch);
73      CloseMecGraph(Anim = 0);
74  END.
```

```
1   program P7_19;
2   {===========================================================================}
3   {   Kinematic analysis of disc cams with flat-face oscillating follower      }
4   {===========================================================================}
5   uses Graph,        {Magenta,Blue,Cyan,White}
6        LibMath,      {_,DEG}
7        LibDXF,       {InitDXFfile}
8        LibInOut,     {IsKeyPressed}
9        LibGraph,     {p_X,p_Y}
10       LibGe2D,      {U2dirs2D90}
11       LibMecIn,     {gCrank}
12       LibCams,      {GetProfileRadii,RotCamOscilFlat}
13       LibMec2D;     {OpenMecGraph,Slider,Offset,L_line}
14                     {Locus,NewFrame,CloseMecDXF,CloseMecGraph}
15   const nPoz = 360;   {number of cam positions}
16         Anim = 0;     {0 = animation OFF, 1 = animation ON}
17         CamXY ='Cam18'; {D2D input file with cam profile points}
18         DTheta= 2*Pi;   {cam rotational amplitude - cannot be 0}
```

```
19      OO1 = 3.5;      {cam-follower center distance - positive only}
20      O1P = 0.4;      {follower offset - positive or negative}
21  var FT:text;  Ch:char;   i,Skip:Integer;  rb,rt, PQ,O1C,xA,yA
22      ,Theta,Phi,Rho,Gamma, xC,yC, xP,yP, xQ,yQ:double;
23  BEGIN
24  if (Anim <> 0) then InitDXFfile('F7_19.DXF');  {..output M3D-DXF file}
25  Assign(FT,'F7_19.TXT'); Rewrite(FT);  {..output ASCII data file}
26  GetProfileRadii(CamXY+'.D2D',rb,rt);  {..extract base & top circle radii}
27  Write('rb=',rb:9:5,'  rt=',rt:9:5,' <CR>..');  ReadLn;
28  If (rt > OO1) OR (Abs(O1P) > rt) then BEGIN
29    Write('Inproper OO1, O1P or PQ values! <CR>..');  ReadLn;
30    Halt;
31    END;
32  WriteLn(FT, CamXY+'.D2D cam with knife-edge translating follower');
33  WriteLn(FT,'DTheta=',DTheta*DEG:6:2,'  OO1=',OO1:6:2,'  O1P=',O1P:6:2);
34  WriteLn(FT,'  Theta  Phi      Gamma  Rho
35  ,'xC   yC    rb    rt');
36  PQ:=1.5*OO1; {..follower face length}
37  OpenMecGraph(-1.1*rt,1.1*OO1,-1.1*rt,1.5*rt);
38  if (Anim > 0) then Skip:=1 else Skip:=nPoz DIV 10;
39  i:=0;
40  repeat
41    if (i > nPoz) then BEGIN
42      i:=0;   CloseMecDXF;
43      END;
44    if (Anim = 1) OR (i MOD Skip = 0) then NewFrame(0);
45    Theta:=-DTheta*i/nPoz;
46    gCrank(Anim*Magenta,0,0,Theta,_,_,xA,yA);
47    Shape('',Anim*Magenta,0,0,xA,yA,_,'_','_',_);  {..draw cam base circle}
48    Shape(CamXY+'.D2D',Anim*Red,0,0,xA,yA);  {..draw cam from file}
49    RotCamOscilFlat(CamXY+'.D2D',OO1,O1P,Theta, Phi,xC,yC,Rho);
```

```
50      O1C:=Dist2Pts2D(xC,yC,OO1,0);
51      xP:=OO1 + O1P*cos(0.5*Pi-Phi);
52      yP:= O1P*sin(0.5*Pi-Phi);
53      xQ:=xP + PQ*cos(Pi-Phi);
54      yQ:=yP + PQ*sin(Pi-Phi);
55      if (Anim <> 0) then  LLine(X_p(OO1),Y_p(0),x_p(xP),Y_p(yP)
56          ,x_p(xQ),Y_p(yQ),JtSz,JtSz);   {..draw follower}
57      PutVector(Anim*Blue,'-',xC,yC,-(yP-yQ)/PQ,(xP-xQ)/PQ, 1.4,'Fc');
58      PutVector(Anim*Cyan,'-',xC,yC, yC/O1C,(OO1-xC)/O1C, 1.2,'Vc');
59      PutPoint(Anim*White,'o', xC,yC,'C ');   {..contact point C}
60      PutGPoint(Anim*White,'/', OO1,0,'');   {..ground pin join O1}
61      Gamma:=U2dirs2D90(-(yP-yQ)/PQ,(xP-xQ)/PQ,yC/O1C,(OO1-xC)/O1C)*DEG;
62      if (i MOD Skip = 0) then PutText(White,_,_,'θ  ='+MySt2(Theta*DEG,7,2)
63      +'°'+'n\γ  ='+MySt2(Gamma,7,2)+'°'n\Rho='+MyStr(Rho,6));
64      RT2D(xC,yC,-Theta,0,xC,yC);   {..rotate backwards point C}
65      if MecOut then  WriteLn(FT,Theta:8:4,' ',Phi*DEG:10:6,' ',Gamma:8:4
66          ,' ',Rho:10:6,' ',xC:10:6,' ',yC:10:6,' ',rb:9:6,' ',rt:9:6);
67      Inc(i);
68      until (NOT MecOut) AND (IsKeyPressed(27) OR (Anim = 0));
69      CloseMecGraph(FALSE);        {..erase all .$2D files}
70      Close(FT);
71  END.


1   program P7_23;
2   {===============================================================================
3   Synthesis of disc cams with arc-faced translating follower
4   ===============================================================================}
5   uses Graph,    {Red,White}
6        LibMath,    {_,RAD}
7        LibGraph,  {p_X,p_Y}
8        LibDXF,    {InitDXFfile}
```

```
 9     LibInOut,   {IsKeyPressed}
10     LibMecIn,   {gCrank}
11     LibCams,    {EnvelofCircles, EndEnvelopes}
12     LibMec2D;   {OpenMecGraph, Slider, Offset,}
13                 {PutAng, PutDist, Locus, NewFrame, CloseMecDXF, CloseMecGraph}
14  const Color = 1;          {0 follower face only, 1 more details}
15      Anim  = 1;            {0 = accumulate frames, 1 = animate}
16      CamXY ='Cam23';       {D2D file name for cam profile}
17      DTheta= 2*Pi;         {cam rotational amplitude - cannot be 0}
18      Gamma =-10*RAD;       {follower face angle - positive or negative}
19      Ds    = 1.0;          {follower amplitude - cannot be 0}
20      s0    = 1.5;          {follower bias - cannot be 0}
21      nPL = 1000;           {polar lines in EnvelofLines - maximum 1000}
22      xPC = 0.0;            {x coord. of polar center in EnvelofLines}
23      yPC = 0.0;            {y coord. of polar center in EnvelofLines}
24      PLL = 1.5*s0+Ds;      {initial polar line length in EnvelofLines}
25      x_1=-2.00; y_1= 0.40; {1st point on follower arc rel. to O1P}
26      x_2= 0.00; y_2= 0.00; {2st point on follower arc rel. to O1P}
27      x_3= 2.00; y_3= 0.40; {3st point on follower arc rel. to O1P}
28  var FT:text;  Ch:char;     dF,dC,Theta, s,xA,yA,xB,yB, DP,
29      x1,y1,x2,y2,x3,y3, xD,yD, xP,yP, xQ,yQ, sG,cG:double;
30  BEGIN
31     InitDXFfile('F7_23.DXF'); {..output M3D-DXF file}
32     if (Anim = 1) OR (Color = 1) then
33       Assign(FT,'dFvdC_L.XY')  {..follower motion file - reduced size}
34     else
35       Assign(FT,'dFvdC.XY'); {..follower motion file - full size}
36     Reset(FT);
37     sG:=sin(Gamma);  cG:=cos(Gamma);
38     DP:=s0;
39     OpenMecGraph(-1.2*PLL,1.2*PLL, -1.5*PLL,1.2*PLL);
```

```
40   repeat
41     if (Anim > 0) then NewFrame(50);
42     gShape(CamXY+'.D2D',Color*Cyan,0,0);  {..draw cam profile from file}
43     ReadLn(FT,dC,dF);
44     Theta:=dC*DTheta+Pi/2;  {..cam rotation angle}
45     s:=s0+dF*Ds;            {..follower displacement}
46     gCrank(-Anim*Color,0,0,Theta,_,_,s0,xA,yA,_,_,_,_);
47     Slider(Anim*Color,0,0,0,0,0, xA,yA,_,_,_, s0, s,_,_
48     ,xD,yD,_,_,_, xB,yB,_,_,_);
49     Offset(0,' ',xD,yD,_,_,_,0,0,_,_,_,_,-DP*sG, DP*cG,xP,yP,_,_,_,_);
50     Offset(0,' ',xD,yD,_,_,_, 0, 0,_,_,_,DP*sG,-DP*cG,xQ,yQ,_,_,_,_);
51     Offset(0,' ',xD,yD,_,_,_,xP,yP,_,_,_,x_1,y_1, x1,y1,_,_,_,_);
52     Offset(0,' ',xD,yD,_,_,_,xP,yP,_,_,_,x_2,y_2, x2,y2,_,_,_,_);
53     Offset(0,' ',xD,yD,_,_,_,xP,yP,_,_,_,x_3,y_3,x3,y3,_,_,_,_);
54     if (Color = 0) then BEGIN
55       EnvelofCircles(-Anim*White,nPL,xPC,yPC, PLL, x1,y1,x2,y2,x3,y3);
56       PutGPoint(Green,'+', 0,0,'');  {..put cam center}
57     END
58     else BEGIN
59       EnvelofCircles(0,nPL,xPC,yPC, PLL, x1,y1,x2,y2,x3,y3);
60       PutGPoint(Anim*White,' ',0,0,'O ');
61       PutPoint(Anim*White,' ',xA,yA,' A');
62       LabelJoint(Anim*White,xD,yD,xB,yB,' B');
63       LabelJoint(0*White,xQ,yQ,xP,yP,' P');
64       LabelJoint(0*White,xP,yP,xQ,yQ,' Q');
65       LabelJoint(Anim*White,0,0,x1,y1,' 1');
66       PutPoint(Anim*Yellow,'.',x2,y2,'');
67       LabelJoint(Anim*White,0,0,x2,y2,' 2');
68       LabelJoint(Anim*White,0,0,x3,y3,' 3');
69       PutAng(Anim*White,0,0,0,xA,yA,6, '<'+#233+'|');  {..#233=Theta}
70       PutDist(Anim*White,0,0,xD,yD,-10, |s>|');
```

```
71      END;
72      SetColor(Red);  SyncDXFColor;
73      x1:=X_p(x1);  y1:=Y_p(y1);
74      x2:=X_p(x2);  y2:=Y_p(y2);
75      x3:=X_p(x3);  y3:=Y_p(y3);
76      PDarc3Pts('',x1,y1,x2,y2,x3,y3);  {..draw arc}
77    until Eof(FT);
78    if (Color = 0) AND (Anim = 0) then
79      EndEnvelopes(CamXY,Cyan)  {..write cam to file CamXY.D2D}
80    else
81      EndEnvelopes(MySt(LastNrLayer,3),Cyan);  {..write cam to last layer}
82    CloseMecDXF;
83    WaitToGo(Ch);
84    CloseMecGraph(Anim = 0);  {..retain cam profile for Anim = 0}
85 END.


1  program P7_25;
2  {=============================================================
3   Systhesis of disk cams with curvilinear-faced translating follower
4  =============================================================}
5  uses Graph,       {Black,DarkGray,Red,White}
6       LibMath,     {_,RAD}
7       LibDXF,      {InitDXFfile}
8       LibInOut,    {IsKeyPressed}
9       LibMecIn,    {gCrank}
10      LibCams,     {EnvelOfPlynes,EndEnvelopes}
11      LibMec2D;    {OpenMecGraph,Slider,Offset,}
12                   {PutDist,Shape,NewFrame,CloseMecDXF,CloseMecGraph}
13  const Color = 1;    {0 cam envelope only, 1 follower animation }
14        Anim  = 1;    {0 = accumulate frames, 1 = animate      }
```

```
15      PlsMns=-1;                       {follower orientation -1:concave, +1:convex }
16      FFace = 'FFace.XY';              {follower face (x,y) input data file }
17      CamXY = 'Cam25';                 {cam profile output file w/ ext. D2D }
18      DTheta= 2*Pi;
19      Gamma =-20*RAD;                  {follower face angle }
20      s0   = 2.0;                      {follower bias - cannot be 0 }
21      Ds   = 1.0;                      {follower amplitude - cannot be 0 }
22      OP   = 0.5;                      {follower offset - positive, 0 or negative }
23      nPL  = 1000;                     {number of polar lines in EnvelOfLines }
24      xPC  = 0.00;                     {x coord. of polar center in EnvelOfCircles }
25      yPC  = 0.00;                     {y coord. of polar center in EnvelOfCircles }
26      PLL  = 1.5*(s0+Ds);              {initial polar line length in EnvelOfLines }
27 var FT:text;  Ch:char;           dF,dC,Theta, s,xA,yA,xB,yB
28      ,xD,yD, xP,yP, xQ,yQ, DP, sG,cG:double;
29 BEGIN
30      InitDXFfile('F7_25.DXF');
31      if (Anim = 1) OR (Color = 1) then
32          Assign(FT,'dFvdC_L.XY') {..follower motion input file - reduced}
33      Else
34          Assign(FT,'dFvdC.XY');  {..follower motion input file - full size}
35      Reset(FT);
36      DP:=s0;          {..half of follower face length}
37      sG:=sin(Gamma);  cG:=cos(Gamma);
38      s:=s0+Ds; OpenMecGraph(-1.8*s,1.8*s,-1.8*s,1.8*s);
39      Repeat
40          if (Anim > 0) then NewFrame(0);
41          Shape(CamXY+'.D2D',Color*Cyan,0,0,1,0);  {..draw cam from file}
42          ReadLn(FT,dC,dF);
43          Theta:=dC*DTheta+Pi/2;  {..cam rotation angle}
```

```
44    s:=s0+dF*Ds;              {..follower displacement}
45    gCrank(-Anim*Color,0,0,Theta,_,_,s0,xA,yA,_,_,_);
46    Slider(Anim*Color,0,0,0,0,0, xA,yA,_,_,_, s0, s,_,_
47    ,xD,yD,_,_,_, xB,yB,_,_,_);
48    Offset(0,'T',xD,yD,_,_,_,0,0,_,_,_,_, -DP*sG, DP*cG,xP,yP,_,_,_,_);
49    Offset(0,'T',xD,yD,_,_,_,0,0,_,_,_,_,  DP*sG,-DP*cG,xQ,yQ,_,_,_,_);
50    Offset(0,'/',xP,yP,_,_,_,xQ,yQ,_,_,_,_,  2*DP,0,_,_,_,_,_,_);
51    if (PlsMns > 0) then BEGIN
52      xP:=xQ; yP:=yQ;
53    END;
54    Shape(FFace,Red,xD,yD,xP,yP);
55    if (Color = 0) then BEGIN
56      EnvelOfPlynes(-Anim*White, nPL,xPC,yPC, PLL, xD,yD,xP,yP, FFace);
57      PutGPoint(Green,'+', 0,0,'');   {..mark cam center}
58    END
59    else BEGIN
60      EnvelOfPlynes(0, nPL,xPC,yPC, PLL, xD,yD,xP,yP, FFace);
61      PutPoint(Anim*White,' ', 0,0,'O ');
62      LabelJoint(Anim*White,0,0,xA,yA,'     A');
63      PutPoint(Anim*White,' ', xB,yB,'B');
64      LabelJoint(Anim*White,0,0,xD,yD,'     D');
65      PutAng(Anim*White,0,0,0,0,xA,yA,6, '<'+#233+'|'); {..#233=Theta}
66      PutDist(Anim*White,0,0,xD,yD,-10, '|s>|');
67    END;
68    until Eof(FT);
69    if (Color = 0) AND (Anim =0) then
70      EndEnvelopes(CamXY,Cyan) {..write cam to file CamXY.D2D}
71    Else
72      EndEnvelopes(MySt(LastNrLayer,3),Cyan); {..write cam to last layer}
```

```
73    CloseMecDXF;
74    WaitToGo(Ch);
75    CloseMecGraph(Anim = 0);   {..retain cam profile file}
76  END.
```

***

```
1   Program F8_01;
2   {==============================================================
3   Generate file F8_01.DXF to plot a circle of radius rb and its involute
4   ==============================================================}
5   uses CRT, LibDXF;
6   const rb   = 30;  {base circle radius}
7         tmax = Pi;  {t upper bound}
8         nt   = 91;  {number of plot points on the involute}
9   var t,x,y: double; it: integer;
10  BEGIN
11    InitDXFfile('F8_01.DXF');
12    DXFcircle('Base_Circle',0,0,0,rb);
13    ExpectDXFplines;
14    for it:=1 to nt do BEGIN {generate the involute curve ..}
15      t:=tmax/(nt-1)*(it-1);
16      x:=rb*(cos(t) + t*sin(t));
17      y:=rb*(sin(t) - t*cos(t));
18      AddVertexPline('Involute',x,y);
19    END;
20    DXFplineEnd('Involute');
21    CloseDXFfile;
22    Write('DXF file output successfully. <CR>..'); ReadLn;
23  END.
```

***

```pascal
1    program P9_01;
2    {====================================================
3    Integrates the Duffing equation and generates files F9_01.TXT & F9_02.D2D
4    ====================================================
5    uses CRT, LibMath;
6    const Poincare = FALSE; {TRUE/FALSE generates or not Poincare map}
7          nPoincare = 1000000;
8          nPhasePath = 20;
9          npintMax = 100;
10         Omega=1.00; Alpha=-1.00;   Beta=1.25;   Gama=0.30;   Delta=0.15;
11   type VecNvar = array[1..2] of double;
12   var  FD: file of double;   FT: text;   x, dxdt: VecNvar;
13        h,tleap,t: double;   np,npMax,npint: Longint;   i: Byte;
14   procedure Write2File(ManyPts: Boolean); {Write t,x & dx/dt to file}
15   BEGIN
16     if ManyPts then Write(FD,x[1],x[2])
17     else BEGIN
18       Write(FT,t:12:8,' ',x[1]:12:8,' ',x[2]:12:8); WriteLn(FT);
19     END;
20   END;
21   procedure Derivs(t: double; x: VecNvar; var dxdt: VecNvar);
22   {Evaluates derivatives dx/dt and d2x/dt2 }
23   BEGIN
24     dxdt[1]:=x[2];
25     dxdt[2]:=-Delta*x[2]-Alpha*x[1]-Beta*Pow(x[1],3)+Gama*cos(Omega*t);
26   END;
27   procedure RK4(var x,dxdt:VecNvar; n:integer; t,h:double);
28   {4th order Runge-Kutta with constant step}
29   var  dxm,dxt,xt: VecNvar;   th,hh,h6: double;  i: integer;
```

```
30    BEGIN
31      hh:=h*0.5;  h6:=h/6.0;   th:=t+hh;
32      for i:=1 to n do xt[i]:=x[i]+hh*dxdt[i];
33      Derivs(th,xt,dxt);
34      for i:=1 to n do xt[i]:=x[i]+hh*dxt[i];
35      Derivs(th,xt,dxm);
36      for i:=1 to n do BEGIN
37        xt[i]:=x[i]+h*dxm[i];    dxm[i]:=dxt[i]+dxm[i]
38      END;
39      Derivs(t+h,xt,dxt);
40      for i:=1 to n do  x[i]:=x[i]+h6*(dxdt[i]+dxt[i]+2*dxm[i]);
41    END;
42    BEGIN
43      ClrScr;
44      x[1]:=0.0;         {initial condition x(0)     }
45      x[2]:=0.000001;    {initial condition dx(0)/dt }
46      if NOT Poincare then BEGIN {phase path points ..}
47        npMax:=nPhasePath;
48        Assign(FT,'F9_01.TXT');   Rewrite(FT);
49        WriteLn(FT,'      t            x1              x2');
50      END
51      else BEGIN {Poincare map points ..}
52        npMax:=nPoincare;
53        Assign(FD,'F9_02.D2D');   Rewrite(FD);
54      END;
55      dxdt[1]:=0.0;  dxdt[2]:=0.0;
56      t:=0;
57      for np:=0 to npMax do BEGIN
58        tleap:=np*(2*Pi/Omega);
59        if Poincare then Write2File(TRUE);
60        for npint:=0 to npintMax do BEGIN
```

```pascal
61            t:=tleap + npint*(2*Pi/Omega)/npintMax;
62            if (NOT Poincare) then Write2File(FALSE);
63            h:=(2*Pi/Omega)/npintMax;
64            RK4(x,dxdt,2,t,h);
65          END;
66          if (np MOD 1000 = 0) then BEGIN {display progress status ..}
67            GoToXY(1,WhereY);   ClrEol;
68            Write('Wait! ',np,'/',npMax);
69          END;
70        END;
71        if Poincare then Close(FD) else Close(FT);
72        GoToXY(1,WhereY);
73        Write('       Integration done! '); ReadLn;
74      END.


1  program P9_03;
2  {=====================================================================
3   Integrates ODE of motion of a spring-mass system and generates data files
4   F9_03LONG.DTA & F9_03SHRT.DTA and animation file F9_03.DXF.
5   =====================================================================}
6  uses Graph,          {Cyan,Red,White}
7       LibMath,         {Max2}
8       LibInOut,        {MySt,IsKeyPressed,CenterMsgT}
9       LibDXF,          {InitDXFfile}
10      LibMec2D;        {PutRefSystem,PutGPoint,PutPoint,Shape,Spring,}
11                       {MecOut,NewFrame,CloseMecDXF,CloseMecGraph}
12  const FleNme  = 'F9_03';  {generic file name         }
13        nPoz    = 500;      {number of simulation positions  }
14        Skip    = 4;        {animate every Skip positions    }
```

```
15     h        = 0.0001;    {[sec] simulation step in Euler subroutine }
16     c        = 0.5;       {damping coefs. 0, 0.5, 6.32456 and 10 Ns/m }
17     L0       = 1.0;       {spring free length in m }
18     k        = 10.0;      {spring constant in N/m }
19     m        = 1.0;       {bob mass in kg }
20     g        = 9.81;      {acceleration due to gravity in m/(s*s) }
21     Rm       = 0.15;      {bob radius in m }
22     Ds       = 0.18;      {spring diameter in m }
23     y0       = 0.5;       {y(t) for t = 0 }
24     dydt0    = 0.0;       {dy/dt(t) for t = 0 }
25     tend     = 15.0;      {final time in sec. }
26   type VecNvar = array[1..2] of double;
27   var FT1,FT2:text; Title,TableHead:string; i1,i2:Word;
28       vy,vdydt,vd2ydt2:VecNvar;   t,Y,dy:double;
29
30   procedure SecondDerivs(t:double; x,dxdt:VecNvar; var d2xdt2:VecNvar);
31   BEGIN
32     Y :=x[1];
33     dY:=dxdt[1];
34     d2xdt2[1]:=(m*g-(y-L0)*k-c*dY)/m;    {..d2v/dt2}
35   END; {..SecondDerivs}
36
37   procedure Euler(var x,dxdt,d2xdt2:VecNvar; n:Byte; t,h:double);
38   {Euler-Taylor integration of the ODE of motion of n variables..}
39   var j: Word;
40   BEGIN
41     SecondDerivs(t, x, dxdt, d2xdt2);
42     for j:=1 to n do BEGIN
43       x[j]:=x[j]+h*(dxdt[j]+0.5*h*d2xdt2[j]);  {..2nd order Taylor}
44       dxdt[j]:=dxdt[j]+h*d2xdt2[j];            {..1st order Taylor}
```

```pascal
45      END;
46    END; {..Euler}
47
48    procedure Write2File(var FT:text);
49    BEGIN
50      Write (FT,t:11:8,' ',vy[1]:12:8);
51      WriteLn(FT,' ',vdydt[1]:12:8,' ',vd2ydt2[1]:12:8);
52    END; {..Write2File}
53
54    BEGIN
55      Title:='k='+MySt(k,6)+'N/m, m='+MySt(m,6)+'kg, c='+MySt(c,6)+'Ns/m';
56      TableHead:='        t              y          dy/dt        d2Y/dt2';
57      Assign(FT1,FleNme+'LNG.DTA'); Rewrite(FT1);
58      Assign(FT2,FleNme+'SRT.DTA'); Rewrite(FT2);
59      WriteLn(FT1,Title);   WriteLn(FT1,TableHead);
60      WriteLn(FT2,Title);   WriteLn(FT2,TableHead);
61      InitDXFfile(FleNme+'.DXF');   {..DXF file for M3D animation}
62      CenterMsgT('','Wait!');
63      t:=0;  i1:=0;  i2:=0;
64      OpenMecGraph(-Max2(Rm,Ds),Max2(Rm,Ds),-4,Max2(Rm,Ds));
65      repeat
66        if (t > tend) then BEGIN
67          t:=0; CloseMecDXF;   {..no effect until OpenMecGraph is called}
68        END;
69        if (t = 0) then BEGIN
70          vy[1]   :=y0;     {..initial y }
71          vdydt[1]:=dydt0;  {..initial dy/dt}
72          SecondDerivs(t,vy,vdydt,vd2ydt2);
73        END
74        else Euler(vy,vdydt,vd2ydt2,1,t,h);
```

```pascal
75              if (Round(t/tend*nPoz) >= i1) then BEGIN
76                Inc(i1);
77                if MecOut then Write2File(FT1);
78              END;
79              if (Round(t/tend*nPoz/Skip) >= i2) then BEGIN
80                Inc(i2);
81                if MecOut then Write2File(FT2);
82                NewFrame(0);
83                PutRefSystem(2,-4,'x','y');    {..reference frame w/ y-axis reversed}
84                PutGPoint(White,'v', 0,0, '');    {..reversed grounded pin joint}
85                Spring(Cyan,0.0,0.0, 0.0,-y, Ds, 6);    {..draw spring}
86                Shape('',Red,0,-y,Rm,-y);    {..draw pendulum bob}
87                PutPoint(White,'O',0,-y,'');    {..mark the center of the bob}
88              END;
89              t:=t+h;
90          until (NOT MecOut) AND IsKeyPressed(27);
91          Close(FT1);  Close(FT2);
92          CloseMecGraph(FALSE);
93     END
```

```pascal
1      program P9_05;
2      {===============================================================
3      Generates the time response curve y(t) of a damped spring-mass system
4      ===============================================================}
5      uses   CRT, LibMath;
6      const  FileName = 'F9_05.D2D';    {output file name}
7             nPts = 200;    {number of data points}
8             tmax = 15.0;    {final time}
9             k    = 10.0;    {spring constant in N/m.}
```

```
10    m     = 1.0;    {mass in kg                    }
11    c     = 0.5;    {damping coefficient in Ns/m   }
12    yInf  = 1.0;    {displacement at equilibrium   }
13    y0    = -0.5;   {initial displacement y(0)     }
14    dydt0 = 0.0;    {initial velocity dy(0)/dt     }
15  var   FD: file of double;  FT: text;
16        t,Y, Zeta,A0,OmegaN,OmegaD,Phi: double;   i,j: Integer;
17  BEGIN
18    Assign(FD,FileName);       Rewrite(FD);
19    Assign(FT,'P9_05.REZ');    Rewrite(FT);
20    OmegaN:=Sqrt(k/m);
21    Zeta:=c/(2*m*OmegaN);
22    OmegaD:=OmegaN*Sqrt(1.0-Sqr(Zeta));
23    A:=Sqrt(Sqr(y0)+Sqr((dydt0+Zeta*OmegaN*y0)/OmegaD));
24    Phi:=Atan2(y0*OmegaD,dydt0+Zeta*OmegaN*y0);
25    WriteLn(FT,'OmegaN =',OmegaN,' rad/s');
26    WriteLn(FT,'OmegaD =',OmegaD,' rad/s');
27    WriteLn(FT,'TauD   =',2*Pi/OmegaD,' s');
28    WriteLn(FT,'A0     =',A0,' m');
29    WriteLn(FT,'Zeta   =',Zeta);
30    WriteLn(FT,'Phi    =',Phi,' rad');
31    for i:=1 to nPts do BEGIN
32      t:=(i-1)*tmax/(nPts-1);
33      y:=yInf + A0*Exp(-Zeta*OmegaN*t)*sin(OmegaD*t + Phi);
34      Write(FD,t,y);
35    END;
36    Close(FD);  Close(FT);
37  END.
```

```pascal
1   program F9_06;
2   {========================================================
3   Finds equilibrium position and fits an exponential through a damped
4   oscillatory response curve y(t)
5   ========================================================}
6   uses   DOS,CRT,
7          LibMath,
8          LibInOut,
9          LibMinEA,
10         LibMinN;
11  const FileName0 = 'F9_05';  {name of .MIN and .MAX input files}
12        FileName1 = 'F9_06';  {name of output .REZ file          }
13        LimAf = 65000;  {maximum calls of the objective function }
14        LimIt = 5000;   {maximum number of iterations            }
15  var vtmin, vYmin, vtmax, vYmax: VDm;   YY, YYbest, YYmin, YYmax: VDn;
16      FT: text;    A0,ZetaOmegaN,yInf, t, vF,vFbest: double;
17      i,j, Nvar, nMin,nMax: Byte; TotalFev: LongInt; s,AuxStr: string;
18  function ExpFunc(t:double): double;
19  BEGIN
20     ExpFunc:=A0*exp(-ZetaOmegaN*t);
21  END;
22  {$F+}
23  function Fobj1(YY: VDn): double;
24  var j:Byte;   MaxError,Error: double;
25  BEGIN
26     Fobj1:=InfD;
27     for j:=1 to Nvar do
28     if (YY[j] < YYmin[j]) OR (YY[j] > YYmax[j]) then EXIT;
29     ZetaOmegaN:=YY[1];  A0:=YY[2]; yInf:=YY[3];
```

```
30      MaxError:=-InfD;
31      for j:=1 to nMin do BEGIN
32          Error:=Abs(yInf - ExpFunc(vtmin[j]) - vYmin[j]);
33          if (MaxError < Error) then MaxError:=Error;
34      END;
35      for j:=1 to nMax do BEGIN
36          Error:=Abs(yInf + ExpFunc(vtmax[j]) - vYmax[j]);
37          if (MaxError < Error) then MaxError:=Error;
38      END;
39      Fobj1:=MaxError;
40  END; {..Fobj1}
41  {$F-}
42  BEGIN
43      ClrScr;
44      Assign(FT,FileName0+'.MIN');  Reset(FT);
45      ReadLn(FT,AuxStr);
46      ReadLn(FT,AuxStr);
47      nMin:=0;
48      while NOT EOF(FT) do BEGIN
49          Inc(nMin);
50          ReadLn(FT,AuxStr);
51          vtmin[nMin]:=Extract1stNo(AuxStr);
52          vYmin[nMin]:=Extract1stNo(AuxStr);
53          if (vtmin[nMin] >= InfD) then Dec(nMin);
54      END;
55      Close(FT);
56      Assign(FT,FileName0+'.MAX');  Reset(FT);
57      ReadLn(FT,AuxStr);
58      ReadLn(FT,AuxStr);
59      nMax:=0;
```

```
60   while NOT EOF(FT) do BEGIN
61     Inc(nMax);
62     ReadLn(FT,AuxStr);
63     vtmax[nMax]:=Extract1stNo(AuxStr);
64     vYmax[nMax]:=Extract1stNo(AuxStr);
65     if (vtmax[nMax] >= InfD) then Dec(nMax);
66   END;
67   Close(FT);
68   Nvar:=3;
69   YYmin[1]:=0.0;  YYmax[1]:=2.0;                      {Zeta*OmegaN}
70   YYmin[2]:=0.0;  YYmax[2]:=vYmax[1]-vYmin[1];        {A0}
71   YYmin[3]:=vYmin[1];  YYmax[3]:=vYmax[1];            {yInf}
72   for i:=1 to Nvar do      {first initial guess ..}
73     YYbest[i]:=YYmin[i]+0.5*(YYmax[i]-YYmin[i]);
74   TotalFev:=0;  vFbest:=InfD;
75   WriteOutN:=FALSE;        {do not display search status info}
76   for j:=1 to 100 do BEGIN  {multistart minimization ..}
77     GoToXY(1,WhereY);  ClrEol;
78     Write('Iteration ',j:3,'  F(X)=',vFbest);
79     for i:=1 to Nvar do BEGIN  {initial guess base on prev. YY}
80       repeat
81         YY[i]:=YYbest[i]+(Random-0.5)*(YYmax[i]-YYmin[i]);
82       until (YYmin[i] <= YY[i]) AND (YY[i] <= YYmax[i]);
83     END;
84     NelderMead('', Fobj1,Nvar, LimAF, 1.0E-16, YYmin,YYmax, vF,YY);
85     TotalFev:=TotalFev + NrFevN;
86     if (vF < vFbest) then BEGIN  {retain the best solution so far ..}
87       vF:=Fobj1(YY);  for i:=1 to Nvar do YYbest[i]:=YY[i];
88       vFbest:=vF;
```

```
89      END;

90    END; {.. multistart minimization}

91    WriteLn;

92    WriteLn('vF            =',vFbest:20:16);

93    WriteLn('Zeta*OmegaN=',YY[1]:20:16);

94    WriteLn('A0           =',YY[2]:20:16);

95    WriteLn('Y(Inf)       =',YY[3]:20:16);

96    Assign(FT,FileName1+'.REZ');  Rewrite(FT);

97    WriteLn(FT,'Error       =',vFbest:20:16);

98    WriteLn(FT,'Zeta*OmegaN=',YY[1]:20:16);

99    WriteLn(FT,'A0          =',YY[2]:20:16);

100   WriteLn(FT,'Y(Inf)      =',YY[3]:20:16);

101   WriteLn(FT,'-----------------------------------------');

102   for i:=1 to nMin do WriteLn(FT,vtmin[i],'   ',vYmin[i]);

103   WriteLn(FT,'-----------------------------------------');

104   for i:=1 to nMax do WriteLn(FT,vtmax[i],'   ',vYmax[i]);

105   WriteLn(FT,'-----------------------------------------');

106   for i:=0 to 200 do BEGIN

107     t:=i*Max2(vtmin[nMin],vtmax[nMax])/200;

108     WriteLn(FT,t:20:16,'    ',yInf+ExpFunc(t):20:16);

109   END;

110   WriteLn(FT,'-----------------------------------------');

111   for i:=0 to 200 do BEGIN

112     t:=i*Max2(vtmin[nMin],vtmax[nMax])/200;

113     WriteLn(FT,t:20:16,'    ',yInf-ExpFunc(t):20:16);

114   END;

115   Close(FT);   ReadLn;

116 END.
```

```pascal
1   program P9_07;
2   {===============================================================
3   Finds the coefficients of Stress-Strain function of an elastomeric material
4   ===============================================================}
5   uses   DOS,CRT,
6          LibInOut,    {MyVal,MyStr}
7          LibMath,     {Pmax,Mmax,VDn,VDp}
8          LibMinN;     {NelderMead}
9   const Nvar=5;    LimAF=10000;
10  var   FT: Text;    Si,Ei: VDp;    X, Xmin, Xmax, Xbest: VDn;
11        TotalFev: LongInt;    nPts,i,j: Word;
12        C1,C2,C3,C4,C5,  vF, vFbest: double;    s: string;
13  function Sigm(E:double): double;    {Sigm(Eps) function ...}
14  var Aux: double;
15  BEGIN
16    Aux:=2*C5*Pow(E,5)+(10*C5+3*C3)*Pow(E,4);
17    Aux:=Aux+(14*C5+4*C4+9*C3+C2)*E*E*E;
18    Aux:=Aux+(6*C5+6*C4+6*C3+3*C2+C1)*E*E;
19    Aux:=Aux+(3*C2+2*C1)*E+C2+C1;
20    Sigm:=2*E*(3+3*E+E*E)*Aux/Pow((1+E),4);
21  END;
22  {$F+}
23  function Fobj2(X: VDn): double;
24  var MaxError,Error: double;    i:Word;
25  BEGIN
26    Fobj2:=InfD;
27    for i:=1 to Nvar do if (X[i]<Xmin[i]) OR (X[i]>Xmax[i]) then EXIT;
28    C1:=X[1]; C2:=X[2]; C3:=X[3]; C4:=X[4]; C5:=X[5];
29    MaxError:=-InfD;
30    for i:=1 to nPts do BEGIN
31      Error:=Abs(Si[i] - Sigm(Ei[i]));
```

```
32        if (MaxError < Error) then MaxError:=Error;
33     END;
34     Fobj2:=MaxError;
35  END; {.. Fobj2}
36  {$F-}
37  BEGIN
38    ClrScr;
39    Assign(FT,'F9_07.DTA');    Reset(FT);
40    nPts:=0;
41    repeat
42      Inc(nPts);
43      ReadLn(FT,Ei[nPts],Si[nPts]);
44      if (nPts MOD 3 <> 0) then
45        Write(Ei[nPts]:10:5,' ',Si[nPts]:10:5,'  |  ')
46      else
47        WriteLn(Ei[nPts]:10:5,' ',Si[nPts]:10:5);
48    until Eof(FT) or (nPts = Pmax);
49    Close(FT);
50    WriteLn(^j);
51    Xmin[1]:=-2;  Xmax[1]:=2;  Xmin[2]:=-2;  Xmax[2]:=2;
52    Xmin[3]:=-2;  Xmax[3]:=2;  Xmin[4]:=-2;  Xmax[4]:=2;
53    Xmin[5]:=-2;  Xmax[5]:=2;
54    WriteOutN:=FALSE;  {do not display search status info}
55    for i:=1 to Nvar do {first initial guess ..}
56      Xbest[i]:=Xmin[i]+0.5*(Xmax[i]-Xmin[i]);
57    TotalFev:=0;
58    vFbest:=InfD;
59    for j:=1 to 100 do BEGIN  {multistart minimization ..}
60      GoToXY(1,WhereY);  ClrEol;
```

```
61      Write('Iteration ',j:3,'   F(X) =',vFbest);
62      for i:=1 to Nvar do BEGIN {initial guess base on previous X[..]}
63        repeat
64          X[i]:=Xbest[i]+(Random-0.5)*(Xmax[i]-Xmin[i]);
65        until (Xmin[i] <= X[i]) AND (X[i] <= Xmax[i]);
66      END;
67      NelderMead('', Fobj2,Nvar, LimAF, 1.0E-32, Xmin,Xmax, vF,X);
68      TotalFev:=TotalFev + NrFevN;
69      if (vF < vFbest) then BEGIN
70        for i:=1 to Nvar do BEGIN
71          Str(X[i]:10:3,s); X[i]:=MyVal(s);
72        END;
73        vF:=Fobj2(X);
74        if (vF < vFbest) then BEGIN
75          for i:=1 to Nvar do Xbest[i]:=X[i];
76          vFbest:=vF;
77        END;
78      END;
79    END; {.. multistart minimization}
80    vF:=Fobj2(Xbest); {call Fobj(..) to re-evaluate C1,C2,C3,C4,C5}
81    WriteLn(^j);
82    WriteLn(' Maximum deviation =',vFbest:12:10);
83    for i:=1 to Nvar do WriteLn(' C',i:1,' =',Xbest[i]);
84    Write('Obj. function calls = ',TotalFev,'     <CR>..');
85    Assign(FT,'F9_07.TXT'); Rewrite(FT);
86    WriteLn(FT,'Max Deviation =',vFbest:12:9);
87    for i:=1 to Nvar do WriteLn(FT,'C',i:1,' =',Xbest[i]:7:4);
88    WriteLn(FT,'  eps      Sigma      Sigm(eps)    Error');
89    for i:=1 to nPts do BEGIN
90      vF:=Sigm(Ei[i]);
```

```
91      WriteLn(FT,Ei[i]:9:6,' ',Si[i]:9:6,' ',vF:9:6,' ',vF-Si[i]:9:5);
92    END;
93    Close(FT);   ReadLn;
94  END.


1   Program P9_09;
2   {=================================================================
3   Generate a T3D file to plot as animation the generalized Rosenbrock's
4   function of three variables
5   =================================================================}
6   uses CRT,DOS,
7        LibInOut,
8        LibMath,
9        LibGE2D;
10  const FileName='F9_09.T3D';
11  var FT: Text;   nX,nY,nZ; Xmin,Xmax, Ymin,Ymax, Zmin,Zmax, X,Y,Z,
12      vF, vFmin,vFmax: double;   iX,iY,iZ: integer;
13  function R3(x1,x2,x3: double): double; {Rosenbrok's function with n=3}
14  var T1,T2: double;
15  BEGIN
16    T1:=100*Sqr(x2-Sqr(x1))+Sqr(1.0-x1);
17    T2:=100*Sqr(x3-Sqr(x2))+Sqr(1.0-x2);
18    R3:=T1+T2;
19  END;  {.. R3()}
20  BEGIN
21    nX:=161;   Xmin:=-2.5;   Xmax:=2.5;
22    nY:=161;   Ymin:=-2.5;   Ymax:=2.5;
23    nZ:=11;    Zmin:=-2.5;   Zmax:=2.5;
24    ClrScr;
25    Assign(FT,FileName);   Rewrite(FT);
```

```
26      for iZ:=1 to round(nZ) do BEGIN
27         Z:=Zmin + (iZ-1)*(Zmax-Zmin)/(nZ-1);
28         Write(FT,'    x3=',MyStr(Z,4):4);
29      END;
30      WriteLn(FT);
31      for iZ:=1 to round(nZ) do Write(FT,MySt(   nX,9):9,'  '); WriteLn(FT);
32      for iZ:=1 to round(nZ) do Write(FT,MySt(   nY,9):9,'  '); WriteLn(FT);
33      for iZ:=1 to round(nZ) do Write(FT,MySt(Xmin,9):9,'  '); WriteLn(FT);
34      for iZ:=1 to round(nZ) do Write(FT,MySt(Xmax,9):9,'  '); WriteLn(FT);
35      for iZ:=1 to round(nZ) do Write(FT,MySt(Ymin,9):9,'  '); WriteLn(FT);
36      for iZ:=1 to round(nZ) do Write(FT,MySt(Ymax,9):9,'  '); WriteLn(FT);
37      vFmin:=InfD;  vFmax:=-InfD;
38      for iX:=1 to round(nX) do BEGIN
39         X:=Xmin+(iX-1)*(Xmax-Xmin)/(nX-1);
40         for iY:=1 to round(nY) do BEGIN
41            Y:=Ymin+(iY-1)*(Ymax-Ymin)/(nY-1);
42            for iZ:=1 to round(nZ) do BEGIN
43               Z:=Zmin+(iZ-1)*(Zmax-Zmin)/(nZ-1);
44               vF:=R3(X,Y,Z);
45               if (vFmin > vF) then vFmin:=vF;
46               if (vFmax < vF) then vFmax:=vF;
47               Write(FT,' ',MyStr(vF,9));
48            END;
49            WriteLn(FT);
50         END;
51         WriteLn(iX:3,' of ',nX:3:0);  {progress report ..}
52      END;
53      WriteLn('vFmin=',vFmin:10:6,'    vFmax=',vFmax:10:6);
54      Close(FT);  ReadLn;
55   END.
```

```pascal
1    Program P9_10;
2    {================================================================
3    Generates the partial minima & maxima of the generalized Rosenbrok's
4    function of n variables for visualizations with D_3D
5    ================================================================}
6    uses CRT,DOS,
7         LibInOut,
8         LibMinN,
9         LibMath;
10   const FileName='F9_10.T3D';
11         n  = 5;    {number of variables of the function to be plotted }
12         ix = 1; mx = 161;    {1st scan variable & number and grid size}
13         iy = 2; my = 161;    {2nd scan variable & number and grid size}
14   var XXmin,XXmax, Xmin,Xmax, Xbest: VDn;
15       XX1, XX2, FVbest, PlsMns: double;
16   function k(k:Integer): Integer; {from k=1..n-2 to k=1..n}
17   BEGIN
18     if (k >= ix) then k:=k+1;
19     if (k >= iy) then k:=k+1;
20     k:=k;
21   END;
22   function k_(k:Integer): Integer; {from k=1..n to k=1..n-2}
23   BEGIN
24     if (k = ix) then BEGIN k_:=n+1; Exit; END;
25     if (k = iy) then BEGIN k_:=n+2; Exit; END;
26     if (k > ix) then k:=k-1;
27     if (k+1 > iy) then k:=k-1;
28     k_:=k;
29   END;
30   {$F+}
```

```pascal
31  function Fobj(vX: VDn): double;    {generalized Rosenbrok function ..}
32  var   x: VDn;   Sum: double;  k: Integer;
33  BEGIN
34     Fobj:=InfD;
35     for k:=1 to n-2 do BEGIN
36        if (vX[k] < Xmin[k]) OR (Xmax[k] < vX[k]) then Exit;
37     END;
38     for k:=1 to n do if (_k(k) <= n) then x[_k(k)]:=vX[k];
39     x[ix]:=XX1;  x[iy]:=XX2;
40     Sum:=0.0;
41     for k:=1 to n-1 do
42        Sum:=Sum+100*Sqr(x[k+1]-Sqr(x[k]))+Sqr(1.0-x[k]);
43     Fobj:=PlsMns*Sum;   {..scale and reverse function value}
44  END;  {.. Fobj}
45  {$F-}
46  function MinMax(XX,YY: double): double;  {partial min/max of Fobj ..}
47  var FV: double;
48      vX: VDn;    TotalFev, LimFC: LongInt;    j,k, jIter: Integer;
49  BEGIN {MinMax ..}
50     TotalFev:=0;
51     FVbest:=InfD;
52     if (PlsMns > 0) then BEGIN LimFC:= 50000; jIter:= 25; END;  {minim.}
53     if (PlsMns < 0) then BEGIN LimFC:=100000; jIter:=100; END;  {maxim.}
54     for j:=1 to jIter do BEGIN  {multistart minimization ..}
55        for k:=1 to n-2 do BEGIN   {initial guess based on previous Xbest..}
56           repeat
57              vX[k]:=Xbest[k]+(Random-0.5)*(XXmax[_k(k)]-XXmin[_k(k)]);
58           until (XXmin[_k(k)] <= vX[k]) AND (vX[k] <= XXmax[_k(k)]);
59        END;
```

```
60        NelderMead('', Fobj, n-2, LimFC, 1.0E-9, Xmin,Xmax, FV,vX);
61        TotalFev:=TotalFev + NrFevN;
62        if (FV < FVbest) then BEGIN
63          FVbest:=FV;
64          for k:=1 to n-2 do Xbest[k]:=vX[k];
65        END;
66      END;
67      MinMax:=FVbest/PlsMns; {.. reverse to actual function value}
68    END; {.. MinMax}
69    var FT: Text;        i,j,k: integer;     FV: double;
70    BEGIN {main ..}
71      if (ix >= iy) then Halt;
72      WriteOutN:=TRUE;    {will NOT display search status info ..}
73      Assign(FT,FileName); ReWrite(FT);
74      for k:=1 to n do BEGIN {set boundaries x1..x5 ..}
75        XXmin[k]:=-1.5; XXmax[k]:=1.5;
76      END;
77      for k:=1 to n-2 do BEGIN    {equalize boundaries ..}
78        Xmin[k]:=XXmin[_k(k)]; Xmax[k]:=XXmax[_k(k)];
79      END;
80      Write(FT,'   Fmin  ');
81      for k:=1 to n do Write(FT,'   xmax',k);
82      Write(FT,'   Fmin');
83      for k:=1 to n do Write(FT,'   xmax',k);
84      WriteLn(FT);
85      for k:=1 to 2*n+2 do Write(FT,'    ',mx:4,'     '); WriteLn(FT);
86      for k:=1 to 2*n+2 do Write(FT,'    ',my:4,'     '); WriteLn(FT);
87      for k:=1 to 2*n+2 do Write(FT,XXmin[ix],XXmax[ix]:13:5); WriteLn(FT);
88      for k:=1 to 2*n+2 do Write(FT,XXmin[ix],XXmax[ix]:13:5); WriteLn(FT);
89      for k:=1 to 2*n+2 do Write(FT,XXmin[iy],XXmax[iy]:13:5); WriteLn(FT);
90      for k:=1 to 2*n+2 do Write(FT,XXmin[iy],XXmax[iy]:13:5); WriteLn(FT);
```

```
 91    ClrScr; StartWatch;
 92    for i:=1 to mx do BEGIN
 93    XX1:=XXmin[ix]+(XXmax[ix]-XXmin[ix])/(mx-1)*(i-1);
 94    for k:=1 to n-2 do    {1st guess ..}
 95    Xbest[k]:=XXmin[_k(k)]+0.5*(XXmax[_k(k)]-XXmin[_k(k)]);
 96    for j:=1 to my do BEGIN
 97    XX2:=XXmin[iy]+(XXmax[iy]-XXmin[iy])/(my-1)*(j-1);
 98    PlsMns:=+1.0;
 99    FV:=MinMax(XX1,XX2);
100    Write(FT,' ',FV:14:8);
101    for k:=1 to n do if (k = ix) OR (k = iy) then
102      Write(FT,'   1.00E100') else Write(FT,' ',Xbest[k_(k)]:12:6);
103    PlsMns:=-1.0E-2;
104    FV:=MinMax(XX1,XX2);
105    Write(FT,' ',FV:12:6);
106    for k:=1 to n do if (k = ix) OR (k = iy) then
107      Write(FT,'   1.00E100') else Write(FT,' ',Xbest[k_(k)]:12:6);
108    WriteLn(FT);
109    END;
110    WriteLn(1.0*i*j:9:0,' of ',1.0*mX*mY:9:0);   {progress report ..}
111    END;
112    Close(FT);   WriteLn(StopWatch);   ReadLn;
113 END.
```

```
1 Program P9_13;
2 {==============================================================
3   Writes to file N uniform random values within the interval [Min..Max]
4 ==============================================================}
5 const FileName = 'F9_13A.DAT';   {output file name}
6       Min = -10.0;   {lower limit }
7       Max =  10.0;   {higher limit }
```

```
 8        N = 10000;        {number of random values }
 9    var FT: Text;     x: double; i: LongInt;
10    BEGIN
11        Assign(FT,FileName);    Rewrite(FT);
12        WriteLn(FT,'Uniform random numbers between ',Min:9:4,'..',Max:9:4);
13        Randomize;
14        for i:=1 to N do BEGIN
15            x:=Min+Random*(Max-Min);    WriteLn(FT,x:16:10);
16        END;
17        Close(FT);
18    END.
```

```
 1  Program P9_14;
 2  {==================================================================}
 3  Writes to file N Gaussian random values
 4  {==================================================================}
 5  const FileName = 'F9_14A.DAT';    {output file name}
 6        Mean    = 0.0;   {mean}
 7        StDev   = 1.0;   {standard deviation}
 8        N       = 10000; {number of random values}
 9  var FT: Text; x1,x2,w: double;  x: double; i: LongInt;
10  procedure MyRandomize;
11  BEGIN
12      Randomize;  x1:=10.0; x2:=10.0; {initialize x1 and x2 in GaussRandom}
13  END; {.. MyRandomize()}
14  function GaussRandom(Mean,StDev: double): double;
15  {Generates Gaussian random numbers with Mean and StDev}
16  BEGIN
17      if (x1+x2 = 20.0) then BEGIN
18      repeat
19          x1:=2.0*Random-1.0;  x2:=2.0*Random-1.0;
```

```
20       w:=x1*x1+x2*x2;
21     until (w < 1.0);
22     w:=Sqrt(-2.0*Ln(w)/w);
23   END;
24   if (x1 < 10.0) then BEGIN
25     GaussRandom:=Mean+w*x1*StDev; x1:=10.0;
26   END;
27   if (x2 < 10.0) then BEGIN
28     GaussRandom:=Mean+w*x2*StDev; x2:=10.0;
29   END;
30 END; {.. GaussRandom()}
31 BEGIN
32   Assign(FT,FileName);    Rewrite(FT);
33   WriteLn(FT,'Mean = ',Mean:9:4,', Standard Deviation = ',StDev:9:4);
34   MyRandomize;
35   for i:=1 to N do BEGIN
36     x:=GaussRandom(Mean,StDev);
37     WriteLn(FT,x:16:10);
38   END;
39   Close(FT);
40 END.


1 program P9_16;
2 {==============================================================================
3  Simulation of a Stephenson III dwell mechanism
4  ==============================================================================}
5 uses Graph,      {Blue,Cyan,Green,Red,Magenta}
6      LibMath,    {_,DEG}
7      LibInOut,   {IsKeyPressed}
8      LibDXF,     {InitDXFfile}
9      LibMecIn,   {gCrank}
```

```
10  LibAssur,  {RRR}
11  LibMec2D; {OpenMecGraph,NewFrame,Locus,SetJointSize,}
12            {AngPVA,CloseMecDXF,MecOut}
13  const nPozDTA = 720; {number of positions in the output file}
14        nPozDXF = 120; {number of positions in the animation }
15  var   FT: text;   i,nPoz: Word;   LR1,LR2: shortint;
16        xO,yO, xA,yA, vxA,vyA, xB,yB, vxB,vyB, xC,yC, xD,yD, vxD,vyD,
17        xE,yE, vxE,vyE, xF,yF, x_D,y_D, OA,AB,BC, DE,EF,
18        Th1,dTh1, Th6,dTh6: double;
19  BEGIN
20    LR1:=Left;    {..orientation of the ABC dyad}
21    LR2:=Right;   {..orientation of the DEF dyad}
22    OA:=60;    AB:=180;    BC:=140;
23    xO:=0.0;   yO:=0.0;
24    xC:=200.00; yC:=0.0;
25    x_D:=79.04; y_D:=40.13; {..coordinates of D relative to coupler AB}
26    xF:=134.6;  yF:=88.1;
27    DE:=84.2;   EF:=79.27;
28    Th6:=Pi;   {..initial value in AngPVA}
29    Assign(FT,'F9_16.DTA'); Rewrite(FT);
30    WriteLn(FT,'Theta1  Theta6   dTheta6');
31    InitDXFfile('F9_16.DXF');   {..DXF file for M3D animation}
32    CenterMsgT('','Wait!');
33    nPoz:=nPozDTA;
34    i:=0;
35    Repeat
36    if (i > nPoz) then BEGIN
37      i:=0;  CloseMecDXF; {..no effect until OpenMecGraph is called}
38      xD:=0.1*(XmaxWS-XminWS); {..multiple uses of variable xD}
39      yD:=0.1*(YmaxWS-YminWS); {..multiple uses of variable yD}
40      OpenMecGraph(XminWS-xD, XmaxWS+xD, YminWS-yD, YmaxWS+yD);
```

```
41        nPoz:=nPozDXF;
42      END;
43      NewFrame(0);
44      Th1:=2*Pi*i/nPoz;
45      dTh1:=1;
46      gCrank(Magenta,xO,yO,Th1,dTh1,_,OA,xA,yA,vxA,vyA,_,_);
47      RRR(Blue,xA,yA,vxA,vyA,_,_,xC,yC,0,0,0,AB,BC,LR1
48          ,xB,yB,vxB,vyB,_,_);
49      Offset(Red,'V',xA,yA,vxA,vyA,_,_,xB,yB,vxB,vyB,_,_
50          ,x_D,y_D,xD,yD,vxD,vyD,_,_);
51      RRR(Green,xD,yD,vxD,vyD,_,_,xF,yF,0,0,0,DE,EF,LR2
52          ,xE,yE,vxE,vyE,_,_);
53      Locus(Cyan, xD,yD,'C');
54      PutGPoint(White,'  ',xO,yO,'O ');
55      PutPoint(White,'  ',xB,yB,'  B');
56      PutGPoint(White,'  ',xC,yC,'  C');
57      PutGPoint(White,'  ',xF,yF,'F ');
58      LabelJoint(White,xO,yO,xA,yA,' A');
59      LabelJoint(White,xE,yE,xD,yD,' D');
60      LabelJoint(White,xF,yF,xE,yE,' E');
61      PutAng(White,xO,yO,xO,yO,xA,yA,8,#233+'1');
62      PutAng(White,xF,yF,xF,yF,xE,yE,8,#233+'6');
63      if (nPoz = nPozDTA) then BEGIN
64          AngPVA(xF,yF,0,0,0,0,xE,yE,vxE,vyE,_,_,  Th6,dTh6,_);
65          WriteLn(FT,Th1*DEG:6:3,' ',Th6*DEG:9:5,' ',dTh6:9:5);
66      END;
67      Inc(i);
68    until (NOT MecOut) AND IsKeyPressed(27);
69    CloseMecGraph(FALSE);
70    Close(FT);
71  END.
```

```
1   program P9_18;
2   {======================================================================
3    Quick-return mechanism simulation with ASCII output using the RT_R
4    and RR_T subroutines
5   ======================================================================}
6   uses Graph,        {Cyan,Red,White}
7        LibMath,      {DEG}
8        LibInOut,     {IsKeyPressed}
9        LibDXF,       {InitDXFfile}
10       LibMecIn,     {gCrank}
11       LibAssur,     {RTR,RRT}
12       LibMec2D;     {OpenMecGraph,NewAnimFrame,Locus,
13                      SetJointSize,CloseMecDXF,MecOutp}
14  const nPoz = 120;  {number of positions}
15  var  i:Word;    FT:text;
16       xO,yO,xA,yA,xB,yB,xC,yC,xD,yD, OA,BC,CD, Theta:double;
17  BEGIN
18    Assign(FT,'F9_18.DTA'); Rewrite(FT);
19    InitDXFfile('F9_18.DXF');
20    xO:=0.0;  yO:= 0.0;    {location of ground joint of crank}
21    xB:=0.0;  yB:=-0.25;   {location of ground joint B of the RTR dyad}
22    OA:=0.1;               {length of input crank OA}
23    BC:=0.50;              {sliding rod length of the RTR dyad}
24    CD:=0.30;              {coupler length of the RRT dyad}
25    yD:=0.20;              {elevation of the RRT dyad slider}
26    OpenMecGraph(-0.3,0.6, -0.3,0.35);
27    SetJointSize(4);
28    i:=0;
29    WriteLn(FT,' Theta    xD');
```

```
30  repeat
31      if (i > nPoz) then BEGIN
32          i:=0;    CloseMecDXF;
33      END;
34      NewFrame(500);
35      Theta:=Pi+2*Pi*i/nPoz;
36      gCrank(Red, xO,yO, Theta,_,_ OA, xA,yA,_,_,_,_);
37      RT_R(Cyan, xA,yA,_,_,_,_, xB,yB,0,0,0,0, 0,0,BC,
38      _,_,_,_,_,_,_,_, xC,yC,_,_,_,_);
39      RRT_(Magenta, xC,yC,_,_,_,_, xO,yD,0,0,0,0, xO,yD,0,0,0,0
40      ,CD,0, +1, xD,yD,_,_,_,_,_,_,_,_);
41      PutGPoint(White,' ', xO,yO ,'O ');
42      PutPoint(White,' ', xA,yA ,'A ');
43      PutGPoint(White,' ', xB,yB ,' B');
44      PutPoint(White,' ', xD,yD-0.02,'D');
45      LabelJoint(White,xD,yD,xC,yC,' C');
46      PutAng(White, xO,yO, xO,yO, xA,yA, 6, #233);
47      PutDist(White, 0,yD, xD,yD, 12, 'xD');
48      if MecOut then WriteLn(FT,Theta*DEG:5:2,' ',xD:9:6);
49      Inc(i);
50  until (NOT MecOut) AND IsKeyPressed(27);
51  Close(FT);
52  CloseMecGraph(FALSE);
53  END.

1  program P9_19;
2  {==================================================================
3  Simulation of a radial engine
4  ==================================================================}
5  uses Graph,        {Brown, Red, White}
```

```pascal
 6       LibInOut,  {IsKeyPressed}
 7       LibDXF,    {InitDXFfile}
 8       LibAssur,  {RR_T}
 9       LibMec2D;  {OpenMecGraph,NewFrame,gCrank,SetJointSize,CloseMecDXF}
10   const nCyl = 3;       {number of cylinders}
11         nPoz = 90;      {number of animation frames}
12         DXFFile = 'F9_19.DXF';  {DXF file name}
13   var Phi, xO,yO, OA,AB, xA,yA,xP,yP,xQ,yQ: double;  i,j: Word;
14   BEGIN
15     xO:=0.0;   yO:=0.0;         {ground joint of crank}
16     OA:=0.20;  AB:=0.50;        {crank length and conrod length}
17     InitDXFfile(DXFFile);
18     OpenMecGraph(xO-OA-AB,xO+OA+AB,yO-OA-AB*1.2,yO+OA+AB*1.2);
19     SetJointSize(-4);
20     i:=0;
21     repeat
22       if (i > nPoz) then BEGIN  i:=0; CloseMecDXF;  END;
23       NewFrame(500);
24       Phi:=2*Pi*i/nPoz;
25       gCrank(Red, xO,yO, Phi,_,_, OA, xA,yA,_,_,_);
26       for j:=1 to nCyl do BEGIN
27         xP:=xO+OA*cos(Pi/2+2*Pi/nCyl*j);
28         yP:=yO+OA*sin(Pi/2+2*Pi/nCyl*j);
29         xQ:=xP+1.1*AB*cos(Pi/2+2*Pi/nCyl*j);
30         yQ:=yP+1.1*AB*sin(Pi/2+2*Pi/nCyl*j);
31         RRT_(Brown, xA,yA,_,_,_,_, xP,yP,0,0,0,  xQ,yQ,0,0,0
32             ,AB,0, +1, _,_,_,_,_ _,_,_,_,_ _);
33         if (nCyl <= 3) then BEGIN
34           PutGPoint(White, ' ', xO,yO,'  0');
```

```
35        PutGPoint(White,  ' ',  xP,yP,' P');
36        PutGPoint(White,  ' ',  xQ,yQ,' Q');
37        PutPoint(White,   ' ',  xA,yA,' A');
38      END;
39    END;
40    Inc(i);
41  until (NOT MecOut) AND IsKeyPressed(27);
42  CloseMecGraph(FALSE); {erase all .$xy files and CloseGraph}
43 END.
```

```
1 program P9_20;
2 {=============================================================================
3 Simulation of Gnome rotary engines
4 =============================================================================}
5 uses Graph,     {Brown,Red,White}
6      LibInOut,  {IsKeyPressed}
7      LibDXF,    {InitDXFfile}
8      LibAssur,  {RR_T}
9      LibMec2D;  {OpenMecGraph,NewFrame,gCrank,SetJointSize,CloseMecDXF}
10 const nCyl = 3;          {number of cylinders}
11       nPoz = 90;         {number of animation frames}
12       DXFile = 'F9_20.DXF';     {DXF file name}
13 var Phi, Ec,cr, xO,yO, PQ, xP,yP,xQ,yQ: double;   i,j: Word;
14 BEGIN
15   PQ:=0.80;              {engine radius                        }
16   Ec:=0.2;  cr:=0.50;    {crank offset and length of conrod    }
17   xP:=0.0;  yP:=0.0;     {engine axis of rotation              }
18   xO:=0.0;  yO:=Ec;      {crankshaft ground-joint location     }
19   OpenMecGraph(xP-PQ,xP+PQ, yP-PQ*1.1,yP+PQ*1.1);
```

```
20    InitDXFfile(DXFfile);
21    SetJointSize(-4);
22    i:=0;
23    repeat
24      if (i > nPoz) then BEGIN         i:=0; CloseMecDXF;         END;
25      NewFrame(500);
26      Phi:=Pi/2 + 2*Pi*i/nPoz;
27      for j:=1 to nCyl do BEGIN
28        gCrank(Red, xP,yP, Phi+2*Pi/nCyl*j,_,_, PQ, xQ,yQ,_,_,_,_);
29        RR_T(-Brown, xO,yO,0,0,0, xP,yP,_,_,_,_, xQ,yQ,_,_,_,_
30          ,CR,0, +1,_,_,_,_,_,_,_,_,_,_,_,_);
31        if (nCyl <= 3) then BEGIN
32          PutGPoint(White, ' ', xO,yO, ' O');
33          PutGPoint(White, ' ', xP,yP, ' P');
34          PutPoint(White, ' ', xQ,yQ, ' Q');
35        END;
36      END;
37      Inc(i);
38    until (NOT MecOut) AND IsKeyPressed(27);
39    CloseMecGraph(FALSE); {erase all .$xy files and CloseGraph}
40  END.
```

```
1  program P9_21;
2  {===================================================================
3   Simulates an iris mechanism with any number of vanes (1st variant)
4   ===================================================================}
5  uses Graph,      {Black,Red,White}
6       LibMath,    {_}
7       LibInOut,   {IsKeyPressed}
8       LibDXF,     {InitDXFfile}
```

```
9      LibMecIn,     {gCrank}
10     LibAssur,     {RR_T}
11     LibMec2D;     {OpenMecGraph,NewFrame,Shape,PutPoint,CloseMecDXF}
12 const DXFile = 'F9_21.DXF';   {DXF file name            }
13     nVane = 3;     {number of vanes    }
14     nPoz  = 30;    {number of animation frames}
15 var i,j, Clr: Word;
16     Phi, xO,yO,    xA,yA, xP,yP, xQ,yQ,    OA,OP,PQ: double;
17 BEGIN
18   xO:= 0.0;   yO:= 0.0;
19   OA:= 6.00;  OP:= 5.25;   PQ:= 9.0;
20   InitDXFfile(DXFile);
21   OpenMecGraph(xO-OA,xO+OA, yO-1.3*OA,yO+1.3*OA);
22   i:=0;
23   if (nVane <= 3) then Clr:=Green else Clr:=Black;
24   repeat
25     if (i > nPoz) then BEGIN
26       i:=0;   CloseMecDXF;
27     END;
28     NewFrame(500);
29     Phi:=-Pi/4 + Pi/6*Sin(2*Pi*i/nPoz);
30     for j:=1 to nVane do BEGIN
31       gCrank(Clr, xO,yO, Phi+2*Pi/nVane*j,_,_, OA, xA,yA,_,_,_,_) ;
32       xP:=OA*Cos(Pi+2*Pi/nVane*j);
33       yP:=OA*Sin(Pi+2*Pi/nVane*j);
34       RR_T(-Clr, xP,yP,0,0,0, xO,yO,0,0,0,0, xA,yA,_,_,_,_
35       ,PQ,0, +1, xQ,yQ,_,_,_,_, xQ,yQ,_,_,_,_ _);
36       Shape('VANE.XY',Red,xP,yP, xQ,yQ);    {..draw vane}
37       if (nVane <= 3) then BEGIN
38         PutGPoint(White,' ', xO,yO,'O         ');
39         LabelJoint(White,xO,yO,xA,yA,' A');
```

```
40      LabelJoint(White,xQ,yQ,xP,yP,'    P');
41      LabelJoint(White,xP,yP,xQ,yQ,'    Q');
42        END;
43      END;
44      Inc(i);
45    until (NOT MecOut) AND IsKeyPressed(27);
46    CloseMecGraph(FALSE);
47 END.


1 program P9_22;
2 {=======================================================================
3 Simulates an iris mechanism with any number of vanes (2nd variant)
4 =======================================================================}
5 uses Graph,        {Red,White}
6   LibInOut,    {IsKeyPressed}
7   LibMath,     {_}
8   LibDXF,      {InitDXFfile}
9   LibMecIn,    {gCrank}
10  LibAssur,    {RR_T}
11  LibMec2D;    {OpenMecGraph,NewFrame,Shape,PutPoint,CloseMecDXF}
12 const DXFile = 'F9_22.DXF';  {DXF file name          }
13   nVane = 3;                 {number of vanes        }
14   nPoz  = 30;                {number of animation frames}
15 var i,j, Clr: Word;
16   Phi, xO,yO,    xA,yA, xP,yP, xQ,yQ,    OA,OP,PQ: double;
17 BEGIN
18  OA:= 6.00;
19  OP:= 5.25;
20  PQ:= 9.0;
```

```
21  xO:= 0.0;   yO:= 0.0;
22  InitDXFfile(DXFfile);
23  OpenMecGraph(xO-OA,xO+OA, yO-1.3*OA,yO+1.3*OA);
24  i:=0;
25  if (nVane <= 3) then Clr:=Green else Clr:=Black;
26  repeat
27    if (i > nPoz) then BEGIN
28      i:=0; CloseMecDXF;
29    END;
30    NewFrame(500);
31    Phi:=Pi + Pi/6*Sin(2*Pi*i/nPoz);
32    for j:=1 to nVane do BEGIN
33      gCrank(Clr, xO,yO, Phi+2*Pi/nVane*j,_,_, OP, xP,yP,_,_,_);
34      xA:=OA*Cos(-Pi/4 + 2*Pi/nVane*j);
35      yA:=OA*Sin(-Pi/4 + 2*Pi/nVane*j);
36      RR_T(Clr, xP,yP,_,_,_, xO,yO,0,0,0, xA,yA,_,_,_
37        ,PQ,0, +1, xQ,yQ,_,_,_, xQ,yQ,_,_,_ _);
38      Shape('VANE.XY',Red,xP,yP, xQ,yQ);
39      if (nVane <= 3) then BEGIN
40        PutGPoint(White, ' ', xO,yO,'O     ');
41        LabelJoint(White,xO,yO,xA,yA,' A');
42        LabelJoint(White,xO,yO,xP,yP,' P');
43        LabelJoint(White,xP,yP,xQ,yQ,' Q');
44      END;
45    END;
46    Inc(i);
47  until (NOT MecOut) AND IsKeyPressed(27);
48  CloseMecGraph(FALSE);
49  END.
```

```pascal
1    program P9_23;
2    {=============================================================
3     Drag-link 4-bar linkage animation showing the fixed and moving centrodes
4     =============================================================}
5    uses CRT,DOS,Graph,
6         LibMath,        {_}
7         LibGraph,       {InitGr}
8         Unit_PCX,       {WritePCX}
9         LibDXF,         {D_X_F,InitDXFfile}
10        LibInOut,       {EraseAll,IsKeyPressed}
11        LibMecIn,       {gCrank}
12        LibAssur,       {RRR}
13        LibMec2D,       {OpenMecGraph,CometLocus,Left,Right,Shape,PutPoint}
14        LibGe2D;        {Int2Lns,RT2D}
15   const DXFile = 'F9_23R.DXF';  {output file name F9_23R.DXF or F9_23R.DXF}
16         LR    = Right;          {Left/Right mechanism closure}
17         nPoz  = 180;            {number of animation frames}
18   var ICm: Text; {..(x,y) ASCII files with moving-centrode points}
19       Phi,xO,yO,xA,yA,xB,yB,xC,yC, OA,AB,BC, xIC,yIC,xICm,yICm: double;
20       i,Loop: Integer; OK: Boolean;        S: string;
21   BEGIN
22     xO:=0.0;   yO:=0.0;     xC:=60;   yC:=0;   {..location of ground joints}
23     OA:=100;   AB:=140;   BC:=120;              {..link lengths}
24     EraseAll('*.PCX');
25     EraseAll('IC*.xy');
26     Assign(ICm,'ICm.xy');        Rewrite(ICm);        Close(ICm);
27     InitDXFfile(DXFile);
28     InitGr(0);    {..make the background white}
29     MecOut:=TRUE;  {..required because of calling InitGr above}
30     OpenMecGraph(xO-OA,xC+BC,yO-1.2*Max2(OA,BC),yO+1.2*Max2(OA,BC));
31     if (LR = Left) then S:=' left ' else S:=' right ';
```

```
32    SetTitle('4-bar'+S+' - fixed & moving centrodes');
33    SetJointSize(3);
34    i:=0; Loop:=0;
35    Repeat
36      if (i > nPoz+1) then BEGIN
37        i:=1;   Inc(Loop);
38        if (Loop = 2) then BEGIN
39          CloseMecDXF;      SetTitle('Press <ESC> to quit..');
40        END;
41      END;
42      NewFrame(0);
43      Phi:=2*Pi/nPoz*i;
44      if (i MOD 5 <> 0) then SuspendDXF else ResumeDXF;
45      gCrank(Red,xO,yO,Phi,_,_, OA, xA,yA,_,_,_);
46      RRR(Red,xA,yA,_,_,_, xC,yC,0,0,0, AB,BC,LR,xB,yB,_,_,_,_);
47      Int2Lns(xO,yO, xA,yA, xC,yC, xB,yB, xIC,yIC, OK);
48      if OK then BEGIN
49        RT2D(xIC-xA,yIC-yA, -Atan2(yB-yA,xB-xA), 0,0, xICm,yICm);
50        if (Loop = 0) then BEGIN      {update moving centrode shape file..}
51          Append(ICm);      WriteLn(ICm, xICm,' ',yICm);
52          Close(ICm);
53        END;
54      END;
55      Shape('ICm.xy',Green,xA,yA, xB,yB);              {..moving centrode}
56      CometLocus(LightBlue,xIC,yIC,'IC');
57      PutPoint(Blue, 'o', xIC,yIC,'');
58      if (Loop < 3) AND D_X_F then WritePCX(ImplicitFileName('IC.PCX'),OK);
59      Inc(i);
60    until IsKeyPressed(27);
61    CloseMecGraph(TRUE);
62 END.
```

```
1   program P9_27;
2   {============================================================
3    Generates D2D files to plot implicit functions within the interval
4    [Xmin..Xmax] x [Ymin..Ymax].
5    ============================================================}
6   uses CRT,        {ClrScr}
7        LibMath;    {Nmax,VDn,InfD}
8   const FileName = 'F9_27A.D2D';  {..'F9_27A.D2D' or 'F9_27B.D2D'}
9   var FD:file of double; vX,vY:VDn;
10      nX,nY, iX,iY:integer; X,Y, Xmin,Xmax, Ymin,Ymax:double;
11  function Fxy(X,Y:double):double;
12  BEGIN
13    if Pos('A',FileName) > 0 then
14      Fxy:=Y*Y*Y-X*X*X-10*X*Y+1;      {..with F9_27A.D2D}
15    if Pos('B',FileName) > 0 then
16      Fxy:=X*Y*Cos(X*X+Y*Y)-1;        {..with F9_27B.D2D}
17  END;
18  {$F+}
19  function Fx(X:double):double; BEGIN Fx:=Fxy(X,Y); END;
20  function Fy(Y:double):double; BEGIN Fy:=Fxy(X,Y); END;
21  {$F-}
22  BEGIN
23    nX:=201; Xmin:=-7.5; Xmax:=7.5;
24    nY:=201; Ymin:=-7.5; Ymax:=7.5;
25    ClrScr;
26    Assign(FD,FileName); Rewrite(FD);
27    for iX:=1 to nX do BEGIN
28      X:=Xmin+(Xmax-Xmin)/(nX-1)*(iX-1);
29      NrFev0:=0;
30      ZeroGrid(Fy, Ymin,Ymax, nY, vY);
31      for iY:=1 to Nmax do
```

```
32    if (vY[iY] <> InfD) then Write(FD,X,vY[iY]);
33   END;
34   for iY:=1 to nY do BEGIN
35     Y:=Ymin+(Ymax-Ymin)/(nY-1)*(iY-1);
36     NrFev0:=0;
37     ZeroGrid(Fx, Xmin,Xmax, nX, vX);
38     for iX:=1 to Nmax do
39        if (vX[iX] <> InfD) then Write(FD,vX[iX],Y);
40   END;
41   Close(FD);
42   Write(FileName,' file output successfully..');   ReadLn;
43  END.
```

```
1  Program P9_28;
2  {============================================================
3  Generate D3D files to plot implicit functions within the interval
4  [Xmin..Xmax] x [Ymin..Ymax].
5  ============================================================}
6  uses CRT;    {ClrScr}
7  const FileName = 'F9_28A.D3D';   {..'F9_28A.D3D' or 'F9_28B.D3D'}
8  var FD:file of double;   i,j:integer;
9      nX,nY, X,Y, Xmin,Xmax, Ymin,Ymax, Z:double;
10 function Fxy(X,Y:double):double;
11 BEGIN
12   if Pos('A',FileName) > 0 then
13     Fxy:=Y*Y*Y-X*X*X-10*X*Y+1;   {..with F9_28A.D2D}
14   if Pos('B',FileName) > 0 then
15     Fxy:=X*Y*Cos(X*X+Y*Y) -1;   {..with F9_28B.D2D}
16 END;
17 BEGIN
```

```
18    nX:=201;        Xmin:=-7.5;  Xmax:=7.5;
19    nY:=201;        Ymin:=-7.5;  Ymax:=7.5;
20    ClrScr;
21    Assign(FD,FileName);      Rewrite(FD);
22    Write(FD,nX,nY,Xmin,Xmax,Ymin,Ymax);
23    for i:=1 to round(nX) do BEGIN
24      X:=Xmin+(Xmax-Xmin)/(nX-1)*(i-1);
25      for j:=1 to round(nY) do BEGIN
26        Y:=Ymin+(Ymax-Ymin)/(nY-1)*(j-1);
27        Z:=Fxy(X,Y);
28        Write(FD,Z);
29      END;
30    END;
31    Close(FD);
32    Write(FileName,' file output successfully..');   ReadLn;
33  END.
```

```
1   program P9_31;
2   {=====================================================================
3    Inverse kinematics of a 5R parallel SCARA robot
4    =====================================================================}
5   uses Graph,    {Blue,Cyan,Green,Red,Magenta}
6        LibGe2D,  {Dist2Pts}
7        LibMath,  {DEG}
8        LibInOut, {IsKeyPressed}
9        LibDXF,   {InitDXFfile}
10       LibAssur, {RRR}
11       LibMec2D; {OpenMecGraph,NewFrame,AngPVA,Ang4PVA,CloseMecDXF}
12                 {,PutRefSystem,LabelJoint,CometLocus,SetTitle}
13  const InpPathXY   = 'RoboPath.XY';
```

```
14        OutJntAngle = 'F9_31.DTA';
15        OutDXF      = 'F9_31.DXF';
16  var FTi,FTo:text;    AuxStr:string;      i: Word;
17    AuxD, AB, BC, xA1,yA1,xA2,yA2, xB1,yB1,xB2,yB2
18    ,xC,yC, xCp,yCp, ThA1,ThB1, ThA2,ThB2, t, vC: double;
19  label Abort;
20  function W(D:double):string;    {shorthand for MyStr()..}
21  BEGIN
22    W:=MyStr(D,10);
23  END;
24  BEGIN
25    Assign(FTi,InpPathXY);      {..input path file}
26    Assign(FTo,OutJntAngle);    {..output joint-angle file}
27    InitDXFfile(OutDXF);        {..output DXF file}
28    vC := 1.00;                 {in/s] speed of point C    }
29    xA1:= 2.25; yA1:=-8.0;      {in] ground-joint A1 coordinates }
30    xA2:= 6.75; yA2:=-8.0;      {in] ground-joint A2 coordinates }
31    AB :=      7.0;             {in] length of link AB     }
32    BC := 10.0;                 {in] length of link BC     }
33    Rewrite(FTo);
34    WriteLn(FTo,W(AB) ,'   ', W(BC));
35    WriteLn(FTo,W(xA1),'   ', W(yA1));
36    WriteLn(FTo,W(xA2),'   ', W(yA2));
37    WriteLn(FTo,W(vC));
38    WriteLn(FTo,'    t        ThetaA1       ThetaB1       ThetaA2       ThetaB2');
39    OpenMecGraph(-6,15, -12,7);
40    SetTitle('Inverse kinematics of an 5R robot');
41    Repeat
42      Reset(FTi);
43      ThA1:=0.0; ThB1:=0.0;  ThA2:=0.0;  ThB2:=0.0;
44      xC:=InfD; yC:=InfD;    t:=0.0;
```

```
45    i:=0;
46    repeat
47    NewFrame(0);
48    if (i MOD 4 <> 0) then SuspendDXF else ResumeDXF;
49    PutRefSystem(3,3,'x','y');
50    xCp:=xC; yCp:=yC;
51    ReadLn(FTi,xC,yC);
52    if (xCp < InfD) then t:=t+Dist2Pts2D(xCp,yCp,xC,yC)/vC;
53    RRR(Red,xA1,yA1,0,0,0,xC,yC,_,_,_,_, AB,BC,-1,xB1,yB1,_,_,_,_,_);
54    RRR(Blue,xA2,yA2,0,0,0,xC,yC,_,_,_,_, AB,BC,+1,xB2,yB2,_,_,_,_,_);
55    AngPVA(xA1,yA1,0,0,0,0,xB1,yB1,_,_,_,ThA1,_,_);
56    Ang4PVA(xA1,yA1,0,0,0,0,xB1,yB1,_,_,_,_,xB1,yB1,_,_,_,_
57    ,xC,yC,_,_,_,_,ThB1,_,_);
58    AngPVA(xA2,yA2,0,0,0,0,xB2,yB2,_,_,_,ThA2,_,_);
59    Ang4PVA(xA2,yA2,0,0,0,0,xB2,yB2,_,_,_,_,xB2,yB2,_,_,_,_
60    ,xC,yC,_,_,_,_,ThB2,_,_);
61    PutAng(White,xA1,yA1,xA1,yA1,xB1,yB1,6,'<'+#233+'_A1|');
62    PutAng(White,2*xB1-xA1,2*yB1-yA1,xB1,yB1,xC,yC,6,'<'+#233+'_B1|');
63    PutAng(White,xA2,yA2,xA2,yA2,xB2,yB2,6,'<'+#233+'_A2|');
64    PutAng(White,2*xB2-xA2,2*yB2-yA2,xB2,yB2,xC,yC,6,'<'+#233+'_B2|');
65    PutGPoint(White,' ',xA1,yA1,'A_1 ');
66    PutGPoint(White,' ',xB1,yB1,'B_1 ');
67    PutGPoint(White,' ',xA2,yA2,'A_2 ');
68    PutGPoint(White,' ',xB2,yB2,'B_2 ');
69    LabelJoint(White,0.5*(xA1+xA2),0.5*(yA2+yA2),xC,yC,'      C');
70    ResumeDXF;
71    Locus(Cyan, xC,yC, 'C');
72    if MecOut then WriteLn(FTo,W(t),' ',W(ThA1*DEG),' '
73    ,W(ThB1*DEG),' ',W(ThA2*DEG),' ',W(ThB2*DEG));
74    if IsKeyPressed(27) then GoTo Abort;
75    Inc(i);
```

```
76      until EOF(FTi);
77      CloseMecDXF;
78    until FALSE;
79  Abort:
80    Close(FTi);  Close(FTo);
81    CloseMecGraph(FALSE);
82 END.


1  program P9_32;
2  {=======================================================================
3   Direct kinematics of a 5R parallel SCARA robot
4   =======================================================================}
5  uses Graph,       {Blue,Red,White}
6    LibMath,        {RAD}
7    LibInOut,       {IsKeyPressed}
8    LibDXF,         {InitDXFfile}
9    LibMecIn,       {gCrank}
10   LibAssur,       {RRR}
11   LibMec2D;       {OpenMecGraph,SetJointSize,NewFrame,
12                    Base,Link,CloseMecDXF,CloseMecGraph}
13 const InAngles = 'F9_31.DTA';
14       OutDXF   = 'F9_32B.DXF';
15       Sticks   = 0;  {0/1; 0 will draw the mechanism as sticks}
16 var FT:text;   AuxStr:string;   i:Word;
                            DD,AuxD,t,vC,AB,BC
17   ,xA1,yA1,xA2,yA2,xB1,yB1,xB2,yB2,xC,yC,ThA1,ThA2:double;
18 label Abort;
19 BEGIN
20   Assign(FT,InAngles);   {input file with joint-angle values}
21   InitDXFfile(OutDXF);   {output DXF file for animation with M_3D.LSP}
22   OpenMecGraph(-6,15, -12,7);
```

```
23   SetTitle('Direct kinematics of an 5R robot');
24   SetJointSize(5);
25   repeat
26     Reset(FT);
27     ReadIn(FT,AB,BC);  ReadIn(FT,xA1,yA1);
28     ReadIn(FT,xA2,yA2);  ReadIn(FT,vC);
29     ReadIn(FT,AuxStr); {..read ` t   ThetaA1   ThetaB1 ` etc.}
30     DD:=0.2*AB;          {..parameter used with Base & Link}
31     i:=0;
32     repeat
33       NewFrame(0);
34       if (i MOD 4 <> 0) then SuspendDXF else ResumeDXF;
35       ReadIn(FT,t,ThA1,AuxD,ThA2,AuxD); {..ignores ThB1 & ThB2}
36       ThA1:=ThA1*RAD;
37       ThA2:=ThA2*RAD;
38       Base(White*Sticks,0.5*(xA1+xA2),0.5*(yA1+yA2)
39       ,0.5*(xA1+xA2),0.5*(yA1+yA2)-1.5*DD,(xA2-xA1)+DD,DD,0);
40       gCrank(-Red,xA1,yA1, ThA1,_,_, AB, xB1,yB1,_,_,_,_);
41       Link(Red,xA1,yA1,xB1,yB1,1.5*DD*Sticks,DD,DD);
42       gCrank(-Red,xA2,yA2, ThA2,_,_, AB, xB2,yB2,_,_,_,_);
43       Link(Red,xA2,yA2,xB2,yB2,1.5*DD*Sticks,DD,DD);
44       RRR(-Blue,xB1,yB1,_,_,_,_,xB2,yB2,_,_,_,_
45       ,BC,BC,-1,xC,yC,_,_,_,_);
46       Link(Blue,xB1,yB1,xC,yC,1.5*DD*Sticks,DD,DD);
47       Link(Blue,xB2,yB2,xC,yC,1.5*DD*Sticks,DD,DD);
48       if (Sticks = 0) then BEGIN
49         PutAng(White,xA1,yA1,xA1,yA1,xB1,yB1,6,'<'+#233+'_A1|');
50         PutAng(White,xA2,yA2,xA2,yA2,xB2,yB2,6,'<'+#233+'_A2|');
51       END;
52       PutGPoint(White,' ',xA1,yA1,'A_1 ');
53       PutGPoint(White,' ',xA2,yA2,'A_2 ');
```

```
54        PutPoint(White,'  ',xB1,yB1,'B_1   ');
55        PutPoint(White,'  ',xB2,yB2,'   B2');
56        LabelJoint(White,0.5*(xA1+xA2),0.5*(yA1+yA2),xC,yC,'      C');
57        CometLocus(Cyan, xC,yC, 'C');
58        if IsKeyPressed(27) then GoTo Abort;
59        Inc(i);
60      until EOF(FT);
61      CloseMecDXF;
62    until FALSE;
63  Abort:
64  Close(FT);
65  CloseMecGraph(FALSE);
66  END.


1  program P9_33;
2  {=========================================================================
3   Direct kinematics of a 2R serial SCARA robot
4   =========================================================================}
5  uses Graph,          {Blue,Red,White}
6       LibMath,        {RAD}
7       LibInOut,       {IsKeyPressed}
8       LibDXF,         {InitDXFfile}
9       LibMecIn,       {gCrank,Crank}
10      LibMec2D;       {OpenMecGraph,SetJointSize,NewFrame,}
11                      {Base,Link,CloseMecDXF,CloseMecGraph}
12 const InAngles = 'F9_31.DTA';
13       OutDXF   = 'F9_33B.DXF';
14       LftRgt   = -1;      {-1 left hand config., +1 right hand config.}
15       Sticks   = 0;       {0/1; 0 will draw the mechanism as sticks}
16 var FT:text;   AuxStr:string;   i:Word;      DD,vC,t,AB,BC,xA,yA
```

```
17        ,xA1,yA1,xA2,yA2,xB,yB,xC,yC,ThA,ThB,ThA1,ThB1,ThA2,ThB2:double;
18   label Abort;
19   BEGIN
20     Assign(FT,InAngles);        {..input joint-angle file}
21     InitDXFfile(OutDXF);        {..output file for M_3D.LSP animation}
22     OpenMecGraph(-6,15, -12,7);
23     if (LftRgt = -1) then
24       SetTitle('Direct kinematics of an 2R robot - left-hand config.')
25     else
26       SetTitle('Direct kinematics of an 2R robot - right-hand config.');
27     SetJointSize(5);
28     repeat
29       Reset(FT);
30       ReadLn(FT,AB,BC);          ReadLn(FT,xA1,yA1);
31       ReadLn(FT,xA2,yA2);        ReadLn(FT,vC);
32       ReadLn(FT,AuxStr);         {..read ` t ThetaA1       ThetaB1 ` etc.}
33       DD:=0.2*AB;
34       i:=0;
35       repeat
36         NewFrame(0);
37         if (i MOD 4 <> 0) then SuspendDXF else ResumeDXF;
38         ReadLn(FT,t, ThA1,ThB1, ThA2,ThB2);
39         if (LftRgt = -1) then BEGIN
40           AuxStr:='1';         xA:=xA1;      yA:=yA1;
41           ThA:=ThA1*RAD;       ThB:=ThB1*RAD;
42         END
43         else BEGIN
44           AuxStr:='2';         xA:=xA2;      yA:=yA2;
45           ThA:=ThA2*RAD;       ThB:=ThB2*RAD;
46         END;
47         gCrank(-Red,xA,yA, ThA,_,_, AB, xB,yB,_,_,_);
```

```
48      Crank(-Blue,xA,yA,0,0,0,xB,yB,_,_,_,ThB,_,_,BC,xC,yC,_,_,_);
49      Link(Red,xA,yA,xB,yB,1.5*DD*Sticks,DD,DD);
50      Link(Blue,xB,yB,xC,yC,1.5*DD*Sticks,DD,DD);
51      if (sticks = 0) then BEGIN {15 is color White..}
52        PutAng(15,xA,yA,xA,yA,xB,yB,6,'<'+#233+'_A'+AuxStr+'|');
53        PutAng(15,2*xB-xA,2*yB-yA,xB,yB,xC,yC,6,'<'+#233+'_B'+AuxStr+'|');
54      END;
55      PutGPoint(White,'   ',xA,yA,'A      ');
56      PutPoint(White,'   ',xB,yB,'B      ');
57      PutPoint(Blue,'o',xC,yC,'');
58      LabelJoint(White,xB,yB,xC,yC,'   C');
59      CometLocus(Cyan, xC,yC, 'C');
60      if IsKeyPressed(27) then GoTo Abort;
61      Inc(i);
62    until EOF(FT);
63    CloseMecDXF;
64  until FALSE;
65  Abort:
66  Close(FT);
67  CloseMecGraph(FALSE);
68  END.


1   program P9_34;
2   {=====================================================================
3    Inverse kinematics of an RTRTR parallel SCARA robot
4   =====================================================================}
5   uses Graph,     {Blue,Cyan,Green,Red,Magenta}
6       LibGe2D,    {Dist2Pts}
7       LibMath,    {DEG}
8       LibInOut, {IsKeyPressed}
```

```
 9       LibDXF,      {InitDXFfile}
10       LibMecIn,{RTRTR}
11       LibMec2D; {OpenMecGraph,NewFrame,AngPVA,Ang4PVA,CloseMecDXF}
12                  {,PutRefSystem,LabelJoint,CometLocus,SetTitle}
13  const InpPathXY = 'RoboPath.XY'; {..input endeffector path          }
14       OutDTA   = 'F9_34.DTA';     {..output t,s1,s2,ds1/dt,ds2/dt}
15       OutDXF   = 'F9_34.DXF';     {..output DXF for animation with M_3D.LSP}
16       vC = 1.0;        {imposed constant speed of point C }
17       xA = 1.00;    yA=-8.0;  {ground-joint A coordinates }
18       xB = 8.00;    yB=-8.0;  {ground-joint B coordinates }
19       A0A= 1.85;    B0B=1.85;  {linear actuator eccentricities}
20       AQ =15.00;
21       PC = 0.00;
22  var FTi,FTo:text;   i: Word;
23       t,dt, xC,yC,xCp,yCp, vxC,vyC, s1,s2, s1p,s2p,ds1,ds2
24       ,xP1,yP1,xP2,yP2, xQ1,yQ1,xQ2,yQ2, xA0,yA0, xB0,yB0: double;
25  label Abort;
26  function W(D:double):string; {shorthand for MyStr()..}
27  BEGIN
28       W:=MyStr(D,8);
29  END;
30  BEGIN
31       Assign(FTi,InpPathXY);
32       Assign(FTo,OutDTA);
33       InitDXFfile(OutDXF);
34       Rewrite(FTo);
35       WriteLn(FTo,'xA=',W(xA),';';         yA=',W(yA),';');
36       WriteLn(FTo,'xB=',W(xB),';';         yB=',W(yB),';');
37       WriteLn(FTo,'A0A=',W(A0A),';';        B0B=',W(B0B),';';');
38       WriteLn(FTo,W(xA),' ',W(yA),'  ',W(xB),' ',W(yB),'  ',W(A0A),'  ',W(B0B));
39       WriteLn(FTo,' t    s1    s2    ds1   ds2');
```

```
40    OpenMecGraph(-4,13, -10,10);
41    SetTitle('RTRTR robot inverse kinematics');
42    repeat
43      Reset(FTi);
44      xC :=InfD; yC :=InfD;
45      s1 :=InfD; s2 :=InfD;
46      ds1:=InfD; ds2:=InfD;
47      t:=0.0;
48      i:=0;
49      repeat
50        NewFrame(0);
51        if (i MOD 4 <> 0) then SuspendDXF else ResumeDXF;
52        s1p:=s1; s2p:=s2;     {..save previous s1 and s2}
53        xCp:=xC; yCp:=yC;     {..save previous xC and yC}
54        ReadIn(FTi,xC,yC);
55        s1:=Sqrt(Sqr(xA-xC)+Sqr(yA-yC) - Sqr(A0A));
56        s2:=Sqrt(Sqr(xB-xC)+Sqr(yB-yC) - Sqr(B0B));
57        if (xCp < InfD) then BEGIN
58          dt:=Dist2Pts2D(xCp,yCp,xC,yC)/vC;
59          t:=t+dt;
60          ds1:= (s1-s1p)/dt;
61          ds2:= (s2-s2p)/dt;
62        END;
63        RTRTR(Red,xA,yA,0,0,0, xB,yB,0,0,0, -A0A,AQ,PC, B0B,AQ,PC
64          ,s1,ds1,_, s2,ds2,_,-1, xC,yC,vxC,vyC,_,_, _);
65        GetA0(xA0,yA0);   GetB0(xB0,yB0);
66        GetP1(xP1,yP1);   GetQ1(xQ1,yQ1);
67        GetP2(xP2,yP2);   GetQ2(xQ2,yQ2);
68        PutDist(White,xA0,yA0,xC,yC, 10,'│ >│');
69        PutDist(White,xB0,yB0,xC,yC,-10,'│ >│');
70        PutGpoint(White,' ',xA,yA,       A');
```

```
71      PutGPoint(White,' ',xB,yB,'B       ');
72      LabelJoint(White,0.5*(xA+xB),0.5*(yA+yB),xC,yC,'   C');
73      if (A0A > 0) then LabelJoint(White,xC,yC,xA0,yA0,' A_0');
74      if (B0B > 0) then LabelJoint(White,xC,yC,xB0,yB0,' B_0');
75      PutVector(LightBlue, '|' ,xC,yC, vxC,vyC, 3,' v_C');
76      ResumeDXF;    Locus(Cyan, xC,yC, 'C');
77      if MecOut then
78          WriteLn(FTo,W(t),' ',W(s1),' ',W(s2),' ',W(ds1),' ',W(ds2));
79      if IsKeyPressed(27) then GoTo Abort;
80      Inc(i);
81    until EOF(FTi);
82    CloseMecDXF;
83  until FALSE;
84  Abort:
85  Close(FTi);  Close(FTo);
86  CloseMecGraph(FALSE);
87  END.
```

```
1   program P9_35;
2   {=============================================================================
3    Direct kinematics of the RTRTR robot with pinion and rake shapes
4   =============================================================================}
5   uses Graph,      {Blue,Cyan,Green,Red,Magenta}
6        LibGe2D,    {S123}
7        LibMath,    {DEG}
8        LibInOut,   {IsKeyPressed}
9        LibDXF,     {InitDXFfile}
10       LibMecIn,
11       LibAssur,   {RTRR}
12       LibMec2D;   {OpenMecGraph,NewFrame,SetJointSize,}
13                   {AngPVA,Ang3PVA,CloseMecDXF,MecOut}
```

```
14  const LabelCol = White;
15        InDTA   = 'F9_34.DTA';    {..input t,s1,ds1,s2,ds2 file        }
16        OutDTA  = 'F9_35.DTA';    {..output t,Thta1,Thta2,Thta1/dt,Thta2/dt}
17        OutDXF  = 'F9_35.DXF';    {..output DXF for animation with M_3D.LSP }
18        rp = 1.57847;  {pitch radius of pinions}
19  var FTi,FTo:text;     AuxStr:string;     i:integer;
20        t,tp, A0A,B0B, xA,yA, xB,yB, xA0,yA0, xB0,yB0, xA_0,yA_0, xB_0,yB_0
21        ,xC,yC,vxC,vyC, s1,s2, ds1,ds2, Phi1,Phi2
22        ,Thta1,Thta2, Thta1p,Thta2p, dThta1,dThta2: double;
23  function W(D:double):string;     {shorthand for MyStr()..}
24  BEGIN
25        W:=MyStr(D,10);
26  END;
27  label Abort;
28  BEGIN
29        Assign(FTi,InDTA);
30        Assign(FTo,OutDTA);
31        InitDXFfile(OutDXF);
32        Rewrite(FTo);
33        WriteLn(FTo,' t Theta1 Theta2 dTheta1/dt dTheta2/dt');
34        OpenMecGraph(-4,13, -18,7);
35        SetJointSize(-2);
36        SetTitle('Direct kinematics of an RTRTR geared robot');
37        repeat
38             Reset(FTi);
39             ReadLn(FTi,AuxStr);    ReadLn(FTi,AuxStr);   ReadLn(FTi,AuxStr);
40             ReadLn(FTi,xA,yA,xB,yB,A0A,B0B);
41             ReadLn(FTi,AuxStr);
42             tp:=0.0;           Phi1:=0.0;   Phi2:=0.0;
43             Thta1:=InfD;        Thta2:=InfD;
44             dThta1:=InfD;       dThta2:=InfD;
```

```
45   i:=0;
46   repeat
47   NewFrame(0);
48   if (i MOD 2 <> 0) then SuspendDXF else ResumeDXF;
49   Thta1p:=Thta1; Thta2p:=Thta2;          {..save previous Thta1 & Thta2}
50   tp:=t;   {..save previous t}
51   ReadIn(FTi,t,s1,s2,ds1,ds2);
52   RTRTR(0,xA,yA,0,0,0, xB,yB,0,0,0,0, -A0A,0.35,0, B0B,0.35,0
53   ,s1,ds1,_, s2,ds2,_,-1, xC,yC,vxC,vyC,_,_, _ );
54   GetA0(xA0,yA0);     GetB0(xB0,yB0);
55   AngPVA(xA,yA,0,0,0, xA0,yA0,_,_,_,_, Phi1,_,_);
56   Thta1:=Phi1 - s1/rp + 0*RAD;
57   gCrank(White,xA,yA, Thta1,_,_, rp, xA_0,yA_0,_,_,_,_);
58   Shape('RTRTR0.XY', LightBlue, xA,yA, xA_0,yA_0);
59   PutAng(White,xA,yA,xA,yA,xA_0,yA_0,6,'<'+#233+'_1|');
60   AngPVA(xB,yB,0,0,0, xB0,yB0,_,_,_,_, Phi2,_,_);
61   Thta2:=Phi2 + s2/rp + 0*RAD;
62   gCrank(White,xB,yB,Thta2,_,_, rp, xB_0,yB_0,_,_,_,_);
63   Shape('RTRTR0.XY', LightBlue, xB,yB, xB_0,yB_0);
64   PutAng(White,xB,yB,xB,yB,xB_0,yB_0,6,'<'+#233+'_2|');
65   Shape('RTRTR1.XY', Red, xC,yC, xA0,yA0);
66   Shape('RTRTR2.XY', Red, xC,yC, xB0,yB0);
67   Shape('RTRTR3.XY', Cyan, xA,yA, xA0,yA0);
68   Shape('RTRTR3.XY', Cyan, xB,yB, xB0,yB0);
69   PutGPoint(White, ',xA,yA,'A       ');
70   PutGPoint(White, ',xB,yB,'B       ');
71   PutPoint(White,'O',xC,yC, 'C');
72   ResumeDXF;     Locus(Cyan, xC,yC, 'C');
73   if (Thta1p < InfD) then BEGIN
74   dThta1:=(Thta1-Thta1p)/(t-tp);
75   dThta2:=(Thta2-Thta2p)/(t-tp)
```

```
76        END;
77        if MecOut then WriteLn(FTo,W(t),' ',W(Thta1*DEG)
78        ,' ',W(Thta2*DEG),' ',W(dThta1),' ',W(dThta2));
79        if IsKeyPressed(27) then GoTo Abort;
80        Inc(i);
81      until EOF(FTi);
82      CloseMecDXF;
83    until FALSE;
84    Abort:
85    Close(FTi); Close(FTo);
86    CloseMecGraph(FALSE);
87 END.


1  program P9_36B;
2  {=============================================================
3  Simulation of an excavator arm motion - includes shapes attached to links
4  =============================================================}
5  uses Graph,      {Brown,Red,White,Magenta}
6       LibMath,    {_}
7       LibInOut,   {IsKeyPressed}
8       LibDXF,     {InitDXFfile,ResetDXFelev,DecDXFelev,SetDXFlayer}
9       LibMecIn,   {RTRR}
10      LibAssur,   {RRR}
11      LibMec2D;   {OpenMecGraph,NewFrame,PutGPoint,gShape,Shape,Offset}
12               {SetJointSize,Locus,PutGPoint,CloseMecDXF,CloseMecDXF}
13 const nPoz = 45;    {..number of positions of the simulation}
14 var i,j:Word; L_R1,L_R2,L_R3,L_R4:shortint;
15     t, s1,s1min,s1max, s2,s2min,s2max, s3,s3min,s3max,
16     xA1,yA1, xB1,yB1, A1Q1,P1C1,B1C1,   xC1,yC1,
17     xA2,yA2, xB2,yB2, A2Q2,P2C2,B2C2,   xC2,yC2,
```

```
18        xA3,yA3,  xB3,yB3,       A3Q3,P3C3,B3C3,    xC3,yC3,
19        xD,yD, C3E,DE, xE,yE,  x_A2,y_A2, x_B2,y_B2, x_A3,y_A3,
20        x_B3,Y_B3, x_D,Y_D, x_F,Y_F, xF,yF: double;
21   BEGIN
22        InitDXFfile('F9_36.DXF');
23        s1min:=0.65; s1max:=0.90;      {..piston #1 motion range}
24        s2min:=0.60; s2max:=0.90;      {..piston #2 motion range}
25        s3min:=0.45; s3max:=0.65;      {..piston #3 motion range}
26        L_R1:=+1;  L_R2:=-1;  {..orientation of loops A1-B1-C1 & A2-B2-C2}
27        L_R3:=-1;  L_R4:=+1;  {..orientation of loops A3-B3-C3 & C3-E-D }
28        xA1 := 0.5000;    yA1 := 0.4239;   {..coord. of ground joint A1 }
29        xB1 := 0.3940;    yB1 := 0.5869;   {..coord. of ground joint B1 }
30        x_A2:= 0.9265;    y_A2:= 0.1753;   {..relative coord. of joint A2 }
31        x_B2:= 1.3477;    y_B2:=-0.4554;   {..relative coord. of joint B2 }
32        x_A3:=-0.0392;    y_A3:=-0.2010;   {..relative coord. of joint A3 }
33        x_B3:=-0.5951;    y_B3:=-0.3253;   {..relative coord. of joint B3 }
34        x_D :=-0.7013;    y_D :=-0.3751;   {..relative coord. of joint D }
35        x_F :=-0.0974;    y_F :=-0.3859;   {..relative coord. of point F }
36        B1C1:=0.7306;     B2C2:= 0.2062;     B3C3:=0.1635;
37        C3E := 0.1584;    DE      := 0.1177;
38        P1C1:=0.8*s1min;  A1Q1:=1.2*(s1max-P1C1);
39        P2C2:=0.8*s2min;  A2Q2:=1.2*(s2max-P2C2);
40        P3C3:=0.8*s3min;  A3Q3:=1.2*(s3max-P3C3);
41        OpenMecGraph(-0.53,2.77,-1.1,2.2);  SetJointSize(-2);
42        i:=0;
43        Repeat
44        if (i > nPoz) then BEGIN
45             i:=0;  CloseMecDXF;
46        END;
47        NewFrame(0);
48        t:=i/nPoz; {..t = time}
```

```
49   s1:=0.5*(s1min+s1max)+0.5*(s1max-s1min)*cos(2*Pi*t-Pi/8);
50   s2:=0.5*(s2min+s2max)+0.5*(s2max-s2min)*cos(2*Pi*t+Pi/4);
51   s3:=0.5*(s3min+s3max)+0.5*(s3max-s3min)*cos(2*Pi*t-Pi/8);
52   gShape('EXbody.XY',Brown,0,0);
53   RTRR(-Magenta,xA1,yA1,0,0,0,xB1,yB1,0,0,0,0,0
54     ,A1Q1,B1C1,s1,_,_,L_R1,xC1,yC1,_,_,_,_,_);
55   Shape('EXboom.XY',Brown,xB1,yB1,xC1,yC_);
56   Offset(0,' ',xB1,yB1,_,_,_,_,xC1,yC1,_,_,_,_
57     ,x_A2,y_A2,xA2,yA2,_,_,_,_);
58   Offset(0,' ',xB1,yB1,_,_,_,_,xC1,yC1,_,_,_,_
59     ,x_B2,y_B2,xB2,yB2,_,_,_,_);
60   RTRR(-Magenta,xA2,yA2,_,_,_,_,xB2,yB2,_,_,_,_
61     ,0,A2Q2,P2C2,B2C2,s2,_,_,L_R2,xC2,yC2,_,_,_,_,_);
62   Shape('EXstick.XY',Brown,xB2,yB2,xC2,yC2);
63   Offset(0,' ',xB2,yB2,_,_,_,_,xC2,yC2,_,_,_,_
64     ,x_A3,y_A3,xA3,yA3,_,_,_,_);
65   Offset(0,' ',xB2,yB2,_,_,_,_,xC2,yC2,_,_,_,_
66     ,x_B3,y_B3,xB3,yB3,_,_,_,_);
67   Offset(0,' ',xB2,yB2,_,_,_,_,xC2,yC2,_,_,_,_
68     ,x_D ,y_D ,xD,yD,_,_,_,_);
69   RTRR(Magenta,xA3,yA3,_,_,_,_,xB3,yB3,_,_,_,_
70     ,0,A3Q3,P3C3,B3C3,s3,_,_,L_R3,xC3,yC3,_,_,_,_,_);
71   RRR(Red,xC3,yC3,_,_,_,_,xD,yD,_,_,_,_,C3E,DE,L_R3,xE,yE,_,_,_,_,_);
72   Offset(0,' ',xD,yD,_,_,_,_,xE,yE,_,_,_,_,x_F,y_F,xF,yF,_,_,_,_);
73   PutGPoint(White,' ',xA1,yA1,'A1 ');
74   PutGPoint(White,' ',xB1,yB1,'B1 ');
75   PutPoint (White,' ',xC1,yC1, C1');
76   PutPoint (White,' ',xA2,yA2,'A2 ');
77   PutPoint (White,' ',xB2,yB2, B2');
78   PutPoint (White,' ',xC2,yC2, C2');
79   PutPoint (White,' ',xA3,yA3, A3');
```

```
80      PutPoint (White,' ',xB3,yB3,'B3      ');
81      PutPoint (White,' ',xC3,yC3,'     C3');
82      PutPoint (White,' ',xD ,yD ,'D      ');
83      PutPoint (White,' ',xE ,yE ,' E');
84      LabelJoint(White,xE,yE,xF,yF,' F');
85      if (i < nPoz) AND MecOut then BEGIN
86        DecDXFelev;    DecDXFelev; DecDXFelev;
87        for j:=i+1 to nPoz do BEGIN
88          SetDXFlayer(MySt(j,3));
89          Shape('EXbucket.XY',LightGray,xD,yD,xE,yE);
90          SetDXFlayer(MySt(LastNrLayer,3));
91        END;
92        ResetDXFelev;
93      END;
94      Shape('EXbucket.XY',DarkGray,xD,yD,xE,yE);
95      Locus(LightBlue,xF,yF,'F');
96      Inc(i);
97    until IsKeyPressed(27);
98    CloseMecGraph(FALSE);
99  END.
```

```
1  program P9_38;
2  {=======================================================================
3    Kinematic simulation of a rope-shovel with shapes read from file
4    =======================================================================}
5  uses Graph,       {Red,White,Green}
6       LibMath,     {_,RAD}
7       LibInOut,    {IsKeyPressed}
8       LibDXF,      {InitDXFfile}
```

```
 9      LibMecIn, {RTRTR}
10      LibMec2D;
11 const nPoz = 90;        {..number of animation frames }
12      xA = 1.8920;  yA=1.5101;  {..ground joint A }
13      xB = 7.3107;  yB=6.8230;  {..ground joint B }
14      x_D= -1.596;  y_D=0.0;    {..local coordinates of point D}
15      A0A=-1.4145;  {..eccentricity A0A            }
16      B0B=-0.600;   {..eccentricity B0B = radius of sheave}
17      P1C= 7.3;     {..rack length measured from C  }
18      rp = 0.35;    {..pitch radius of pinion       }
19 var i:Word;   t,s1,s2,xA0,yA0,xB0,yB0,xC,yC,xD,yD,xP1,yP1
20     ,xA0p,yA0p,xB0p,yB0p,Phi1,Theta1,Phi2,Theta2: double;
21 BEGIN
22    InitDXFfile('F9_38.DXF');
23    OpenMecGraph(-6.4,10.3, -3.4,7.4);
24    SetJointSize(-2);
25    i:=0;
26    Phi1:=0.0;  Phi2:=0.0;  {..for AngPVA continuity}
27    repeat
28      if (i > nPoz) then BEGIN
29        i:=0;      CloseMecDXF;
30      END;
31    NewFrame(0);
32    gShape('RSbody.XY',Brown, 0,0);
33    t:=i/nPoz; {..t = time}
34    s1:=4.5 + 2.0*sin(2*Pi*t - Pi/6);  {..input 1}
35    s2:=5.5 + 3.0*cos(2*Pi*t + Pi/9);  {..input 2}
36    RTRTR(0, xA,yA,0,0,0, xB,yB,0,0,0, A0A,0,P1C
37        ,B0B,0,0, s1,_,_, s2,_,_, +1, xC,yC,_,_,_,_, );
38    GetA0(xA0,yA0);
```

```
39   GetB0(xB0,yB0);
40   GetP1(xP1,yP1);
41   Shape('RSboom.XY', LightBlue, xC,yC,xA0,yA0);
42   Shape('RSbrack.XY', Cyan, xA,yA,xA0,yA0);
43   AngPVA(xA0,yA0,_,_,_, xC,yC,_,_,_,Phi1,_,_);
44   Theta1:=Phi1 - s1/rp + 7.632*RAD;
45   gCrank(Red, xA,yA, Theta1,_,_, rp, xA0p,yA0p,_,_,_);
46   Shape('RSpinion.XY',Blue, xA,yA,xA0p,yA0p);
47   AngPVA(xA0,yA0,_,_,_, xC,yC,_,_,_,Phi2,_,_);
48   Theta2:=Phi2 + s2/B0B;
49   gCrank(Magenta, xB,yB, Theta2,_,_, B0B,  xB0p,yB0p,_,_,_);
50   Shape('', Magenta, xB,yB,xB+B0B,yB);   {..plot sheave}
51   Offset(0,'',xC,yC,_,_,_,xP1,yP1,_,_,_,x_D,y_D, xD,yD,_,_,_,_);
52   Locus(Magenta, xD,yD,'D');
53   PutDist(White,xB0,yB0, xC,yC, 0.0, '');  {..this is the rope}
54   PutGPoint(White,'',xA,yA,'A ');
55   PutGPoint(White,'',xB,yB,'B ');
56   PutPoint (White,'',xD,yD,' D');
57   PutPoint (White,'.',xA0,yA0,' A_0');
58   PutPoint (White,'.',xB0,yB0,' B_0');
59   PutPoint (White,'.',xP1,yP1,' P_1');
60   PutGPoint(White,'O',xC,yC,'');
61   LabelJoint(White,xB,yB,xC,yC,'C      ');
62   Inc(i);
63   until IsKeyPressed(27);
64   CloseMecDXF;
65   CloseMecGraph(FALSE);
66 END.
```

```pascal
1   program P9_49;
2   {========================================================================
3    Crank-slide punch press simulation
4   ========================================================================}
5   uses Graph,         {Red,White}
6        LibMath,       {_}
7        LibInOut,      {IsKeyPressed}
8        LibDXF,        {InitDXFfile}
9        LibMecIn,      {gCrank}
10       LibAssur,      {RR_T}
11       LibMec2D;      {OpenMecGraph,NewFrame,SetJointSize,CloseMecDXF}
12  const FName ='F9_49'; {DXF and DAT file names}
13        nPoz = 1800;   {# of simulation points; nPoz/10 are # of animation frames}
14        RPM  = 80;     {crank RPM}
15        s_f  = 0.75;   {[m] end of punch}
16        h    = 0.02;   {[m] stock thickness}
17        Fmax = 5.72E5; {maximum punch force}
18  var FT:text;    i,ViewOn: Word;
19      time,Theta,dTheta, F,T, xO,yO, OA,AB, BP, s,s_s,
20      xA,yA,vxA,vyA, xB,yB, xP,yP,vxP,vyP, xQ,yQ: double;
21  BEGIN
22    dTheta:=Pi*RPM/30; {angular velocity in rad/s}
23    OA:=0.15;  {[m] crank length }
24    AB:=0.50;  {[m] conrod length }
25    BP:=0.15;  {punch length }
26    xO:=0.0; yO:=0.65;  {crank ground joint}
27    s_s:=s_f-h;
28    xQ:=xO; yQ:=yO-0.5*(s_s+s_f);  {coordinates of punch guide center}
29    Assign(FT,FName+'.DAT'); Rewrite(FT);
30    WriteLn(FT,' time      Theta      Theta      s        ds/dt        F          T');
31    WriteLn(FT,'          [s]       [RAD]      [DEG]      [m]       [m/s]       [N]        [N-m]');
```

```
32   InitDXFfile(FName+'.DXF');
33   OpenMecGraph(-OA,OA,yO-OA-1.5*AB-BP,yO+1.5*OA);
34   OpaqueJoints:=FALSE;
35   i:=0;
36   T:=InfD;
37   repeat
38     if (i > nPoz) then BEGIN
39       i:=1; CloseMecDXF;
40     END;
41     if (i MOD 10 = 0) then BEGIN
42       NewFrame(0); ViewOn:=1;
43     END
44     else ViewOn:=0;
45     time:=i/nPoz*(2*Pi/dTheta);
46     Theta:=dTheta*time;
47     gCrank(ViewOn*Brown, xO,yO, Theta,dTheta,_, OA, xA,yA,vxA,vyA,_,_);
48     RR_T(ViewOn*Red,xA,yA,vxA,vyA,_,_, xQ,yQ,0,0,0,0, 0.5*Pi,0,0,
49     AB,0,BP, -1, xB,yB,vxB,vyB,_,_, xB,yB,vxB,vyB,_,_, xP,yP,vxP,vyP,_,_, _);
50     s:=Abs(yP-yO);
51     if (s >= s_s) AND (s <= s_f) AND (vyP < 0) then BEGIN
52       F:=Fmax*(s_f - s)/(s_f - s_s);
53       PutVector(ViewOn*Blue,'=',xP,yP,0,F/Fmax,-0.2,'F');
54     END
55     else F:=0;
56     T:=-F*vyP/dTheta;
57     PutGPoint(ViewOn*White, '', xO,yO,'O  ');
58     PutPoint(ViewOn*White, '', xB,yB,'B  ');
59     PutPoint(ViewOn*White, '', xP,yP,'P  ');
60     LabelJoint(ViewOn*White,xO,yO, xA,yA,' A');
61     PutGPoint(ViewOn*White, '.', xO,yQ+h/2,'');
62     PutGPoint(ViewOn*White, '.', xO,yQ-h/2,'');
```

```
63      if MecOut then WriteLn(FT,MyStr(time,9),' ',MyStr(Theta,9),' ',MyStr(Theta*DEG,9)
64      ,' ',MyStr(s,9),' ',MyStr(vyP,9),' ',MyStr(F,9),' ',MyStr(T,9));
65      Inc(i);
66    until (NOT MecOut) OR IsKeyPressed(27);
67    Close(FT);
68    CloseMecGraph(FALSE);
69  END.


 1  program Purge;
 2  {=============================================================
 3   Deletes without confirmation all files of extensions $XY, $2D, $3D, OLD, BAK,
 4   all files ~POLY*.TMP and files acad.err and acadstk.dmp if present.
 5   Erases with confirmation all files of extension BMP,PCX and SCR.
 6   =============================================================}
 7  uses CRT, LibInOut;
 8  var Ch: char;
 9  BEGIN
10    ClrScr;
11    EraseAll('acad.err');
12    EraseAll('acadstk.dmp');
13    EraseAll('~POLY*.TMP');
14    EraseAll('*.$XY');
15    EraseAll('*.$2D');
16    EraseAll('*.$3D');
17    EraseAll('*.OLD');
18    EraseAll('*.BAK');
19    Write(^j^j^m'       Erase all BMP files in current directory <Y/N>? ');
20    WaitToGo(Ch);
21    if (Ch = #27) then Exit;
22    if UpCase(Ch) = 'Y' then EraseAll('*.BMP');
23    Write(^j^j^m'       Erase all PCX files in current directory <Y/N>? ');
```

```
24    GoToXY(WhereX-1,WhereY);
25    WaitToGo(Ch);
26    if (Ch = #27) then Exit;
27    if UpCase(Ch) = 'Y' then EraseAll('*.PCX');
28    Write(^j^j^m'    Erase all SLD files in current directory <Y/N>? ');
29    GoToXY(WhereX-1,WhereY);
30    WaitToGo(Ch);
31    if (Ch = #27) then Exit;
32    if UpCase(Ch) = 'Y' then BEGIN
33      EraseAll('*.SLD');
34      Write(^j^j^m'  Erase all SCR files in current directory <Y/N>? ');
35      GoToXY(WhereX-1,WhereY);
36      WaitToGo(Ch);
37      if (Ch = #27) then Exit;
38      if UpCase(Ch) = 'Y' then EraseAll('*.SCR');
39    END;
40  END.
```

***

Index

Note: A complete list of Turbo Pascal procedures and functions is available from the book website at www.crcpress.com/product/isbn/9781482252903/.

A

Acceleration
  angular, 155, 166, 168, 209, 215
  Coriolis, 169, 171–172, 480, 482
  gravitational, 97, 543
  normal, 172–173
  tangential, 97, 172–173, 482
  vector, 169, 171–173, 480
Ackermann, *see* Steering
Actuator
  linear, 158, 161, 274, 283, 356, 584; *see also* Motor
  RTRR, 153, 185–186, 206–210, 358–359
  RTRTR, 153, 185–186, 190–204–205, 207, 209, 213, 355–358, 360–361
Addendum, *see* Gear
Algorithm; *see also* Optimization, algorithm
  Brent, 132–133, 149, 455
  Euler–Taylor, 322
  evolutionary, 129, 141–142
  Nelder–Mead, 134–137, 139, 328, 458, 459, 462, 464, 469, 472
  Newton, 392, 399
  painter's, 47, 56; *see also* Polygon, scan conversion
  ranking, 335
  Zero, 129–131, 452; *see also* Brent
Aliasing, 104
Angle
  between two lines, 378
  between two vectors, 276
  value rectification, 103
Animation
  accumulated frame, 3, 34, 36–38, 152, 525, 529, 534, 536
  multiple frame, 34, 36
Approximation, 8–9, 130, 173, 271, 327; *see also* Curve fitting; Interpolation
  linear, 9, 32, 41, 401

  minimax, 328, 373
  parabolic, 9, 132, 276, 324, 326, 386
  polygonal, 28–30, 40–41
  Taylor, 399
Archimedean spiral, 28–32, 151–153, 418, 419, 421, 424–426, 475
Area, 60, 369; *see also* Integral
  plot, 2, 11
  under a curve, 12, 372, 400
Arnhold–Kennedy Theorem, 305
Arrow; *see also* **PutAng** in Unit **LibMec2D**;
      PutDist in Unit **LibMec2D**
  head, 176
  marker, 2, 94, 319–321
  vector, 44, 69, 123, 126–127
Assur group, 151, 183, 185–186, 210, 216, 245, 266
AutoCAD block, 13, 66, 108, 122, 124–126, 364, 450
AutoCAD command
  align, 108
  appload, 122, 313
  _ddptype, 28
  dxfin, 11
  dxfout, 108
  extend, 113
  hide, 11, 14–15, 25, 46, 56, 66, 82–84, 127
  ltscale, 14
  mirrtext, 67, 122
  move, 114
  offset, 272, 274, 277, 289
  purge, 108
  render, 82, 84–85, 125, 127
  scale, 322, 359
  script, 121–122
  shade, 10–11, 25, 46, 80, 82–83, 85
  spline, 270
  3dmesh, 119
  trim, 66, 113, 138, 140
  upload, 116

AutoCAD layer, 67
AutoCAD region, 11, 46, 66
AutoCAD slide, 121–122, 597
AutoCAD spline, 92, 106, 111, 114, 178, 270–271
AutoLISP, 51, 91–92, 116, 121, 126, 128, 153, 297, 313, 316; *see also* `G_3D.LSP`; `Gears. LSP`; `M_3D.LSP`
Axis
    label, *see* Category
    reversal, 67–68
    secondary, 9, 93

B

Background, 23, 34, 36, 110, 121, 139
    color, 43, 56, 572
    curve, 23, 34, 37–38, 140, 142, 323
    drawing, 123–124, 126
    image, 33, 152
    layer, 151
    plot, 35, 38
Backlash, *see* Gear, backlash
Ball joint, 124, 361–363
Base circle, *see* Gear, base circle
BGI
    Black constant 156, 162, 194, 206, 211, 343
    graphic mode, 92
BMP, 121–123, 142, 373, 597–598
Branching configuration, 202
Brent, *see* Algorithm, Brent
Bridge-like defects, *see* Defects, bridge-like
B-spline, *see* Spline, B-
Bump steer, 124–126; *see also* Steering

C

Cam
    dwell, 269–271, 290
    follower contact, 276, 279
    pitch profile, 273–282, 289, 291, 516, 520
Camber angle, 361, 363–365; *see also* Suspension
Cantilever beam, 144–147, 149
Cartesian robot, 163; *see also* Robot, Cartesian
Category
    axis, 3, 7, 23, 94–95
    name, 10–11
Caterpillar, 358
Center of curvature, *see* Curvature, center of
Centrode, 319, 345–346, 374, 572–573
Circular pitch, 300, 309, 317
Coincidence parameter, 5, 30, 46, 112
Colinearity parameter, 46, 112, 345
Combined plot, 43, 70, 85–86, 99, 110, 327

Comet, 2, 35–38, 179, 276; *see also* Locus
    animation 2, 28–29, 33
    plot, 2–3, 28–29, 33–38, 179–180, 319, 321
Concatenate
    files, 104
    lines, 32–33, 46, 65, 95
Cone, 121, 123, 127–128
Constrain, 44, 134
    equality, 139
    inequality, 137, 145
    side, 137, 140–141
Constraint equation, 190, 192, 206, 211
Contact
    force, 269–270, 273, 278, 281–282, 287
    patch, 363, 369
    ratio, 308, 310, 317
Control
    line, 17–19
    point, 100–102, 270
    polygon, 101, 106
Coriolis acceleration, *see* Acceleration, Coriolis
CPU time, 89
Cramer rule, 170, 209, 226
Crossover, 141–143; *see also* Evolutionary algorithm
Cubic
    extrapolation, 98
    parabola, 387–388; *see also* Parabola
    spline, *see* Spline, cubic
Curtain, 16–18, 76–78, 81, 87
Curvature, 14, 60–61, 74–75
    cam, 273
    center of, 14, 176, 273, 298, 339
    involute, 298, 304, 306
    radius of, 275–278, 280–281, 284–286, 288–289, 298
    3D surfaces, 60–61, 74–75
    2D curves, 14, 339, 401
Curve fitting, 133, 319, 324, 326, 328
Cusp, 277, 283, 286

D

Damping, 96, 321, 324
    coefficient, 323, 543, 546
    ratio, 20, 324–325, 441
Decimation, 31, 44, 98, 104–105, 179; *see also* Scatter
Dedendum, *see* Gear, dedendum
Defects
    bridge-like, 58, 60–61, 68, 73–74, 351
    connectivity, 61
    visibility, 3, 72, 113
Derivative
    first, 9, 102, 173, 215, 272, 390
    numerical, 91, 101–102, 271, 398

second, 91, 98, 101–102, 156, 213, 222–223, 244, 271–272, 401
zero of, 10
Design space, 141, 144–147, 149, 466, 468
Diametral pitch, 298–301, 305, 311, 313, 317
Division line, 7–8, 14, 46, 52, 57, 59, 66, 81
Duffing oscillator, 319–321, 374, 540
Dwell linkage, 319, 338–340, 374, 561; *see also* Cam
DXF layer, 33–34, 152, 177, 179–180, 322–323, 346, 370
Dyad
    RRR, 184–186, 210–213, 339, 353–354, 358–359; *see also* Procedure **RRR** and **RRRc** in Unit **LibAssur**
    RRT, 184–186, 214–222, 270; *see also* Procedure **RRT_** and **RR_T** in Unit **LibAssur**
    RTR, 186, 222–228, 341, 501, 564; *see also* Procedure **RT_R** Unit **LibAssur**
    RTT, 185–186, 228, 245–267; *see also* Procedure **R_T_T**, **RT_T_**, **R_TT_** and **RT__T** in Unit **LibAssur**
    TRT, 185–186, 228–245, 505–506; *see also* Procedure **T_R_T**, **T_RT_** and **_TRT_** in Unit **LibAssur**
Dyadic isomer; *see also* **Turbo Pascal** Unit, **LibAssur**
    R_T_T, *see* Dyad, RTT
    R_TT_, *see* Dyad, RTT
    RR_T, *see* Dyad, RRT
    RRT_, *see* Dyad, RRT
    RT__T, *see* Dyad, RTT
    RT_R, *see* Dyad, RTR
    RT_T_, *see* Dyad, RTT
    T_R_T, *see* Dyad, TRT
    T_RT_, *see* Dyad, TRT
    TRT, *see* Dyad, TRT

E

Ellipse, 2, 111, 114, 146, 293
Empty plot, 81
End effector path, 353, 356, 360; *see also* Robot
Energy, 367, 371–372, 375
Engine, 281, 286, 342
    radial, 342, 565
    rotary, 342–343, 567
    cycle, 367
Envelope, 9, 313, 326, 338, 536
Epicycloid, 157–158
Euclidean norm, 331
Euler
    equation, 166
    integration, 402
    method, 97, 402
    Taylor algorithm, *see* Algorithm, Euler–Taylor

Evolutionary algorithm, 129, 141–142
Excavator, 319, 358–360–361
Excel, 1, 70–71, 98, 173, 407, 434
Expansion coefficient, 142
Export to
    ASCII, 2, 8–9
    BMP, *see* **M_3D.LSP**
    D_2D, 91, 110, 136
    DXF, 3, 5–6, 10–11, 13–14, 110, 114, 139
    DXF one-to-one, 3, 17, 32, 44, 65, 107, 110–111
Extrema, 8, 12, 111, 129, 133, 139, 324, 388
Extrusion, 118–119

F

Family
    curves, 20, 47–48
    lines, 284, 290, 293
    solutions, 144
Feasible
    individual, 141; *see also* Evolutionary algorithm
    region, *see* Feasible, space
    space, 129, 138–139, 141–142, 145–146, 148–149
    value, 139
Finite difference, 44, 69, 172–173, 272, 358
First recurrence map, *see* Poincaré map
Flywheel, 319, 367–368, 371–372, 375
Follower; *see also* Procedures
    EnvelOfCircles in Unit **LibCams**; Procedures **EnvelOfLines** in Unit **LibCams**; Procedures **EnvelOfPlynes** in Unit **LibCams**
    amplitude, 272, 286, 516, 520, 525, 529, 534, 537
    bounce, 269–270
    eccentricity, 272, 276, 281, 286–287
    envelope, 193, 269, 282, 283, 287, 289, 291
    fall, 270, 289, 291
    jerk, 269–270
    rise, 269–270, 289, 291
Formula language, *see* **Working Model 2D (WM 2D)**
Four-bar linkage, *see* Mechanism, four-bar
Frame rate, 34–35, 122; *see also* Animation
Free-body diagram, 97, 322, 371
Frequency, 20, 104, 324, 336–337
    circular, 325, 327, 441–442
    damped, 325, 327, 441–442
    natural, 325, 327, 441–442
    ratio, 20
    relative, 337–338; *see also* Histogram
Friction, 301–303, 319, 368

Function; *see also* Hyperfunction
  banana, *see* Rosenbrock
  four-hump, 50–51, 58, 62–65, 68–70, 72, 82–84,
      119, 134
  harmonic, 195, 209, 319, 361
  implicit, 319, 350–351, 574–575
  multimodal, 131, 141, 146
  orange-squeezer, 49–50, 61, 71, 75, 86
  Rosenbrock, 329–335, 554

G

G_3D.LSP, 51, 65, 92, 116–120, 128
Gaussian distribution, 337–338, 560–561
Gear; *see also* Pinion
  addendum
      circle, 300, 306, 311–317
      modification, 306, 311–315; *see also* Gear,
          profile shift
  backlash, 300–302, 306, 308–310, 317
  base circle, 273–275, 277–280, 282, 285–286,
      288–289, 298, 300, 302–304, 306, 311,
      317, 516, 519, 521, 523, 525, 528, 530,
      532, 539
  clearance, 300
  contact point, 303–306, 310
  dedendum, 300, 313, 316–317
  interference, 310, 312
  nonstandard, 317
  profile shift, 306, 310–312, 314–315, 317
  reducer, 178, 487
  standard, 306–307, 311–312, 315
      center distance, 302, 307
      profile, 209, 300, 306–308, 310–312,
          315, 317
  tooth
      full depth, 300–301, 307–308, 311–314,
          316–317
      land, 313, 316
      stub, 300, 313–314, 316–317
  transmission, 305
  zero profile shift, *see* Gear, standard
Gears.LSP, 179, 313–316
Generating pinion, *see* Pinion, cutter
Generating rack, *see* Rack, cutter
Geneva wheel, 338
GIF Animator, 3, 40
Global maximum, 58, 136, 140
Global minimum, 58, 136, 140–141, 147, 329–330,
      333, 335
Glyph, 2, 5, 14, 78; *see also* Marker
Gnome, *see* Engine, rotary

Gradient, 43–44, 47, 68–70, 81, 331, 333–335
Grid
  line, 67, 443
  point, 86–87, 145
  size, 28, 47, 51, 55, 58, 81–82, 88, 131, 329, 336,
      350, 415, 417, 556
Gruebler–Kutzbach criterion, 184

H

Helical spring, 323; *see also* Procedure **Spring** in
      Unit **LibMec2D**
Helix, 92, 107, 117, 121, 124, 448–449
Hemisphere, 85, 438
Hewlett-Packard Graphics Language (HP–GL),
      91–92, 111–112, 128
Histogram, 6, 25–27, 40, 319, 336–338
Hob tool, 289; *see also* Gear
Hydraulic cylinder, 153, 158; *see also* Actuator,
      linear
Hyperbolic paraboloid, 73, 137
Hyperboloid, 27
Hyperfunction, 329
Hypersurface, 330, 335–336, 374

I

Image space, 47, 69, 87, 89
Inequality, 247, 253, 259, 263
  constraint, 137–138, 145–146
  plot, 27–28, 80, 417, 437
Infeasible
  area, 77
  region, 141, 149
  space, 138, 141
Initial
  conditions, 97, 129, 323–324, 402, 541
  guess, 8, 129, 131–132, 135, 140, 328, 335, 399,
      465, 471, 474, 549, 552, 557
Instant center of rotation, 305, 345–346
Integral, 11–13, 27, 370, 372, 400–401; *see also*
      Numerical, integration
Interpolation
  linear, 91, 99–100, 115–116, 271, 324, 352
  parabolical, 132, 324, 326
  spline, 98, 100–101
Intersection
  circle-circle, 164, 186–190
  circle-line, 273, 290, 382–383
  circle-parabola, 279, 391
  line-line, 282–285, 345, 393
  line-parabola, 276

surface–plane, 44, 47, 60–61, 70–75, 84, 89
surface–surface, 27, 120
with *x*-axis, *see* Zero, of functions
Inverted motion, 269, 278, 282, 284–287, 290, 292, 294
Involute, 179, 297–318, 348–349, 488, 539; *see also* Gear
Iris, *see* Mechanism, iris
Isomer, *see* Dyadic isomer
Isometric view, 50, 123
Isotropic, 2–3, 8, 17, 93

J

Jerk, 269–270
John Deere, 358
Joint
    pin, 184, 193–194, 207, 242, 251
        center, 193, 206, 229, 236–237, 243
        grounded, 154, 174, 194, 211, 523, 545
        size, 162, 195; *see also* Procedure **SetJointSize** in Unit **LibMec2D**
    potential, 220, 256, 261
    prismatic, *see* Joint, sliding
    sliding, 153, 183–184, 201, 220, 245, 281; *see also* Actuator, linear
        eccentricity, 204–205, 214, 216, 219, 222, 496, 498
    symbol, 153
    turning, 229, 245; *see also* Joint, pin

K

Kinematics
    direct, 158, 351, 353–355, 579, 581–582, 586–587
    inverse, 319, 348, 353–355, 577, 583–584
Kinetic energy, 371–372, 375; *see also* Energy

L

Land of 3D plots; *see also* Gear, tooth land
    bottom, 72, 76, 78, 80–81, 85
    top, 72, 77, 80–81, 84, 87, 313, 337
Law of gearing, 302
Length of action, 310; *see also* Gear
Level curves, 44, 46, 74, 89, 114, 138, 142, 329–330
    color, 43, 57, 66, 82, 110
    defects, 60–61, 68, 75
    elevation, 46, 58–59, 62, 72, 351, 433
    equally spaced, 51, 61, 73
    layer, 114
    logarithmically spaced, 62–65, 75, 433

mapped on surface, 72, 81
number, 44, 60–62, 82
projected, 46, 66, 558
raised, 43, 49, 53, 59, 66, 72, 82, 113
thickness, 57, 82
Line
    of action, 304, 306
    breaker, 16–18, 20–21, 32–34, 36–37
    separator, 109, 116; *see also* Line, breaker
Linear
    approximation, *see* Approximation, linear
    interpolation, *see* Interpolation, linear
Linkage, 122, 124–125, 374; *see also* Mechanism
Lissajous curve, 162–163, 477
Local maximum, 58, 110, 136
Local minimum, 9, 58, 110, 136, 335
Locus, 96–98, 157, 163, 171, 228, 244, 251, 266; *see also* End effector path; Procedures **CometLocus** and **Locus** in Unit **LibMec2D**
    cam design, 274, 278–279, 283, 293
    file, 283, 293, 484, 495–497, 499, 501–502, 504, 506, 508–510, 512, 515
    gears, 298, 302, 305
    layer, 152
    linkages, 157–158, 163, 171, 228, 244, 251, 266
Logarithmic
    decrement, 324–326
    spaced level-curves, *see* Level curves
Lower bound path, 331; *see also* Upper bound path

M

M_3D.LSP, 92, 121–128, 142, 152–153, 157, 179, 185, 345–346, 352, 364, 579, 582, 584, 587
Manipulator, 158, 355–357, 374; *see also* Robot
Marker; *see also* Glyph
    arrow, 2, 94, 319–321
    size, 2, 10, 12, 94
    spacing, 5–6
    type, 2–3, 7, 10, 14, 18, 23–24, 94, 142
MathCAD, 1
Mathematica, 1, 70
MATLAB, 1, 44, 61, 70, 98, 134, 407
Mechanism
    dwell, 319, 338–340, 374, 561
    four-bar, 202, 319, 339, 345
    iris, 342–345, 352, 568, 570
    quick-return, 319, 341, 564
    slider–crank, 202, 204, 341
    Stephenson, 338–339, 561

Mesh
color, 45, 55
line, 45
node, 45
surface, 43, 46, 51
Mini–max, 146, 374
Mitsubishi, 351–352
Motion inversion, 274, 279, 283, 287, 290–294, 346
Motor
grounded, 157, 174, 194, 211, 545
linear, 151, 153, 158–159, 161–162, 185
moving, 156
representation, 153, 156, 163
rotary, 151, 153, 185, 303; *see also* Engine, rotary
MP4, 304–308, 311–312, 314, 347–349
Multibody software, 297; *see also* **Working Model 2D (WM 2D)**
Multicriteria optimization 129, 144–146, 149–150; *see also* Optimization

N

Newton, 390, 399; *see also* Algorithm, Newton
Normal
acceleration, *see* Acceleration, normal
distribution, *see* Gaussian distribution
force, 306, 519, 524
line, 273, 275–276, 278, 280–282, 287, 302, 304
Notepad, 18, 22, 57, 59, 109, 122–123
Numerical; *see also* Algorithm
differentiation, 101, 128, 271
integration, 128, 322
NURBS, 270–271; *see also* Spline, NURBS

O

Objective function, *see* Optimization
Optimization, 70, 129, 138, 144–147, 149, 327
algorithm, 132–134, 146–147, 149, 329, 335–336
constrained, 140
polyline vertices, 46, 112
Oscillating
follower, 278, 293–294, 297, 520, 522, 529, 531
slide, 185–186, 196, 204–205, 209, 495; *see also* actuator, RTRR
Oscillation, 93–94, 125–126, 321, 325, 443, 444, 464

P

Paint, 70, 74, 82, 84–86
Painter's algorithm, *see* Algorithm, painter's
Pan, 107
Parabola, 9, 276, 279, 284, 288, 386–391

Parabolic interpolation, *see* Interpolation, parabolic
Parallel processing, 335
Parallel robot, 185, 353, 374; *see also* Robot, parallel
Parametric equation, 118, 138, 377
Pareto
front, 144–149
point, 471, 474
solution, 144, 146
Partial mini–max, 329–330, 333, 335–336, 374, 556–557
Pascal, *see* **Turbo Pascal** Unit
Pause, 139
PCX stream export, 3, 15, 34–36
Penalty function, 138; *see also* Optimization
Pendulum, 96–98, 151, 323, 545
Performance space, 145–150, 466, 468
Period of motion, 324–325; *see also* Frequency
Phase angle, 163, 325, 360; *see also* Oscillation
Phase path, 319–321, 540–541; *see also* Duffing oscillator
Phasor, 179; *see also* Vector
Pinion, 178, 297, 300, 317, 355, 357–358, 361, 487–488, 586, 593–594; *see also* Gear; Rack, and Pinion
cutter, 310–312
generating, *see* Pinion, cutter
Piston bump, 194–195, 495, 497–498; *see also* constant **BumpPiston** in Unit **LibMec2D**
Pitch
angle, 387
cam profile, *see* Cam
circle, 300, 310–311, 317
helix, 117
profile, 381, 385
radius, 356, 587, 593
Pivot joint, *see* Joint, pin
Planetary gear, 319, 346–349, 374
Plot
animated, 3, 180
area, 2, 11–13
bottom land, *see* Land of 3D plots
box
orientation, 44–45, 48, 50, 69, 87, 118
size, 2, 4–5, 8, 17, 29, 44–45, 65, 120
colored, 2–3, 7, 10, 17, 39, 81–85
combined, 85–86, 99, 110, 327
comet, 2–3, 28–29, 33–37, 179–180, 319, 321
contour, *see* Level curves
crosshatch, 49, 52, 55, 57
dynamic, 96, 151, 179
histogram, 6, 25–27, 40, 319, 336–338
line breaker, 16–18, 20–21

mesh, 51–52, 55, 79
node, 58, 78, 80–82
stem, 2, 12, 43, 56–58, 80
surface, 43, 47, 72, 79, 81–82, 85, 92, 112, 119–120
top land, *see* Land of a 3D plot
wireframe, 45–46, 55, 57, 59, 65–66, 82
Plotter, 20, 163
Poincaré map, 321, 540–542
Point; *see also* Marker
    cloud, 58, 350
    end, 32, 98, 126, 157
    extremum (minimum or maximum), *see* Optimization
    ground, 152; *see also* Procedure **PutGPoint** in Unit **LibMec2D**
    label, 14, 174, 493, 516, 520
    offset, 151, 164, 167, 184, 273, 280, 358–359, 479
    saddle, 58
    scan, 34, 36, 96, 151, 179–180
    singular, 9, 15, 17–18, 70, 78, 131, 382–383
Polar
    angle, 299
    array, 31–32, 152, 272, 313
    coordinate, 97
    curve, 28, 482, 482
    lines, 282, 290, 293, 525, 529, 534, 537
    point, 282, 284, 287, 290
    radius, 299
Polygon
    approximation, 30
    scan conversion, 47, 55
    spiraling, 37–38, 41, 426–428
Polygonalization, 29
Polyline
    DXF, 5, 30, 33, 46, 104, 109, 116
    3D, 104, 107, 116
    2D, 104, 106–107, 112
Polyline optimization, *see* Optimization, polyline vertices
Preference function, 146–147
Pressure angle, 202
    cams, 273, 275–282, 286–289, 294–295
    gears, 302, 306–308, 317
Principle of virtual work, 368, 370
Prismatic pair, *see* Joint, sliding
Probability density, 338
Profile shift, *see* Gear, Profile shift
Projection, 47, 379, 402
    on an axis, 252–263, 280
    minimax, 330, 335–336
    oblique, 47–48
    parallel, 44–45

perspective, 44–45, 50
    on a plane, 21, 44, 117
Punch press, 319, 367–373, 595; *see also* Slider-crank
Purge
    command, *see* **AutoCAD** command, purge
    files, 122, 319, 373, 597

Q

Quadrant, 49, 55–56, 197, 407
Quadratic B-spline, *see* Spline, B-
Quadratic equation, 188, 216, 383, 385, 389

R

Rack; *see also* Gear; Pinion
    cutter, 310–311, 313, 315–317
    generating, *see* Rack, cutter
    and pinion, 317, 355
Radial engine, *see* Engine, radial
Radial line, 282, 284–285, 287, 289–291, 294; *see also* Polar, lines
Random
    number, 319, 336–337, 374, 428, 560
    perturbation, 140; *see also* Optimization
    point, 141, 145, 469
Raster image, 3, 28, 40, 82, 91, 104, 108–109, 359
Reaction
    force, 202, 205
    moment, 281
Recombination, *see* Evolutionary algorithm
Reduction ratio, 125
Robot
    Cartesian, 28, 163
    5R, 351–354, 576–577, 579
    parallel, 185, 353, 374, 576, 579, 583
    SCARA, 158, 319, 351–352, 355, 576, 579, 581, 583
    serial, 158, 351, 353, 355, 374, 581
    2R, 351–353, 355, 581–582
Roll angle, 397–398
Roller follower, 272, 274, 277, 281, 289–291
Root finding, 18, 129–132, 350, 385, 399, 452, 453
Rope shovel, 185, 190, 319, 358, 360–361, 592
Rosenbrock, *see* Function, Rosenbrock
Rotation; *see also* Translation
    angle, 125, 284, 450, 526, 530, 535, 537
    axis of, 107, 275, 567
    3D, 68, 107, 124–126, 362, 397, 451; *see also* Axis, reversal
    2D, 154, 156, 165, 269–270, 272–273, 275
Rotational velocity, *see* Velocity, angular
Roto-translation 126; *see also* Rotation
Round off error, 402

S

Saddle point, 58, 64, 73–76, 134
Sampling, 9, 15, 18, 28, 40, 61, 74
Scale, 109, 114, 116–117, 128, 322, 359, 557; *see also*
        AutoCAD command, scale
    color, 45, 57, 81, 110
    factor, 3, 44, 66, 419–420
    one-to-one, 17, 65
    plot, 66
    rate, 13
Scan
    line, 2, 34, 36, 96, 151, 179–180, 316, 489
    point, 34, 36, 96, 151, 179–180
    variable, 330–336, 556
SCARA robot, *see* Robot, SCARA
Scatter, 3, 21–22, 25, 27, 44, 52, 54–55, 65, 69; *see also*
        Decimation
Scilab, 44, 407
Search, *see* Optimization
Secant method, 9, 41
Secondary axis, 9, 93
Sewing machine, 184, 493
Shear
    area, 369
    strength, *see* Strength, shear
    stress, *see* Stress
    transformation, 45, 47–48, 50, 89
Side view, 21, 44, 107
Simplex, 134–137, 139–140, 458, 459, 462, 464,
        469, 472
Single-valued function, 28, 43, 85, 89, 319, 329
Singularity, 78
Slider–crank, *see* Mechanism, slider–crank
Sliding; *see also* Joint, sliding
    block, 228–229, 235–236, 241, 251–252, 257–258,
        262, 509, 512
    rod, 220, 222–223, 228, 235, 241,
        500–501, 564
Sorting, 141, 143; *see also* Evolutionary
        algorithm
Speed
    fluctuation, 367–369, 372
    ratio, 303, 368
Spline; *see also* **AutoCAD** command, spline;
        AutoCAD, spline
    B-, 91, 98, 100–102, 270
    cubic, 98, 100
    follower motion, 270
    interpolation, spline
    NURBS, 270
Spring-mass, 321–324, 327, 542, 545
Standard deviation, 560

T

Steering, 122–125
    Ackermann, 124, 363, 365, 374
    angle, 125–126, 363, 365
Stephenson linkage, *see* Mechanism, Stephenson
Stopping criteria, 134, 142
Strength
    shear, 367, 369
    yield, 144–145
Stress
    bending, 145
    contact, 283, 289
    raiser, 145
    stress, 367, 369
    strain curve, 108, 110, 327, 551
Sum of squares, 328
Sun gear, 346–348; *see also* Planetary gear
Surface, 43, 51, 65, 116, 119–120, 128, 329; *see also*
        Projection
Suspension, 361–362, 364–367, 374
Synthesis
    cam profile, 151, 272, 274, 277, 279, 281, 284, 286,
        289, 291–292, 374
    follower motion, 269

Tangent
    circle, 288, 310, 382–385
    circle–circle, 302, 304, 393–395
    circle–parabola, 288, 391
    curve, 350
    end, 270–271
    parabola, 284, 388–389
Taper transformation, 45, 47–48, 50
Taylor series, 399, 543; *see also* Euler–Taylor algorithm
Toe angle, 361, 363–365; *see also* Suspension
Toroidal helix, 449
Toroidal spiral, 118
Torus, 92, 118, 121, 123, 128, 449
Tractor, 122–123, 127–128
Translation; *see also* Rotation
    2D, 165, 395
    3D, 126, 362, 396–397, 451
TRT dyad, *see* Dyad, TRT
Truncated plot, 3, 44, 47, 70–72, 76–77, 79, 84–85, 350
Truncated surface, 47, 85
Turbo Pascal Unit
    LibAssur, 153, 179, 183, 185–186, 211, 213, 218,
        221, 227–228, 234, 240, 244, 250, 255, 260,
        265–266, 271, 319, 342, 351, 358, 493, 499,
        501, 503, 505–506, 507, 508, 510, 512, 514,
        562, 564–566, 567–569, 570, 572, 576, 579,
        586, 589, 595

LibCams, 271, 273, 275–276, 279, 282–284, 288, 293, 518, 522, 524, 527, 529, 531, 534, 536

LibDXF, 92, 163, 299, 441–442, 444, 445–447, 475–476, 482, 485, 486–487, 489, 492, 493, 495, 497, 499, 501, 503, 505–506, 508, 510, 512, 514, 516, 518, 520, 522, 524, 527, 529, 531, 533, 536, 539, 542, 561, 564, 566–568, 570, 572, 576, 579, 581, 584, 586, 589, 592, 595

LibGe2D, 30, 188–189, 248, 276, 284, 288, 346, 421, 485, 497, 518, 522, 527, 531, 554, 572, 576, 583, 586

LibGe3D, 126, 450

LibGIntf, 93, 442, 445, 489

LibGraph, 92, 441–443, 445, 489, 492, 522, 527, 529, 531, 533, 572

LibInOut, 93, 140, 438–439, 442–443, 445, 464, 469, 472, 475–476, 477–479, 480, 482, 485–486, 487, 489, 492, 493, 495, 497, 499, 501, 503–505, 506, 508, 510, 512, 514, 516, 518, 520, 522, 524, 527, 529, 531, 534, 536, 542, 547, 551, 554, 556, 561, 564–566, 567, 568, 570, 572, 576, 579, 581, 583, 586, 589, 592, 595, 597

LibMath, 16, 30, 51, 92, 103, 129, 131, 138, 154, 185, 197, 244, 343, 350, 412, 414–415, 419, 421, 424–425, 426, 428, 429, 441–442, 443, 445, 47, 450–452, 453, 456, 458, 459–461, 462, 464, 467–468, 469, 472–476, 477–479, 480, 482, 485–486, 487, 489, 493, 495, 497, 499, 501, 503–505, 506, 508, 510, 512, 514, 515, 518, 520, 522, 524, 527, 529, 531, 533, 536, 540, 542, 545, 547, 551, 554, 556, 561, 564, 568, 570, 572, 574, 576, 579, 581, 583, 586, 589, 592, 595

LibMec2D, 96, 151, 153, 156, 166, 169, 172, 174–175, 180, 188, 190, 194–195, 207, 219, 222, 227, 241, 251, 271, 345, 351, 358, 475–476, 477–479, 480, 485, 486–487, 489, 492, 493, 495, 497, 500, 501, 503, 505–507, 508, 510, 512, 514, 516, 518, 520, 522, 524, 527, 529, 531, 534, 536, 542, 562, 564, 566–567, 569–570, 572, 576, 579, 581, 584, 586, 589, 593, 595

LibMecIn, 151, 153, 156, 179, 185, 193, 358, 361, 477, 479, 480, 482–485, 486–487, 493, 495, 497, 499, 501, 503, 505–506, 508, 510, 512, 514, 516, 518, 520, 522, 524, 527, 529, 531, 534, 536, 561, 564, 569–570, 572, 579, 581, 584, 586, 589, 593, 595

LibMin1, 132, 454–456

LibMinN, 134, 139–140, 458, 459, 462, 464, 469, 472, 547, 551, 556

LibMinEA, 547

LibPlots, 91–94, 128, 151, 441–444, 445, 489

Unit_D2D, 1, 40

Unit_D3D, 43, 87

Unit_PCX, 92, 441–444, 445, 489, 572

Unit_TXT, 98

Turning joint, *see* Joint, turning

U

Unconstrained function, 133

Undercut
    cam, 277, 283, 289
    gear, 312–313, 316–317

Upper bound path, 330, 332, 335

Upper envelope

Util~DXF, 20, 91, 98, 104, 106–112, 114, 116, 128, 140, 178–179, 269, 271, 293, 345, 352, 358, 360

Util~PLT, 66, 92, 111–115, 128, 178, 343, 352, 358–359

Util~TXT, 91, 98–100–104, 106, 108–110, 115–116, 128, 269, 271, 352

V

Vector
    acceleration, *see* Acceleration, vector
    data, 92–96, 131, 134, 323, 441–445, 451, 489
    displacement, 281
    equation, 155, 159–160, 165, 169, 197–199, 207–208, 212, 215, 217, 224, 247, 299, 304
    field 68; *see also* Gradient
    force, 273, 278
    format 3, 40, 44, 110
    graphics, *see* Vector, format
    loop
        equation, 230, 236, 242, 253, 258, 263, 267
        method, 196, 207, 210
    normal 276; *see also* Normal, line
    2D, 155, 159, 165, 168, 174, 179–180, 196, 204, 210, 213, 273, 276, 379, 397, 482, 489, 491

Velocity
    angular, 155, 165, 168, 208, 212, 215, 297, 303, 339–340, 358, 367, 371, 595
    linear, 155, 160, 164
    vector, 173, 273, 276, 278, 281–282, 287, 306, 323, 356

Vertex, 73, 76, 108, 114, 119, 283; *see also* Procedure
    AddVertexPline in Unit **LibDXF**
    elimination, 32, 112, 134
    extraction, 106, 179
    file, 107, 110, 116, 293, 447

merger, 73
    numbering, 2, 14, 23, 37
Vibration, 128, 269, 319, 324, 374
Video clip, 122; *see also* **MP4**
View point, 49, 119, 123
View window, 48, 106, 112, 175

W

Weak Pareto solution, *see* Pareto, solution
Weighted sum 146–147, 149; *see also* Multicriteria
    optimization
Weighting coefficient, 147
Wheel, 122, 125, 361; *see also* Gear
    base, 363
    carrier, 362–363
    friction, 301
    jounce, 363–365
    rebound, 363–365
    recessional motion, 364
    steering, 125–126
    track, 361, 363–364
Width of space, 300, 310, 312; *see also* Gear
Windows, 6, 46
Wireframe, 45–46, 55, 57, 59, 65–66, 82
Working Model 2D (WM 2D) 295, 297, 310, 346,
    368, 373
    formula language, 305, 372
    simulations
        gear teeth cutting, 310–313, 349
        planetary gear, 346–349
    punch press, 372
    of two gears, 307–308
    of two involutes, 303–305
Workspace, 152, 175, 339, 360, 484, 495

Y

Yaw
    angle, 398
    motion, 358

Z

Zero; *see also* Algorithm, Zero; Level curves;
    Procedures **ZeroGrid** in Unit
    LibMath; Procedures **ZeroStart** in
    Unit **LibMath**
    backlash, 302, 306; *see also* Backlash
    color, 279, 282, 290–292; *see also* **BGI**, **Black**
        constant
    division by, 15, 17, 30, 166, 188, 382, 385, 412,
        414, 452–453
    DOF, 183
    elastic range, 369
    elevation, 58, 113, 351; *see also* Level curves
    of functions, 29, 41, 129–132; *see also* Root
        finding
    of graphs, 2, 8–11
    line, 66
    profile shift, *see* Gears, zero profile shift